AF823913

Ecological Solutions to Agricultural Land Degradation

Subhash Babu • Anup Das
Sanjay Singh Rathore
Raghavendra Singh • Rajbir Singh
Editors

Ecological Solutions to Agricultural Land Degradation

Springer

Editors
Subhash Babu
Division of Agronomy
ICAR-Indian Agricultural Research Institute
New Delhi, India

Sanjay Singh Rathore
Division of Agronomy
ICAR-Indian Agricultural Research Institute
New Delhi, India

Rajbir Singh
Division of Natural Resource Management
ICAR
New Delhi, Delhi, India

Anup Das
ICAR Research Complex for Eastern Region
Patna, Bihar, India

Raghavendra Singh
Cropping Systems and Resource Management
ICAR-Indian Institute of Farming Systems Research
Meerut, Uttar Pradesh, India

ISBN 978-981-96-3391-3 ISBN 978-981-96-3392-0 (eBook)
https://doi.org/10.1007/978-981-96-3392-0

© The Editor(s) (if applicable) and The Author(s), under exclusive license to Springer Nature Singapore Pte Ltd. 2025

This work is subject to copyright. All rights are solely and exclusively licensed by the Publisher, whether the whole or part of the material is concerned, specifically the rights of translation, reprinting, reuse of illustrations, recitation, broadcasting, reproduction on microfilms or in any other physical way, and transmission or information storage and retrieval, electronic adaptation, computer software, or by similar or dissimilar methodology now known or hereafter developed.
The use of general descriptive names, registered names, trademarks, service marks, etc. in this publication does not imply, even in the absence of a specific statement, that such names are exempt from the relevant protective laws and regulations and therefore free for general use.
The publisher, the authors and the editors are safe to assume that the advice and information in this book are believed to be true and accurate at the date of publication. Neither the publisher nor the authors or the editors give a warranty, expressed or implied, with respect to the material contained herein or for any errors or omissions that may have been made. The publisher remains neutral with regard to jurisdictional claims in published maps and institutional affiliations.

This Springer imprint is published by the registered company Springer Nature Singapore Pte Ltd.
The registered company address is: 152 Beach Road, #21-01/04 Gateway East, Singapore 189721, Singapore

If disposing of this product, please recycle the paper.

This book is dedicated to Late Dr. Gulab Singh Yadav, whose pioneering work and dedication to advancing agricultural sciences have left an enduring legacy. His insights and commitment to addressing challenges faced by farming community and promoting sustainable agricultural practices have inspired countless others. May his legacy continue to guide and inspire future generations in the pursuit of frontiers in agronomy for soil and environmental sustainability.

Preface

Agricultural land degradation is a pressing challenge that threatens food security, biodiversity, and the livelihoods of millions of populaces globally. Land degradation is not only affecting the environment but also affecting the socioeconomic condition of global society. As per an estimate of UN, about 100 Mha of productive land lost every year due to land degradation. If current trends continue, the world will need to restore a staggering 1500 Mha of degraded land by 2030 to reach land degradation neutrality (LDN) targets enshrined in the Sustainable Development Goals (SDGs). The economic consequences of land degradation are enormous, as up to 40% of the earths land is degraded, directly affecting half of humanity, and threatening roughly half of global GDP (US$44 trillion). As global demand for food rises, agricultural lands face unbearable pressure, leading to severe degradation. Hence, a shift from conventional agricultural production practices toward more sustainable farming is crucial for soil health restoration, enhancing resilience, and securing food and nutritional security. Embracing ecological solutions can halt/reverse land and environmental degradation and ensure a sustainable future for agriculture and humanity. Ecological practices aim to restore the health of degraded lands, enhance its resilience to climate change, and promote the sustainable use of natural resources. This book explores a wide array of strategies and interventions that have been designed and implemented across the various regions to combat land degradation, offering a comprehensive view on the futuristic challenges and opportunities. Very few resource and reference books are available in the global and regional platforms that address specific issues on agricultural land degradation. The book "Ecological Solutions to Agricultural Land Degradation" explores the conservation agriculture and regenerative farming practices and approaches that focus on the soil health. The role of agroforestry in developing resilient farming systems, the integration of legumes and cereals in cropping systems to enhance soil fertility and environmental sustainability, and the role of precision farming to optimize resource use and minimize negative environmental outcomes along with other ecofriendly and feasible practices are included in this book. The book also emphasizes the importance of nature-based solutions, crop diversification, watershed management, organic farming, and biomass recycling to restore degraded lands and secure food production for a sustainable future. The strategies discussed in this book are not only about restoring agricultural land but also about improving food-nutrition and livelihoods security of diverse population. We would like to express our profound gratitude to the contributors who have generously shared their knowledge and expertise to address the noble cause of rehabilitating degraded agricultural lands. Through research findings and practical examples, the book aims to provide a comprehensive understanding of the challenges and opportunities associated with agricultural land restoration. We hope that this book serves as a valuable resource for researchers, students, and policymakers to develop strategies for restoring the degraded agricultural lands. Moreover, this publication will serve as a handy resource book for preparing a roadmap for policy planning and advancing knowledge in sustainable and resilient farming, helping to address futuristic agricultural land restoration challenges. This book is a call to action—a call to rethink the way we approach agriculture, to embrace the potential of ecological solutions, and to commit ourselves to the restoration and protection of agricultural lands and achieving the goal of LDN.

New Delhi, India — Subhash Babu
Patna, India — Anup Das
New Delhi, India — Sanjay Singh Rathore
Modipuram, India — Raghavendra Singh
New Delhi, India — Rajbir Singh

Contents

Editors and Contributors

About the Editors

Subhash Babu, PhD (IARI, New Delhi), Fellow National Academy of Agricultural Sciences (FNAAS), Associate Fellow, Indian National Science Academy (IAF), AF-NAAS, Member Indian National Young Academy of Sciences (INYAS), and ISA Associate Fellow, currently working as a senior scientist at the Division of Agronomy, ICAR-IARI, New Delhi, India. He is recipient of the Dr. Mahatim Singh Memorial Gold Medal and BHU medals during his MSc (Ag.) Agronomy. He is known for his work in integrated soil health improvement, resource conservation, and the integration of innovative climate resilient technologies in agriculture. His research and initiatives aim to improve food-nutritional security and environmental sustainability. Dr. Babu is actively collaborating with various institutions and organizations to promote knowledge sharing and the adoption of best practices in farming, making a substantial impact on the agricultural sector. He has been conferred with the Best Scientist-2016 of ICAR Research Complex, Sikkim Centre, and NESA Young Scientist of the Year Award-2020, PS Deshmukh Young Agronomist Award by ISA. He is serving as an associate editor of the *Frontiers of Agronomy*, *Scientific Reports*, *Indian Journal of Agronomy*, *Indian Journal of Hill Farming*, and editor in *Proceeding of Indian National Science Academy*. He is also serving as an advisory board member of Heliyon. He served as a Councillor ISA for NE states (two times) and Indian Society of Weed Sciences for Sikkim and as Joint Secretary of the Indian Association of Hill Farming, Meghalaya.

Anup Das, PhD (IARI, New Delhi), FNAAS, FISA, FNABS, and FIAHF, currently serving as Director, ICAR Research Complex for Eastern Region, Patna, India, and Vice President, ISA. His extensive research works led to the development of the package of practices for production of pulses and oilseeds in rice and maize fallow areas; organic production packages of important crops in system approach and developed many climate resilient integrated farming system & land use models. He has conceptualized and executed an integrated organic farming system in a cluster approach for safe and healthy food production in northeast India. Greening vast rice fallow areas of Eastern India, multiple use of water for enhancing water productivity and sustainable intensification of rice-wheat system in Eastern IGPs are his current areas of research. His research focuses primarily on developing innovative agronomic approaches for managing soil health and environmental security and providing practical solutions to challenges faced by small farmers. Guided/ co-guided 17 masters and two PhD students. He has published 250 research articles and 45 books and bulletins. He is recipient of several national awards like ICAR-Lal Bahadur Shastri Outstanding Young Scientist Award, Swami Sahajanand Saraswati Outstanding Extension Scientist Award, Fakhruddin Ali Ahmed Award, Inter-disciplinary Team Research Award, and First Dr. H.K. Jain–CAU, Imphal Award, among many others. He has served as Vice President, Co-chief editor, Secretary, and Joint Secretary of IAHF. He has organized several national and international conferences.

Sanjay Singh Rathore, PhD (IARI, New Delhi), FNAAS, FISA, FRMSI, FIAHF, FSRMR, FSEE, currently working as a Head, Division of Agronomy, ICAR-IARI, New Delhi, and National Coordinator of World Bank funded National Agricultural Higher Education Project (NAHEP). Earlier, Dr Rathore has served as Executive (PE) in the National Dairy Development Board (2001–2002), ARS-Scientist at ICAR Research Complex for NEH Region, Nagaland, Jharnapani (2003–2008), Senior Scientist (Agronomy) at ICAR-DRMR, Bhartapur (2008–2015), and Principal Scientist at the Division of Agronomy, ICAR-IARI, Pusa, New Delhi (2015–2023). He has published 160 international and national research papers and 8 books. He has developed IFS models under jhum cultivation, efficient irrigation water management, and resource conservation technologies for mustard-based cropping systems, suitable agri horti, and potential crop diversification options for maize-based systems under limited water conditions. He has developed climate resilient agronomy for oilseed-based systems. Dr. Rathore's research focuses on enhancing crop yields, improving resource use efficiency, and promoting environmentally sustainable farming techniques. He has received ICAR Junior and Senior Research Fellowship during master's and doctorate programmes. He has also received the Young Scientist Associate 2012 Award, Distinguished Scientist Award 2014, Dhiru Morarji Memorial Award 2007, 2019, and Sri Ram Purskar 2018 and 2020 from Fertilizer Association of India (FAI). He has guided 18 students for master and PhD program.

Raghavendra Singh, PhD (BHU, Varanasi), FISA, and FIAHF is currently working as a Head, Division of Cropping Systems and Resource Management, ICAR-Indian Institute of Farming Systems Research, Modipuram, Meerut, UP (India). He is a prominent agronomist recognized for his extensive work on sustainable agriculture and crop management. He has made significant contributions to soil health, nutrient management, and the development of innovative farming practices. Dr. Singh's research often focuses on enhancing crop productivity, resource use efficiency, and environmental sustainability. Dr Singh is associated with the development of high yielding varieties of rice, maize, and rajmash for organic conditions of Sikkim. He has published 120 research articles and 3 books. He received several national awards like Pandit Deen Dayal Upadhyay Krishi Vigyan Rashtriya Protshahan Puraskar-2018 Zonal Awards, the Best Scientist Award, and the Best Extensionist Award. He has served as Joint Secretary of the Indian Society of Agronomy, Varanasi Chapter, and as a Councilor for both the ISA and the IAHF.

Rajbir Singh, PhD (HAU, Hisar), FNAAS, FISA, FHSI, currently serving as the Deputy Director General (Agricultural Extension) at Indian Council of Agricultural Research (ICAR), New Delhi. Earlier as Assistant Director General (ADG) for Agronomy, Agroforestry, and Climate Change (ICAR), he played instrumental role in planning, coordinating, and monitoring R&D programmes aimed at sustainable agricultural production and resource conservation amid climate change. Before joining ADG, Dr. Singh was the Director at ICAR-ATARI, Ludhiana (2015–2023). He revitalized KVKs and advanced conservation agriculture and natural farming to tackle climate change, soil health issues, and low farm profitability. Known for his team-building skills and effective communication, Dr. Singh is recognized for his problem-solving abilities. Dr. Singh has received several prestigious awards, including the Rafi Ahmed Kidwai Award (2021), Nanaji Deshmukh multidisciplinary Team ICAR Award (2019), and Swami Sahajanand Saraswati Outstanding Scientist Award (2016). His expertise lies in conservation agriculture, extension research, and natural farming. Dr. Singh has published over 100 research papers, alongside numerous technical bulletins, popular articles, books, and book chapters. With over 29 years of experience, he has led impactful projects, including a central scheme for agricultural mechanization that significantly reduced crop residue burning and expanded sustainable practices in Punjab, Haryana, and Uttar Pradesh.

Contributors

Gunturi Alekhya Division of Agronomy, ICAR-Indian Agricultural Research Institute, New Delhi, India

Sk Asraful Ali Division of Agronomy, ICAR-Indian Agricultural Research Institute, New Delhi, India

Anjana J. Atapattu Agronomy Division, Coconut Research Institute, Lunuwila, Sri Lanka

Subhash Babu Division of Agronomy, ICAR-Indian Agricultural Research Institute, New Delhi, India

Subrata Bag ICAR-Central Research Institute for Dryland Agriculture, Hyderabad, Telangana, India

Shreyas Bagrecha Agronomy Section, ICAR-National Dairy Research Institute, Karnal, Haryana, India

Ruchika Choudhary Department of Agronomy, Maharana Pratap University of Agriculture and Technology, Udaipur, Rajasthan, India

Devendra Kumar Dadhich ICAR-National Dairy Research Institute, Karnal, Haryana, India

Anup Das ICAR Research Complex for Eastern Region, Patna, Bihar, India

T.K. Das Division of Agronomy, ICAR-Indian Agricultural Research Institute, New Delhi, India

Tanmay Das ICAR-Indian Agricultural Research Institute, New Delhi, India

Khushboo Devi Division of Agronomy, ICAR-Indian Agricultural Research Institute, New Delhi, India

Nuwandhya S. Dissanayaka Agronomy Division, Coconut Research Institute, Lunuwila, Sri Lanka

Suman Dutta Ramkrishna Mission Vivekananda Educational and Research Institute, Kolkata, West Bengal, India

Ananya Gairola Division of Agronomy, ICAR-Indian Agricultural Research Institute, New Delhi, India

Alekhya Gunturi ICAR-Indian Agricultural Research Institute, New Delhi, India

Twinkle Jena Division of Agronomy, ICAR-Indian Agricultural Research Institute, New Delhi, India

Kannoj Forage Section, Department of Genetics and Plant Breeding, CCS Haryana Agricultural University, Hisar, Haryana, India

Sabyasachi Koley Banaras Hindu University, Varanasi, Uttar Pradesh, India

Amit Kumar ICAR Research Complex for NEH Region, Sikkim Centre, Gangtok, Sikkim, India

Gautam Kumar ICAR Research Complex for NEH Region, Sikkim Centre, Gangtok, Sikkim, India

Sandeep Kumar ICAR-National Bureau of Soil Survey and Land Use Planning, Jorhat, Assam, India

Sanjeev Kumar Department of Soil Science & Agricultural Chemistry, Sardar Vallabhbhai Patel University of Agriculture and Technology, Meerut, Uttar Pradesh, India

Shamal Shasang Kumar Crop Research Division, Ministry of Agriculture & Waterways (MOA & W), Suva, Fiji

Vipin Kumar Division of Agronomy, ICAR-Indian Agricultural Research Institute, New Delhi, India

Megha Kumari Division of Agronomy, ICAR-Indian Agricultural Research Institute, New Delhi, India

Rupa Kumpatla Department of Soil Science, G. B. Pant University of Agriculture & Technology, Pantnagar, Uttarakhand, India

S.S. Mali ICAR Research Complex for Eastern Region, Farming System Research Centre for Hill and Plateau Region, Ranchi, Jharkhand, India

S.K. Naik ICAR Research Complex for Eastern Region, Farming System Research Centre for Hill and Plateau Region, Ranchi, Jharkhand, India

Tony Manoj Kumar Nandipamu Department of Agronomy, G. B. Pant University of Agriculture & Technology, Pantnagar, Uttarakhand, India

Kadagonda Nithinkumar Division of Agronomy, ICAR-Indian Agricultural Research Institute, New Delhi, India

Tharindu D. Nuwarapaksha Agronomy Division, Coconut Research Institute, Lunuwila, Sri Lanka

Abhishek Panchadi Department of Plant and Soil Sciences, Mississippi State University, Starkville, MS, USA

Aastika Pandey Division of Agronomy, ICAR-Indian Agricultural Research Institute, New Delhi, India

Anjali Patel ICAR-Indian Agricultural Research Institute, New Delhi, India

Sunil Kumar Prajapati Division of Agronomy, ICAR-Indian Agricultural Research Institute, New Delhi, India

Radha Prasanna Division of Microbiology, ICAR-Indian Agricultural Research Institute, New Delhi, India

Rishi Raj Division of Agronomy, ICAR-Indian Agricultural Research Institute, New Delhi, India

Sanjay Singh Rathore Division of Agronomy, ICAR-Indian Agricultural Research Institute, New Delhi, India

Kadapa Sreenivasa Reddy Division of Agronomy, ICAR-Indian Agricultural Research Institute, New Delhi, India

Sougata Roy International Potato Centre (CIP-CGIAR), Patna, Bihar, India

Trisha Roy ICAR-Indian Institute of Soil and Water Conservation, Dehradun, Uttarakhand, India

Priyanka Saha ICAR-Krishi Vigyan Kendra, West Garo Hills, ICAR-Research Complex for NEH Region, Umiam, Meghalaya, India

Seema Sangwan Division of Microbiology, ICAR-Indian Agricultural Research Institute, New Delhi, India

G.D. Sanketh ICAR-Central Island Agricultural Research Institute, Port Blair, Andaman and Nicobar, India

Ayan Sarkar ICAR-Indian Agricultural Research Institute, New Delhi, India

P.K. Sarkar ICAR Research Complex for North Eastern Hill Region, Tripura Centre, Lembucherra, West Tripura, India

Suchismita Satapathy Banaras Hindu University, Varanasi, Uttar Pradesh, India

Garima Saxena Division of Microbiology, ICAR-Indian Agricultural Research Institute, New Delhi, India

Diksha Sharma Division of Agronomy, ICAR-Indian Agricultural Research Institute, New Delhi, India

Kapila Shekhawat ICAR-Indian Agricultural Research Institute, New Delhi, India

Reshma Shinde ICAR Research Complex for Eastern Region, Farming System Research Centre for Hill and Plateau Region, Ranchi, Jharkhand, India

Deo Narayan Singh Department of Agronomy, Udai Pratap College, Varanasi, Uttar Pradesh, India

Raghavendra Singh ICAR-Indian Institute of Farming Systems Research, Meerut, Uttar Pradesh, India

Rajiv K. Singh ICAR-Indian Agricultural Research Institute, New Delhi, India

Raman Jeet Singh ICAR-Indian Institute of Soil and Water Conservation, Dehradun, Uttarakhand, India

Manish Tomar Department of Agronomy, CCS Haryana Agricultural University, Hisar, Haryana, India

Shashi S. Udumann Agronomy Division, Coconut Research Institute, Lunuwila, Sri Lanka

Arvind Verma Department of Agronomy, Maharana Pratap University of Agriculture and Technology, Udaipur, Rajasthan, India

Owais Ali Wani Division of Soil Science and Agricultural Chemistry, Sher-e-Kashmir University of Agricultural Sciences and Technology of Kashmir, Srinagar, India

Devideen Yadav ICAR-Indian Institute of Soil & Water Conservation Research, Dehradun, Uttarakhand, India

Sarla Yadav ICAR-Central Potato Research Station, Patna, Bihar, India

Land Degradation: A Global Challenge to Environmental Sustainability and Livelihood Security

1

S. K. Naik, Reshma Shinde, S. S. Mali, P. K. Sarkar, and Anup Das

Abstract

Land degradation poses a serious threat to environmental sustainability and the livelihoods of millions, especially in developing countries. Land degradation refers to the deterioration of land quality and productivity, often resulting from human activities such as deforestation, overgrazing, and unsustainable agricultural practices. These actions disrupt essential ecosystem functions and services, leading to negative consequences such as a decline in agricultural productivity, loss of biodiversity, reduced food security, and compromised water availability. Furthermore, degraded lands are more vulnerable to the adverse effects of climate change. From an economic perspective, land degradation leads to substantial losses by depleting soil fertility and increasing farming costs, further intensifying food insecurity. As the global population grows, the pressure on land resources intensifies, exacerbating the challenges faced by already degraded lands. To combat these challenges, implementing sustainable land management (SLM) and restoration strategies is crucial. Key SLM approaches include agroforestry, conservation agriculture, Integrated Nutrient Management (INM), integrated water resource management, and reforestation. Agroforestry promotes the integration of trees into agricultural landscapes, enhancing soil fertility, reducing erosion, and increasing biodiversity. Conservation agriculture, with practices like minimal tillage and crop rotation, preserves soil health and improves water

S. K. Naik (✉) · R. Shinde · S. S. Mali
ICAR Research Complex for Eastern Region, Farming System Research Centre for Hill and Plateau Region, Ranchi, Jharkhand, India

P. K. Sarkar
ICAR Research Complex for North Eastern Hill Region, Tripura Centre, Lembucherra, West Tripura, India

A. Das
ICAR Research Complex for Eastern Region, Patna, Bihar, India

© The Author(s), under exclusive license to Springer Nature Singapore Pte Ltd. 2025

S. Babu et al. (eds.), *Ecological Solutions to Agricultural Land Degradation*,
https://doi.org/10.1007/978-981-96-3392-0_1

retention, leading to higher agricultural productivity. Effective water resource management, including rainwater harvesting and drip irrigation, is vital, particularly in water-scarce regions. Restoration efforts aim to rehabilitate degraded lands, restoring their productivity and ecological balance. Techniques such as terracing and contour plowing are essential for rebuilding soil structure and arresting erosion. Reforestation and afforestation initiatives, especially in degraded areas, are substantial in carbon sequestration, improving soil health, and reinstating ecosystem services. The reclamation of saline soils through methods like leaching and the cultivation of salt-tolerant crops is another key strategy in land restoration. Government policies are instrumental in addressing land degradation, though challenges persist in scaling up effective practices and ensuring coordinated efforts. The future success of land degradation mitigation in India hinges on a sustained commitment to SLM practices, community involvement, and integration of modern technology with traditional techniques plays a vital role in restoring degraded lands and promoting a sustainable future.

Keywords

Land degradation · Livelihood security · Soil health · Food security

Abbreviations

GHG	Green house gas
INM	Integrated nutrient management
SLM	Sustainable land management
SDGs	Sustainable Development Goals
SOC	Soil organic carbon
CA	Conservation agriculture
ZTDSR	Zero-till direct-seeded rice

1.1 Introduction

Land is one of the most vital natural resources, essential for supporting life and sustaining ecosystems. It comprises a complex interplay of soil, water, and biodiversity. These three elements working together generate goods and services, which provide a foundation for sustainable livelihoods. However, land degradation knocking the health, livelihood, and security of human beings at risk. There is high pressure on global land due to agriculture and land use change. Land degradation refers to the decline in land's ability to generate benefits under a specific land use and management system (FAO 1999). It has both onsite impacts, such as reduced productivity and increased input needs, and offsite effects, including sedimentation of water bodies, declining water quality, and changes in water and wind regimes. The various land degradation viz., water erosion, wind erosion, soil fertility decline, waterlogging, salinization, lowering of the water table, deforestation, forest degradation, and rangeland degradation affects livelihood security and is a global challenge to environmental sustainability. Land degradation is linked to poverty, as it reduces crop and livestock yields, leaving people with less food and fewer resources to sell.

The foremost drivers of global land degradation include unsustainable management or overexploitation of resources, clearance of natural vegetation, nutrient depletion, overgrazing, improper irrigation, excessive use of agrochemicals, pollution, urbanization, mining, and excavation. The problem is more severe in the driest part of the world. The vast majority of the people, who depend on dryland reside in developing countries, where children and adults are susceptible to the influences of land degradation and drought. Land degradation is a major contributor to climate change, driving greenhouse gas (GHG) emissions and reducing the land's ability to function as a carbon sink. Enhancing the productivity of land and water is critical for achieving sustainable agricultural production, ensuring food security, and meeting the Sustainable Development Goals (SDGs). Climate change intensifies land degradation through factors such as heavier rainfall, increased flooding, rising temperatures, prolonged droughts, extended dry spells, and sea-level rise. The combined effects of land degradation and climate change pose serious challenges to the sustainability of natural resources and the livelihoods of millions who depend on them. Land degradation impends the security of 3.2 billion people and is one of the crucial reasons along with climate change, forcing 50–700 million people to migrate by 2050 (UNCCD 2019).

Land degradation, driven primarily by unsustainable land use and deforestation, is one of the most persistent challenges faced globally. Unsustainable land management and land-use changes, particularly in agriculture, are key human-driven causes of land degradation, with far-reaching effects on social, economic, and environmental systems. Its impacts extend beyond the degraded areas, disrupting marine and freshwater ecosystems and affecting distant communities and biodiversity (Olsson et al. 2019).

1.2 Land Degradation Status in the World and India

The IPCC (2019) highlights that since 1961, rising population and per capita consumption have altered land use patterns by intensifying commercial agriculture and forestry, which, while supporting food supply, have contributed to these changes. Human activities and extensive land exploitation have degraded a significant portion of the planet's ice-free land. The "World Atlas of Desertification" reports degradation across 75% of the land, while the Global Land Outlook 2022 places this figure at 40% (UNCCD 2022). In India, addressing land degradation is crucial given its significant effects on the economy and the well-being of millions of people who depend on agriculture. The population in India exceeds 1.4 billion which is ~18% of the global population whereas it occupies only 2.4% of the world's land area, putting immense pressure on limited land resources. The latest assessment by ISRO's Space Applications Centre (2018–2019) reveals that 97.85 million hectares (Mha), or 29.77% of India's land area, have been degraded, marking an increase of 3.32 Mha since 2003–2005. Major contributors to this degradation include water erosion (11.01% of total land), vegetation degradation (9.15%), and wind erosion (5.46%). Over time, areas impacted by water erosion, vegetation degradation, waterlogging, frost, and human activity have risen, while wind erosion and salinity-affected areas have declined. Approximately 80% of India's degraded lands are in nine states: Rajasthan, Maharashtra, Gujarat, Ladakh, Karnataka, Jharkhand, Odisha, Madhya Pradesh, and Telangana.

1.3 Sources and Causes of Land Degradation

Land degradation, caused by both natural and human factors, leads to reduced soil fertility, fewer plants and animals, and diminished natural benefits. This results in lower agricultural productivity and challenges to environmental sustainability and livelihood security. The major causes/sources of land degradation are:

1.3.1 Unsustainable Agriculture

The various reasons for unsustainable agriculture are attributed to (1) monoculture farming, (2) overuse of pesticides, (3) poor water management, (4) imbalanced fertilization, (5) excessive tillage and use of heavy machinery, (6) crop residue burning and inadequate organic matter inputs.

1.3.2 Deforestation

Land degradation is driven by deforestation for agriculture, logging, urbanization, and practices such as excessive fuelwood collection, overgrazing, and forest fires. These factors disrupt ecosystems, increase soil erosion, affect the water cycle, and reduce biodiversity.

1.3.3 Overgrazing

Excessive livestock grazing leads to vegetation degradation. Overgrazing aggravates soil compaction, erosion, and soil fertility loss. Overgrazing by large livestock populations in arid regions reduces soil infiltration, accelerates runoff, and leads to increased soil erosion.

1.3.4 Urbanization

It is the expansion and cities and infrastructure. Urbanization converts productive land into impervious surfaces, gradually removing large areas from agriculture, forestry, grasslands, pastures, and wild vegetation while causing soil sealing and disrupting water cycles.

1.3.5 Mining

It involves the extraction of minerals and resources, leading to habitat destruction and deforestation. It pollutes the soil and water with harmful chemicals. Opencast mining alters the biological, chemical and physical properties of the soil, vis-à-vis affecting the socioeconomic conditions of the area. Harmful impacts include water shortage due to a low water table, soil pollution, biodiversity loss, air and water pollution, and acid drainage. The removal of overburden during mining causes significant loss of vegetation and fertile topsoil.

1.3.6 Land Pollution

Land pollution occurs when industrial waste, pesticides, heavy metals, and other pollutants are disposed of on land, making it incompatible for agriculture and posing significant health risks.

1.3.7 Natural and Social Sources

Natural causes of land degradation include events such as earthquakes, tsunamis, droughts, avalanches, landslides, volcanic eruptions, floods, tornadoes, and wildfires. Social factors contributing to land degradation include limited land resources, declining per capita land access, economic strain, land tenure challenges, poverty, and population growth.

1.3.8 Climate Change

It alters the temperature and precipitation patterns. Climate change increases the frequency of extreme weather events. It results in desertification and soil erosion.

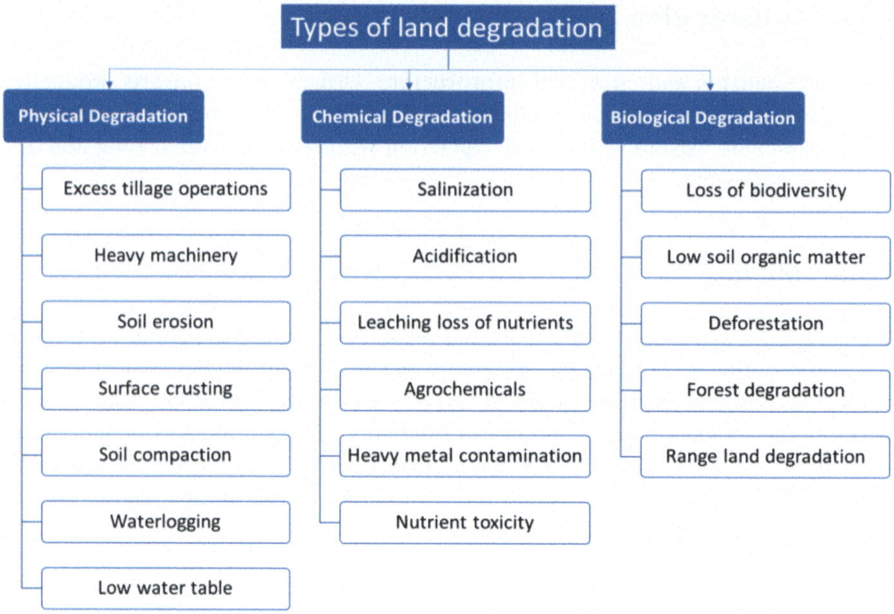

Fig. 1.1 Types of land degradation

1.3.9 Soil Degradation

Soil degradation results from ineffective land use planning, insufficient soil protection measures, absence of land use policies, and heightened erosion coupled with the destruction of habitats.

1.3.10 Desertification

The conversion of fertile land into desert is typically driven by climate change, overgrazing, deforestation, and unsustainable agricultural practices. Land degradation primarily occurs in arid, semi-arid, and sub-humid regions.

Land degradation can be classified as physical, chemical, biological, and desertification with its sources of degradation depicted below in Fig 1.1.

1.4 Land Degradation and Soil Health

Land degradation leads to soil degradation, and ultimately decline in soil quality. There are four types of soil degradation *viz.*, physical, chemical, biological, and ecological (Fig. 1.2.).

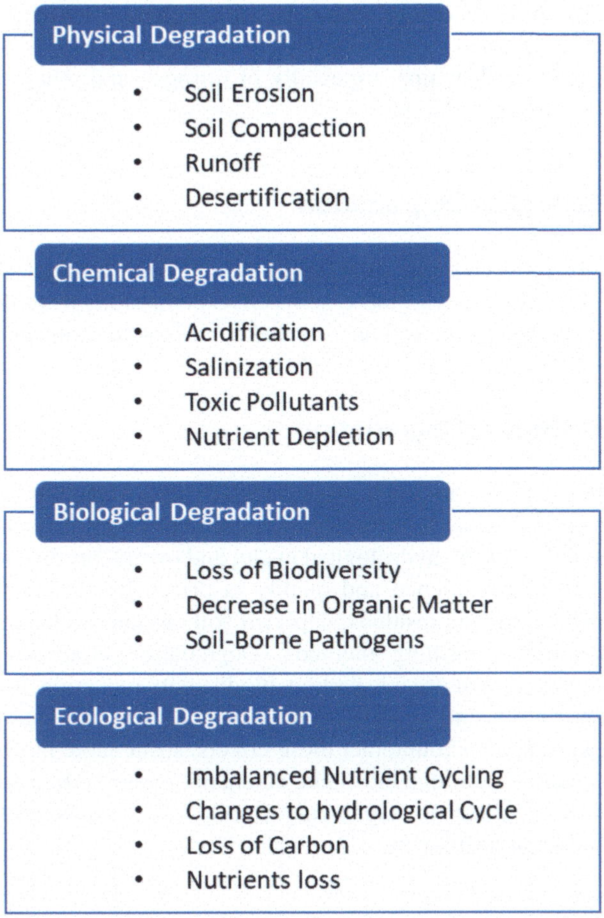

Fig. 1.2 Types of soil degradation

1.4.1 Soil Physical Degradation

This occurs due to changes in soil structure, particularly the disruption of pore connectivity and geometry. It makes soil more prone to compaction, crusting, reduced water absorption, and increased surface runoff. Such changes lead to enhanced wind and water erosion, extreme soil temperature variability, and accelerated desertification processes.

1.4.2 Soil Chemical Degradation

Chemical degradation stems from issues such as soil acidification, salinity build-up, reduced nutrient-holding capacity, and the depletion of essential nutrients.

Additional contributors include aluminum and manganese toxicity, calcium and magnesium deficiency, and contamination caused by industrial pollutants or by-product waste. Nutrient leaching, especially of nitrogen and other vital elements, further intensifies the degradation.

1.4.3 Soil Biological Degradation

This is typified by reduced organic carbon in the soil, biodiversity loss, diminished capacity to store carbon, and an increase in GHGs emissions from the soil. These factors collectively undermine soil fertility and contribute to ecological imbalances.

1.4.4 Soil Ecological Degradation

This is a multifaceted issue arising from physical, chemical, and biological degradation. It disrupts critical ecosystem services viz., nutrient recycling, water absorption and filtration, stability of the hydrological cycle, and overall productivity of biome.

The decline in soil resilience and quality is driven by factors such as overplowing, removal of organic residues, depleted soil carbon stocks, nutrient imbalances, and exploitative farming methods. Degradation processes include soil structure breakdown, loss of organic carbon, biodiversity reduction, increased crusting, compaction, runoff, and erosion, all of which deplete essential nutrients and water from ecosystems. To counteract these effects, sustainable practices focus on enhancing soil resilience through the management of organic carbon pools. Effective solutions involve integrated management of soil, water, crops, and livestock to restore and sustain soil health (Fig. 1.3).

1.5 Land Degradation and Human Health, Climate Change and Livelihood Security

Land degradation has extensive consequences, leading to food and water insecurity, economic hardship, and social challenges, including migration and conflict. DLDD also exacerbates climate change, which in turn intensifies its impacts on the environment, human health, and livelihoods. Degraded lands lose their ability to sequester carbon dioxide, increasing greenhouse gas emissions and exacerbating global warming. The reduction in vegetation cover raises surface temperatures and disrupts precipitation patterns, intensifying climate change effects (Lal 2015). This cycle contributes to worsening soil erosion, more frequent extreme weather events, and the spread of invasive species and disease vectors, along with shifts in species distribution that raise the risk of emerging and re-emerging diseases, including malaria, dengue, and Zika, as well as zoonotic infections. Rising temperatures and humidity create conditions favorable to disease-carrying vectors such as mosquitoes,

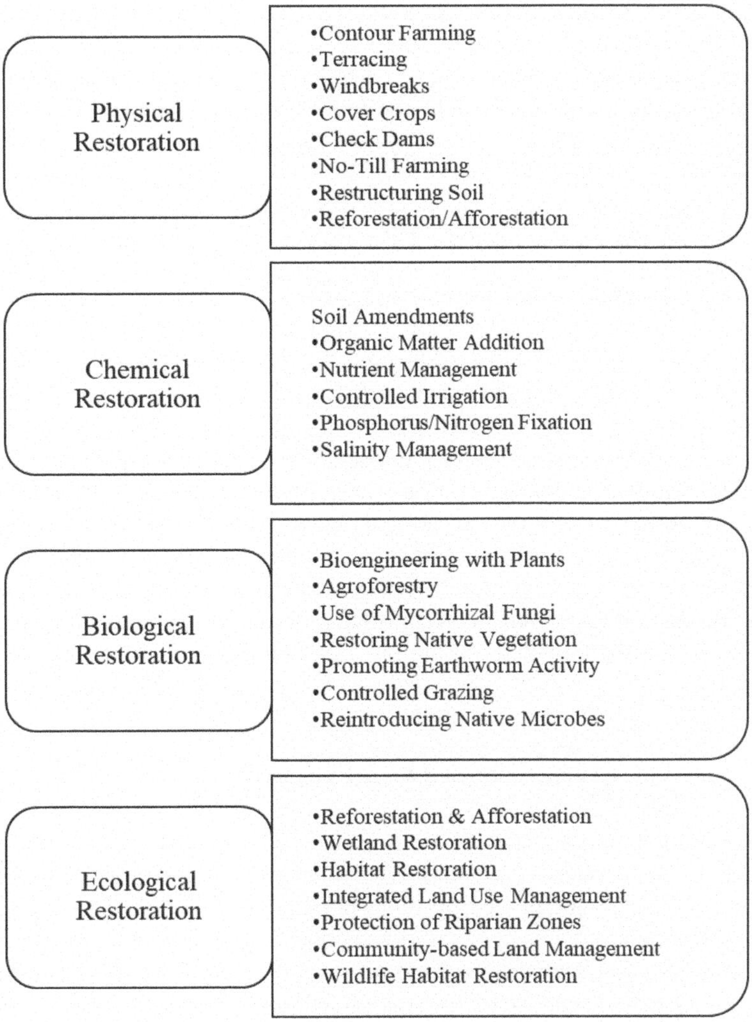

Fig. 1.3 Strategies for restoring soil quality

increasing the transmission of vector-borne diseases like malaria and dengue (Campbell et al. 2015; Sena and Ebi 2021).

Land degradation associated with climate change also increases fire risks, and water availability, reduces air quality, and contributes to human health issues due to water, land, dust, and air pollution from degraded soils. Frequent extreme events like droughts, floods, dust storms, and heat waves driven by both land degradation and climate change pose numerous health risks, including respiratory, skin, heart problems, infectious diseases, and injuries. Vulnerable populations, particularly those in low-income or unstable socioeconomic conditions, face the greatest risks from these effects. Sena and Ebi (2021) highlights the complex relationship between desertification, land degradation, drought (DLDD), and health outcomes across WHO's primary disease categories. Factors such as water availability and quality, sanitation, food security, air quality, soil health, and access to healthcare services mediate these health impacts. DLDD, through various exposure pathways, contributes to a wide range of health effects, encompassing up to 19 distinct health outcomes across these disease categories. This complex web of interconnections emphasizes the urgent need for an integrated management approach that addresses not only environmental factors but also social, economic, and health vulnerabilities.

The land degradation impacts are particularly severe in India for smallholder farmers who rely on the land for their livelihoods, it exacerbates poverty, and reduced agricultural productivity leads to lower incomes and increased food prices. This, in turn, can lead to migration, unemployment, social unrest, and conflicts over scarce resources. More than 58% of India's population depends on agriculture for livelihood. Land degradation directly impacts this large demographic, particularly smallholder farmers and rural communities. Reduced soil fertility translates to lower agricultural productivity, shrinking the incomes of farmers. Land degradation leads to a reduction in arable land, making cultivation economically unviable. In many cases, this forces migration to urban areas, contributing to increased urban poverty. The loss of ecosystem services, such as freshwater provision and biodiversity, also diminishes livelihood options, particularly in small and marginal communities.

1.6 Land Degradation and Food Security

Land degradation significantly threatens global food security. In 2024 the Global Hunger Index, India ranks 105th out of 127 countries, highlighting substantial food security challenges compounded by rapid population growth and climate change impacts. Although India is the world's second-largest producer of agricultural goods, continued soil degradation could necessitate future food imports, especially as the country has to support 18% of the global population with only 2.4% of the world's land area. By 2050, the demand for food grains is expected to rise to 400 million tonnes compared to the present 328 million tonnes, intensifying the need for SLM practices to support food production.

To address these challenges, SLM strategies like conservation agriculture, agroforestry, and reforestation play a critical role in improving soil health, water retention, and biodiversity (Das et al. 2022; Panwar et al. 2022). Conservation agriculture employs techniques such as minimum tillage, crop residue management, and crop rotation to reduce soil erosion and boost productivity. (Grace et al. 2013; Babu et al. 2023; Naik et al. 2023). Agroforestry integrates trees into agricultural systems, enhancing soil fertility and protecting against erosion (Sarkar et al. 2017; Babu et al. 2020; Kumar et al. 2022b).

Additionally, reforestation and afforestation in degraded areas contribute to soil restoration and carbon sequestration, while integrated water resource management,

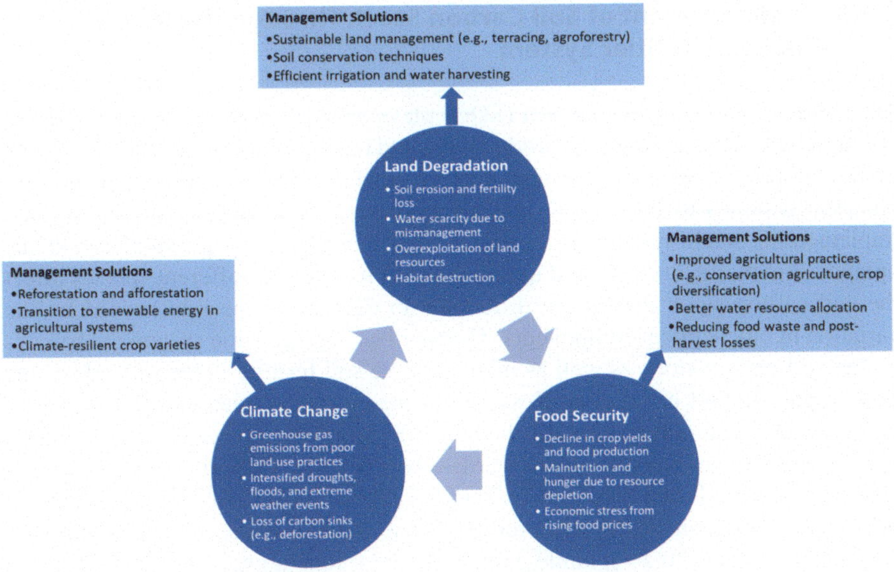

Fig. 1.4 Land degradation—food security—climate change nexus

including rainwater harvesting and drip irrigation, supports efficient water use in arid regions. Reclaiming degraded and problematic soils is also crucial, using techniques like liming, leaching, organic amendments, salt-tolerant crops to restore productivity, etc (Shinde et al. 2020). These practices align with the United Nations Sustainable Development Goals, as sustainable land management and soil restoration directly impact food security and climate resilience. Addressing the interlinked challenges of land degradation, food security, and climate change requires a holistic approach (Fig. 1.4), prioritizing soil health, moisture retention, and efficient water management (Abrol and Gupta 2019). Adopting these strategies can help farmers cultivate resilient crops, maintain soil health, and secure India's food and nutrition security for future generations.

1.7 Sustainable Land Management: Strategies for Restoring Degraded Soil Health

Sustainable land management plays a vital role in restoring degraded soil health, ensuring crop productivity, and supporting environmental resilience. Degraded soils pose important challenges to food security and ecosystem stability, making it essential to adopt effective restoration strategies. This approach integrates sustainable practices, conservation techniques, and community participation to enhance soil fertility, structure, and function for long-term agricultural and environmental sustainability.

1.7.1 Enhancement of Soil Carbon Stock Through the Mango-Based Multitier System

The storage of soil organic carbon (SOC) plays a crucial role in mitigating global climate change, enhancing crop productivity, and maintaining soil fertility. Cropping systems combined with management practices that promote the addition of substantial crop residues to the soil are key to increasing SOC stocks. Identifying and implementing such systems is essential for sustaining crop productivity in the uplands. Again, the quantity and quality of crop residues of different species affect the soil carbon stock due to the difference in decomposition rate caused by the variation in biochemical components. The tree-based production system contributes higher soil carbon stock as well as maintains the soil fertility in the uplands of the eastern plateau and hill region. Research in the eastern plateau and hill region of India revealed that mango orchards had the highest total soil organic carbon (SOC) at 62.5 Mg ha^{-1} among various orchards. Over 6 years, mango, guava, and litchi orchards increased total SOC, which stands 17.2%, 12.6%, and 11%, respectively, in comparison to control. Mango orchards showed the most significant increases in very labile carbon (20.7%), labile carbon (13.5%), less labile carbon (17.4%), and non-labile carbon (Naik and Bhatt 2017) and considered the best tree production system for enhancing soil carbon stock (Fig. 1.5).

The fruit-based multitier cropping system involves integrating fruit trees with large canopies, such as mango and litchi, as the primary crop, complemented by precocious, dwarf-canopy fruit species like guava, peach, or aonla as filler crops planted between rows and individual trees. Intercrops were cultivated in the remaining interspaces to optimize land use. In this system, mango trees (cv. Amrapali) were spaced at 10 m × 20 m (plant-to-plant and row-to-row) as the main crop, with fast-growing mahogany timber species planted at the same spacing. The mango and mahogany plants were alternated within rows, maintaining a 5-m distance between them. The filler trees like aonla (Var. NA-7) and peach (Var. Florida prince) were

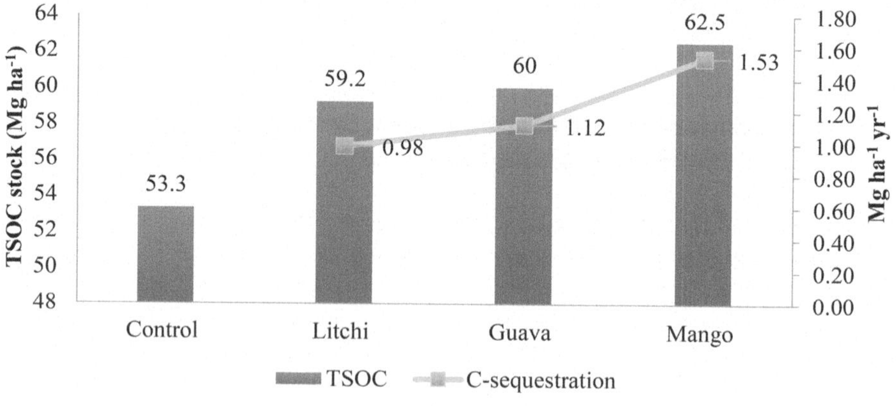

Fig. 1.5 Soil organic carbon stock and C-sequestration in 6-year-old fruit orchards of eastern plateau and hill region

planted 5 m apart between and within the rows of main plants of mango, while 10 m apart within the same rows. The intercrops like rice and finger millet were grown in different combinations in the multitier production system. The combination of Mango + Mahogany + Aonla + Rice and Mango + Mahogany + Peach + Rice demonstrated higher levels of total organic carbon, active and passive carbon pools, improved soil biological properties, and elevated carbon management indices. These combinations are considered the most effective production systems for carbon sequestration and maintaining soil health in the eastern plateau and hill regions of India.

1.7.2 Carbon Sequestration in Fruit Crops in the Rainfed Agroecosystems

Orchard fruit trees were classified as trees outside forests, are vital for carbon sequestration and offer significant potential for economic opportunities through carbon trading. Given that a large portion of India's population depends on agriculture, the adoption of fruit based multitier systems serves as a long-term, sustainable strategy to enhance carbon credit generation. These systems also contribute additional revenue streams through the production of timber, fuelwood, fodder and food, while supporting agricultural livelihoods. The biomass of below-ground and above-ground carbon stock in mango, guava and litchi orchards of varying ages was assessed in the hot and dry sub-humid climate of the eastern hill and plateau region. In litchi orchards, the biomass carbon stock ranged from 0.10 to 1.85 Mg ha^{-1} in 2–10-year-old orchards (Naik et al. 2018). Carbon mitigation in litchi, guava, and mango orchards varied with age (Fig. 1.6). In guava orchards, biomass carbon stock ranged from 0.24 to 4.55 Mg ha^{-1} in 2–10-year-old orchards, with carbon mitigation between 0.13 and 3.14 Mg ha^{-1}, sequestering 0.50–11.54 Mg CO_2 ha^{-1} from the atmosphere (Naik et al. 2021). In mango orchards, the total stored biomass carbon (branches, bole, roots) ranged from 0.2 to 3.87 Mg ha^{-1} in 2–10-year-old orchards, with the bole contributing 0.09–1.86 Mg ha^{-1}, accounting for 45–48% of the total carbon stored. Mango orchards' CO_2 mitigation potential ranged from 0.57 to 11.04 Mg ha^{-1} in 2–10-year-old orchards (Fig. 1.7) (Naik et al. 2019). The study highlights that CO_2 mitigation is directly linked to biomass production in these components. The developed carbon-stock models can help fruit growers estimate biomass carbon at various growth stages by the measurement of collar diameter and facilitate opportunities for supplementary revenue through carbon credits.

1.7.3 Basin Enrichment in the Orchards of Rainfed Agroecosystem

The rising need for organic produce, coupled with the increasing costs of manure and fertilizers, has created a need for alternative, cost-effective strategies for viable nutrient management. Cultivating biomass-producing plants in the alley areas of

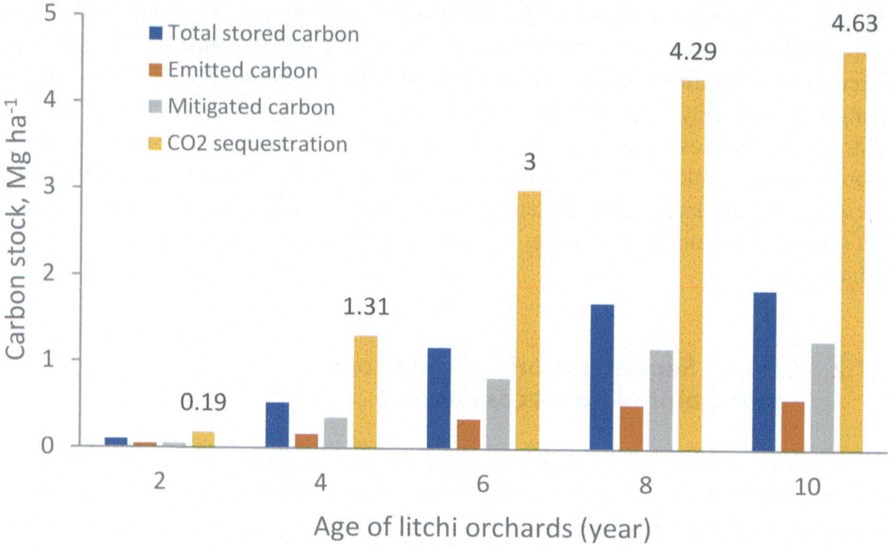

Fig. 1.6 Biomass carbon stock in litchi orchards of the eastern plateau and hill region

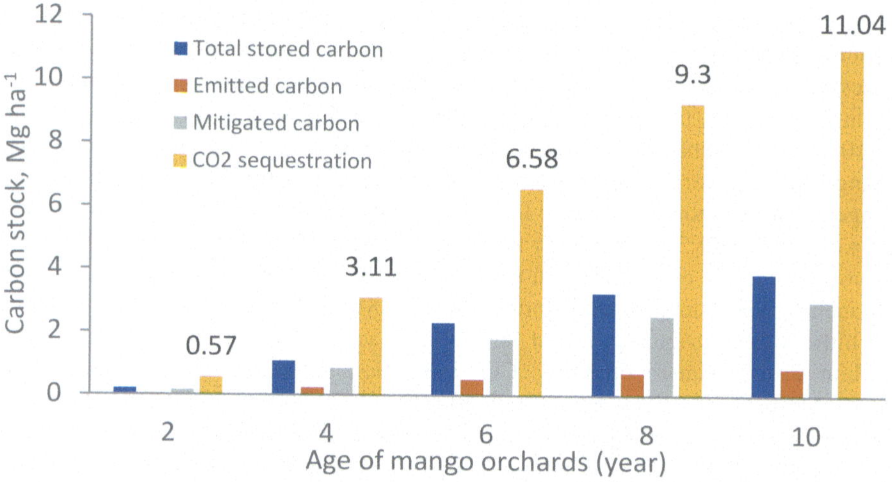

Fig. 1.7 Biomass carbon stock in mango orchards of eastern hill and plateau region of India

orchards and reutilizing the harvested biomass into the plant basins of fruit tree has proven effective in enhancing soil fertility and supporting the growth of various fruit tree species. Biological nutrient recycling using Tephrosia candida in fruit orchards is a promising and affordable method for enhancing soil fertility and boosting fruit crop productivity, especially in low-fertility regions. A 6-year study was investigated in the eastern hill and plateau region of India using a Bael (*Aegle marmelos* Correa) as test crop (Fig. 1.8). The study focused on enriching the plant basin

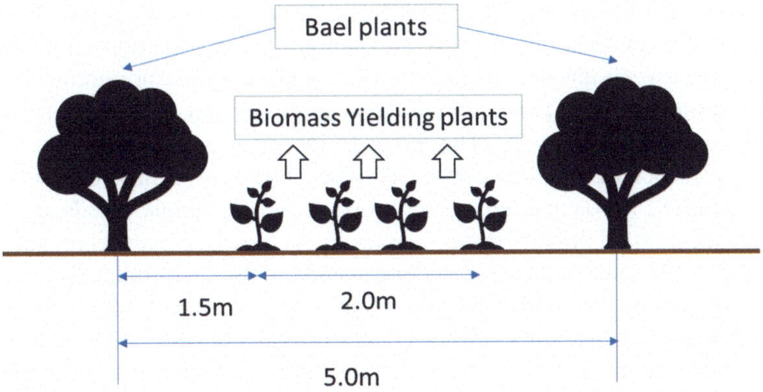

Fig. 1.8 Schematic representation of the basin enrichment of orchards

through recycling of biomass from various biomass-producing plants (Das et al. 2021). A significant increase in available N, P, and K levels across various treatments was noted compared to the control. Additionally, mulching *Tephrosia candida* notably improved soil moisture levels during the drier months. This study demonstrated that integrating *Tephrosia candida* into the alley areas of Bael orchards is an effective strategy for enhancing soil fertility and promoting plant growth.

1.7.4 Management of Yield-Limiting Nutrients in Fruit Orchards in a Rainfed Agroecosystem

Accurate information on leaf nutrient concentration is crucial for effective fertilizer application, leading to the development of leaf analysis-based Diagnosis and Recommendation Integrated System (DRIS) indices for mango fruit yield in the eastern hill and plateau region (Naik and Bhatt 2017). DRIS analysis revealed a relative deficit of boron, zinc and magnesium, alongside a relative abundance of sulphur, phosphorus, calcium, potassium and nitrogen in 9-year-old mango orchards. In young orchards (6–7 years), the nutrient demand hierarchy was identified as calcium > sulfur > potassium > boron > nitrogen > phosphorus > magnesium > zinc. Boron consistently emerged as the most yield-minimizing nutrient across all orchard age groups. Soil pH also influenced nutrient limitations: calcium and magnesium were critical below pH 5.5, whereas boron, zinc and nitrogen were most limiting above pH 5.5. To achieve optimal yields, a balanced application of N, S, Zn, B, Ca, Mg and K is essential for mango orchards in acidic soils within this region.

1.7.5 Planting Systems with Polythene Mulching

Fertigation applies nutrients directly to the root zone through drip emitters, significantly enhancing fertilizer-use efficiency compared to traditional methods. Frequent nutrient applications via drip irrigation have shown excellent crop response and improved efficiency (Mali et al. 2016). Combining drip irrigation with polythene mulching conserves vital resources like water and nutrients, promoting crop production and productivity. This approach also enhances soil moisture retention and increases topsoil temperature. The process involves covering the soil with a 50-micron bicolor LLDPE plastic sheet. The sheet is tightly secured without creases, with edges buried 7–10 cm deep in small furrows at a 45° angle. Seeds are sown or seedlings transplanted directly through holes in the mulch, spaced as recommended for the crop. Field trials in the eastern plateau and hill region of India evaluated various planting systems, including flat-bed, broad-bed, and ridge-furrow, with or without polythene mulching, for cauliflower and brinjal. Results highlighted that the ridge-furrow system with polythene mulching significantly boosted crop productivity while enhancing soil quality in the hot, sub-humid climate of the region (Naik et al. 2022).

1.7.6 Improving Vegetable Yields with Precision Nutrient Management

To boost crop yields, excessive fertilizer application has become common, often resulting in reduced efficiency and environmental harm due to nutrient runoff into water resources. Additionally, rising irrigation infrastructure costs and escalating fertilizer prices highlight the need to improve water and nutrient use efficiency at the field level. Beyond ecological and management factors, crop productivity under drip fertigation are also shaped by planting geometry and plant spacing. Research indicates that applying the right amount of fertilizer at the appropriate growth stage yields better results than using a uniform dosage. Vegetable crops having different planting systems in combination with fertigation doses recommended in the eastern plateau and hill region of India are,

Chilli: A triangulated paired-row planting system (60 cm spacing between paired rows, with 40 × 30 cm crop spacing) combined with stage-specific nutrient application significantly enhances chilli yields. The recommended schedule involves applying 6% of the recommended fertilizer (RF) during weeks 3–8, 8% during weeks 9–10, 10% during weeks 11–13, 4% during weeks 14–15, and 2% during weeks 16–20 after transplanting (Mali et al. 2019).

Broccoli: A planting setup similar to chilli (60 cm between paired rows, 40 × 30 cm crop spacing), combined with precise fertigation timing, resulted in a 46.7% higher yield compared to conventional practices. The nutrient application consisted of 7% RF during weeks 1–4, 8% during weeks 5–7, 12% during weeks 8–9, 9% during weeks 10–11, and 2% during weeks 12–14 after transplanting (Mali et al. 2016).

Tomato: For tomato crops, triangulated paired-row planting (60 cm spacing between rows, with a crop spacing of 40 × 70 cm) combined with mid-stage fertigation showed the highest yield potential. Fertilizer was applied at 2% RF during weeks

1–6, 4% during weeks 7–9, 8% during weeks 10–12, 10% during weeks 13–16, and 6% during weeks 17–18 (Mali et al. 2016).

Sweet Corn: A rectangular planting system (50 × 15 cm spacing, 1,33,300 plants per hectare) with late-stage fertigation achieved maximum cob yields. Fertilizer application followed a schedule of 4% RF during weeks 1–3, 6% during weeks 4–6, 8% during weeks 7–9, 10% during weeks 10–12, and 8% during weeks 13–14 (Jha et al. 2015).

1.7.7 Rainwater Harvesting Using Doba Technology

Rainwater harvesting using plastic-lined *doba* pits has proven effective for providing lifesaving irrigation to newly planted mango trees in upland areas of Jharkhand (Naik et al. 2016). This method involves digging pits of 3.0 m × 1.5 m × 1.0 m dimensions and lining them with UV-stabilized black polythene (250 microns). During the rainy season, the collected rainwater is stored, and the pit is covered with locally sourced thatch to prevent evaporation. Each *doba* can store up to 4500 l of water, sufficient to sustain 10 fruit plants during dry periods. The *doba* can also be utilized to store water from seasonal streams, aiding the establishment of fruit orchards. This technology has gained widespread adoption among farmers in Jharkhand, Chhattisgarh, West Bengal, and Odisha, significantly benefiting mango, litchi, and guava cultivation.

1.7.8 Conservation Agriculture to Sustain Soil Health and Food Security

Rice-fallow areas, predominantly in Eastern India, account for 84% (9.7 Mha) of the country's total rice-fallow land (11.7 Mha). These areas face productivity challenges due to poor soil health and limited moisture during the dry season, hindering the cultivation of subsequent crops. Conservation agriculture (CA)-based practices have demonstrated significant improvements in soil health, carbon sequestration, and productivity, enabling the inclusion of second and third crops in rice-fallow systems. The Direct Seeded Rice (DSR)–Mustard–Blackgram cropping system under CA in the eastern plateau and hill region emerged as highly efficient, offering enhanced soil organic carbon (SOC) stabilization, biological activity, and reduced carbon footprints (Naik et al. 2023). Integrating legumes into cereal-based systems further improved soil quality and SOC levels, especially under no-till practices. Legume cover crops, with their resistant carbon compounds, increased soil carbon stability and diversity (Ghosh et al. 2012).

In eastern India, zero-till direct-seeded rice (ZTDSR) reduced carbon footprints by 2.8% compared to transplanted puddled rice systems, particularly when paired with residue retention and winter crops like chickpea, lentil, safflower, linseed, and mustard (Kumar et al. 2022a). Reduced tillage with 30% residue retention in rice–maize systems recorded the highest SOC stock (51.0 Mg ha^{-1}) after 3 years (Mondal et al. 2020), emphasizing the benefits of no-till and residue management for sustainable intensification. This optimized approach offers a pathway to enhance soil carbon dynamics, reduce environmental impacts, and increase productivity in rice-fallow ecosystems.

1.8 Reclaiming Coalmine Affected Barren Lands

Land degradation stands among the most critical global environmental issues, severely impacting food security, biodiversity, water resources, and livelihoods, particularly in developing countries. It involves the reduction of land productivity caused by factors viz., deforestation, desertification, soil erosion, and unsustainable farming practices. In regions like Jharkhand, India, the problem is exacerbated by extensive coal mining, leading to significant deforestation and loss of fertile land.

Land degradation in Jharkhand not only threatens the agricultural livelihoods of farmers but also exacerbates rural poverty and environmental degradation. Coal mining has devastated both natural ecosystems and agricultural productivity. Without restoration efforts, these degraded lands are locked into cycles of low productivity, soil erosion, and loss of biodiversity, leading to severe environmental and socio-economic consequences. The innovative interventions of the Indian Council of Agricultural Research (ICAR) have reversed land degradation caused by coal mining in Jharkhand's Phusri village. Through sustainable agricultural and agroforestry practices, degraded land has been reclaimed, offering new hope for environmental sustainability and livelihood security. Jharkhand is rich in mineral resources, particularly having more than 180 coal mines which has long been a major contributor to India's coal industry. However, intensive mining activities have left vast tracts of land barren, unfit for agricultural use, and prone to soil erosion. The village of Phusri, situated near Charhi, Mandu in Ramgarh district, exemplifies the challenges faced by local communities living in these mine-affected areas (Sarkar et al. 2019). Before the interventions, the villagers of Phusri relied on quarrying raw coal for income, and their agricultural activities were limited to cultivating rainfed lowland paddy during the monsoon season. Water from abandoned mining pits, which dried up by March, was the only source for washing, bathing, and occasionally irrigating crops. With only a few trees and no sustainable farming options, the village suffered from food insecurity, poor nutrition, and economic instability.

1.8.1 The ICAR Interventions: A Sustainable Approach

Recognizing the urgent need for reclamation, ICAR- Research Complex for Eastern Region, Farming System Research Centre for Hill and Plateau Region (FSRCHPR), Ranchi, Jharkhand initiated a project in Phusri in 2014 (Anonymous 2019; Sarkar et al. 2019). This project aimed to restore the region's agricultural productivity while enhancing environmental sustainability. The goal was to transform barren, mining-affected lands into fertile agricultural land using innovative, low-cost, and sustainable practices. The following techniques were implemented to improve soil and water retention and overall ecosystem health:

Plastic-Lined Dobhas: Small, plastic-lined water harvesting pits were created to capture rainwater during the monsoon season, ensuring a year-round water supply.

Full Moon and Half Moon Terracing: These innovative terracing techniques were designed to capture and retain water, reduce soil erosion, and improve soil moisture content. The curvature of the terraces allowed water to infiltrate the soil slowly, supporting plant growth during dry spells.

Staggered Contour Trenching: This method was employed to break the flow of water, reducing runoff, and allowing moisture to seep deeper into the soil.

Mulching: Fresh as well as dried crop residues were used as mulch to cover the soil, retaining moisture, reducing weed growth, and adding organic matter as the mulch decomposed.

1.8.1.1 Agroforestry: A Model for Land Reclamation

At the core of the reclamation strategy was the implementation of an agri-horti-silvi-pastoral system, a multi-tier approach integrating agriculture, horticulture, silviculture, and pastoral farming. This model was crucial for ensuring sustainable land use and providing diverse benefits to local communities.

Species Selection for Agroforestry

- *Agroforestry Trees*: Multipurpose Tree species (MPTs) were carefully selected for their adaptability to the local environment and their ability to provide multiple benefits. Some of the key species planted included:
  - *Aegle marmelos* (Bael)
  - *Artocarpus heterophyllus* (Jackfruit)
  - *Citrus limon* (Lemon)
  - *Mangifera indica* (Mango)
  - *Punica granatum* (Pomegranate),
  - *Psidium guajava* (Guava),
  - *Tectona grandis* (Teak)
  - *Swietenia mahogany* (Mahogany)
  - *Dalbergia latifolia* (Kala Shisham)
  - *Melia azedarach* (Bakain)
  - *Pongamia pinnata* (Karanj),
- *Fodder Crops*: In addition to tree species, *Tephrosia candida* (Bhoomi sudha), bamboo species such as *Dendrocalamus asper* and *Saccharum spontaneum* (Kans grass) were introduced to provide fodder for livestock, along with the dual benefits of nitrogen fixation and nutrient cycling in the soil.

The agroforestry model created a self-sustaining ecosystem that generated economic value for local farmers while restoring environmental health. Fruit trees provided income through the sale of produce, while timber and fodder trees contributed to household needs and livestock farming. The inclusion of hardy forest trees like Teak, Mahogany, Bakain, Karanj, and Kala Shisham also helped protect the land from erosion and acted as carbon sinks, contributing to climate change mitigation.

1.8.1.2 Empowering Local Communities: Women in Agroforestry

In rural Jharkhand, women have been instrumental in promoting sustainable agriculture, particularly through organic farming techniques that reduce reliance on chemical pesticides and fertilizers. This shift has enhanced soil health and product quality. In Gumla District, women tackled pest infestations in mango orchards by using innovative, low-cost methods such as covering fruits with paper envelopes and employing pheromone traps like Methyl eugenol to manage fruit flies. These practices eliminated the need for toxic pesticides, yielding healthier, pesticide-free produce. Beyond improving agricultural outcomes, these initiatives empowered women economically and elevated their roles in household decision-making, with many becoming advocates for eco-friendly farming, inspiring their communities.

1.8.1.3 Reclaiming Mine Lands: A Success Story

The transformation of Phusri's barren lands into productive agricultural fields was nothing short of a remarkable success. By 2019, the village had achieved significant milestones in both environmental restoration and livelihood security:

- *Soil Health Improvements*: The adoption of soil and water conservation measures, including techniques like Dobha pits, full-moon and half-moon terraces, staggered contour trenching, mulching, and zero-tillage practices, has significantly curbed soil erosion and improved land productivity (Fig 1.9). These prac-

Fig. 1.9 Measurement of sediment discharge under developed agroforestry model over control at Phusri coal mine-affected area

tices fostered vegetation establishment, with root systems effectively stabilizing the soil. As a result, SOC content increased from 0.23% to 0.58%. Improvements in soil pH and electrical conductivity further enhanced conditions for crop growth. Mulching with fresh and dried vegetation, coupled with the decomposition of organic matter and agricultural activities, boosted microbial populations and their activity.

- *Increased Productivity*: Villagers successfully established paddy nurseries on previously barren lands, marking a significant achievement (Fig. 1.10). Additionally, the cultivation of diverse vegetables and fruits, including *tomatoes, chilli, brinjal, and mustard,* led to an approximate 30% improvement in the nutritional intake of local families (Fig. 1.11). After 8 years of plantation, *Bakain* exhibited the highest *plant height (10.98 ± 0.40 m)* and *trunk diameter (40.13 ± 1.80 cm)* among the multipurpose tree species (MPTs). Among fruit crops, *Bael* recorded the maximum *plant height (5.24 ± 0.38 m), trunk diameter (37.87 ± 4.29 cm),* and *plant spread.* Furthermore, a total of *628 kg ha*$^{-1}$ *green biomass* was reaped from *Kans grass (Saccharum spontaneum), Melia azedarach,* and *Bhoomi sudha (Tephrosia candida).*
- *Economic Benefits*: The adoption of the agroforestry model has proven to be a highly profitable strategy for enhancing farmers' livelihoods. Under the agroforestry system implemented in coal mine-affected areas, *MPTs* were planted at a spacing of 5 m × 5 m (equivalent to 400 trees ha^{-1}). The study highlighted the economic potential of individual species, with *Bakain* showing the highest economic return of Rs. 349,899 per ha after 8 years, followed by Teak and Mahogany. Additionally, villagers achieved a minimum net return of Rs. 42,683 ha^{-1} from

Fig. 1.10 Paddy nursery- a component of the developed agroforestry model at the Phusri coal mine-affected area

Fig. 1.11 Cultivation of seasonal vegetables in the alleys of developed agroforestry model

the crops growing in rainy season and Rs. 18,293 ha^{-1} from winter crops, while surplus vegetables and fruits were sold in local markets. This increase in income substantially improved household livelihoods and reduced the need for migration to urban areas in search of employment opportunities.

- *Saving water, labor, time, and energy*: Water conservation practices effectively reduced water usage, labor, time, and energy demands. A standardized summer irrigation schedule was established to maintain consistency. In 2014, it was noted that soil moisture levels dropped to 6–8% within 24 h of irrigation and continued to decline over time. However, the adoption of the agroforestry model combined with intercropping has significantly enhanced the soil's water retention capacity. This improvement is primarily attributed to the decomposition of crop residues, which increases organic matter, promotes microbial populations, and enhances their activity in the soil.
- *Biodiversity Revival*: The introduction of agroforestry species and sustainable land management practices led to a significant increase in biodiversity. Grey patches of land are turning green. The number of weed species in the area grew from 7 in 2014 to 23 in 2023, while wildlife such as birds, wild boars, and reptiles began to reappear in the regenerated landscapes.
- *Carbon sequestration and environmental sustainability*: One of the most significant achievements of the project was its positive impact on the environment. By reintroducing vegetation and implementing soil conservation techniques, the project helped mitigate the effects of climate change and restored the natural ecosystem. The increase in vegetation cover also contributed to carbon sequestration, while the growing microbial population in the soil improved soil fertility and resilience against climate variability. The annual carbon sequestration by

tree biomass (above and below ground) of the system after 8 years of plantation was recorded as 0.266 ± 0.040 Mg C ha^{-1}, where maximum potential for annual carbon sequestration was observed in Bakain tree (0.084 ± 0.006 Mg C ha^{-1}).

1.8.2 Lessons for Global Sustainability

The reclamation of degraded lands in Jharkhand serves as a model for other regions facing similar challenges. The success of this project demonstrates that land degradation is reversible with the right combination of sustainable land management, community engagement, and technological innovation. By focusing on agroforestry, organic farming, and water conservation, communities can restore degraded lands, improve biodiversity, and enhance food security.

The key lessons are,

Community Participation: Engaging local communities, especially women, is vital for the success of land reclamation projects.

Agroforestry Integration: Combining agriculture, horticulture, and forestry offers a holistic solution to land degradation.

Sustainable Practices: Eco-friendly farming techniques such as mulching, organic pest control, and water conservation can significantly enhance soil health and agricultural productivity.

The success story of Phusri village offers hope that land degradation, even in the harshest conditions, can be reversed. Through community involvement, scientific innovation, and sustainable practices, degraded lands can be transformed into productive ecosystems, supporting both human livelihoods and environmental sustainability. As the global community continues to grapple with land degradation and its impacts, this story from Jharkhand serves as a blueprint for creating resilient landscapes capable of supporting future generations.

1.9 Indian Government Policies and Future Prospects

To address land degradation and encourage SLM, the Indian government has implemented a series of significant policies and programs:

- Desert Development Programme (DDP): Established in 1995, this initiative focuses on mitigating the effects of drought and rehabilitating natural resources in desert regions to support ecosystems and local communities.
- United Nations Convention to Combat Desertification (UNCCD): India joined the UNCCD in 1994 and ratified it in 1996. As part of its commitment to Sustainable Development Goal (SDG) target 15.3, India aims to restore 26 million hectares of degraded land by 2030, thereby achieving Land Degradation Neutrality (LDN) to support ecosystems, food security, and ecosystem services.

- National Afforestation Programme (NAP): Since 2000 and overseen by the Ministry of Environment, Forest and Climate Change, this program focuses on reforesting degraded lands, promoting biodiversity, carbon sequestration, and increasing forest cover.
- National Action Programme to Combat Desertification (NAPCD): Launched in 2001, this program implements proactive strategies to combat desertification and improve resilience in degraded areas.
- Mahatma Gandhi National Rural Employment Guarantee Act (MGNREGA): Introduced in 2006, MGNREGA combines soil and water conservation with rural employment opportunities by supporting ecological interventions like water conservation and land restoration.
- Watershed Development Component of Pradhan Mantri Krishi Sinchayee Yojana (PMKSY): Launched in 2015, this program focuses on restoring 26 million hectares of degraded land through sustainable water management, afforestation, irrigation, and other conservation activities.
- National Initiatives for Climate-Resilient and Sustainable Agriculture: Initiated in 2010 under the National Action Plan on Climate Change (NAPCC), this initiative promotes agricultural productivity, sustainability, and climate resilience.
- National Mission for Sustainable Agriculture (NMSA): Launched in 2010 as part of the National Action Plan on Climate Change, NMSA aims to enhance agricultural productivity while addressing climate change by promoting sustainable agricultural practices.
- National Mission for a Green India (GIM): Established in 2014, GIM focuses on increasing forest cover and ecosystem resilience by promoting large-scale afforestation efforts across both forested and non-forest regions.
- Soil Health Card Scheme: Introduced in 2014–2015, this initiative aims to combat declining soil health by promoting soil testing and encouraging targeted, test-based fertilizer applications.

The focus on integrated land management highlights the need to combat land degradation and advance sustainable development. The government's efforts in afforestation, watershed management, and climate-resilient agriculture are significant steps toward achieving land degradation neutrality. However, challenges such as population growth, climate change, and intensive farming will continue to strain land resources, impacting soil health, rural livelihoods, and food security. Sustainable measures like organic farming, agroforestry and conservation agriculture offer promising solutions by improving soil quality and resilience. Technological tools like GIS and AI support effective land monitoring and planning, while government policies and global commitments strengthen restoration efforts. With community participation and eco-friendly farming, India can address climate and food security challenges through sustainable land use. Long-term success depends on collaboration among the government, civil society, and local communities to implement these practices.

1.10 Conclusions

Land degradation is a foremost threat to environmental sustainability and livelihood security, especially in developing countries such as India. Its effects on climate, human being and the environment are far-reaching, posing serious risks to food security and economic growth. Combating land degradation demands a holistic approach that incorporates sustainable land management, effective restoration methods, and strong government support. By engaging local communities and combining traditional knowledge with modern technologies, it is possible to rehabilitate degraded lands and promote a sustainable future. Implementing these strategies successfully will be essential for meeting both global and national objectives related to land degradation neutrality and sustainable development.

References

Abrol I, Gupta R (2019) Climate change-land degradation-food security nexus: addressing India's challenge. J Agron Res 2(2):17–35. https://doi.org/10.14302/issn.2639-3166.jar-19-3015

Anonymous (2019) Annual report 2019. Patna, ICAR Research Complex for Eastern Region, ICAR Publication, pp 12–30

Babu S, Mohapatra KP, Yadav GS, Lal R, Singh R, Avasthe RK, Das A, Chandra P, Gudade BA, Kumar A (2020) Soil carbon dynamics in diverse organic land use systems in the North Eastern Himalayan ecosystem of India. Catena 194(2020):104785. https://doi.org/10.1016/j.catena.2020.104785

Babu S, Raghavendra S, Avasthe RK, Rathore SS, Kumar S, Das A, Layek J, Sharma V, Wani OA, Singh VK (2023) Conservation tillage and diversified cropping enhance system productivity and ecoefficiency and reduce greenhouse gas intensity in organic farming. Front Sustain Food Syst 7:1114617. https://doi.org/10.3389/fsufs.2023.1114617

Campbell LP, Luther C, Moo-Llanes D, Ramsey JM, Danis-Lozano R, Peterson AT (2015) Climate change influences on global distributions of dengue and chikungunya virus vectors. Philos Trans R Soc Lond Ser B Biol Sci 370:20140135. https://www.ncbi.nlm.nih.gov/pmc/articles/PMC4342968

Das B, Sarkar PK, Dhakar MK, Naik SK, Maurya S, Singh AK, Kumar S, Bhatt BP (2021) Basin enrichment of bael plants (Aegle marmelos Correa) through alley cropping of biomass producing plants: effects on plant growth and soil properties. Fruits 76(2):61–71

Das A, Babu S, Datta MK, Sanjeev SR, Avasthe RK, Rathore SS, Yadav SK, Singh VK (2022) Restoring soil carbon in marginal land of Indian Himalayas: impact of crop intensification and conservation tillage. J Environ Manag. https://doi.org/10.1016/j.jenvman.2022.115603

FAO (1999) Poverty alleviation and food security in Asia. In: Land degradation, Chapter 3. Food and Agricultural Organization of the United Nations (FAO). Accessed 18 Oct 2020

Ghosh PK, Venkatesh MS, Hazra KK, Kumar N (2012) Long-term effect of pulses and nutrient management on soil organic carbon dynamics and sustainability on an Inceptisol of Indo-Gangetic Plains of India. Exp Agric 48(4):473–487

Grace JK, Sharma KL, Sumachandrika D, Srinivas K, Mandal UK, Raju BMK, Korwar GR, Venkateswarlu B, Kumar S, Sankar GM, Kumar TS (2013) Effect of long-term use of tillage, residues, and n levels in sorghum (sorghum vulgare (l))—castor (ricinus communis) cropping system under rainfed conditions—crop responses and economic performance—part I. Exp Agric 49(3):395–415

IPCC (2019) Special report on climate change and land. https://unfccc.int/sites/default/files/resource/20191204_COP_SRCCL_SBSTA_V10.pdf

Jha BK, Mali SS, Naik SK, Singh AK, Kumar A (2015) Optimal planting geometry and growth stage based fertigation in vegetable crops. Technical bulletin no. R-56/RAN-25. Research Centre/ICAR-Research Complex for Eastern Region, Ranchi/Patna, pp 1–34

Kumar R, Mishra JS, Mali SS, Mondal S, Meena RS, Lal R, Jha BK, Naik SK, Biswas AK, Hans H, Sundaram PK (2022a) Comprehensive environmental impact assessment for designing carbon-cum-energy efficient, cleaner and eco-friendly production system for rice-fallow agro-ecosystems of South Asia. J Clean Prod 331:129973

Kumar SS, Wani OA, Mir SA, Babu S, Sharma V, Chesti MUH, Baba ZA, Sofi PA, Wani FJ, Dar SR, Singh R, Yadav D (2022b) Soil carbon dynamics in the temperate Himalayas: impact of land use management. Front Environ Sci 10:1009660. https://doi.org/10.3389/fenvs.2022.1009660

Lal R (2015) Restoring soil quality to mitigate soil degradation. Sustainability 7(5):5875–5895

Mali SS, Jha BK, Naik SK, Singh AK, Kumar A (2016) Effect of fertigation pattern and planting geometry on growth, yield and water productivity of tomato (Solanum lycopersicum). Indian J Agric Sci 86(9):1208–1213

Mali SS, Naik SK, Jha BK, Singh AK, Bhatt BP (2019) Planting geometry and growth stage linked fertigation patterns: Impact on yield, nutrient uptake and water productivity of Chilli pepper in hot and subhumid climate. Sci Hortic 249:289–298

Mondal S, Naik SK, Haris AA, Mishra JS, Mukherjee J, Rao KK, Bhatt BP (2020) Effect of conservation tillage and rice-based cropping systems on soil aggregation characteristics and carbon dynamics in Eastern Indo-Gangetic Plain. Paddy Water Environ 18:573–586

Naik SK, Bhatt BP (2017) Diagnostic leaf nutrient norms and identification of yield-limiting nutrients of mango in eastern plateau and hill region of India. Commun Soil Sci Plant Anal 48(13):1574–1583

Naik SK, Mali SS, Das B, Bhatnagar PR, Kumar S, Sikka AK (2016) Rainwater harvesting using plastic-lined Doba technology for orchard establishment in the eastern plateau and hill region of India. Curr Sci 111(11):1751–1753

Naik SK, Sarkar PK, Das B, Singh AK, Bhatt BP (2018) Predictive models for dry biomass and carbon stock estimation in Litchi chinensis under hot and dry sub-humid climate. Arch Agron Soil Sci 64(10):1366–1378

Naik SK, Sarkar PK, Das B, Singh AK, Bhatt BP (2019) Biomass production and carbon stocks estimate in mango orchards of hot and sub-humid climate in eastern region, India. Carbon Manag 10(5):477–487

Naik SK, Sarkar PK, Das B, Singh AK, Bhatt BP (2021) Biomass production and carbon stock in Psidium guajava orchards under hot and sub-humid climate. Curr Sci 120(10):1627–1635

Naik SK, Jha BK, Singh AK (2022) Drip fertigated planting systems with polythene mulching on cauliflower–eggplant cropping systems in hot and subhumid climate: impact on soil health and crop yield. Commun Soil Sci Plant Anal 53(10):1261–1276

Naik SK, Mali SS, Jha BK, Kumar R, Mondal S, Mishra JS, Singh AK, Biswas AK, Choudhary AK, Choudhary JS, Hans H (2023) Intensification of rice-fallow agroecosystem of South Asia with oilseeds and pulses: impacts on system productivity, soil carbon dynamics and energetics. Sustainability 15(2):1054

Olsson L, Barbosa H, Bhadwal S et al (2019) Land degradation. In: Shukla PR, Skea J, Calvo Buendia E et al (eds) Climate change and land: an UNCCD special report on climate change, desertification, land degradation, sustainable land management, food security, and greenhouse gas fluxes in terrestrial ecosystems. Accessed 18 Oct 2020

Panwar AS, Ansari MA, Ravisankar N, Babu S, Prusty AK, Ghasal PC, Chaudhary J, Shamim M, Singh R, Raghavendra KJ, Dutta D, Meena AL, Chauhan GV, Ansari MH, Singh R, Aulakh CS, Singh DK, Sharma PB (2022) Restoration of soil quality, ecosystem services and productivity of rice-wheat system: 15- years impact of organic farming under diverse agro-ecosystems. Front Environ Sci 10:972394

Sarkar PK, Bishnoi SK, Shinde R, Das B (2017) Improvement in agroforestry system. Indian Farm 67(7):19–20

Sarkar PK, Dhakar MK, Das B, Kumar PK, Maurya S, Mali SS, Shinde R, Choudhary JS, Naik SK, Kherwar D, Chakrabarti A, Raghav DK, Singh AK, Bhatt BP (2019) Healthy plant: foundation for nutritional and environmental sustainability. LEISA, New Delhi, pp 6–10
Sena A, Ebi K (2021) When land is under pressure health is under stress. Int J Environ Res Public Health 18:136. https://doi.org/10.3390/ijerph18010136
Shinde R, Thombare N, Sarkar PK, Verma S (2020) Management of acid soils for crop production. ENVIS JHARKHAND NEWS, Govt. of Jharkhand, Issue Feb–March 2017–18, published in the year 2020, pp 9–12
UNCCD (2019) UNCCD 14th session of the Conference of the Parties, UNCCD, 13 September 2019, New Delhi. https://www.unccd.int/cop14
UNCCD (2022) The global land outlook, 2nd edn. Bonn

Agricultural Land Restoration: Strategies and Challenges

2

Vipin Kumar, Sunil Kumar Prajapati, Ananya Gairola,
G. D. Sanketh, Aastika Pandey, and Devideen Yadav

Abstract

The degradation of agricultural land presents a serious global challenge, compromising food security, economic stability, and environmental sustainability. Prolonged unsustainable land management, accelerates soil degradation, resulting in the potential release of around 2/3rd of their initial carbon stores into the atmosphere as CO_2. Land degradation impacted nearly one fourth of the global land surface and the lives of ~3.2 billion people. Soil degradation encompasses nutrient depletion, diminished microbial activity, and degradation of physical structure, all of which contribute to decreased fertility and moisture retention capacity, often leading to the land abandonment. This condition is more aggravated in middle- and low-income nations, where agricultural intensification has been pursued to bolster crop yields swiftly. The loss of fertile arable land is accelerating annually with a rate of ~24 Mha and 23 hectare per minute and economic losses amounting to 10% of the global GDP annually. Furthermore, if appropriate land and crop management practices not upscaled, up to 95% of global land could be degraded by 2050. Hence, a comprehensive approach is highly warranted to arrest and/or restore the land degradation. Restoration efforts span various practices aimed at rejuvenating ecosystems, preserving biodiversity, and fostering economic development. Aligned with international objectives such as achieving Land Degradation Neutrality by 2030, restoring agricultural lands is crucial for securing food stability and sustainable agricultural practices.

V. Kumar (✉) · S. K. Prajapati · A. Gairola · A. Pandey
Division of Agronomy, ICAR-Indian Agricultural Research Institute, New Delhi, India

G. D. Sanketh
ICAR-Central Island Agricultural Research Institute, Port Blair, Andaman and Nicobar, India

D. Yadav
ICAR-Indian Institute of Soil & Water Conservation Research, Dehradun, Uttarakhand, India

© The Author(s), under exclusive license to Springer Nature Singapore Pte Ltd. 2025
S. Babu et al. (eds.), *Ecological Solutions to Agricultural Land Degradation*,
https://doi.org/10.1007/978-981-96-3392-0_2

Moreover, immediate and concerted action is imperative for land restoration to ensure a sustainable future for forthcoming generations.

Keywords

Biodiversity · Crop diversification · Precision farming · Organic matter

Abbreviations

~	Approximate
₹	Rupees
C	Carbon
CA	Conservation Agriculture
DCR	Diversified Crop Rotation
FAO	Food and Agriculture Organization
FGDG	Flue Gas De-Sulphurised Gypsum
FYM	Farm Yard Manure
GDP	Gross Domestic Product
GEF	Global Environment Facility
GLASOD	Global Assessment of Soil Degradation
GPS	Global Positioning System
GHG	Green House Gases
IPBES	Intergovernmental Science-Policy Platform on Biodiversity and Ecosystem Services
IPCC	Intergovernmental Panel on Climate Change
kg	kilogram
km^3	Cubic kilometre
LDN	Land degradation neutrality
m	Metre
Mg	Megagram
Mha	Million hectares
mS/cm	Milli Simon per centimetre
NAAS	National Academy of Agricultural Sciences
PGPR	Plant Growth Promoting Rhizobacteria
SDGs	Sustainable Development Goals
SLM	Sustainable Land Management
SOC	Soil Organic Carbon
Tg	Tera gram
UN	United Nations
UNCCD	United Nations Convention to Combat Desertification
USA	United States of America
USD	United States Dollar
VNIR	Visible and Near-Infrared
WOCAT	World Overview of Conservation Approaches and Technologies

2.1 Introduction

The degradation of agricultural land has emerged as a formidable global challenge, threatening food security, economic stability, and environmental sustainability. The repercussions of this pervasive issue are far-reaching, affecting nearly a quarter of the Earth surface and impacting livelihoods of ~3.2 billion people worldwide, especially smallholder farmers and rural communities (Montanarella 2016; Olsson et al. 2019). Economic toll of land degradation amounts to an astonishing 10% of the annual gross domestic product globally, emphasizing the critical need for concerted restoration efforts (IPBES 2018). A remarkable 25% of the global land area has undergone degradation, with an alarming annual loss of nearly 24 billion Mg of fertile soil (Chasek 2022). If current trends persist, ~95% of the gross land surface could face degradation by 2050 (GEF 2019). The implications are dire, particularly for the 1.5 billion human being who rely directly on degraded land for their livelihoods (UNCCD 2020a). Globally the economic costs of land degradation are ~85.8 trillion USD per annum (Reinl 2019). Land degradation contributes significantly to food insecurity, escalating food prices, environmental hazard, and depletion of bio-diversity and soil ecosystem services (Yadav et al. 2024). Hence a multifaceted approach centred on land restoration is highly warranted. *Land restoration is the ecological process aimed at reviving landscapes to support human settlements, wildlife, and plant communities* (UNCCD 2024). Land restoration presents an opportunity to protect ecosystems, foster economic development, mitigate natural disasters, enhance soil productivity, and secure food supplies (Das et al. 2022a; Yadav et al. 2023).

2.2 Concept of Restoration

The concept of restoration encompasses any deliberate action aimed at initiating or expediting the recovery of a degraded ecosystem (IPBES 2018). Efforts for restoration should be comprehensive and implemented at the landscape level, aiming to ecological integrity restoration, and support human well-being (Maginnis and Jackson 2002; Das et al. 2022b). Restoring degraded agricultural land encompasses a wide array of practices, including soil water conservation, afforestation, biodiversity promotion, and the preservation of natural vegetation (Yadav et al. 2021a). Furthermore, beyond reversing land degradation, restoration initiatives create opportunities to improve livelihoods and prepare for future challenges. Economically, investing in land restoration yields substantial returns, every rupee invested in restoration can generate up to ₹2509.5 benefits (UNEP 2022). This aligns with the United Nations Convention to Combat Desertification's mission, emphasizing actions that protect and revitalize land resources as crucial for achieving Land Degradation Neutrality (LDN) by 2030. Restoring arable and pasture land stands as a critical component of ensuring food security and sustainable agricultural practices (Wani et al. 2022). Meeting restoration targets requires intensified efforts focused on agricultural lands. The engagement of agricultural sectors in restoration efforts is imperative, given their direct involvement in communities and food production. Restoring degraded agricultural lands not only unlocks economic benefits but also ensures the resilience of food systems, essential for meeting the needs of a growing global population (Rathore et al. 2022a). Hence, there is need to explore various approaches, challenges, and the transformative potential of agricultural land restoration in mitigating the pervasive impacts of land degradation on ecosystems and human well-being.

LDN Goal- 2030

UNCCD, 3/COP.12 of the, Land Degradation Neutrality (LDN) is defined as the condition wherein the quantity and calibre of land resources essential for sustaining ecosystem functions and services, as well as fortifying food security, either remain constant or elevate within predetermined temporal and spatial scales across various ecosystems (UNCCD 2015). Presently, 128 nations have pledged to establish targets for achieving LDN. The UNCCD Secretariat approximates that approximately 2.5 billion individuals in rural areas derive their sustenance from agriculture and natural resources (UNCCD 2020b).

2.3 Historical Perspectives on Agricultural Land Degradation

2.3.1 Overview of Agricultural Land Degradation

The deterioration of arable land, termed as *Agricultural land degradation*, signifies a decline in the inherent ability of soil to yield economic resources and services, which is attributable to both natural processes and the human activities combined (Hossain et al. 2020). The soil productivity has been consistently threatened by the pervasive issue of land degradation since the advent of agricultural practices (Sullivan 2004; Utuk and Daniel 2015). There are several evidences supporting that extensive degradation of soil has taken place globally for the last 5000 years (Scherr 1999). For instance, an approximate of around 2 billion hectares of formerly biologically productive land have undergone irreversible degradation over past millennium (Nkonya et al. 2016).

The degradation of agricultural land was a major global concern in the twentieth century and is also expected to remain a critical issue in the twenty-first century (Table 2.1). This issue has emerged as a major hurdle to achieve the sustainable advancement of agriculture. Current estimations indicate its impact on ~23% of the planet's terrestrial area, expanding by 5–10 Mha annually, profoundly affecting the livelihoods of ~1.5 billion individuals globally. In India, ~120.4 Mha of area face degradation due to various factors, out of which water erosion accounts for 82.6 Mha, wind erosion impacts 12.0 Mha, chemical degradation affects 24.8 Mha, while physical degradation affects 1.0 Mha (NAAS 2010). The losses due to water erosion alone causes a staggering crop output worth ₹292 billion (Sharda et al. 2019). Such extensive nutrient depletion and diminished crop output have profound

Table 2.1 Worldwide land degradation statistics

Categories	Share of land (%)
Most degraded land	25
Moderately degradation trend	36
Moderately and slightly degradation trend	8
Under improvement	8
Bare lands	18
Embodied by water	2

Source: Reynolds et al. (2007), FAO (2011), Bindraban et al. (2012), and Gomiero (2016)

implications for food security, livelihood sustainability, and even impacts environmental stability (Babu et al. 2020a; Singh et al. 2023). Furthermore, the consequences of land degradation extend to the loss of biodiversity and alterations in species abundance. Considering that land represents a scarce and non-renewable natural resource, it becomes imperative to curtail and reverse the process of land degradation to safeguard the integrity of ecosystems, ensure food security, and preserve environmental stability (Kumar et al. 2022; Wani et al. 2023).

2.3.2 Factors Contributing to Degradation of Arable Land

A multitude of factors contribute to both the land degradation and restoration of land, exhibiting intricate interactions among them (Fig. 2.1). It is essential to note that both natural and anthropogenic forces can act as catalysts for not only land degradation but also for land improvement (Kiage 2013; Bisaro et al. 2014). Land degradation also encompasses erosive processes caused by water and wind, chemical deterioration leading to soil acidity, salinity, and nutrient leaching, among other factors. On the other hand, physical soil degradation involves phenomena like crusting, compaction, hard-setting, and related issues (Eswaran et al. 2001).

The project entitled global assessment of soil degradation (GLASOD) has introduced a comprehensive map highlighting intensity of land degradation on a worldwide scale. Within this assessment, four major causes of land degradation were identified *viz.*, deforestation resulting from clearing off natural vegetation, the excessive utilization of vegetation meant for the domestic purposes, overgrazing, and adoption of the intensive agricultural practices (Oldeman et al. 1991). Nations such as Central African Republic, Egypt, Ghana, Pakistan, Paraguay and Tajikistan have encountered substantial degradation of land, subsequently leading to escalating food prices (Braimoh 2015). Furthermore, within the framework of developing nations, the gender issue assumes paramount significance, considering that impoverished women typically face heightened vulnerability to land degradation. Additionally, women often possess diminished authority and reduced decision-making capabilities concerning land-related matters.

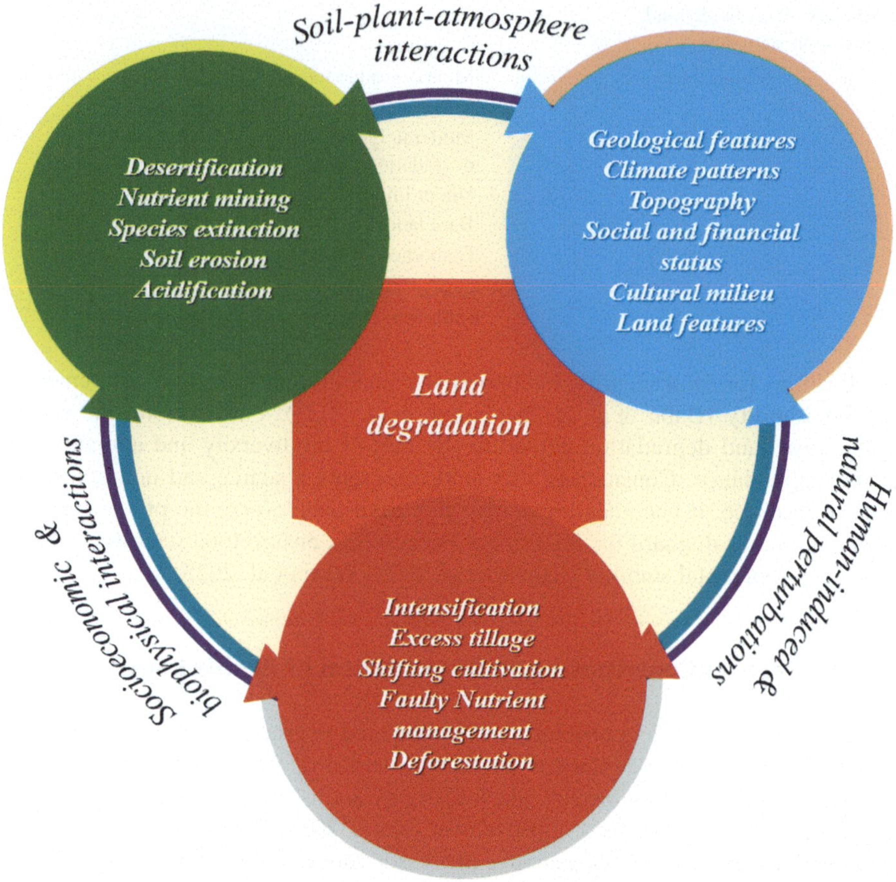

Fig. 2.1 Mechanisms-agents-activities nexus as a drivers of agricultural land degradation

Population Pressure The escalating demands of an expanding global population exert immense pressure on the finite resources of land and natural reserves, significantly reducing their accessibility and efficacy. This heightened demand, especially for food and habitation, leads to rapid degradation of arable land, posing substantial threats to global food security. Additionally, within developing nations, heightened population growth among groups reliant on ecosystem services intensifies environmental stress and degradation, disproportionately affecting regions highly dependent on these services (UN 2012).

Impact of Socioeconomic Deprivation Socioeconomic challenges, encompassing low-income levels and limited educational opportunities, are significantly linked with the overexploitation of various resources, leading to widespread ecological degradation. This vulnerability is further compounded by the exacerbation of environmental fragility due to natural calamities like droughts and floods. These disasters contribute to the depletion of vegetation as well as soil, consequently limiting the land's capacity to sustain life and provision of essential ecosystem services (UN 2012).

Consequences of Unsustainable Practices Unsustainable anthropogenic activities such as inadequate irrigation application techniques, excessive usage of mineral fertilizers, and unchecked overgrazing result in severe consequences. These practices foster the salinization of land, acidification of soil, and degradation of vegetation, disrupting the soil fertility and productivity to naturally rejuvenate its ecosystem services (UN 2012).

Challenges in Land Ownership Systems Land tenure systems characterized by inequality, gender bias, and insecurity create barriers to essential investments and protective decision making required in order to safeguard land from getting degraded. Consequently, marginalized groups, particularly the females, find themselves entrenched in the never-ending cycles of impoverishment, lacking ownership and control over their own land and its productive capabilities (FAO 2012).

Poor Governance Prevailing top-down governance models prevalent in many developing nations significantly contribute to environmental degradation. Conversely, inclusive governance models hold the potential to yield more favourable outcomes, particularly in terms of fostering sustainable land management practices (FAO 2012).

Each of these areas requires comprehensive, collaborative action to mitigate degradation, promote sustainable resource management, and ensure a more equitable and resilient future.

2.4 Impact of Soil Degradation on Agricultural Productivity

The persistent issue of soil degradation has posed a significant global concern throughout the twenty-first century owing to its negative repercussions on the soil functioning, along with profound implications on the food security challenges amid a growing population. Moreover, the yearly economic loss due to reduced agricultural productivity from land degradation is ~36 million tons of cereal equivalent. This economic loss has been assessed at USD 5400 million attributed to the water erosion and USD 1800 million due to the wind erosion (Nosberger et al. 2001). In the context of crop productivity, a vital contributor to several SDGs, distinct consequences of soil degradation can be grouped into on-site and off-site impacts. On-site degradation results in the direct decline of soil and its potential in terms of productivity, undermining the different types of ecosystem services and the capacity of farmers and livestock managers to yield food and animal products (Singh et al. 2021a). Off-site impacts may manifest positively, characterized by sediment deposition that enhances soil quality, or negatively, attributable to pollution, sedimentation, and various other processes.

The off-site effects of soil erosion may not always be as apparent as the direct on-site impacts. Pesticides and fertilizers, often carried along with the eroded soil, have the potential to contaminate the downstream sources of water as well as the areas meant for recreation (Morgan 1999). The major off-site consequence of erosion is transportation of the sediments and accompanying agricultural pollutants into watercourses (Abebaw 2019).

2.4.1 Declining Treasure of Nutrient in Soil

As soil erosion occurs, nutrients that are completely essential for plant growth such as nitrogen, phosphorus, potassium, and calcium are concurrently lost. It was found that the eroded soil typically retains around three times more nutrients compared to the soil that remains on the eroded land (Lat 2001). A Mg of fertile soil generally contains ~1–6 kilograms (kg) of nitrogen, 1–3 kg of phosphorus, and 2–30 kg of potassium. Conversely, eroded soil commonly exhibits lower levels of nitrogen, ranging from 0.1 to 0.5 kg per Mg (Troch et al. 2002). The rate of annual SOC loss across the world is estimated at 21.80 ± 0.91 g C per m^2 per year (Abdalla et al. 2020). As the organic matter layer proceeds to diminish, the ecosystem's productivity, as gauged by crop yield, diminishes significantly (Yadav et al. 2020a, b; Babu et al. 2023a).

2.4.2 Impact on Crop Yield

Soil erosion causes reduction in biomass productivity within ecosystem, exerting a profound impact on the floral and faunal diversity existing within the ecosystem. However, a significant challenge in numerous estimations of land degradation lies in the lacuna of a definitive cause-and-effect relationship existing between extent of degradation and the subsequent decline in agricultural productivity. Sometimes, the variation in soil loss due to erosion in different agronomic practices in field did not translate into notable changes in the crop productivity. It indicates that factors other than soil degradation or additive effect with erosion might have significant effect on crop productivity (Maglinao et al. 2000, 2003).

The global potential land productivity had decreased by 15–30% due to soil erosion (IPCC 2019). Percentage-wise, estimations revealed a decline in yield of approximately 15%, 35%, and 75% for crop land soil categorized as lightly, moderately, and severely degraded, respectively. Conversely, rangeland soils were estimated to incur losses of ~5%, 18%, and 50% within the same respective degradation categories (Crosson 1997). Land degradation has more negative impact on the crop land than the pasture land. The croplands had a potential productivity of 12.7%, lower than pastures due to land degradation (Oldeman 1998).

2.4.3 Impact on Reservoir Sedimentation

Soil erosion is the main cause for reservoir sedimentation. Approximately 0.5–1% of total volume (6800 km^3) of water which is stored in reservoirs globally is annually lost due to sedimentation (Morris et al. 2008). The pace at which sediment accumulates in reservoirs hinges on several factors including the soil's susceptibility to erosion in the contributing watershed, its topography, surrounding climate, agricultural practices, and also the ratio estimated between the watershed size and the reservoir volume (Gray 2002). Broadly, the process of reservoir siltation can significantly impact agricultural productivity in a given area, either directly or indirectly.

2.5 Ecological Implications of Soil Degradation

The effect of soil degradation that takes place on ecology can be explained through a term '*Biological degradation*', which includes reduction in vegetation cover on soils, habitat depletion, decline in biomass, fire ramifications, deterioration in quality and diversity of species composition, imbalanced soil life, insurgence of pests or diseases, loss of predators. Land degradation leads to a reduction in supply of the diverse terrestrial ecosystem services, resulting in the immediate economic ramifications like elevated food costs, along with extensive socioeconomic implications like food and water insecurity, malnutrition etc. The ecological consequences directly impact the species/type of flora inhabiting the particular area. Additionally, it leads to the decreased access to potable water, reduction in surface water levels, aquifer depletion due to insufficient recharge, and a decline in biodiversity. Soil degradation disrupts the regulating and provisioning services of soil ecosystems (GEF 2006; Babu et al. 2020b) (Fig. 2.2).

2.6 Soil Health as the Bedrock of SDGs

Land degradation has continued to exist as an alarming global concern throughout the twenty-first century due to its detrimental effects on the land, environment, and ultimately the food security, especially amidst an expanding global population. Additionally, it holds significance concerning the attainment of some of the Sustainable Development Goals (SDGs), notably SDG 1 (no poverty), 2 (ensuring food security), 3 (clean water and sanitation), 13 (taking action against climate change), 14 (preserving life that exist below water), and 15 (protecting life on the land).

A central objective in the pursuit of SDGs lies in Goal 15, denoted as "Life on land," alongside Target 15.3 centred on the LDN (Fig. 2.3; UNCCD 2016). The SDG 15 strives to endorse the judicious utilization of land resources, ensuring sustained food production, adoption of resilient agricultural practices, and optimal use of the existing natural resources. Consequently, accomplishment of this goal through

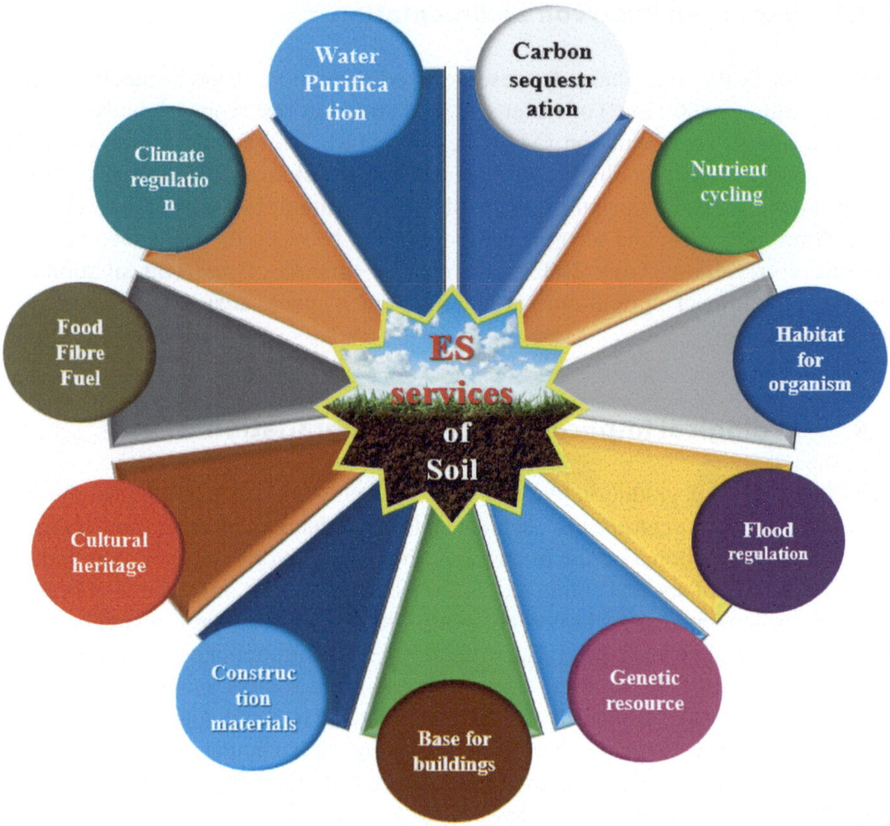

Fig. 2.2 Ecosystem services provided by soil

the practice of Sustainable Land Management (SLM) substantiates attainment of the various other SDGs. The implementation of SLM techniques, for example, can augment both water productivity and quality, fostering revival of the existing wetland ecosystems, thus contributing notably to SDG 6, addressing 'Clean water and sanitation'. Moreover, SLM encompasses the maintenance of robust lands capable of bolstering agricultural output, fortifying the food security, and fostering the agricultural economies. This directly aligns with SDG 1, addressing 'No poverty', SDG 2, aiming for 'Zero hunger', and SDG 5, advocating for the good health and well-being of all. The rehabilitation of land resources assumes a pivotal role in combating climate change and advancing Goal 13, targeting 'Climate action', considering climate change to be a substantial threat to sustainable development. Moreover, the sustainable management of land holds the potential to annually sequester up to 3 billion tonnes of carbon (UNCCD 2013).

Fig. 2.3 SDGs relation with Target 15.3 on LDN

The land sector is responsible for 25% of the world's total greenhouse gas emissions (Smith et al. 2014). Implementing enhanced land-use practices and proficient management approaches, like low-emission agricultural techniques, agroforestry, and the conservation and restoration of ecosystems, holds the potential to narrow the gap caused by rest of the emissions by as much as 25%. SDG 15 holds crucial importance in the pursuit of SDG 11 concerning 'Sustainable cities and communities.' Forecasts suggest that between the year of 2000 and 2030, ~1.6–3.3 Mha of primarily agricultural land annually will ultimately succumb to urbanization due to the migration of the global population into urban areas (UNCCD 2017). The fulfilment of SDG 15 and Target 15.3 through SLM practices could therefore establish enduring linkages with other goals, necessitating comprehensive policy strategies like forest landscape restoration. These strategies are pivotal in aligning global sustainability commitments, warranting synchronized implementation, and translating them into actionable measures at the grassroots level.

2.7 Approaches to Agricultural Land Restoration

2.7.1 Agronomic Management Techniques

2.7.1.1 Crop Management Practices

Implementing effective crop management practices significantly mitigates wind and water erosion. Preserving soil coverage stands as a foundational principle in the rejuvenation of deteriorated agricultural land. Shielding the soil against erosion involves retaining the various kinds of crop residues on the soil surface post-harvest, proving to be a beneficial approach.

2.7.1.2 Crop Rotation

Crop rotation enhances productivity and financial gains when introducing legumes into rotation and it will not boost only cropping frequency but also maintain overall soil health as well as sustainability (Babu et al. 2023b; Singh et al. 2021b). Soil capacity for supplying of essential nutrients improves by the incorporation of legumes in succession with cereals (Yadav et al. 2021b). The soil structure and fertility can be improved with the alternation of shallow-rooted and deep-rooted plants, which is an additional avenue for improvement facilitated by the strategic implementation of crop rotation (Yadav et al. 2021c). The enhanced potential of soil ecosystem is achieved through an increase in crop residues, the development of diverse root systems, and augmentation of microbial activity (Studdert and Echeveria 2000; Yadav et al. 2020a, b; Yadav et al. 2023). The diversified crop rotation (DCR) contributes to soil water conservation, equilibrium in wet-soil aggregates, and increased soil enzymatic activities (Babu et al. 2023a; Kumawat et al. 2023). The DCR plays quite a vital role in managing soil erosion and improving the crop production (Alhameid et al. 2017; Yadav et al. 2024), leading to a significant enhancement in the soil productivity. Moreover, it efficiently utilizes water as well as nutrients within the soil profile to sustain profitability (Miller et al. 2003; Das et al. 2022a, b; Walia et al. 2022).

2.7.1.3 Intercropping

Inter-cropping, involving the simultaneous growing of two or more than two crops in the same field in a given year, emerges as an agronomic practice that strengthens agricultural systems ecologically. The structural density and vegetation cover in intercropping not only overcome the intense volume of rainfall that directly erodes the soil but it also reduces the intense impact of raindrop on the soil surface (Ma et al. 2014; Ansari et al. 2023). Furthermore, legume inter-cropping aligns with the United Nations' sustainable development goals by boosting the agricultural production, improving the nutritional quality of food, and promoting soil health. As intercropping involves inclusion of two or more crops in a definite row pattern, it increases soil cover and improves resource-use efficiency compare to single crop. For instance, solo cropping of maize can only cover 30% of area during kharif season, leaving more bare soil vulnerable to soil erosion while maize + sweet potato intercropping can cover up to 60% of soil which can reduce the soil erosion efficiently (Yadav et al. 2023). Higher biomass production capacity of intercrops enhances soil organic matter which improves soil structure and infiltration in soil.

2.7.1.4 Rotational Grazing

Employing perennial forage management in the form of pastures for grazing is a crucial approach to restore degraded lands. Unlike existing grazing practices that often contribute to soil degradation, primarily due to overgrazing, hence efficient grazing management strategy can sustain the perennial forage production and landscape. Rotational grazing practices like, simple rotational grazing, low-intensity animal movements with around 2–14 pastures per herd (Briske et al. 2011; Roche et al. 2015), management-intensive grazing with higher intensities and 16+ pastures per herd, and adaptive multi-paddock grazing that may include as high as 40+ pastures per herd (Teague et al. 2015; Barton et al. 2020; Mosier et al. 2021) are important to minimize the soil erosion. By rotating animals, soil as well as the perennial vegetation benefit from grazing-free periods, promoting soil fertility along with perennial productivity (Kemp et al. 2000). Rotational grazing in perennial pastures improves soil fertility and also enhance soil physical, chemical, and biological indicators (Teague et al. 2011; Byrnes et al. 2018), leading to improved water retention and infiltration (McCallum et al. 2004; Teague et al. 2011).

2.7.2 Agroforestry

Agroforestry encompassing woody perennials, such as various species of trees and shrubs, cultivated alongside herbaceous plants like crops or pastures, and/or livestock. These components are organized spatially, involve rotational practices, or a combination of both, fostering intricate ecological and economic interplay between the arboreal and non-arboreal elements within the system (Young 1995). Trees and shrubs serve several purposes in this context: they mitigate splash erosion by diminishing the impact of raindrops on the soil (Fig. 2.4). Through shading, they regulate soil temperature, reducing water evaporation. Additionally, trees act as barriers, curtailing wind erosion. Moreover, they play a pivotal role in the nutrient recycling within the soil depths; certain trees, specifically leguminous varieties, have the capacity to fix nitrogen, thereby benefiting food crops (Bashir et al. 2017). The agroforestry has the capacity to sequester approximately 83.6 Mg of carbon per hectare up to a soil depth of 30 cm, indicating a 26% increase in carbon storage compared to conventional cultivation (Jha et al. 2001). If, 630 Mha of land could be allocated for agroforestry, presenting the potential to sequester ~586 million metric tons of carbon per year by 2040, equivalent to 2150.60 Tg per year of CO_2 absorption (Watson et al. 2000).

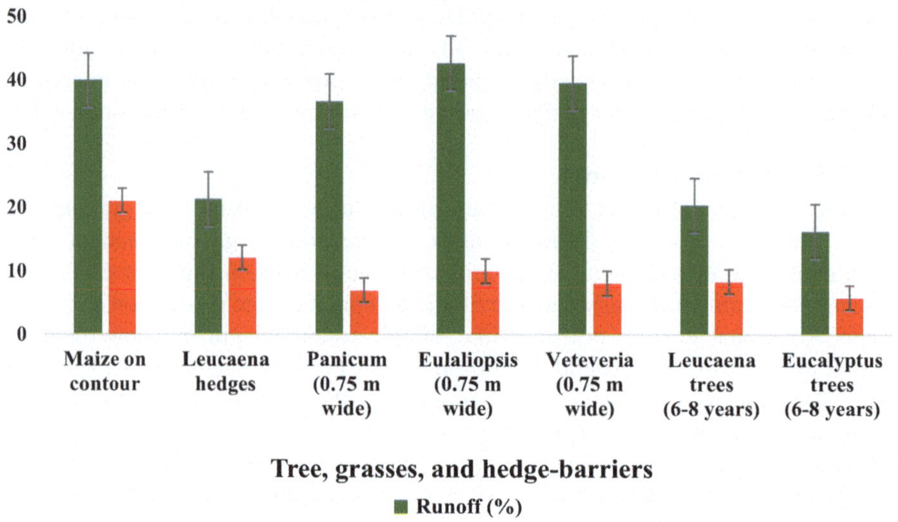

Fig. 2.4 Potential of different trees, grasses, and hedge barriers for erosion control

2.7.3 Strip Cropping and Shelter Belts

Strip cropping involves cultivating crops in various alternating strips, comprising row crops or forage/grasses. This agricultural system serves as a very effective method to mitigate soil erosion by segmenting the existing sloping terrains into broad sections with diverse vegetative cover thereby reducing nutrient loss and stabilize soil carbon (Table 2.2). Diverse covers under strip cropping intercepts runoff and fosters water infiltration, consequently reducing both runoff and soil erosion. Frequently, strip cropping is intertwined with crop rotations, where distinct crops are sown in strips each year. These belts also used to improve microclimate (optimum temperature, heat induced transpiration, moisture loss, and wind speed), especially boon for sandy drylands during winter and spring. Generally, the distance effectively controlled by structures are 15–20 times of the tree height. In a strip of variables width, there are 2–5 lines of trees (1–3 m gap) and individual trees planted at intervals of 1–2 m within each row (WOCAT 2010).

Shelterbelts serve as protective measures for both irrigated and rain-fed farms, primarily safeguarding agricultural land and irrigation channels from encroaching sands. These belts effectively reduce wind speed, enhance the microclimate, and augment livestock yields. Under arid regions crop production can potentially increase of up to 300%, with average annual increments often ranging between 30% and 50% (Saturday 2018). Moreover, by improving the microclimate, shelterbelts frequently enhance fruit production by a third or more. These protective barriers also significantly impact livestock production by shielding animals from strong winds and excessive heat, thereby elevating lambing percentages by around 30% (Elhouri Ahmed 1989). Predominantly, species like Eucalyptus, *Casuarina sp.*,

Table 2.2 Importance of strip cropping

Strip cropping	Services provided	References
Indigofera + GFS (Grass filter strip)	Sequester carbon @ 0.4–1.5 t/ha/year Reduces runoff, nutrients and soil erosion losses	Lenka et al. (2012)
Mize-soybean/Maize-bean (Contour farming)	The maize-soybean sequence in combination with alternating strips reduces erosion up to 40% besides a 10–20% yield advantage over sole crop	Francis et al. (1986)
Maize-soybean	Strip crops result in more profit (US$ 568 ha^{-1}; US$163 ha^{-1} for maize and soybean, respectively) over sole cropping	Al-Amin et al. (2024)
Faba bean (*Vicia faba*) -tomato (*Solanum lycopersicum*)	Higher resource use efficiency Low economic risk Higher carbon return (>1)	Campanelli et al. (2023)
Common wheat (*Triticum aestivum*)—zucchini (*Cucurbita pepo*)	Higher land equivalent ratio (>1) Higher above-ground biomass (+63%)	Campanelli et al. (2023)
Finger millet + groundnut (6:4)	Finger millet + groundnut had a 53.33% yield advantage with 22.4% less runoff over sole finger millet	Jakhar et al. (2015)
Narrow buffer strip of *Fescue arundinacea* in maize	Reduced quantity of nitrogen in sub-surface lateral flow, protection of groundwater resources	Salazar et al. (2015)
Finger millet + groundnut (10:4)	A 19.6% more conservation of runoff compared to finger millet sole cropping	Jakhar et al. (2015)
Rotational strip intercropping of maize and peanut	Erosion intensity in continuous maize and rotational strip intercropping (maize + groundnut) was reduced by 38.88% and 45.45% over continuous groundnut, respectively	Zou et al. (2021)
Stipa bungeana, *Lespedeza davurica* and *Artemesia capillaries* with wheat	Effective trapping of sediments of eroded soils, enhance root growth, improves nutrient cycling	Yuan et al. (2009)

Populus, and Prosopis are employed for this purpose, with seedlings typically used for establishing these shelterbelts.

2.7.4 Conservation Agriculture and Cover Cropping

The land deterioration has undergone a marked exponential escalation within the preceding 200 years (Sanderman et al. 2017). Therefore, rebuilding the soil organic carbon is of utmost requirement for agricultural sustainability in context of the current climate change (Rumpel and Chabbi 2021; Yadav et al. 2019a, b).

Recurrent disturbances from tillage and harvesting practices, transition from the various types of deep-rooted perennial plants to the limited shallow-rooted annual plants has ultimately resulted in decline of 20–60% of SOC content in agricultural

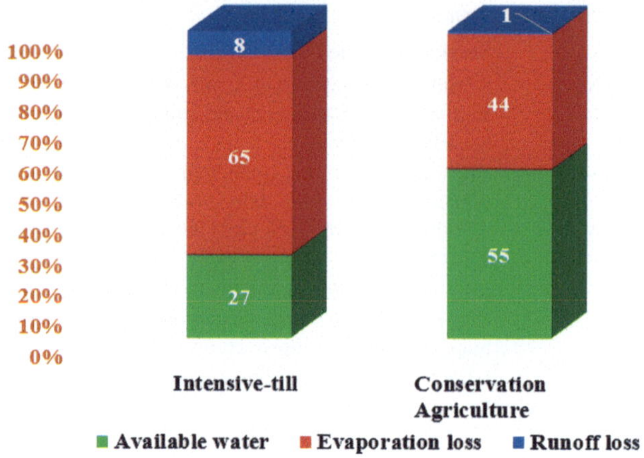

Fig. 2.5 Effect of tillage practices and mulching on moisture and runoff losses

soils (Crews and Rumsey 2017). Figure 2.5 depicted that the adoption of reduced tillage, and /or no-tillage, stands as a significant strategy aimed at mitigating soil erosion and curbing nutrient losses through the wind and water erosion (Van Pelt et al. 2017; Borrelli et al. 2016; Panagos et al. 2015). In Brazil, the transition from conventional tillage to no-till system led to a substantial reduction in soil erosion from 3.4 to 8.0 Mg per ha to 0.4 Mg per ha, while in Paraguay, it decreased from 23 to 0.53 Mg per ha per year (Sorrenson et al. 1997). Additionally, the adoption of no-till practices, coupled with enhanced vegetation cover, resulted in a 22% reduction in sediment load from 1994 to 1998 in a watershed in Brazil. Furthermore, the net soil carbon stock exhibited an annual increase from 0.5 to 0.7 Mg per ha in the USA and Brazil. The adoption of CA reduces soil loss by 7.5% in 4 years in Mediterranean agricultural soils (Petito et al. 2022). Likewise mulching of diverse crop residue reduced soil erosion and preserved moisture (Bashir et al. 2017). In traditional farming practices, raindrops impacting bare soil profiles lead to the degradation of soil, resulting in detached soil particles obstructing the soil pores and impeding water infiltration (Srinivasarao et al. 2022). The impact of erosivity factor on crop yields depends on soil types and crop varieties. Cover crops (leguminous or non-leguminous species) such as cowpea, beans, sweet potato crops are best candidates to reduce soil degradation (Yadav et al. 2024; Singh et al. 2021b). The presence of surface or crop cover substantially lowers the amount of soil detachment, resulting in the reduction of overall soil erosion (Li et al. 2019). The nitrogen fixing capacity and biomass of legume helps to eliminate vulnerability of soil to erosion and deficiency of nutrient in soil (Singh et al. 2021a, 2023; Walia et al. 2022). The cover crops are an excellent tool to control the unsustainable soil and organic matter depletion measured in vineyards (Lopez-Vicente et al. 2020; Yadav et al. 2013, 2021b). Cover crops cultivated between primary cropping seasons, have a unique potential to augment soil carbon stocks at around 0.22–0.4 Mg C per ha per year (Poeplau and Don 2015; Kaye and Quemada 2017).

2.7.5 Watershed Management

A watershed denotes the geographic region encompassed and drained by a particular watercourse. The services and functions provided by watersheds might facing challenges due to deforestation, unregulated timber extraction, alterations in agricultural practices, excessive grazing, construction of roads, pollution, and the encroachment of non-native plant species. Additionally, natural events like wildfires, severe winds, and diseases could impact these watershed functions. The decline in watershed functions can result in severe adverse effects, including soil erosion, diminished soil fertility, sedimentation of water bodies like streams, reservoirs, and coastlines, heightened surface runoff leading to sudden floods, decreased infiltration into groundwater sources, compromised water quality, and the degradation of aquatic habitats and biodiversity (FAO 2017; Koralay and Kara 2018). Watershed management, characterized by human interventions aiming to ensure sustainable utilization of the natural resources within a watershed, aims at addressing these imminent threats. Implementation of Improved Watershed Management practices significantly contributed to enhancing soil health (Gebremeskel et al. 2019).

2.7.6 Bioremediation and Phytoremediation Approaches

The rapid progression of industrialization, urban expansion, and intensified agricultural practices in the recent decades has led to extensive degradation of land resources on a global scale. About 1.5 billion individuals are adversely impacted by land degradation (www.unccd.int) and about 12.2 billion ha of land worldwide has suffered severe degradation primarily caused by pollution and various other factors.

2.7.6.1 Bioremediation

Of the diverse array of remediation methods, bioremediation is gaining traction due to its cost-effectiveness and non-invasive nature when compared to traditional techniques. Bioremediation is recognized as a promising technological approach aimed at the restoration of contaminated and degraded agricultural lands.

Recent advancements in biotechnology, such as the development of highly effective microbial communities, utilization of enzymes with enhanced specificity, and the creation of plants tailored to engage specific microbial associations, are introducing novel opportunities in the realm of remediation technology (Fig. 2.6). The development of new omic revolutions, nanotechnology with vault packaging, novel catabolic genes can enhance the unique potential to reclaim the soil and remove the soil pollution efficiently (Floris et al. 2017; Kang et al. 2016; Tripathi et al. 2016).

2.7.6.2 Phytoremediation

Phytoremediation is an emergent field within environmental science experiencing rapid growth, demonstrating substantial potential in detoxifying both inorganic and

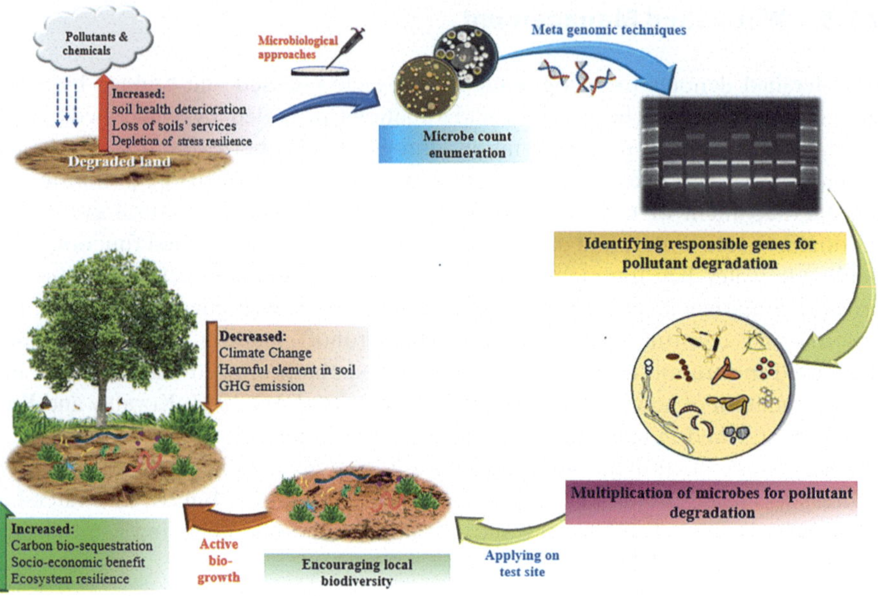

Fig. 2.6 Monitoring biological systems and bioremediation of contaminated soil

organic pollutants within the soil environment. The phytoremediation approaches encompass metal accumulation within plant tissues (phytoextraction), mitigation of risks through gentle soil remediation (phyto-stabilization), and enhancement of soil fertility and improvement in its ecosystem services. The cultivation of perennial crops on degraded agricultural land presents an auspicious practice for the restoration of soil fertility and improvement of soil ecosystem services (Barbosa et al. 2015; Arco-Lazaro et al. 2017). The phytoremediation strategy for restoration of the degraded land is presented in Fig. 2.7. This approach entails a reduction in soil tillage, an augmentation of soil coverage, and the subsequent establishment of a beneficial litter effect. Profound and densely structured root systems, coupled with enduring vegetative cover can greatly mitigate runoff and soil erosion (Fagnano et al. 2015; Fernando et al. 2015). Several tree and herbaceous species can be employed for the rehabilitation of agriculturally degraded lands (Table 2.3).

2.7.7 Precision Agriculture

Precision agriculture is an innovative and unique approach to agricultural land management that leverages technology and data-driven techniques to optimize the use of resources and enhance overall productivity (Choudhary 2023). While precision agriculture is commonly associated with increasing crop yields and minimizing input wastage, its application for agricultural land restoration is an emerging and

Fig. 2.7 Phytoremediation strategies on degraded lands

promising area (Fig. 2.8). A pivotal facet of precision agriculture concerning land restoration revolves around the meticulous targeting of interventions predicated on intricate information pertaining to the precise conditions of the land. This encompasses factors like soil health, nutrient concentrations, moisture content, and other pertinent parameters. The incorporation of advanced technologies such as Revised Universal Soil Loss Equation (RUSLE) with GPS, sensors, VNIR, and remote sensing, a real-time data pertaining to soil health can be collected that will facilitate more precise and informed decision-making processes in remediation (Zhang et al. 2013; Borrelli et al. 2021). The expeditious imperative lies in the rehabilitation of degraded soil resources, necessitating a thorough assessment that takes into account a variety of factors. Remote sensing technology effectively caters to these requirements, encompassing tasks such as identifying and delineating the temporal attributes of land degradation, forecasting the distribution of deteriorated soil and spatial disparities in soil quality, and discerning alterations induced by natural phenomena or human activities (Khaledian et al. 2017; Meinen and Robinson 2020; Wei et al. 2021).

2.7.8 Green Manuring

The maintenance of soil health poses a substantial challenge given the intricate interplay between inputs and costs, the expensive nature of fertilizers, and the persistent problem of soil degradation. In the face of such challenges, green or organic manures emerge as an economical and effective source of plant nutrients and soil improvement (Babu et al. 2020c).

Consequently, the incorporation of various green manuring crops into the existing cropping systems not only exerts a favourable impact on the different soil properties but also augments crop productivity, preserves soil moisture, and serves as a deterrent against weed infestations (Das et al. 2020, 2022b). Legume green

Table 2.3 Species suitable for the remediation and rehabilitation of degraded lands

Species for saline and alkali areas	Species for degraded soils	Species for silvi-pasture system	Vegetation strips
Grasses and legumes	**Trees**	**Trees**	*Panicum maximum*
Dicanthium annulatum	*Azadirachta indica*	*Albizia spp.*	*Eulaliopsis binata*
Sporobolus helvolus	*Prosopis juliflora*	*Acacia nilotica*	*Vetivera zizinoides*
Leptochloa fusca	*Prosopis cineraria*	*Albizia lebbeck*	*Cenchrus ciliares*
Paspalum vaginatum	*Acacia nilotica*	*Dalbergia sissoo*	*Chrysopogon zizanioides*
Cenchrus ciliares	*Acacia radianoa*	*Leucaena leucocephala*	*Gliricidia sepium*
Sesbania sesban	*Zizyphus mauritiana*	*Azadirachta indica*	*Calliandra calothyrsus*
Medicago sativa	**Grasses**	*Moringa oleifera*	*Macroptilium atropurpureum*
Crotalaria juncea	*Lasiuruss indicus*	*Prosopis cineraria*	*Clitoria ternatea*
Trees and shrubs	*Panicum turgidum*	*Eucalyptus tereticornis*	**Bamboo**
Tamarix (Tamarisk)	*Panicum antidotale*	*Hardiwickia binata*	*Dendrocalamus strictus (Male Bamboo)*
Eucalyptus camaldulensis	**Shrubs**	**Grasses**	*Bambusa bambos (Indian Thorny Bamboo)*
Casuarina equisetifolia	*Calotropis procera*	*Chloris gayana*	*Bambusa vulgaris*
Salvadora persica	*Calligaonum polygonoides*	*Chrysopogon fulvus*	
Acacia tortilis	*Dodonaea viscosa*	*Cynodon dactylon*	
Dichrostachys cinerea	*Aervaja vanica*	*Dichanthium annulatum*	
Leucaena leucocephala	*Cassia siamea*	*Cenchrus ciliaris*	
Dendrocalamus strictus	*Exotolaria burhia*	*Panicum maximum*	
Delbergia sissoo	*Zizyphus nummularia*	*Panicum pedicellatum*	
Haloxylon ammodendron		*Heteropogon*	
Albizia lebbeck		*Bothrochloa*	
Prosopis juliflora		**Legumes**	
Azadirachta indica		*Stylosanthes hamata*	
		Desmodium intortum	
		Sirata spp.	
		Macroptilium atropurpureum	
		Clitoria ternatea	

manuring amplifies the overall pore space by reducing soil bulk density, consequently fostering improved root development, increased soil water content, and enhanced nutrient use efficiency (Anderson et al. 1997; Yadav et al. 2020a, b, 2021b). Legumes play a dual role in the soil enrichment by contributing nitrogen and facilitating the recycling of various nutrients on. Green manure crops, including legumes, accumulate essential plant nutrients (such as nitrogen (N), phosphorous (P), potassium (K), calcium (Ca), magnesium (Mg), sulphur (S), and micronutrients) throughout their growth season. Hence, the choice of green manures may be tailored based on their distinct functions (Fig. 2.9) and their nutrient compositions (Table 2.4).

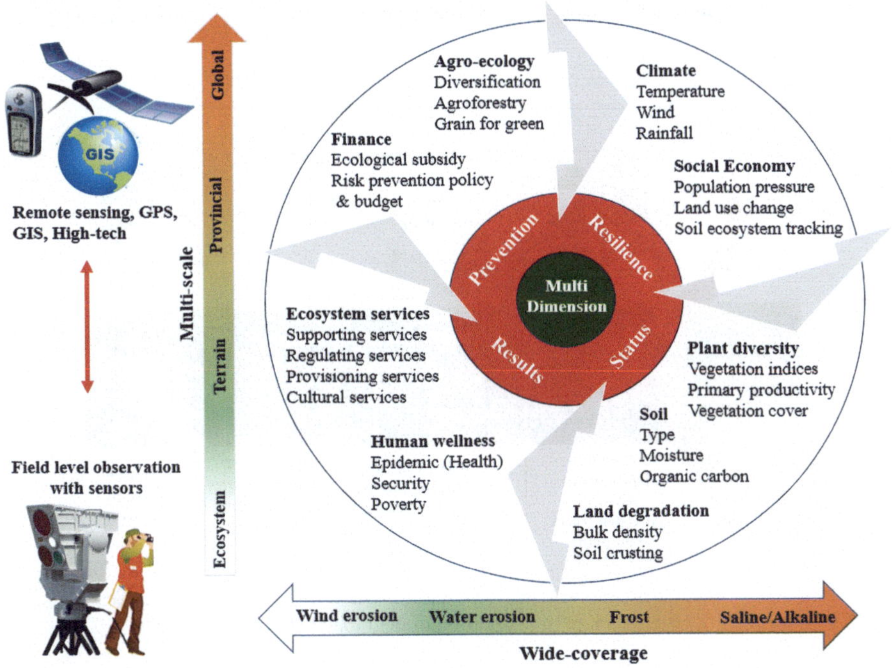

Fig. 2.8 Land degradation monitoring with remote sensing technology

2.7.9 Soil Reclamation with Organic and Inorganic Materials

Reclamation of degraded agricultural lands often involves use of various materials to improve soil fertility, structure, and overall health. These materials are strategically applied to address specific soil deficiencies and promote sustainable agricultural practices.

- Organic sources: FYM, compost, Biochar, Vermicompost, PGPR
- Inorganic sources: Flue gas desulphurisation gypsum (FGDS), Gypsum, lime, Elemental sulphur, fertilizers, $MgCO_3$.

The application of compost amendments lowers the soil bulk density and also increases permeability by promoting soil particle aggregation, thereby augmenting leaching down of salts (Murtaza et al. 2019). Soil pH significantly influences nutrient availability for plants; even a slight pH variation can hinder plant growth (Babu et al. 2020a). Compost possesses pH-buffering capabilities, and its addition can lead to pH buffering in the soil due to the nitrification effects (Zhao et al. 2014; Singh et al. 2023). Likewise, biochar incorporation into saline soils significantly improves the water retention capacity, water percentage, and hydraulic conductivity and nutrient dynamics (Babu et al. 2022, 2023c). Earthworms, notably *Pheretima posthuma, Eutyphoeus incommodus,* and *Lampito mauritii* are more suitable for salt-affected

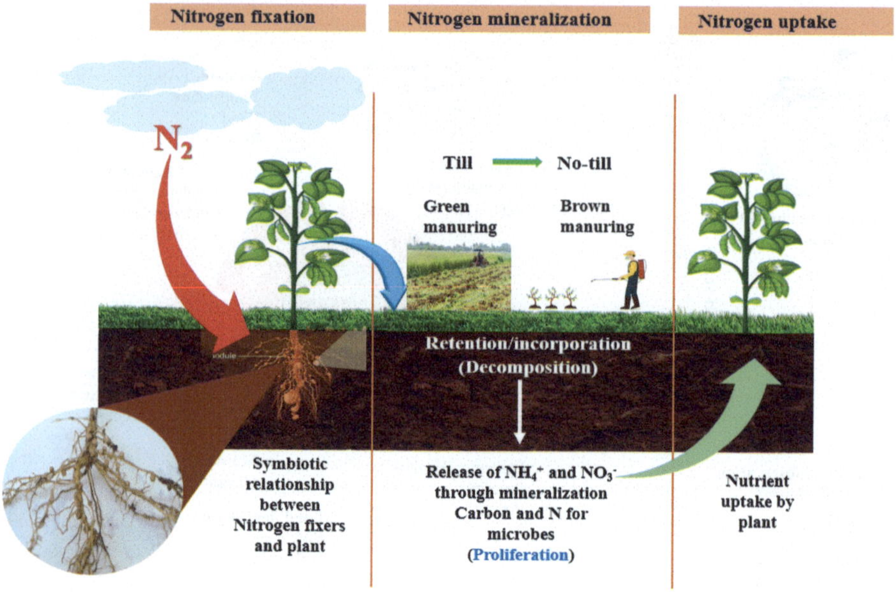

Fig. 2.9 Functions of green manure in restoration of degraded land

Table 2.4 Nutrient composition of green manure and green leaf manure crops

Plant species	Average nutrient content (%) on air dry basis		
	N	P	K
Green manure crop species			
Crotalaria juncea	2.3	0.218	1.494
Sesbania aculeata	3.5	0.262	0.996
Sesbania speciosa	2.7	0.218	1.826
Tephrosia purpurea	3.1	0.218	0.996
Green leaf manure species			
Azadirachta indica	2.80	0.131	0.332
Leucaena leucocephala	3.50	0.218	0.664
Pongamia glabra	3.30	0.175	1.992
Gliricidia sepium	2.80	0.131	3.818
Delonix regia	2.80	0.218	0.415
Parthenium hysterophorus	2.70	0.306	1.245
Eichhornia crassipes	3.00	0.393	0.166
Cassia fistula	1.60	0.087	0.996

soils due to their tolerance capacity (up to 10.48 mS/cm) and robust survival rates in degraded soils (Sharif et al. 2016). Adoption of balance fertilization with slow-release fertilizers to ensure an optimum nutrient supply, reducing the risk of nutrient leaching and improves biological activity in soil is crucial for maintaining the soil ecosystem functionality. Gypsum ($CaSO_4.2H_2O$) is cheapest and readily available source for reclaiming the degraded alkali soils. A new material i.e., FGDG is

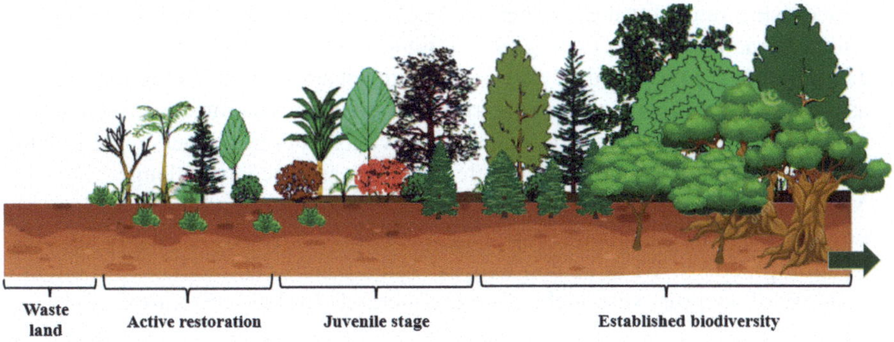

Fig. 2.10 Land restoration through native tree species

excellent source of calcium for sodic soil and can be spread with lime and litter spreaders. It maintains calcium and magnesium levels in soil and offers readily available form of sulphur, that benefit the soil and crop (Rai et al. 2023).

2.7.10 Biodiversity Restoration

After extended periods of unsustainable cultivation, agricultural soils undergo degradation, leading to the potential release of up to two-thirds of their initial soil carbon stores into the atmosphere in the form of carbon dioxide (Lal 2004). Generally, the degraded and abandoned farmlands go through a process of natural succession and then recovery, which is often referred to as passive restoration (McIver and Starr 2001). This process holds the potential to lead the land towards a state resembling its pre-disturbance vegetation. Purposeful active restoration, which deliberately encourages the diversity of plant species, has the potential to significantly enhance total carbon storage on degraded land. Both passive and active methods play crucial roles in restoring degraded lands, fostering an increase in carbon stocks, and augmenting various soil ecosystem services (Fig. 2.10). The success of restoration efforts is influenced by various factors, including historical cultivation practices, intrinsic soil characteristics, and the overall composition of the surrounding vegetation (Yang et al. 2020).

2.8 Challenges and Constraints in Agricultural Land Restoration

2.8.1 Economic and Technological Challenges

- **Financial constraints:** Agricultural land restoration often requires significant investments in technology, equipment, and inputs. Farmers may face challenges

in accessing funds for these expenses. High costs associated with modern agricultural practices and technologies can hinder widespread adoption.
- ***Technological gaps:*** Limited access to advanced agricultural technologies, such as precision farming or sustainable irrigation methods, can impede effective land restoration efforts. Lack of training and knowledge among farmers regarding the use of modern technologies and implements is a significant challenge.
- **Market access and value chains:** Limited market opportunities for sustainably produced agricultural goods may discourage farmers from investing in land restoration practices. Weak or non-existent value chains can hinder the economic viability of adopting sustainable farming methods for land restoration.

2.8.2 Policy and Regulatory Barriers

- **Lack of supportive policies:** Inadequate policies that incentivize or support agricultural land restoration may hinder progress in this area or to farmers. Policies that favour conventional farming practices over sustainable alternatives may discourage farmers from adopting restoration techniques.
- **Land tenure issues:** Unclear land tenure arrangements and insecure land rights can deter farmers from making long-term investments in restoring land. Lack of policies addressing land tenure issues often lead to conflicts and hinder sustainable land management practices.
- **Regulatory compliance challenges:** Stringent regulations without adequate support mechanisms can burden farmers and limit their ability to adopt sustainable practices. Conflicting regulations or unclear guidelines may create confusion and hinder the implementation of land restoration measures.

2.8.3 Socio-economic and Cultural Considerations

- **Traditional farming practices:** Deep-rooted cultural and traditional farming practices may resist changes, making it challenging to introduce new, sustainable methods. Resistance to abandoning conventional methods and adopting new, ecologically friendly practices that have been passed down through generations is a most common challenge in promoting sustainable agriculture development.
- **Social dynamics and gender Issues:** Gender-based disparities in access to resources and decision-making can affect the successful implementation of land restoration initiatives. Socio-cultural norms may limit the involvement of certain groups in decision-making processes related to land management.
- **Community engagement:** Limited community involvement and participation in land restoration projects may hinder their success. Lack of awareness and involvement in decision-making processes can lead to resistance or indifference within local communities.

- **Attitudinal barriers:** Negative perceptions or misconceptions about the benefits of sustainable agriculture. Lack of awareness about the positive impacts of land restoration in local communities.

2.8.4 Limited Awareness and Capacity

- **Educational gaps:** Limited awareness and understanding among farmers about the importance and benefits of land restoration practices. Insufficient training and extension services to educate farmers about sustainable farming.
- **Capacity building challenges:** Inadequate infrastructure for training and capacity building programs related to land restoration. Shortage of skilled personnel to provide training and support to farmers in implementing sustainable land restoration and farming practices.
- **Information accessibility:** Limited access to relevant information and data regarding sustainable farming practices and their benefits. Lack of communication channels for disseminating information about successful land restoration initiatives.
- **Communication barriers:** Ineffective communication between farmers, stakeholders, and extension workers hindering the exchange of knowledge and experiences. Lack of awareness campaigns to bridge the information gap for land restoration.

2.9 Conclusion

Agricultural land restoration is a multifaceted and essential aspect of sustainable land management. It not only rejuvenates degraded soils but also contributes to global efforts for food security, environmental conservation, and the promotion of resilient agricultural systems. The detrimental effects of factors either natural or human-induced such as soil erosion, nutrient depletion, salinization, desertification, deforestation and altered soil structure have created severe degradation of soil and restricted supply of its ecosystem services. Key approaches to agricultural land restoration include agronomic and soil management practices like cover cropping, crop rotation, conservation agriculture, agroforestry, organic and inorganic amendments, and sustainable water management. Successful agricultural land restoration contributes to improved soil fertility, increased water retention, and enhanced nutrient cycling. It plays helps in mitigating the impacts of land degradation, preserving biodiversity, and fostering resilient and productive agricultural ecosystems. Moreover, the restoration of agricultural lands is closely linked to broader environmental goals.

References

Abdalla K, Mutema M, Hill T (2020) Soil and organic carbon losses from varying land uses: a global meta-analysis. Geogr Res 58(2):167–185

Abebaw WA (2019) Review on impacts of land degradation on agricultural production in Ethiopia. J Resour Dev Manag 57

Al-Amin AA, Lowenberg DeBoer J, Erickson BJ, Evans JT, Langemeier MR, Franklin K, Behrendt K (2024) Economics of strip cropping with autonomous machines. Agron J 116(2):572–589

Alhameid A, Ibrahim M, Kumar S, Sexton P, Schumacher TE (2017) Soil organic carbon changes impacted by crop rotational diversity under no-till farming in South Dakota, USA. Soil Sci Soc Am J 81(4):868–877

Anderson I, Buxton DR, Karlen DL, Cambardella C (1997) Cropping system effects on nitrogen removal, soil nitrogen, aggregate stability, and subsequent corn grain yield. Agron J 89:881–886

Ansari MA, Ravisankar N, Ansari MH, Babu S, Layek J, Panwar AS (2023) Integrating conservation agriculture with intensive crop diversification in the maize-based organic system: impact on sustaining food and nutritional security. Front Nutr 10:1137247. https://doi.org/10.3389/fnut.2023.1137247

Arco-Lazaro E, Martínez-Fernández D, Bernal MP, Clemente R (2017) Response of Piptatherum miliaceum to co-culture with a legume species for the phytostabilisation of trace elements contaminated soils. J Soils Sediments 17:1349–1357

Babu S, Singh R, Avasthe RK, Yadav GS, Das A, Singh VK, Mohapatra KP, Rathore SS, Chandra P, Kumar A (2020a) Impact of land configuration and organic nutrient management on productivity, quality and soil properties under baby corn in Eastern Himalayas. Sci Rep 10(1):1–14

Babu S, Singh R, Avasthe RK, Yadav GS, Mohapatra KP, Selvan T, Das A, Singh VK, Valente D, Petrosillo I (2020b) Soil carbon dynamics in Indian Himalayan intensified organic rice-based cropping sequences. Ecol Indic. https://doi.org/10.1016/j.ecolind.2020.106292

Babu S, Mohapatra KP, Yadav GS, Lal R, Singh R, Avasthe RK, Das A, Chandra P, Gudade BA, Kumar A (2020c) Soil carbon dynamics in diverse organic land use systems in North Eastern Himalayan ecosystem of India. Catena. https://doi.org/10.1016/j.catena.2020.104785

Babu S, Rathore SS, Singh R, Kumar S, Singh VK, Yadav SK, Yadav V, Raj R, Yadav D, Shekhawat K, Owais AW (2022) Exploring agricultural waste biomass for energy, food and feed production and pollution mitigation: a review. Bioresour Technol. https://doi.org/10.1016/j.biortech.2022.127566

Babu S, Singh R, Avasthe RK, Kumar S, Rathore SS, Singh VK, Ansari MA, Valente D, Petrosillo I (2023a) Soil carbon dynamics under organic farming: impact of tillage and cropping diversity. Ecol Indic. https://doi.org/10.1016/j.ecolind.2023.109940

Babu S, Singh R, Avasthe RK, Kumar S, Rathore SS, Singh VK, Ansari MA, Valente D, Petrosillo I (2023b) Conservation tillage and diversified cropping enhance system productivity and eco-efficiency and reduce greenhouse gas intensity in organic farming. Front Sustain Food Syst 7:1114617. https://doi.org/10.3389/fsufs.2023.1114617

Babu Subhash, Singh R, Kumar S, Rathore SS, Yadav D, Yadav SK, Yadav V, Ansari MA, Das A, Rajanna GA, Wani OA, Raj R, Yadav DK, Singh VK (2023c) Biochar implications in cleaner agricultural production and environmental sustainability. Environ Sci Adv. https://doi.org/10.1039/D2VA00324D

Barbosa B, Boléo S, Sidella S, Costa J, Duarte MP, Mendes B, Cosentino SL, Fernando AL (2015) Phytoremediation of heavy metal-contaminated soils using the perennial energy crops Miscanthus spp. and Arundo donax L. Bioeng Res 8:1500–1511

Barton E, Bennett DE, Burnidge W (2020) Holistic perspectives—understanding rancher experiences with holistic resource management to bridge the gap between rancher and researcher perspectives. Rangelands 42(5):143–150

Bashir S, Javed A, Bibi I, Ahmad N (2017) Soil and water conservation. University of Agriculture, Faisalabad, Faisalabad, pp 263–286

Bindraban PS, van der Velde M, Ye L, Van den Berg M, Materechera S, Kiba DI, Tamene L, Ragnarsdóttir KV, Jongschaap R, Hoogmoed M, Hoogmoed W (2012) Assessing the impact of soil degradation on food production. Curr Opin Environ Sustain 4(5):478–488

Bisaro A, Kirk M, Zdruli P, Zimmermann W (2014) Global drivers setting desertification research priorities: insights from a stakeholder consultation forum. Land Degrad Dev 25:5–16. https://doi.org/10.1002/ldr.2220

Borrelli P, Panagos P, Ballabio C, Lugato E, Weynants M, Montanarella L (2016) Towards a pan-European assessment of land susceptibility to wind erosion. Land Degrad Dev 27(4):1093–1105

Borrelli P, Alewell C, Alvarez P, Anache JAA, Baartman J, Ballabio C, Bezak N, Biddoccu M, Cerdà A, Chalise D, Chen S (2021) Soil erosion modelling: a global review and statistical analysis. Sci Total Environ 780:146494

Braimoh A (2015) The role of climate-smart agriculture in addressing land degradation. Solutions J 6(5):48–57

Briske DD, Sayre NF, Huntsinger L, Fernández-Giménez M, Budd B, Derner JD (2011) Origin, persistence, and resolution of the rotational grazing debate: integrating human dimensions into rangeland research. Rangel Ecol Manag 64(4):325–334

Byrnes RC, Eastburn DJ, Tate KW, Roche LM (2018) A global meta-analysis of grazing impacts on soil health indicators. J Environ Qual 47(4):758–765

Campanelli G, Iocola I, Leteo F, Montemurro F, Platani C, Testani E, Canali S (2023) Strip cropping in organically managed vegetable systems: agronomic and environmental effects. Renew Agric Food Syst 38

Chasek P (2022) From land degradation to land restoration. Policy Briefs 29(2)

Choudhary VK (2023) Applications of emerging smart technologies in farming systems: a review. Glob J Comput Sci Technol 23:49–64

Crews TE, Rumsey BE (2017) What agriculture can learn from native ecosystems in building soil organic matter: a review. Sustainability 9:1–18

Crosson PR (1997) Will erosion threaten agricultural productivity? Environment 39(8):4–9, 29–31

Das K, Biswakarma N, Zhiipao R, Kumar A, Ghasal PC, Pooniya V (2020) Significance and management of green manures. Soil Health:197–217

Das A, Babu S, Datta M, Kumar S, Singh R, Avasthe RK, Rathore SS, Yadav SK, Singh VK (2022a) Restoring soil carbon in marginal land of Indian Himalayas: impact of crop intensification and conservation tillage. J Environ Manag. https://doi.org/10.1016/j.jenvman.2022.115603

Das A, Babu S, Singh R, Kumar S, Rathore SS, Singh VK, Datta M, Yadav SK, Wani OA, Yadav D (2022b) Impact of live mulch-based conservation tillage on soil properties and productivity of summer maize in Indian Himalayas. Sustainability 14:12078. https://doi.org/10.3390/su141912078

Elhouri Ahmed A (1989) Land restoration and revegetation. FAO Conservation Guide (FAO).

Eswaran H, Lal R, Reich PF (2001) Land degradation: an overview. Responses to land degradation. In: Bridges E, Hannam I, Oldeman L, Penning de Vries F, Scherr S, Sompatpanit S (eds) Proceedings of the 2nd international conference on land degradation and desertification. Khon Kaen, Thailand. Oxford Press, New Delhi

Fagnano M, Impagliazzo A, Mori M, Fiorentino N (2015) Agronomic and environmental impacts of giant reed (Arundo donax L.): results from a long-term field experiment in hilly areas subject to soil erosion. Bioenergy Res 8:415–422

FAO (The Food and Agriculture Organization of the United Nations) (2011) The state of the world's land and water resources for Food and Agriculture (SOLAW)—managing systems at risk. Food and Agriculture Organization of the United Nations/Earthscan, Rome/London. Available online: http://www.fao.org/docrep/017/i1688e/i1688e.pdf. Accessed 29 Aug 2019

FAO (2012) Voluntary guidelines on the responsible governance of tenure of land, fisheries and forests in the context of national food security

FAO (2017) Watershed management in action—lessons learned from FAO field projects. Rome

Fernando AL, Boléo S, Barbosa B, Costa J, Duarte MP, Monti A (2015) Perennial grass production opportunities on marginal Mediterranean land. Bioenergy Res 8:1523–1537

Floris B, Galloni P, Sabuzi F, Conte V (2017) Metal systems as tools for soil remediation. Inorg Chim Acta 455:429–445

Francis C, Jones A, Crookston K, Wittler K, Goodman S (1986) Strip cropping corn and grain legumes: a review. Am J Altern Agric 1(4):159–164

Gebremeskel K, Teka K, Birhane E, Negash E (2019) The role of integrated watershed management on soil-health in northern Ethiopia. Acta Agric Scand,Section B Soil Plant Sci 69(8):667–673

Global Environment Facility (2019) Land degradation. Available online: https://thegef.org/sites/default/files/publications/gef_land_degradation_bifold_2019.pdf

Global Environmental Facility (GEF) (2006) Land Degradation as a Global Environmental Issue: a synthesis of three studies commissioned by the global environment facility to strengthen the knowledge base to support the land degradation focal area. Scientific and Technical Advisory Panel, Washington, DC

Gomiero T (2016) Soil degradation, land scarcity and food security: reviewing a complex challenge. Sustainability 8(3):281. https://doi.org/10.3390/su8030281

Gray JR (2002) The need for sediment surrogate technologies to monitor fluvial-sediment transport. In: Proceedings of the turbidity and other sediment surrogates workshop, April 30–May2, 2002, Silver Legacy Hotel, Reno, Nevada, U.S. Geological Survey Circular 1250, 4 p. Available online: http://water.usgs.gov/osw/techniques/TSS/listofabstracts.htm

Hossain A, Krupnik TJ, Timsina J, Mahboob MG, Chaki AK, Farooq M, Bhatt R, Fahad S, Hasanuzzaman M (2020) Agricultural land degradation: processes and problems undermining future food security. In: Environment, climate, plant and vegetation growth. Springer, Cham, pp 17–61. https://doi.org/10.1007/978-3-030-49732-3_2

Intergovernmental Science-Policy Platform on Biodiversity and Ecosystem Services (IPBES) Secretariat (2018) Summary for policymakers of the IPBES assessment report on land degradation and restoration. IPBES Secretariat, Bonn. https://doi.org/10.1016/0025-326x(95)90325-6

IPCC (2019) Final government distribution: land degradation (IPCC SRCCL). Available online: https://www.ipcc.ch/site/assets/uploads/2019/08/2e.-Chapter-4_FINAL.pdf. Accessed 29 Aug 2019

Jakhar P, Adhikary PP, Naik BS, Madhu M (2015) Finger millet (Eleusine coracana)-groundnut (Arachis hypogaea) strip cropping for enhanced productivity and resource conservation in uplands of Eastern Ghats of Odisha. Indian J Agron 60(3):365–371

Jha MN, Gupta MK, Raina AK (2001) Carbon sequestration: forest soil and land use management. Ann For Sci 9:249–256

Kang CH, Kwon YJ, So JS (2016) Bioremediation of heavy metals by using bacterial mixtures. Ecol Eng 89:64–69. https://doi.org/10.1016/j.ecoleng.2016.01.023

Kaye JP, Quemada M (2017) Using cover crops to mitigate and adapt to climate change. A review. Agron Sustain Dev 37:1–17

Kemp DR, Michalk DL, Virgona JM (2000) Towards more sustainable pastures: lessons learnt. Aust J Exp Agric 40(2):343–356

Khaledian Y, Kiani F, Ebrahimi S, Brevik EC, Aitkenhead-Peterson J (2017) Assessment and monitoring of soil degradation during land use change using multivariate analysis. Land Degrad Dev 28:128e141

Kiage LM (2013) Perspectives on the assumed causes of land degradation in the rangelands of Sub-Saharan Africa. Prog Phys Geogr 37:664–684. https://doi.org/10.1177/0309133313492543

Koralay N, Kara Ö (2018) Effects of soil erosion on water quality and aquatic ecosystem in a watershed. Scientific Committee-Bilim Kurulu, p 20

Kumar SS, Mir SA, Wani OA, Babu S, Yeasin M, Bhat MA, Hussain N, Wani AA, Kumar R, Yadav D, Dar SR (2022) Land-use systems regulate carbon geochemistry in the temperate Himalayas, India. J Environ Manag. https://doi.org/10.1016/j.jenvman.2022.115811

Kumawat A, Yadav D, Srivastava P, Babu S, Kumar D, Singh D, Vishwakarma DK, Sharma VK, Madhu M (2023) Restoration of agroecosystems with conservation agriculture for food security to achieve UN-Sustainable Development Goals. Land Degrad Dev. https://doi.org/10.1002/ldr.4677

Lal R (2004) Soil carbon sequestration impacts on global climate change and food security. Science 304:1623–1627

Lat R (2001) Loss of plant nutrient in runoff and eroded soil. In: Rosswall I (ed) Nitrogen cycling in West Africa ecosystem. Uppsala, Rekian and kata logtryck, pp 31–38

Lenka NK, Dass A, Sudhishri S, Patnaik US (2012) Soil carbon sequestration and erosion control potential of hedgerows and grass filter strips in sloping agricultural lands of eastern India. Agric Ecosyst Environ 158:31–40

Li G, Wan L, Cui M, Wu B, Zhou J (2019) Influence of canopy interception and rainfall kinetic energy on soil erosion under forests. Forests 10(6):509

López-Vicente M, Calvo-Seas E, Álvarez S, Cerdà A (2020) Effectiveness of cover crops to reduce loss of soil organic matter in a rainfed vineyard. Land 9(7):230

Ma B, Yu X, Ma F, Li Z, Wu F (2014) Effects of crop canopies on rain splash detachment. PLoS One 9(7):e99717

Maginnis S, Jackson W (2002) Building assets for people and nature: an introduction to forest landscape restoration. IUCN, Gland

Maglinao AR, Santoso D, Penning F (2000) Network approach in soil management research: IWMI's experience in southeast Asia. Available online: http://www.infoagro.net/sites/default/files/migrated_documents/attachment/Maglinao.pdf. Accessed 29 Aug 2019

Maglinao AR, Santoso D, Penning F (2003) Network approach in soil management research: IWMI's experience in Southeast Asia. International conference on impacts of agricultural research and development: why has impact assessment research not made more of a difference? In: International conference on impacts of agricultural research and development: why has impact assessment research not made more of a difference? A. Watson, DJ, A Mexico, DF (Mexico)^ BCIMMYT^ C2003 2003 (No. 338.91 WAT. CIMMYT)

McCallum MH, Kirkegaard JA, Green TW, Cresswell HP, Davies SL, Angus JF, Peoples MB (2004) Improved subsoil macroporosity following perennial pastures. Aust J Exp Agric 44(3):299–307

McIver J, Starr L (2001) Restoration of degraded lands in the interior Columbia River basin: passive vs. active approaches. For Ecol Manag 153:15–28

Meinen BU, Robinson DT (2020) Mapping erosion and deposition in an agricultural landscape: optimization of UAV image acquisition schemes for SfM-MVS. Remote Sens Environ 239:111666

Miller PR, Gan Y, McConkey BG, McDonald CL (2003) Pulse crops for the northern Great Plains: II. Cropping sequence effects on cereal, oilseed, and pulse crops. Agron J 95(4):980–986

Montanarella L (2016) The importance of land restoration for achieving a land degradation–neutral world. In: Land restoration. Academic, pp 249–258

Morgan L (1999) Pesticides and groundwater in the State of Washington. Environ Manag Health 10(1):7–17

Morris GL, Annandale G, Hotchkiss R (2008) Reservoir sedimentation, in sedimentation engineering: processes, measurements, modeling, and practice. American Society of Civil Engineers, Reston

Mosier S, Apfelbaum S, Byck P, Calderon F, Teague R, Thompson R, Cotrufo MF (2021) Adaptive multi-paddock grazing enhances soil carbon and nitrogen stocks and stabilization through mineral association in southeastern US grazing lands. J Environ Manag 288:112409

Murtaza B, Zaman G, Imran M, Shah GM, Amjad M, Ahmad N, Naeem MA, Zakir A, Farooq A, Ahmad S, Murtaza G (2019) Municipal solid waste compost improves crop productivity in saline-sodic soil: a multivariate analysis of soil chemical properties and yield response. Commun Soil Sci Plant Anal 50(8):1013–1029. https://doi.org/10.1080/00103624.2019.1604016

NAAS (2010) Degraded and wastelands of India, status and spatial distribution. ICAR, New Delhi

Nkonya E, Mirzabaev A, Von Braun J (eds) (2016) Economics of land degradation and improvement: a global assessment for sustainable development. Springer Open, Cham. https://doi.org/10.1007/978-3-319-19168-3

Nösberger J, Geiger HH, Struik PC (2001) Crop science: progress and prospects/edited by p. cm. In: Papers presented at the third international crop science congress in Hamburg, Germany. Available from: https://www.researchgate.net/publication/344485158_Agricultural_Land_

Degradation_Processes_and_Problems_Undermining_Future_Food_Security. Accessed 13 Jan 2024

Oldeman LR (1998) Soil degradation: a threat of food security. Report 98/01. International Soil Reference and Information Centre, Wageningen. Available from: https://www.isric.org/documents/document-type/isric-report-199007-world-map-status-human-induced-soil-degradation

Oldeman RL, Hakkeling RTA, Sombroek WG (1991) World map of the status of human-induced soil degradation. 2nd rev. edn. International Soil Reference and Information Centre, Wageningen

Olsson L, Barbosa H, Bhadwal S, Cowie A, Delusca K, Flores-Renteria D, Hermans K, Jobbagy E, Kurz W, Li D, Sonwa DJ (2019) Land degradation: IPCC special report on climate change, desertification, land degradation, sustainable land management, food security, and greenhouse gas fluxes in terrestrial ecosystems. In: IPCC special report on climate change, desertification, land degradation, sustainable land management, food security, and greenhouse gas fluxes in terrestrial ecosystems. Intergovernmental Panel on Climate Change (IPCC), p 1

Panagos P, Borrelli P, Meusburger K, Alewell C, Lugato E, Montanarella L (2015) Estimating the soil erosion cover-management factor at the European scale. Land Use Policy 48:38–50

Petito M, Cantalamessa S, Pagnani G, Degiorgio F, Parisse B, Pisante M (2022) Impact of conservation agriculture on soil erosion in the Annual Cropland of the Apulia Region (Southern Italy) based on the RUSLE-GIS-GEE framework. Agronomy 12(2):281. https://doi.org/10.3390/agronomy12020281

Poeplau C, Don A (2015) Carbon sequestration in agricultural soils via cultivation of cover crops—a meta-analysis. Agric Ecosyst Environ 200:33–41. https://doi.org/10.1016/j.agee.2014.10.024

Rai AK, Basak N, Meena RL, Sundha P, Khandkar UR, Bangar KS, Jha SK, Sharma DK, Sharma PC, Yadav RK, Chandra P, Patel S, Kumar S, Kaur H, Bedwal S, Jasra RV, Chintansinh SC, Sidhpuria K, Kumar N (2023) Reclamation of sodic soils through flue gas desulphurization gypsum. Technology Folder/2023/04. ICAR–Central Soil Salinity Research Institute, Karnal

Rathore SS, Babu S, El-Sappah AH, Shekhawat K, Singh VK, Singh RK, Upadhyay PK, Singh R (2022a) Integrated agroforestry systems improve soil carbon storage, water productivity, and economic returns in the marginal land of the semi-arid region. Saudi J Biol Sci. https://doi.org/10.1016/j.sjbs.2022.103427

Rathore SS, Babu S, Shekhawat K, Singh R, Yadav SK, Singh VK, Singh C (2022b) Designing energy cum carbon-efficient environmentally clean production system for achieving green economy in agriculture. Sustain Energy Technol Assess. https://doi.org/10.1016/j.seta.2022.102190

Reinl J (2019) Desertification costs world economy up to 15 trillion dollars. UN Inter Press Service. Available from: https://ipsnews.net/2019/09/desertification-costs-world-economy-15-trillion-dollars-u-n/

Reynolds JF, Smith DM, Lambin EF, Turner BL, Mortimore M, Batterbury SP, Downing TE, Dowlatabadi H, Fernández RJ, Herrick JE, Huber-Sannwald E (2007) Global desertification: building a science for dryland development. Science 316:847–851

Roche LM, Cutts BB, Derner JD, Lubell MN, Tate KW (2015) On-ranch grazing strategies: context for the rotational grazing dilemma. Rangel Ecol Manag 68(3):248–256

Rumpel C, Chabbi A (2021) Managing soil organic carbon for mitigating climate change and increasing food security. Agronomy 11(8):1553. https://doi.org/10.3390/agronomy11081553

Salazar O, Rojas C, Avendaño F, Realini P, Nájera F, Tapia Y (2015) Inorganic nitrogen losses from irrigated maize fields with narrow buffer strips. Nutr Cycl Agroecosyst 102:359–370. https://doi.org/10.1007/s10705-015-9707-4

Sanderman J, Hengl T, Fiske GJ (2017) Soil carbon debt of 12,000 years of human land use. Proc Natl Acad Sci 114(36):9575–9580

Saturday A (2018) Restoration of degraded agricultural land: a review. J Environ Health Sci 4(2):44–51

Scherr SJ (1999) Soil degradation: a threat to developing-country food security by 2020? Volume 27 of Food, agriculture, and the environment discussion paper. International Food Policy Research Institute, p 63

Sharda VN, Mandal D, Dogra P (2019) Assessment of cost of soil erosion and energy saving value of soil conservation measures in India. Indian J Soil Conserv 47(1):1–6

Sharif F, Danish MU, Ali AS, Khan AU, Shahzad L, Ali H, Ghafoor A (2016) Salinity tolerance of earthworms and effects of salinity and vermi amendments on growth of Sorghum bicolor. Arch Agron Soil Sci 62. https://doi.org/10.1080/03650340.2015.1132838

Singh R, Babu S, Avasthe RK, Meena RS, Yadav GS, Das A, Mohapatra KP, Rathore SS, Kumar A, Singh C (2021a) Conservation tillage and organic nutrients management improve soil properties, productivity, and economics of the maize-vegetable pea system in the Eastern Himalayas. Land Degrad Dev. https://doi.org/10.1002/ldr.4066

Singh R, Babu S, Avasthe RK, Yadav GS, Das A, Mohapatra KP, Kumar A, Singh VK, Chandra P (2021b) Crop productivity, soil health, and energy dynamics of Indian Himalayan intensified organic maize-based systems. Int Soil Water Conserv Res 9(2):260–270. https://doi.org/10.1016/j.iswcr.2020.11.003

Singh D, Mishra AK, Patra S, Dwivedi AK, Ojha CSP, Singh VP, Mariappan S, Babu S, Singh N, Yadav D, Ojasvi PR, Kumar G, Madhu M, Sena DR, Chand L, Kumar S (2023) Effect of long-term tillage practices on runoff and soil erosion in sloping croplands of Himalaya, India. Sustainability 15:8285. https://doi.org/10.3390/su15108285

Smith P, Bustamante M, Ahammad H, Clark H, Dong H, Elsiddig EA, Haberl H, Harper R, House J, Jafari M, Masera O (2014) Agriculture, forestry and other land use (AFOLU). In: Climate change 2014: mitigation of climate change. Contribution of Working Group III to the Fifth Assessment Report of the Intergovernmental Panel on Climate Change. Cambridge University Press, pp 811–922

Sorrenson WJ, Lopez Portillo J, Nunez M (1997) The economics of no-tillage and crop rotations in Paraguay. Policy and investment implications. FAO, Final Report to the MAG/GTZ Soil Conservation Project, 215 pp

Srinivasarao C, Naik MR, Kumar GR, Ravula M, Kundu S, Swamy GN, Nataraj KC, Prasad JVNS (2022) Cover-crop technology for soil health improvement, land degradation neutrality, and climate change adaptation. Indian J Fertil 18(5):440–460

Studdert GA, Echeverria HE (2000) Crop rotations and nitrogen fertilization to manage soil organic carbon dynamics. Soil Sci Soc Am J 64(4):1496–1503

Sullivan P (2004) Sustainable soil management: soil systems guide. Appropriate Technology Transfer for Rural Areas (ATTRA) FairettevilleA.R.72702, National Centre for AppropriateTechnology. Available from: https://www.researchgate.com/publication/344485158_Agricultural_Land_Degradation_Processes_and_Problems_Undermining_Future_Food_Security. Accessed 13 Jan 2024

Teague WR, Dowhower SL, Baker SA, Haile N, DeLaune PB, Conover DM (2011) Grazing management impacts on vegetation, soil biota and soil chemical, physical and hydrological properties in tall grass prairie. Agric Ecosyst Environ 141(3–4):310–322

Teague WR, Grant B, Wang HH (2015) Assessing optimal configurations of multi-paddock grazing strategies in tallgrass prairie using a simulation model. J Environ Manag 150:262–273

Tripathi V, Edrisi SA, Abhilash PC (2016) Towards the coupling of phytoremediation with bioenergy production. Renew Sust Energ Rev 57:1386–1389

Troch FR, Hobbs JA, Donahue KL (2002) Soil and water conservation. Prentice Hall. UN Center on Trans-national, Englewood Cliffs

United Nations (UN) (2012) The future we want. Available from: http://www.uncsd2012.org/thefuturewewant.html. Accessed 1 Jan 2015

United Nations Convention to Combat Desertification (UNCCD) (2013) Land matters for climate: reducing the gap and approaching the target. UNCCD, Bonn

United Nations Convention to Combat Desertification (2015) Decision 3/COP.12. 20831 (January): 80. UNCCD, Bonn

United Nations Convention to Combat Desertification (2016) Achieving land degradation neutrality at the country level: building blocks for LDN target setting. UNCCD, Bonn

United Nations Convention to Combat Desertification (2017) Global land outlook. UNCCD, Bonn

United Nations Convention to Combat Desertification (2020a) Land tenure. unccd.int/actions/land-tenure

United Nations Convention to Combat Desertification (2020b) Operationalization of the Land Degradation Neutrality Fund. Report by the Global Mechanism. ICCD/CRIC(19)/3. unccd.int/sites/default/files/sessions/documents/2020

United Nations Convention to Combat Desertification (2024) Land management & restoration. UNCCD

United Nations Environment Programme (UNEP) (2022) Beyond GDP: making nature count in the shift to sustainability. Beyond GDP: making nature count in the shift to sustainability (unep.org)

Utuk IO, Daniel EE (2015) Land degradation: a threat to food security: a global assessment. J Environ Earth Sci IISTE 5:13–21

Van Pelt RS, Hushmurodov SX, Baumhardt RL, Chappell A, Nearing MA, Polyakov VO, Strack JE (2017) The reduction of partitioned wind and water erosion by conservation agriculture. Catena 148:160–167

Walia SS, Babu S, Gill RS, Kaur T, Kohima N, Panwar AS, Yadav DK, Ansari MA, Ravishankar N, Kumar S, Kaur K, Ansari MH (2022) Designing resource-efficient and environmentally safe cropping systems for sustainable energy use and economic returns in Indo-Gangetic Plains, India. Sustainability 14:14636. https://doi.org/10.3390/su142114636

Wani OA, Kumar S, Hussain N, Wani AA, Babu S, Alam P, Rashid M, Popescu SM, Mansoor S (2022) Multi-scale processes influencing global carbon storage and land-carbon-climate nexus: a critical review. Pedosphere. https://doi.org/10.1016/j.pedsph.2022.07.002

Wani OA, Sharma V, Kumar SS, Babu S, Sharma KR, Rathore SS, Marwaha S, Ganai NA, Dar SR, Yeasin M, Singh R, Tomar J (2023) Climate plays a dominant role over land management in governing soil carbon dynamics in North Western Himalayas. J Environ Manag. https://doi.org/10.1016/j.jenvman.2023.117740

Watson R, Noble I, Bolin B, Ravindranath NH, Verardo D, Andrasko K, Apps M, Brown S, Farquhar G, Goldbert D, Hamburg S (2000) Summary to policy makers: land use, land-use change, and forestry. In: Land use, land-use change and forestry

Wei L, Zhang Y, Lu Q, Yuan Z, Li H, Huang Q (2021) Estimating the spatial distribution of soil total arsenic in the suspected contaminated area using UAVBorne hyperspectral imagery and deep learning. Ecol Indicators:133108384

World Overview of Conservation Approaches and Technologies (2010) Shelterbelts for farmland in sandy areas [China]. WOCAT SLM Technologies

Yadav SK, Singh Y, Yadav MK, Babu S, Singh K (2013) Effect of organic nitrogen sources on yield, nutrient uptake and soil health under rice (*Oryza sativa*) based cropping sequence. Indian J Agric Sci 83(2):170–175

Yadav GS, Babu S, Das A, Bhowmik SN, Datta M, Singh R (2019a) Soil carbon dynamics and productivity of rice-rice system under conservation tillage in submerged and unsubmerged ecologies of eastern Indian Himalaya. Carbon Manag 10(1):51–62. https://doi.org/10.1080/17583004.2018.1545518

Yadav GS, Das A, Lal R, Babu S, Datta M, Meena RS, Patil SB, Singh R (2019b) Impact of no-till and mulching on soil carbon sequestration under rice (*Oryza sativa* L.)-rapeseed (Brassica campestris L. var. rapeseed) cropping system in hilly agro-ecosystem of the Eastern Himalayas, India. Agric Ecosyst Environ 275:81–92

Yadav GS, Das A, Babu S, Mohapatra KP, Lal R, Rajkhowa DJ (2020a) Potential of conservation tillage and altered land configuration to improve soil properties, carbon sequestration and productivity of maize based cropping system in eastern Himalayas. International Soil and Water Conservation Research, India. https://doi.org/10.1016/j.iswcr.2020.12.003

Yadav GS, Babu S, Das A, Mohapatra KP, Singh R, Avasthe RK, Roy S (2020b) No till and mulching enhance energy use efficiency and reduce carbon footprint of a direct-seeded upland rice production system. J Clean Prod. https://doi.org/10.1016/j.jclepro.122700

Yadav GS, Kandpal BK, Das A, Babu S, Mohapatra KP, Devi AG, Devi HL, Chandra P, Singh R, Barman KK (2021a) Impact of 28-year-old agroforestry systems on soil carbon dynamics in Eastern Himalayas. J Environ Manag. https://doi.org/10.1016/j.jenvman.2021.111978

Yadav GS, Das A, Kandpal BK, Babu S, Lal R, Datta M, Das B, Singh R, Singh VK, Mohapatra KP, Chakraborty M (2021b) The food-energy-water-carbon nexus in a maize-maize-mustard cropping sequence of the Indian Himalayas: an impact of tillage-cum-live mulching. Renew Sust Energ Rev. https://doi.org/10.1016/j.rser.2021.111602

Yadav GS, Babu S, Das A, Datta M, Mohapatra KP, Singh R, Singh VK, Rathore SS, Chakraborty M (2021c) Productivity, soil health, and carbon management index of Indian Himalayan intensified maize-based cropping systems under live mulch-based conservation tillage practices. Field Crop Res. https://doi.org/10.1016/j.fcr.2021.108080

Yadav D, Singh D, Babu S, Madegowda M, Singh DV, Mandal D, Rathore AC, Sharma VK, Singhal V, Kumawat A, Yadav DK, Yadav RK, Kumar S (2023) Intensified cropping reduces soil erosion and improves rainfall partitioning and soil properties in the marginal land of the Indian Himalayas. Int Soil Water Conserv Res. https://doi.org/10.1016/j.iswcr.2023.10.002

Yadav D, Singh D, Madhu M, Singh DV, Islam S, Kumar G, Gupta AK, Mandal D, Babu S, Barh A, Kumawat A, Dubey R, Pati K, Bhargavi B, Yadav DK (2024) Interplanted crops and rainfall variability regulate soil erosion in rainfed maize of the Indian Himalayas. Land Degrad Dev. https://doi.org/10.1002/ldr.5221

Yang Y, Hobbie SE, Hernandez RR, Fargione J, Grodsky SM, Tilman D, Zhu YG, Luo Y, Smith TM, Jungers JM, Yang M (2020) Restoring abandoned farmland to mitigate climate change on a full earth. One Earth 3(2):176–186

Young A (1995) Agroforestry for soil conservation. CTA, Wageningen, p 194

Yuan Y, Bingner RL, Locke MA (2009) A review of effectiveness of vegetative buffers on sediment trapping in agricultural areas. Ecohydrol Ecosyst Land Water Process Interact Ecohydrogeomorphol 2(3):321–336

Zhang H, Yang Q, Li R, Liu Q, Moore D, He P, Ritsema CJ, Geissen V (2013) Extension of a GIS procedure for calculating the RUSLE equation LS factor. Comput Geosci 52:177–188

Zhao X, Wang S, Xing G (2014) Nitrification, acidification, and nitrogen leaching from subtropical cropland soils as affected by rice straw-based biochar: laboratory incubation and column leaching studies. J Soils Sediments 14:471–482. https://doi.org/10.1007/s11368-013-1149-0803-2

Zou XX, Shi PX, Zhang CJ, Si T, Wang YF, Zhang XJ, Yu XN, Wang HX, Wang ML (2021) Rotational strip intercropping of maize and peanuts has multiple benefits for agricultural production in the northern agropastoral ecotone region of China. Eur J Agron 129:126304

Agroforestry for Restoring Degraded Land

3

Nuwandhya S. Dissanayaka, Tharindu D. Nuwarapaksha, Shashi S. Udumann, and Anjana J. Atapattu

Abstract

Land degradation poses a critical challenge to environmental sustainability and human welfare. Agroforestry has multifaceted opportunities to restore degraded lands. The diverse causes of land degradation like deforestation, overgrazing, unsustainable agricultural practices, and climate change need to be adequately addressed along with their far-reaching environmental, economic, social, and geopolitical impacts. Agroforestry, which deliberately integrates woody perennials with agricultural crops and/or animals, is a promising approach to land restoration. There is an urgent need to understand various agroforestry systems and their mechanisms for rehabilitating degraded lands, focusing on soil improvement, biodiversity enhancement, water management, and carbon sequestration. An in-depth analysis of how these systems contribute to ecosystem restoration while simultaneously offering economic benefits to farmers is very important. The challenges and limitations in implementing agroforestry systems, including technical complexities, socioeconomic barriers, and policy constraints need adequate attention. Further, future research needs, possible policy recommendations, and strategies for scaling up successful agroforestry practices should be emphasized. Thus, an attempt is made to understand the role of agroforestry in the growing body of knowledge on sustainable land management practices and their potential to address global environmental challenges.

Keywords

Deforestation · Biodiversity · Climate change · Land restoration · Sustainability

N. S. Dissanayaka · T. D. Nuwarapaksha · S. S. Udumann · A. J. Atapattu (✉)
Agronomy Division, Coconut Research Institute, Lunuwila, Sri Lanka

© The Author(s), under exclusive license to Springer Nature Singapore Pte Ltd. 2025
S. Babu et al. (eds.), *Ecological Solutions to Agricultural Land Degradation*,
https://doi.org/10.1007/978-981-96-3392-0_3

3.1 Introduction

The land is a vital natural resource that supports the sustainable livelihood of human populations and natural surroundings. Forests serve as the basis for agriculture, timber, and urban planning which are essential for food, fiber, non-timber products, bio-diversity, and maintaining the balance in the ecological system. Land resources conservation and management are essential for sustaining soil productivity, supporting water cycles, and storing carbon, all to ensure that the environment is stable for mitigating climate change. However, land resource degradation is on the rise readily giving way to various issues and challenges (Cowie et al. 2018). According to the United Nations, globally more than 23% of the land is considered as moderately to highly degraded, escalating beyond humankind and affecting over 1.5 billion global population (Fig. 3.1) (Stavi and Lal 2015). This can be described as the state whereby the biophysical environment is altered by a series of human activities operating on land (Mohamed et al. 2019). It is characterized by the deterioration of the quality and or the carrying capacity of resources such as land, water, and other assets largely attributed to poor utilization of land assets (Gomiero 2016). The biophysical environment refers to the biotic and abiotic components of the natural world. It includes soil, water, air, plants, animals, and microorganisms.

Signs of land degradation include soil erosion, evidenced by the loss of topsoil, formation of gullies, and sediment accumulation in water bodies (Chalise et al. 2019). The salinity of the soil an be observed as a white layer on the soil surface and poor plant growth resulting from the intake of salts becomes toxic to the plant (Acharya and Kafle 2009). Soil compaction can also occur and lead to hard compacted layers of the soil, water repellency, and animal's unhealthy condition (Nawaz et al. 2013). Loss of vegetation cover happens due to land quality depletion, leading to sparse vegetation, invasive species, and loss of biodiversity (Hossain et al. 2020).

Fig. 3.1 Status of the top 10 land degraded countries of the world. (Source: Ulain et al. 2022)

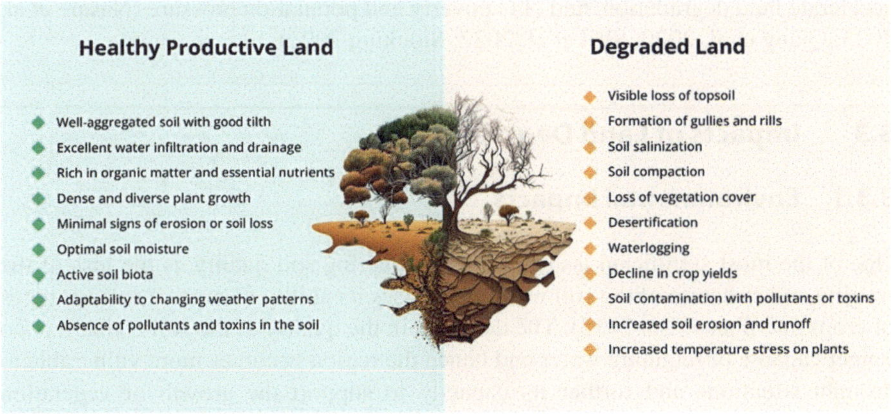

Fig. 3.2 Signs of healthy and productive land vs degraded land

Desertification is also a sign of land degradation (Siril et al. 2022). It is marked by the expansion of barren areas, increased dust storms, and a decline in agricultural productivity (Al-Kulabi 2022). Waterlogging causes excess water on the soil surface and stunted plant growth (Weeraratna 2022). Nutrient depletion due to poor farming practices leads to declining crop yields and yellowing plants, necessitating more fertilizers (Gomiero 2016). Chemical degradation involves soil contamination, acidification, and reduced microbial activity (Lehman et al. 2015). Deforestation and land clearing results reduced forest cover, enhanced soil erosion, and habitat destruction. A summary of signs of healthy, productive land and degraded land is shown in Fig. 3.2.

3.2 Causes of Land Degradation

Some natural processes can cause land degradation over the years, however, very often land degradation is promoted by anthropogenic activities as the primary drivers (Talukder et al. 2021). The causes for this process are diverse and often interconnected. Importantly, land degradation is a complex process that can occur over long periods and may not always be immediately visible. The main causes for this irreversible phenomenon are (1) deforestation activities involving clearing forests for agriculture, or logging, (2) overgrazing land beyond its carrying capacity, (3) unsustainable farming practices like prolonged monoculture, excessive use of synthetic fertilizers and pesticides, poor irrigation practices causing salinization, and failure to implement crop rotation, (4) conversion of fertile land to urban areas, (5) industrial pollution contaminating soil and water resources, (6) open-pit mining destroying top soil and vegetation, (7) poor soil and water management like lack of soil conservation techniques and inefficient irrigation systems leading to waterlogging or salinization, (8) land fragmentation, (9) inappropriate land-use policies, (10) climate change and natural disasters like floods, landslides, and wildfires can accelerate land degradation, and (11) poverty and population pressure (Nasare et al. 2023; Qiang et al. 2020; Siril et al. 2022; Stocking 2001).

3.3 Impacts of Land Degradation

3.3.1 Environmental Impacts

One of the most significant issues, directly affecting soil quality, is the loss of the fertility and structure of the soil which decreases its ability to support various types of crops and plants (Lal 2015). The decrease in the quality of the soil is that it is no longer capable of retaining water and hence the region becomes more vulnerable to drought situations and further its capacity to support the growth of vegetation declines (Nawaz et al. 2013; Singh et al. 2021). This degradation also enhances the chance of the soil to be eroded, an aspect that further aggravates the degradation (Chalise et al. 2019). Recent studies have suggested soil erosion rates in the range of 75 Pg year^{-1} for croplands globally (Borelli et al. 2017). El-Gammal et al. (2015) revealed that the high risk of physical (i.e. soil compaction) and chemical vulnerability (i.e. salinization) covered an area of 86.02 km^2 (12.83%) and 2.28 km^2 (0.34%), respectively in the surface soil layers in Egypt. Biological diversity is then reduced especially in the measures of habitats by small or large-scale developers causing species decline and, at worst, extinction (Pueyo et al. 2006). This loss prevents such factors as pollination and natural control of pests that are so essential to both nature and crops. The result is the harming of water resources major one being the effect on the quantity and quality of water in rivers, lakes, and aquifers among others (Bera et al. 2022). Climate change affects the existing hydrological cycles to an extent that it changes the rates of groundwater recharge and affects the long-term availability of water (Gomiero 2016; Avasthe et al. 2023). Also, land degradation affects climate change since some of the emissions, such as carbon from the soils and vegetation, are influenced by the process (Lal 2004). As soil organic matter is exposed, it releases greenhouse gases, contributing to global warming (Babu et al. 2023b). These interconnected impacts create a cascade effect, where each aspect of environmental degradation reinforces and amplifies the others, posing a weighty threat to the overall health and stability of ecosystems.

3.3.2 Economic Impacts

Land degradation exerts significant economic impacts that ripple through local and global economies. At the forefront is the decline in agricultural productivity, manifested through decreased crop yields and diminished quality of produce. Marginal degradation can reduce the crop yield by 10%, moderate somewhere between 10% and 50% while severely degraded soil can reduce the crop yield by over 50% (Gupta 2019). Farmers often find themselves caught in a vicious cycle, as degraded land requires more inputs (such as fertilizers and water) to maintain productivity, thereby enhancing cost of production and minimizing profit margins (Lal 2015). Food security becomes one of the noteworthy issues arising mainly in developing nations where subsistence farming is rife (Nuwarapaksha et al. 2024b). According to Gupta (2019), the loss of cropland is likely to cause a 13–18% decline in China's food production capacity by 2030–2050. That is why when production reduces and when it becomes expensive to produce, the global food basket is at risk of having its prices raised with a corresponding increase in the number of hungry people and the incidences of malnutrition. Regarding the economic impacts, poverty levels rise, and most of the effects are felt in rural areas where social groups depend on land resources for their existence (Abu Hammad and Tumeizi 2012). These communities may end up being completely displaced from the regions or being forced to look for other means of livelihood since they may end up having no land to till, further increasing their poverty levels and consequently the rate of rural-urban migration (Barbier and Hochard 2018). These multiple intertwined economic effects show the importance of developing and implementing the practical strategies of sustainable land management to ensure both the health of the environment and the solidity of economic future.

3.3.3 Social Impacts

Land degradation triggers a cascade of social impacts that profoundly affect communities and societal structures. Forced migration emerges as a critical issue when land can no longer support local populations, leading to the displacement of people from their ancestral homes (McLeman 2017). This pushes people from their homes and displaces them and although it affects their lives, this might also cause conflicts in the areas that people are relocated to and in assimilation for resources and culture. Environmental problems make people develop diseases as a way of showing the impacts of a damaged environment (Lal 2009). Food insecurity rises due to the onset of scarcity and whereby quality food may be inaccessible, disease prevalence is high specifically among children and other susceptible groups. Moreover, degraded lands often experience increased pollution, exposing inhabitants to various contaminants that cause health problems (Manisalidis et al. 2020). The escalating land degradation hampers positive social transformation as it has negative impacts on the life quality of vulnerable and marginalized groups of people who cannot afford to find ways of adapting to the effects or resettling. This leads to social and economic marginalization where only some individuals have access to productive land and are hence excluded from this privilege, deepening social and economic inequalities in societies. The cultural dimension of land degradation is often overlooked but is equally significant. As communities are forced to abandon their lands or radically change their practices, traditional land management knowledge and techniques are lost. This erosion of cultural heritage extends to the physical landscape itself, with unique cultural landscapes and heritage sites being altered or destroyed.

3.3.4 Geopolitical Impacts

As land degradation progresses, it leads to increased competition for diminishing land and water resources. These scarcities can provoke conflict along ethnic, geographical, and sometimes national lines (Baechler 1999). For instance, along international water bodies, such as rivers, water quality and quantity degradation by the upstream nation causes diplomatic flare-ups with the downstream nation. Likewise, with limited accessible and productive land for agriculture being available, there could be more efforts made to gain a piece of land in another country which may then land one in accusations of "land grab." The bidding war over resources intensifies and may transform into a full-on war mostly in areas that are politically unstable or have historical beefs (Baechler 1999). The stress on resources caused by land degradation can exacerbate existing political and social tensions, potentially destabilizing entire regions (Barbut and Alexander 2016). This instability can be evidenced in civil unrest in a particular country or even conflict with neighboring countries. For example, migration due to degradation can stretch the capacity and cohesion of receiving regions, and this can pose a problem of xenophobia between indigenous persons and settlers. Due to its effects on food and water security, land degradation remains a major threat to national security (Abrol and Gupta 2019). As agricultural productivity declines due to soil degradation, countries may struggle to feed their populations or maintain economic stability in their agricultural sectors. This can again result in reliance on imported food products, and in view of the fact that any country can be subjected to food embargoes by other countries or face unfavorable market conditions in the international market, this can be disadvantageous (Atapattu et al. 2024c). Water scarcity resulting from land degradation can similarly threaten national security. Lack of access to clean water can lead to health crises, economic disruption, and social unrest. In extreme cases, water scarcity can become a strategic asset in conflicts. Furthermore, the nations with their lands severely degraded may become politically or economically unstable, and therefore, become nexuses for extremism or need intervention by the international community members (Gomiero 2016).

3.4 Long-Term Sustainability

Soil degradation is significant in its impact on long-term sustainability of ecosystems and, therefore, on the prospective generations' potential (Imeson 2011). Furthermore, it severely limits the ecological resilience, that is the capacity of the ecosystems to absorb disturbances and return to the original state. Absolute stability in ecosystems depicts their core functionality while self-healing, which means ecosystems possess preconditions that help them maintain their stability. However, as land degrades, this resilience is eroded. Soil loses its structure and nutrient content, vegetation cover diminishes, and biodiversity declines. These changes make ecosystems more vulnerable to disturbances such as extreme weather events, pest outbreaks, or invasive species (Dale et al. 2005; Gibbs and Salmon 2015). The decline of resilience also means that ecosystems get mechanically inclined to future environmental changes, which include climate change (Babu et al. 2023a). For example, eroded forests easily fall under the pressure of influence such as drought or fire influence more than eroded agricultural lands which are less capable of resisting alterations in precipitation (Lal 2012). This makes the ecosystem less capable of recovering and thus there is a vicious cycle that results to further reduces the capability of the ecosystem to recover from environmental stresses. The effects of land degradation are long-term reaching down to future generations and therefore have negatively affects the human capacity to achieve their needs. There are still many people who continue to use land resources poorly or directly consume them, that is, the world is living at the expense of the future. Thus, degraded soils yield lower hence the next generation would be faced with the problem of producing enough food, timber, and other crop outputs (Sharma et al. 2018). This has proven to be a big challenge towards the achievement of sustainable development in the world. The United Nations' Agenda for Sustainable Development, the realization of which is to lead to a better future for human beings and the planet as a whole, depends to a great extent on their healthy state and productivity (Atapattu et al. 2024a). Of these, land degradation puts at risk many of the SDGs such as poverty, food, water and climate by and for 2030 (Chew et al. 2024; Jiang et al. 2022; Sims et al. 2019).

The previous approaches to land rehabilitation are commonly known to be single-goal activities, typically including afforestation or soil protection (Debie 2020; Song et al. 2022). Agroforestry technology is almost a win-win strategy that can at the same time restore the ecological potential, on the one hand, and bring material dividends to the farmers with the owners of the plots of land (Nuwarapaksha et al. 2023c). By integrating trees with crops or livestock, agroforestry systems create a synergistic environment that combats various forms of land degradation (Dissanayaka et al. 2024a). The deep root systems of trees improve soil structure, enhance water infiltration, and reduce erosion, effectively countering soil degradation (Dissanayaka et al. 2023a). The continuous addition of organic matter through leaf litter and root decomposition enhances soil fertility and carbon content, reversing nutrient depletion. Agroforestry's diverse vegetation structure promotes biodiversity, supporting natural pest control and pollination services, which helps restore ecological balance in degraded landscapes (Udawatta et al. 2019). Analyzing the role of agroforestry in land restoration aims to contribute to the growing body of knowledge on sustainable land management practices and their potential to address global environmental challenges.

3.5 Agroforestry Systems for Land Restoration

Agroforestry is the deliberate integration of woody perennials (trees, shrubs, palms, bamboos, etc.) with agricultural crops and/or animals on the same land management unit (Mosquera-Losada et al. 2018). It is a dynamic, ecologically based natural resource management system that diversifies and sustains production for increased social, economic, and environmental benefits (Mukhlis et al. 2022).

3.5.1 Agroforestry System Classification

Agroforestry systems are classified using various approaches, primarily based on their structural, functional, socioeconomic, and ecological characteristics, as well as the components involved. Structurally, systems are categorized by spatial arrangements (e.g., mixed dense plantings, boundary plantings, alternate rows planting, and widely spaced trees) and temporal arrangements (coincident, concomitant, or sequential) (Nair et al. 2021). Functionally, they are classified based on productive (food, fodder, wood, or/and other products like medicinal plants, and resins) and protective (as for soil conservation, biodiversity, and water conservation) roles (Atreya et al. 2021). Socioeconomic classification considers the scale of production and management level, while ecological classification is based on environmental settings as tropical agroforestry systems, temperate agroforestry systems, and arid and semi-arid agroforestry systems (Dissanayaka et al. 2023d). Commonly used organization in scientific literature and practice is component-based, which can be summarized in the following Fig. 3.3.

Additional practices like windbreaks, riparian buffer strips, and forest farming can fit into one or more of these categories depending on their specific implementation. These classification systems provide a framework for understanding the diverse range of agroforestry practices, although many real-world systems may incorporate elements from multiple categories, reflecting their complex and integrated nature.

Fig. 3.3 Coconut-based different agroforestry systems (**a**) Agrisilvicultural system, (**b**) Silvopastoral system, (**c**) Agrosilvopastoral system (Source: Sharma et al. 2022)

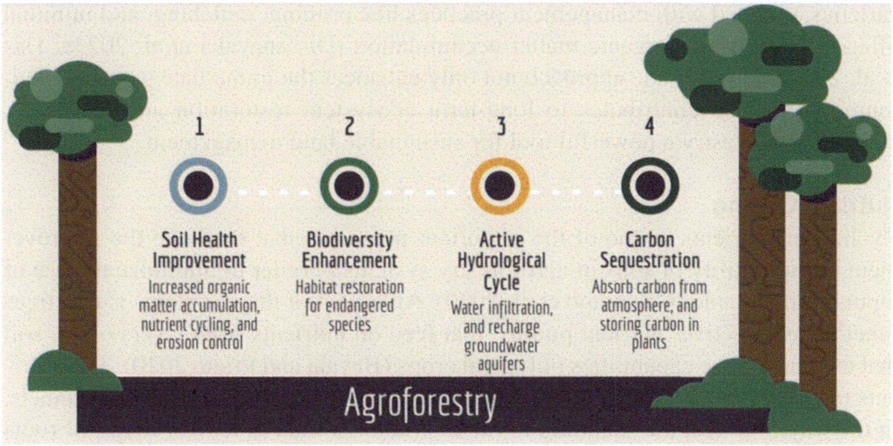

Fig. 3.4 Main mechanisms of land restoration through agroforestry

3.5.2 Mechanisms of Land Restoration Through Agroforestry

Agroforestry systems is a comprehensive approach offer several mechanisms for land restoration, addressing various aspects of ecosystem health and function (Fig. 3.4).

3.5.2.1 Soil Improvement

Organic Matter Accumulation in Agroforestry

Organic matter accumulation in agroforestry systems is a multifaceted process that significantly enhances soil quality and ecosystem health (Leite et al. 2014; Babu et al. 2020). These systems contribute diverse organic inputs through leaf and livestock farm litter, root decomposition, pruning residues, understory vegetation, and crop residues (Dissanayaka et al. 2022; Nuwarapaksha et al. 2023d). Deep-rooted trees play a crucial role by accessing nutrients from depths of 10–20 m or more, mobilizing elements from weathering bedrock, and even facilitating hydraulic lift of deep water to surface layers (Lebrazi and Fikri-Benbrahim 2022; Yadav et al. 2021a). This organic matter is deposited on the soil surface through leaf fall, root exudates, and mycorrhizal associations, gradually upgrading soil structure, water-holding capacity, and nutrient availability (Fahad et al. 2022; Wani et al. 2023). The increased organic matter content acts as a binding agent for soil particles, enhancing aggregation and water retention, while also functioning as a slow-release fertilizer and boosting the soil's cation exchange capacity (Lorenz and Lal 2014). Moreover, these organic inputs stimulate diverse microbial populations, further improving nutrient cycling and contributing to the formation of stable soil organic carbon (Wang et al. 2017). Over time, this process creates a more resilient and productive soil ecosystem, with potential for significant carbon sequestration. The selection of appropriate plant components, such as fast-growing nitrogen-fixers or deciduous varieties, coupled with management practices like pruning, mulching, and minimal tillage, can optimize organic matter accumulation (Dissanayaka et al. 2023a; Das et al. 2022). This holistic approach not only enhances the immediate growing environment but also contributes to long-term ecosystem restoration and resilience, making agroforestry a powerful tool for sustainable land management.

Nutrient Cycling

Cycling of nutrients is one of the important processes that facilitate the improvement of the fertility of soils in agroforestry systems in order to minimize the use of inputs from outside (Celentano et al. 2020). An aspect of these systems is that trees function as effective 'nutrient pumps' that feed on nutrients from layers of the soil that are outside the capabilities of typical crops (Bayala and Prieto 2020). The nutrients that might not otherwise be accessible are mobilized through several channels. In the way of a tree, the shedding of these structures such as leaves, twigs, or roots happens on the terrain surface, where they break down and release elements to the upper soil layers (Nuwarapaksha et al. 2022; Pardon et al. 2017). Such a process is of great value in recycling soluble nutrients that would otherwise be removed from the cycle. Further, nitrogen-fixing trees like legumes that have the potential of sampling nitrogen from the air and fixing it in a form that is useful to plants are often adopted in most agroforestry systems (Dissanayaka et al. 2024b; Nuwarapaksha et al. 2023a). This biological nitrogen fixation significantly enhances soil nitrogen content, a crucial element for plant growth (Soumare et al. 2020). The combination of nutrient pumping and nitrogen fixation creates a more self-sustaining system, dramatically reducing the need for external fertilizer inputs (Yadav et al. 2021b). This will minimize the environmental impacts associated with synthetic fertilizer use, such as water pollution and greenhouse gas emissions (Sileshi et al. 2014). This process of cycling grab enhances the nutrient of soil for a long time, hence countering the sustainable productivity of farming.

Erosion Control

Erosion control is an important role of agroforestry systems, and it can mitigate soil loss at various levels (Fahad et al. 2022). In this process, trees are highly advantageous due to their extensive root systems that physically entwine soil particles to become one body. Such roots form a system of branches below the ground, which helps to modify the structure and cohesion of the soil and minimize its tendency to be washed away by water and wind (Masi et al. 2021; Yadav et al. 2023). The impact is especially observed on the elevated land and on the territories with geomorphologically unstable surfaces. In the above-ground space, the tree canopy layer shields the soil surface from direct exposure. During rainfall, the water droplets' kinetic energy is also tamed by the foliage and branches before any impact on the ground (Yan et al. 2021). This interception minimizes the effects of raindrops detaching soil particles and thus the ability to erode them (Luo et al. 2020; Yadav et al. 2024). Furthermore, the canvas of fallen leaves that develops on the forest floor also creates a mulch layer. This layer contributes to an accumulation of organic matter, as well as a physical barrier against erosion (Quintarelli et al. 2022). It also helps in slowing down the water on the surface and gives ample time for infiltration hence minimizing the runoff. It also shields the soil from wind hence greatly reducing wind erosion (van Ramshorst et al. 2022). In the agricultural ecosystems, this means that trees are planted with crops or pastures and this results in a system whereby vegetation cover is maintained throughout the year even when the crop plants are not able. These continuous protections are particularly helpful in areas that experience heavy downpours or strung season winds. These mechanisms include; root binding, canopy protection, and surface cover, and through the application of all these mechanisms, agroforestry systems ease the problem of soil erosion thus securing soil resources and overall land productivity in the long run.

3.5.3 Biodiversity Enhancement

Agroforestry systems (Fig. 3.5) are a vital source of improving biological diversity as well as providing overall value-addition to the production systems (Jose 2012). These systems combine trees with crops or livestock and establish a variety of habitats that foster various plant and animal species. The tree species and their architecture form a ground cover to canopy structure that supports a wide variety of insects, birds, mammals, and microorganisms (Udawatta et al. 2019). It is not simply a greater number of species; it enhances the capabilities of the ecosystem many times. Some of the agroforestry plots may act as important habitats and connectors for species living in fragmented natural habitats in agricultural and urbanized landscapes (Montagnini 2020). These corridors also allow animals to migrate and breed genetically with other populations that would otherwise potentially disappear entirely from that region. The increased spread of species also leads to increased resistance in ecosystems (Oliver et al. 2015). A diverse system is capable of coping and bailing with some disturbances like; pests, diseases or harsh climatic conditions. This robustness can be partially attributable to the abundance of naturally occurring predators and parasitoids in the area (Pumariño et al. 2015). The distinct relations between species in agroforestry systems make pest control natural hence no extensive use of chemicals. Some of the key advantages of these systems include

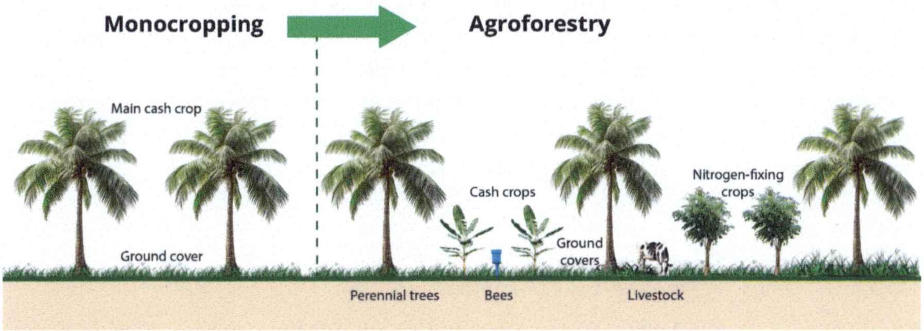

Fig. 3.5 Components in a typical agroforestry system

providing pollinators needed for crop production and the availability of floral resources for foraging throughout the year (Bentrup et al. 2019). Besides the increased biodiversity in animals, the improvement of soil microorganisms improves nutrient cycling and the health of the soils. In this way, aside from promoting conservation, agroforestry systems increase yields, make them more stable, and ensure agricultural productivity in the long term. For example, crops like *Tithonia diversifolia, Gliricidia sepium,* and grasses are identical for soil rehabilitation in tropical zones (Nuwarapaksha et al. 2023b; Senarathne et al. 2019).

3.5.4 Water Management

Agroforestry systems have numerous advantages when it comes to water and how they interact with the hydrological cycle (Kaushal et al. 2021). Trees mixed with crops or pasture significantly enhance the water infiltration on the lands (Zhu et al. 2019). This is because the tree roots make channels in the soil and through these structures, the porosity of the soil increases and water penetrates deeper levels of the soil easily. This enhanced participation together with a better soil structure due to the incorporation of organic amendments significantly decrease the runoff rate (Pavlidis and Tsihrintzis 2018). Agroforestry minimizes the tendency of soil erosion and aids in the recharge of the groundwater aquifers. The tree canopy and understory vegetation in the agroforestry systems have another importance, in reducing transpiration (Kaushal et al. 2021). The cover offered by trees reduces the surface temperature of the soil while total vegetative cover helps maintain a more humid microclimate in the area (de Carvalho et al. 2021). Trees reduce the effects of wind by serving as windbreaks, thus minimizing the loss of water through evaporation (Rosenfeld et al. 2010). This is especially useful in areas with low rainfall rates such as the arid and semi-arid areas because it prolongs the availability of water for the crops and reduces the amount of irrigation water used. Agroforestry systems within riparian zones were most effective in enhancing water quality. These buffer strips contribute to a vegetative cover that has the ability to inhibit the flow of sediments, nutrients, and other pollutants into water bodies (Stutter et al. 2012). The root systems help to combat stream bank erosion while the aerial growths act as a barrier to slow down the force of flood waters hence discouraging erosion. Such filtering is important for sustaining the health of water bodies and a supply of clean water for consumption by other users. Due to increased water infiltration, checking of water evaporation, and water filtering effects within the agroforestry systems, this practice plays a crucial role in the management of water in a sustainable manner by increasing and improving agricultural production as well as conservation.

3.5.5 Carbon Sequestration

Agroforestry is a good strategy for promoting carbon sequestration, a critical process in addressing climate change and improving the soil condition (Atapattu and

Udumann 2024; Ghale et al. 2022). Trees act as significant carbon sinks, storing substantial amounts of carbon in their biomass both above and below ground (Duque et al. 2021; Kumar et al. 2022a). As trees grow, they continuously absorb carbon dioxide from the air through photosynthesis, turning it into woody biomass and root structures (Lambers and Oliveira 2019). This process of carbon storage is long-term, with mature trees holding considerable amounts of carbon for decades or even centuries. Below ground, tree roots contribute to carbon sequestration by depositing carbon-rich organic matter deep in the soil profile (Ranasinghe and Thimothias 2012; Kumar et al. 2022b). Furthermore, practices of agroforestry enhance the soil organic carbon since this type of plant system constantly supplies the soil with fresh input derived from the shedding of leaves, root secretions, and decaying plant residues. There is a virtuous cycle in this process of bettering the soil health due to higher levels of soil organic carbon—better plant growth that in turn supports carbon sequestration (Wani et al. 2022). Furthermore, agroforestry systems lead to decreased release of greenhouse gases since practices discourage the usage of artificial fertilizers and mechanical farming techniques compared to conventional techniques (Torres et al. 2017; Rathore et al. 2022). This diverse nature also increases their resilience to climate change impacts, ensuring long-term carbon storage potential (Gnonlonfoun et al. 2019). This multifaceted approach to carbon sequestration demonstrates how agroforestry can effectively address climate change mitigation while providing numerous co-benefits for agricultural productivity and ecosystem health.

3.6 Challenges and Limitations

3.6.1 Technical Challenges

There are various technical constraints inherent in the implementation and management of agroforestry systems, as discussed below. The first potential problem involves tree-crop interactions which are rather complicated and need deep understanding and skills to manage properly (Camilli et al. 2018). Stakeholders should know how various species interfere with each other about the utilization of factors such as light, water, and nutrients in plants. The choice of trees and crops that are compatible and well-adapted to the site is important but challenging at times (Nuwarapaksha et al. 2022). However, with trees, results are not always immediate and there is a long-term growth, thus it needs patience and planning in the long run. Interactions between pests and diseases in diversified systems may pose more challenges than those found in monoculture systems; therefore, integrated pest management may be required (Sollen-Norrlin et al. 2020). There's also a need for specialized equipment and techniques for managing trees in agricultural settings, which may not be readily available or familiar to traditional farmers (Udumann et al. 2024b). Moreover, there is a dearth of extensive literature and locally developed examples of various forms of agroforestry systems across diverse ecosystems, which creates difficulties in offering specific suggestions to farmers (Kiyani et al. 2017).

3.6.2 Socioeconomic Barriers

Several outcomes of agroforestry systems' adoption and successful implementation could encounter daunting socioeconomic challenges. A major challenge reported is related to the startup cost, both in capital and time (Sagastuy and Krause 2019). Tree planting may be a challenge to many farmers, especially the smallholder farmers as they may lack the capital to invest in it and also may be highly discouraged due to the long initial gestation period to get returns on the investment (Atreya et al. 2021). This is because in many regions there is insecurity of land tenure, which means farmers cannot be certain of the benefits of planting a tree that will start bearing many years later. The culture and tradition also play a role by limiting farmers and encouraging them to stick with the tradition instead of taking up with the new methods they have not seen before. There is the question of market development for the products through agroforestry where in most cases, the component of output has no well-developed market structure where farmers can easily turn their diversification efforts into profit (Baig et al. 2021; Do et al. 2020). Further, education, and extension services are critical; this is because the diffuse and often knowledge-intensive nature of agroforestry implies these services are either inadequate, unavailable, or unreachable in many instances in rural areas (Kiyani et al. 2017). Scarcity of labor especially in areas where people have moved from the rural areas to the urban centers, to look for employment, it becomes difficult to manage a more complex system of agroforestry. In addition, the perceived risk that is often associated with new systems will discourage farmers, especially where there are no alternative examples of farmers who have implemented the systems or government support.

3.7 Policy Constraints

Government policies are another key factor and constrain the use of agroforestry systems as well as the regulates the success in most areas. The policies of many countries concerning agroforestry are still general because most of the countries do not differentiate or classify agroforestry as separate from both agriculture and forestry but as a part of either of them (Dhyani et al. 2021). This classification issue can lead to conflicts in land use regulations and access to governmental support programs. Agricultural subsidies and incentives are frequently biased towards conventional monoculture systems, disadvantaging agroforestry practices. There's often a lack of coordination between different governmental departments (e.g., agriculture, forestry, environment) in policies affecting agroforestry (Ndlovu and Borrass 2021; Paudel et al. 2022). Research funding and extension services may be inadequately allocated to agroforestry compared to conventional agricultural systems. Land use policies in many areas do not recognize the multifunctional nature of agroforestry, failing to account for its environmental services like carbon sequestration or biodiversity conservation. Additionally, there can be legal barriers to tree planting on agricultural land or harvesting trees from farms. The absence of payment mechanisms for ecosystem services provided by agroforestry systems further diminishes their economic attractiveness. Lastly, the long-term nature of agroforestry investments often conflicts with short-term political cycles and policy-making processes, making it challenging to implement and maintain supportive policies over the timeframes needed for agroforestry systems to mature and deliver their full range of benefits.

3.8 Future Directions

3.8.1 Research Needs

Agroforestry as a vital and promising land use practice can only have a certain future if these critical research gaps are resolved. In this perspective, more long-term research is needed to unravel the intricate dynamics of the interactions within multifunctional and agroforestry systems and their temporal dynamics in relation to ecosystem services. Research should focus on optimizing tree-crop combinations for different agroecological zones, considering factors such as climate change resilience. Development of improved varieties of multipurpose trees that can thrive in diverse agroforestry settings is crucial. There's also a need for more comprehensive economic analyses of agroforestry systems, including the valuation of ecosystem services, to better demonstrate their full benefits to policymakers and farmers (Huang et al. 2023; Udumann et al. 2024a). Research into innovative techniques for rapid establishment and management of agroforestry systems could help overcome adoption barriers (Dissanayaka et al. 2023a; Dissanayake et al. 2023a). Furthermore, research on the application of agroforestry in other areas that have not yet been explored like urban agriculture, as a technique of weed management, or climate-smart landscapes, among others should be conducted (Atapattu et al. 2024b; Udumann et al. 2023). Furthermore, novel applications in the field of soil and moisture conservation including biochar and ash amendment, the improved techniques of mulching processes, and the agri-waste management systems also deserve more attention and research work (Dissanayaka et al. 2023b, c; Dissanayake et al. 2023b; Ekanayaka et al. 2023; Nuwarapaksha et al. 2024a; Shamila et al. 2024). Interdisciplinary research combining agricultural, ecological, and social sciences is essential to address the multifaceted nature of agroforestry. Lastly, user-friendly models and decision support tools that would help the farmers and the extension agents in designing and managing the appropriate site-specific agroforestry systems locally would further enhance adoption significantly.

3.9 Policy Recommendations

In order to promote the development of agroforestry several policy actions should be considered (Duguma et al. 2023). Firstly, there is a concern that the legislation and policies of the country do not have separate provisions for recognizing agroforestry as a formal land management system with corresponding support and legal regulation. Future legislation to be enacted should offer farmers the potential economic benefit for taking up agroforestry practices form of tax reductions, grants, or compensation for public services. Whichever systems of land tenure reforms are chosen, there is the need to ensure that the farmers holding trees on their farms have long-term leasehold tenure systems. It is therefore imperative to embrace and incorporate agroforestry into national climate change mitigation and adaptation policies and strategies to maximize on climate finance for the noble cause. It is also recommended that more policy support should be directed towards enhancing market connection of the produce of agroforestry systems such as certification on sustainable production (Shah 2023). Education policies need revision to incorporate agroforestry into agricultural and forestry curricula at various levels. Cross-sectoral coordination mechanisms should be established to ensure coherent policies across agriculture, forestry, and environmental departments. Additionally, policies supporting research and extension services specifically for agroforestry are essential (Mosquera-Losada et al. 2023). Finally, international trade policies should be reviewed to ensure they don't disadvantage agroforestry products and instead promote sustainable land use practices.

3.10 Scaling Up Successful Practices

Expansion of effective practices in the agroforestry system is necessary to achieve the demonstrated potential of this practice in the fight against various global problems. Thus, this process has to be addressed through a multifaceted approach (Dissanayaka et al. 2023a). Firstly, identifying and documenting successful agroforestry models across different ecological and socioeconomic contexts is essential. These success stories should be widely disseminated through various media channels to inspire and inform potential adopters. There is a need to strengthen the extension services and farmer-to-farmer knowledge exchange to improve the dissemination and adoption of best practices fundamentally tailored to the local context (Taylor and Bhasme 2018). Leveraging digital technologies, such as mobile apps and online platforms, can help in the widespread dissemination of information and provide real-time support to farmers. Creating demonstration sites and model farms in different regions can provide tangible examples of successful agroforestry implementation. Public-private partnerships should be fostered to invest in value chain development for agroforestry products, ensuring market demand drives adoption (Bettles et al. 2021). It has been noted that developing the capacity of farmers, extensionists and local administrators is significant in implementing and managing agroforestry innovations. Eradicating initial investments and implementing the economical and effective methods for creating agroforestry systems might be effective. Finally, the incorporation of agroforestry into large-scale restoration and climate change mitigation projects can attract the necessary funding and momentum, to upscale (Buck et al. 2020; Romijn et al. 2019). By combining these approaches, it's possible to significantly expand the adoption of agroforestry practices, maximizing their environmental, economic, and social benefits on a global scale.

3.11 Conclusion

Agroforestry is a viable approach for restoring degraded areas and it's also has a direct impact on solving most of the global issues regarding the environment, humanity, and economy. Because agroforestry systems are embedded into production systems of trees with crops or animals, the practice provides a holistic approach towards land remediation and improvement next to interventions such as growing cover crops, adding fertilizers, or contending with invasive plant species. Apart from giving assistance in the rehabilitation of damaged ecosystems, these systems also support income generation among farmers, and lessen the effects of climate change.

However, the widespread adoption and success of agroforestry face several challenges, including technical complexities, socioeconomic barriers, and policy constraints. Overcoming these obstacles requires a multi-faceted approach involving further research, policy reforms, and strategies for scaling up successful practices.

In the future, the emphasis should be placed on determining the best tree-crop associations depending on the agroecological regions, the generation of new and better multipurpose trees, and the integrated cost of agroforestry systems. Some of the policy guidelines can be specific through acknowledgment of agroforestry as a separate land use type, grants towards adoption of agroforestry practices, including it as a component in the country's climate change frameworks. To increase the application of good models for the growth of the practice, regional best practices need to be looked for, extension services need to be empowered, digital technologies fostered and PPPs encouraged. In this regard, addressing these aspects opens a great perspective in employing agroforestry in the attempts to rehabilitate degraded terrains, make agriculture production more sustainable, and counteract climate change.

Moreover, agroforestry has great potential to restore the degraded land, although more concentrated and definite measures are guaranteed by researchers, policymakers, and practitioners. Thus, the development of the subject of agroforestry which is bringing balanced and sustainable solutions to the increasing environmental problems, should gain more attention and funds.

References

Abrol I, Gupta R (2019) Climate change-land degradation-food security nexus: addressing India's challenge. J Agron Res 2(2):17–35. https://doi.org/10.14302/issn.2639-3166.jar-19-3015

Abu Hammad A, Tumeizi A (2012) Land degradation: socioeconomic and environmental causes and consequences in the eastern Mediterranean. Land Degrad Dev 23(3):216–226. https://doi.org/10.1002/ldr.1069

Acharya AK, Kafle N (2009) Land degradation issues in Nepal and its management through agroforestry. J Agric Environ 10:133–143. https://doi.org/10.3126/aej.v10i0.2138

Al-Kulabi AKJ (2022) The concept of desertification, its causes and effects, and treatments. J La Lifesci 3(1):1–13. https://doi.org/10.37899/journallalifesci.v3i1.550

Atapattu AJ, Udumann SS (2024) Leveraging agroforestry principles for nature-based climate-smart solutions for coconut cultivation. In: Leal Filho W, Nagy GJ, Ayal DY (eds) Handbook of nature-based solutions to mitigation and adaptation to climate change. Springer, Cham, pp 1–28. https://doi.org/10.1007/978-3-030-98067-2_166-1

Atapattu AJ, Ranasinghe CS, Nuwarapaksha TD, Udumann SS, Dissanayaka NS (2024a) Sustainable agriculture and sustainable development goals (SDGs). In: Garwi J, Masengu R, Chiwaridzo O (eds) Emerging technologies and marketing strategies for sustainable agriculture. IGI Global, pp 1–27. https://doi.org/10.4018/979-8-3693-4864-2.ch001

Atapattu AJ, Perera LK, Nuwarapaksha TD, Udumann SS, Dissanayaka NS (2024b) Challenges in achieving artificial intelligence in agriculture. In: Chouhan SS, Saxena A, Singh UP, Jain S (eds) Artificial intelligence techniques in smart agriculture. Springer, Singapore, pp 7–34. https://doi.org/10.1007/978-981-97-5878-4_2

Atapattu AJ, Udumann SS, Nuwarapaksha TD, Dissanayaka NS (2024c) Upcycling coconut husk by-products: transitioning from traditional applications to emerging high-value usages. In: Neelancherry R, Gao B, Wisniewski A Jr (eds) Agricultural waste to value-added products. Springer Nature, Singapore, pp 249–273. https://doi.org/10.1007/978-981-97-2535-9_12

Atreya K, Subedi BP, Ghimire PL, Khanal SC, Charmakar S, Adhikari R (2021) Agroforestry for mountain development: prospects, challenges, and ways forward in Nepal. Arch Agric Environ Sci 6(1):87–99. https://doi.org/10.26832/24566632.2021.0601012

Avasthe RK, Devi SH, Bhupenchandra I, Kumar A, Chongtham SK, Babu S, Singh R, Das A, Gudade BA, Bora SS (2023) Agro-tactics for reducing carbon footprint in agricultural production systems: a review. Indian J Agron 68(2):115–125. https://doi.org/10.59797/ija.v68i2.332

Babu S, Mohapatra KP, Yadav GS, Lal R, Singh R, Avasthe RK, Das A, Chandra P, Gudade BA, Kumar A (2020) Soil carbon dynamics in diverse organic land use systems in North Eastern Himalayan ecosystem of India. Catena. https://doi.org/10.1016/j.catena.2020.104785

Babu S, Das A, Singh R, Mohapatra KP, Kumar S, Rathore SS, Yadav SK, Yadav P, Ansari MA, Panwar AS, Wani OA, Singh M, Ravishankar N, Layek J, Chandra P, Singh VK (2023a) Designing an energy efficient, economically feasible, and environmentally robust integrated farming system model for sustainable food production in the Indian Himalayas. Sustain Food Technol. https://doi.org/10.1039/D2FB00016D

Babu S, Singh R, Avasthe RK, Kumar S, Rathore SS, Singh VK, Ansari MA, Valente D, Petrosillo I (2023b) Soil carbon dynamics under organic farming: impact of tillage and cropping diversity. Ecol Indic. https://doi.org/10.1016/j.ecolind.2023.109940

Baechler G (1999) Environmental degradation in the south as a cause of armed conflict. In: Carius A, Lietzmann KM (eds) Environmental change and security. Springer, Berlin/Heidelberg, pp 107–129. https://doi.org/10.1007/978-3-642-60229-0_7

Baig MB, Burgess PJ, Fike JH (2021) Agroforestry for healthy ecosystems: constraints, improvement strategies and extension in Pakistan. Agrofor Syst 95(5):995–1013. https://doi.org/10.1007/s10457-019-00467-4

Barbier EB, Hochard JP (2018) Land degradation and poverty. Nature Sustain 1(11):623–631. https://doi.org/10.1038/s41893-018-0155-4

Barbut M, Alexander S (2016) Land degradation as a security threat amplifier: the new global frontline. In: Land restoration, pp 3–12. https://doi.org/10.1016/B978-0-12-801231-4.00001-X

Bayala J, Prieto I (2020) Water acquisition, sharing and redistribution by roots: applications to agroforestry systems. Plant Soil 453(1–2):17–28. https://doi.org/10.1007/s11104-019-04173-z

Bentrup G, Hopwood J, Adamson NL, Vaughan M (2019) Temperate agroforestry systems and insect pollinators: a review. Forests 10(11):981. https://doi.org/10.3390/f10110981

Bera A, Dutta TK, Bhattacharya M, Patra BC, Kumar Sar U (2022) Anthropogenic stress on river health: with special reference to Kangsabati River, West Bengal, India. In: Patra BC, Shit PK, Bhunia GS, Bhattacharya M (eds) River health and ecology in South Asia. Springer, pp 39–61. https://doi.org/10.1007/978-3-030-83553-8_3

Bettles J, Battisti DS, Cook-Patton SC, Kroeger T, Spector JT, Wolff NH, Masuda YJ (2021) Agroforestry and non-state actors: a review. Forest Policy Econ 130:102538. https://doi.org/10.1016/j.forpol.2021.102538

Borrelli P, Robinson DA, Fleischer LR, Lugato E, Ballabio C, Alewell C, Meusburger K, Modugno S, Schütt B, Ferro V, Bagarello V (2017) An assessment of the global impact of 21st century land use change on soil erosion. Nat Commun 8(1):1–13. https://doi.org/10.1038/s41467-017-02142-7

Buck L, Scherr S, Trujillo L, Mecham J, Fleming M (2020) Using integrated landscape management to scale agroforestry: examples from Ecuador. Sustain Sci 15(5):1401–1415. https://doi.org/10.1007/s11625-020-00839-1

Camilli F, Pisanelli A, Seddaiu G, Franca A, Bondesan V, Rosati A, Moreno GM, Pantera A, Hermansen JE, Burgess PJ (2018) How local stakeholders perceive agroforestry systems: an Italian perspective. Agrofor Syst 92(4):849–862. https://doi.org/10.1007/s10457-017-0127-0

Celentano D, Rousseau GX, Paixão LS, Lourenço F, Cardozo EG, Rodrigues TO, Silva HR, Medina J, de Sousa TMC, Rocha AE, de Oliveira Reis F (2020) Carbon sequestration and nutrient cycling in agroforestry systems on degraded soils of Eastern Amazon, Brazil. Agrofor Syst 94(5):1781–1792. https://doi.org/10.1007/s10457-020-00496-4

Chalise D, Kumar L, Kristiansen P (2019) Land degradation by soil erosion in Nepal: a review. Soil Syst 3(1):12. https://doi.org/10.3390/soilsystems3010012

Chew BH, Maxwell L, Anyiam FE, Menouni A, Kurniawan TA, Dimobe K, Sharma TPP, Ali GA, Shah RDT, Saleem R, Mashroofa MM, Nasr M, Abbas B, Atapattu AJ, Mahmoud M, Singh N, Sarker MR (2024) Statements on open science for Sustainable Development Goals. Data Sci J 23(1):49. https://doi.org/10.5334/dsj-2024-049

Cowie AL, Orr BJ, Castillo Sanchez VM, Chasek P, Crossman ND, Erlewein A, Louwagie G, Maron M, Metternicht GI, Minelli S, Tengberg AE, Walter S, Welton S (2018) Land in balance: the scientific conceptual framework for Land Degradation Neutrality. Environ Sci Pol 79:25–35. https://doi.org/10.1016/j.envsci.2017.10.011

Dale V, Archer S, Chang M, Ojima D (2005) Ecological impacts and mitigation strategies for rural land management. Ecol Appl 15(6):1879–1892. https://doi.org/10.1890/03-5330

Das A, Babu S, Datta M, Kumar S, Singh R, Avasthe RK, Rathore SS, Yadav SK, Singh VK (2022) Restoring soil carbon in marginal land of Indian Himalayas: impact of crop intensification and conservation tillage. J Environ Manag. https://doi.org/10.1016/j.jenvman.2022.115603

de Carvalho AF, Fernandes-Filho EI, Daher M, Gomes LDC, Cardoso IM, Fernandes RBA, Schaefer CEGR (2021) Microclimate and soil and water loss in shaded and unshaded agroforestry coffee systems. Agrofor Syst 95(1):119–134. https://doi.org/10.1007/s10457-020-00567-6

Debie E (2020) Stakeholder perceptions on multiple ecosystem services of soil conservation practices in Tiwa Watershed, Northwest Ethiopia. J Environ Assess Policy Manag 22:2250010. https://doi.org/10.1142/S1464333222500107

Dhyani S, Murthy IK, Kadaverugu R, Dasgupta R, Kumar M, Adesh Gadpayle K (2021) Agroforestry to achieve global climate adaptation and mitigation targets: are South Asian countries sufficiently prepared? Forests 12(3):303. https://doi.org/10.3390/f12030303

Dissanayaka D, Nuwarapaksha T, Udumann S, Dissanayake D, Atapattu AJ (2022) A sustainable way of increasing the productivity of coconut cultivation using cover crops: a review. Circ Agric Syst 2(1):1–9. https://doi.org/10.48130/CAS-2022-0007

Dissanayaka DMNS, Dissanayake DKRPL, Udumann SS, Nuwarapaksha TD, Atapattu AJ (2023a) Agroforestry—a key tool in the climate-smart agriculture context: a review on coconut cultivation in Sri Lanka. Front Agron 5. https://doi.org/10.3389/fagro.2023.1162750

Dissanayaka DMNS, Udumann SS, Dissanayake DKRPL, Nuwarapaksha TD, Atapattu AJ (2023b) Review on aquatic weeds as potential source for compost production to meet sustainable plant nutrient management needs. Waste 1(1):264–280. https://doi.org/10.3390/waste1010017

Dissanayaka DMNS, Udumann SS, Nuwarapaksha TD, Atapattu AJ (2023c) Effects of pyrolysis temperature on chemical composition of coconut-husk biochar for agricultural applications: a characterization study. Technol Agron 3(1). https://doi.org/10.48130/TIA-2023-0013

Dissanayaka NS, Udumann SS, Nuwarapaksha TD, Atapattu AJ (2023d) Agroforestry: an avenue for resilient and productive farming through integrated crops and livestock production. In: Kiba DI (ed) Transitioning to zero hunger. MDPI Books, pp 115–136. https://doi.org/10.3390/books978-3-03897-863-3-5

Dissanayaka DMNS, Udumann SS, Atapattu AJ (2024a) Synergies between tree crops and ecosystems in tropical agroforestry. In: Raj A, Jhariya MK, Banerjee A, Jha RK, Singh KP (eds) Agroforestry, pp 49–87. https://doi.org/10.1002/9781394231164.ch3

Dissanayaka DMNS, Udumann SS, Nuwarapaksha TD, Atapattu AJ (2024b) Harnessing the potential of Mucuna cover cropping: a comprehensive review of its agronomic and environmental benefits. Circ Agric Syst 4(1). https://doi.org/10.48130/cas-0024-0001

Dissanayake DKRPL, Dissanayaka DMNS, Udumann SS, Nuwarapaksha TD, Atapattu AJ (2023a) Is biochar a promising soil amendment to enhance perennial crop yield and soil quality in the tropics? Technol Agron 3(1). https://doi.org/10.48130/TIA-2023-0004

Dissanayake DKRPL, Udumann SS, Dissanayaka DMNS, Nuwarapaksha TD, Atapattu AJ (2023b) Effect of biochar application rate on macronutrient retention and leaching in two coconut growing soils. Technol Agron 3(1). https://doi.org/10.48130/TIA-2023-0005

Do H, Luedeling E, Whitney C (2020) Decision analysis of agroforestry options reveals adoption risks for resource-poor farmers. Agron Sustain Dev 40(3):20. https://doi.org/10.1007/s13593-020-00624-5

Duguma LA, Minang PA, Watson C, Nath AJ, Muthee KW, van Noordwijk M, Mutune JM, Sileshi GW (2023) Agroforestry as a key intervention to achieve nationally determined contribution (NDC) targets. In: Dagar JC, Gupta SR, Sileshi GW (eds) Agroforestry for sustainable intensification of agriculture in Asia and Africa. Sustainability sciences in Asia and Africa, pp 641–664. https://doi.org/10.1007/978-981-19-4602-8_19

Duque A, Peña MA, Cuesta F, González-Caro S, Kennedy P, Phillips OL, Calderón-Loor M, Blundo C, Carilla J, Cayola L, Farfán-Ríos W, Fuentes A, Grau R, Homeier J, Loza-Rivera MI, Malhi Y, Malizia A, Malizia L, Martínez-Villa JA et al (2021) Mature Andean forests as globally important carbon sinks and future carbon refuges. Nat Commun 12(1):2138. https://doi.org/10.1038/s41467-021-22459-8

Ekanayaka EMGN, Dissanayake DKRPL, Udumann SS, Dissanayaka DMNS, Nuwarapaksha TD, Herath HMSK, Atapattu AJ (2023) Sustainable utilization of king coconut husk as a feedstock in biochar production with the highest conversion efficiency and desirable properties. IOP Conf Ser Earth Environ Sci 1235(1):012009. https://doi.org/10.1088/1755-1315/1235/1/012009

El-Gammal MI, Ali RR, Abou Samra RM (2015) GIS-based land degradation risk assessment of Damietta governorate, Egypt. Egypt J Basic Appl Sci 2(3):183–189. https://doi.org/10.1016/j.ejbas.2015.01.001

Fahad S, Chavan SB, Chichaghare AR, Uthappa AR, Kumar M, Kakade V, Pradhan A, Jinger D, Rawale G, Yadav DK, Kumar V, Farooq TH, Ali B, Sawant AV, Saud S, Chen S, Poczai P (2022) Agroforestry systems for soil health improvement and maintenance. Sustainability 14(22):14877. https://doi.org/10.3390/su142214877

Ghale B, Mitra E, Sodhi HS, Verma AK, Kumar S (2022) Carbon sequestration potential of agroforestry systems and its potential in climate change mitigation. Water Air Soil Pollut 233(7):228. https://doi.org/10.1007/s11270-022-05689-4

Gibbs HK, Salmon JM (2015) Mapping the world's degraded lands. Appl Geogr 57:12–21. https://doi.org/10.1016/j.apgeog.2014.11.024

Gnonlonfoun I, Assogbadjo AE, Gnanglè CP, Glèlè Kakaï RL (2019) New indicators of vulnerability and resilience of agroforestry systems to climate change in West Africa. Agron Sustain Dev 39(2):23. https://doi.org/10.1007/s13593-019-0566-2

Gomiero T (2016) Soil degradation, land scarcity and food security: reviewing a complex challenge. Sustainability 8(3):281. https://doi.org/10.3390/su8030281

Gupta GS (2019) Land degradation and challenges of food security. Rev Eur Stud 11(1):63. https://doi.org/10.5539/res.v11n1p63

Hossain A, Krupnik TJ, Timsina J, Mahboob MG, Chaki AK, Farooq M, Bhatt R, Fahad S, Hasanuzzaman M (2020) Agricultural land degradation: processes and problems undermining future food security. In: Fahad S, Hasanuzzaman M, Alam M, Ullah H, Saeed M, Ali Khan I, Adnan M (eds) Environment, climate, plant and vegetation growth. Springer, pp 17–61. https://doi.org/10.1007/978-3-030-49732-3_2

Huang IY, James K, Thamthanakoon N, Pinitjitsamut P, Rattanamanee N, Pinitjitsamut M, Yamklin S, Lowenberg-DeBoer J (2023) Economic outcomes of rubber-based agroforestry systems: a systematic review and narrative synthesis. Agrofor Syst 97(3):335–354. https://doi.org/10.1007/s10457-022-00734-x

Imeson A (ed) (2011) Desertification, land degradation and sustainability. Wiley, Hoboken, pp 1–326

Jiang L, Bao A, Jiapaer G, Liu R, Yuan Y, Yu T (2022) Monitoring land degradation and assessing its drivers to support sustainable development goal 15.3 in Central Asia. Sci Total Environ 807:150868. https://doi.org/10.1016/j.scitotenv.2021.150868

Jose S (2012) Agroforestry for conserving and enhancing biodiversity. Agrofor Syst 85(1):1–8. https://doi.org/10.1007/s10457-012-9517-5

Kaushal R, Mandal D, Panwar P, Rajkumar, Kumar P, Tomar JMS, Mehta H (2021). Soil and water conservation benefits of agroforestry. In: Forest resources resilience and conflicts, pp 259–275. https://doi.org/10.1016/B978-0-12-822931-6.00020-4

Kiyani P, Andoh J, Lee Y, Lee DK (2017) Benefits and challenges of agroforestry adoption: a case of Musebeya sector, Nyamagabe District in southern province of Rwanda. For Sci Technol 13(4):174–180. https://doi.org/10.1080/21580103.2017.1392367

Kumar SS, Mir SA, Wani OA, Babu S, Yeasin M, Bhat MA, Hussain N, Wani AA, Kumar R, Devideen Y, Dar SR (2022a) Land-use systems regulate carbon geochemistry in the temperate Himalayas, India. J Environ Manag. https://doi.org/10.1016/j.jenvman.2022.115811

Kumar SS, Wani OA, Mir SA, Babu S, Sharma V, Chisti MH, Baba ZA, Sofi PA, Wani F, Dar SR, Singh R, Yadav D (2022b) Soil carbon dynamics in the temperate Himalayas: impact of land use management. Front Environ Sci. https://doi.org/10.3389/fenvs.2022.1009660

Lal R (2004) Soil carbon sequestration to mitigate climate change. Geoderma 123(1–2):1–22. https://doi.org/10.1016/j.geoderma.2004.01.032

Lal R (2009) Soil degradation as a reason for inadequate human nutrition. Food Secur 1(1):45–57. https://doi.org/10.1007/s12571-009-0009-z

Lal R (2012) Climate change and soil degradation mitigation by sustainable management of soils and other natural resources. Agric Res 1(3):199–212. https://doi.org/10.1007/s40003-012-0031-9

Lal R (2015) Restoring soil quality to mitigate soil degradation. Sustainability 7(5):5875–5895. https://doi.org/10.3390/su7055875

Lambers H, Oliveira RS (2019) Photosynthesis, respiration, and long-distance transport: photosynthesis. In: Lambers H, Oliveira RS (eds) Plant physiological ecology. Springer, pp 11–114. https://doi.org/10.1007/978-3-030-29639-1_2

Lebrazi S, Fikri-Benbrahim K (2022) Potential of tree legumes in agroforestry systems and soil conservation. IN: Advances in legumes for sustainable intensification, pp 461–482. https://doi.org/10.1016/B978-0-323-85797-0.00004-5

Lehman R, Cambardella C, Stott D, Acosta-Martinez V, Manter D, Buyer J, Maul J, Smith J, Collins H, Halvorson J, Kremer R, Lundgren J, Ducey T, Jin V, Karlen D (2015) Understanding and enhancing soil biological health: the solution for reversing soil degradation. Sustainability 7(1):988–1027. https://doi.org/10.3390/su7010988

Leite LFC, Iwata BDF, Araújo ASF (2014) Soil organic matter pools in a tropical savanna under agroforestry system in Northeastern Brazil. Revista Árvore 38(4):711–723. https://doi.org/10.1590/S0100-67622014000400014

Lorenz K, Lal R (2014) Soil organic carbon sequestration in agroforestry systems. A review. Agron Sustain Dev 34(2):443–454. https://doi.org/10.1007/s13593-014-0212-y

Luo J, Zhou X, Rubinato M, Li G, Tian Y, Zhou J (2020) Impact of multiple vegetation covers on surface runoff and sediment yield in the small basin of Nverzhai, Hunan Province, China. Forests 11(3):329. https://doi.org/10.3390/f11030329

Manisalidis I, Stavropoulou E, Stavropoulos A, Bezirtzoglou E (2020) Environmental and health impacts of air pollution: a review. Front Public Health 8. https://doi.org/10.3389/fpubh.2020.00014

Masi EB, Segoni S, Tofani V (2021) Root reinforcement in slope stability models: a review. Geosciences 11(5):212. https://doi.org/10.3390/geosciences11050212

McLeman R (2017) Migration and land degradation: recent experience and future trends. In: The global land outlook working paper; United Nations Convention to Combat Desertification (UNCCD), Bonn, p 2018–06

Mohamed E, Belal A-A, Ali RR, Saleh A, Hendawy EA (2019) Land degradation. In: El-Ramady H, Alshaal T, Bakr N, Elbana T, Mohamed E, Belal AA (eds) The soils of Egypt, pp 159–174. https://doi.org/10.1007/978-3-319-95516-2_9

Montagnini F (2020) The contribution of agroforestry to restoration and conservation: biodiversity islands in degraded landscapes. In: Dagar JC, Gupta SR, Teketay D (eds) Agroforestry for degraded landscapes. Springer, Singapore, pp 445–479. https://doi.org/10.1007/978-981-15-4136-0_15

Mosquera-Losada MR, Santiago-Freijanes JJ, Rois-Díaz M, Moreno G, den Herder M, Aldrey-Vázquez JA, Ferreiro-Domínguez N, Pantera A, Pisanelli A, Rigueiro-Rodríguez A (2018) Agroforestry in Europe: a land management policy tool to combat climate change. Land Use Policy 78:603–613. https://doi.org/10.1016/j.landusepol.2018.06.052

Mosquera-Losada MR, Santos MGS, Gonçalves B, Ferreiro-Domínguez N, Castro M, Rigueiro-Rodríguez A, González-Hernández MP, Fernández-Lorenzo JL, Romero-Franco R, Aldrey-Vázquez JA, Sobrino CC, García-Berrios JJ, Santiago-Freijanes JJ (2023) Policy challenges for agroforestry implementation in Europe. Front For Glob Change 6. https://doi.org/10.3389/ffgc.2023.1127601

Mukhlis I, Rizaludin MS, Hidayah I (2022) Understanding socio-economic and environmental impacts of agroforestry on rural communities. Forests 13(4):556. https://doi.org/10.3390/f13040556

Nair PKR, Kumar BM, Nair VD (2021) Classification of agroforestry systems. In: Nair PKR, Kumar BM, Nair VD (eds) An introduction to agroforestry. Springer, pp 29–44. https://doi.org/10.1007/978-3-030-75358-0_3

Nasare LI, Rahman LA, Darko FD (2023) Indigenous knowledge systems for land condition assessment and sustainable land management in the Talensi District, Ghana. J Degrad Mining Lands Manag 10(2):4209. https://doi.org/10.15243/jdmlm.2023.102.4209

Nawaz MF, Bourrié G, Trolard F (2013) Soil compaction impact and modelling. A review. Agron Sustain Dev 33(2):291–309. https://doi.org/10.1007/s13593-011-0071-8

Ndlovu NP, Borrass L (2021) Promises and potentials do not grow trees and crops. A review of institutional and policy research in agroforestry for the Southern African region. Land Use Policy 103:105298. https://doi.org/10.1016/j.landusepol.2021.105298

Nuwarapaksha T, Udumann S, Dissanayaka D, Dissanayake D, Atapattu AJ (2022) Coconut based multiple cropping systems: an analytical review in Sri Lankan coconut cultivations. Circ Agric Syst 2(1):1–7. https://doi.org/10.48130/CAS-2022-0008

Nuwarapaksha TD, Dissanayaka DMNS, Udumann SS, Atapattu AJ (2023a) Gliricidia as a beneficial crop in resource-limiting agroforestry systems in Sri Lanka. Indian J Agrofor 25(1):12–18. https://www.researchgate.net/publication/372232549

Nuwarapaksha TD, Dissanayake WK, Gunathilaka WS, Udumann SS, Dissanayaka NS, Atapattu AJ (2023b) Assessing the optimum harvesting stage of *Tithonia diversifolia* as climate smart soil amendment for coconut plantations. In: The 2nd international online conference on agriculture, p 1. https://doi.org/10.3390/IOCAG2023-15967

Nuwarapaksha TD, Rajapaksha UN, Ekanayake J, Weerasooriya SA, Atapattu AJ (2023c) Exploring the economic viability of integrating Jamnapari goat into underutilized pastures under coconut cultivations in Coconut Research Institute, Sri Lanka. IECAG 2023:27. https://doi.org/10.3390/IECAG2023-14990

Nuwarapaksha TD, Udumann SS, Dissanayaka NS, Atapattu JA (2023d) Coconut-based livestock farming: a sustainable approach to enhancing food security in Sri Lanka. In: Kiba DI (ed) Transitioning to zero hunger. MDPI Books, pp 197–213. https://doi.org/10.3390/books978-3-03897-863-3-8

Nuwarapaksha TD, Dissanayaka DMNS, Udumann SS, Vinujan S, Atapattu AJ (2024a) Exploring the impact of pyrolysis temperature on nutrient composition of *Gliricidia sepium* biochar: a comprehensive study. Technol Agron 4(1). https://doi.org/10.48130/tia-0024-0014

Nuwarapaksha TD, Udumann SS, Atapattu AJ (2024b) Fostering food and nutritional security through agroforestry practices. In: Raj A, Jhariya MK, Banerjee A, Jha RK, Singh KP (eds) Agroforestry, pp 285–318. https://doi.org/10.1002/9781394231164.ch10

Oliver TH, Heard MS, Isaac NJB, Roy DB, Procter D, Eigenbrod F, Freckleton R, Hector A, Orme CDL, Petchey OL, Proença V, Raffaelli D, Suttle KB, Mace GM, Martín-López B, Woodcock BA, Bullock JM (2015) Biodiversity and resilience of ecosystem functions. Trends Ecol Evol 30(11):673–684. https://doi.org/10.1016/j.tree.2015.08.009

Pardon P, Reubens B, Reheul D, Mertens J, De Frenne P, Coussement T, Janssens P, Verheyen K (2017) Trees increase soil organic carbon and nutrient availability in temperate agroforestry systems. Agric Ecosyst Environ 247:98–111. https://doi.org/10.1016/j.agee.2017.06.018

Paudel S, Baral H, Rojario A, Bhatta KP, Artati Y (2022) Agroforestry: opportunities and challenges in Timor-Leste. Forests 13(1):41. https://doi.org/10.3390/f13010041

Pavlidis G, Tsihrintzis VA (2018) Environmental benefits and control of pollution to surface water and groundwater by agroforestry systems: a review. Water Resour Manag 32(1):1–29. https://doi.org/10.1007/s11269-017-1805-4

Pueyo Y, Alados CL, Barrantes O (2006) Determinants of land degradation and fragmentation in semiarid vegetation at landscape scale. Biodivers Conserv 15(3):939–956. https://doi.org/10.1007/s10531-004-2936-5

Pumariño L, Sileshi GW, Gripenberg S, Kaartinen R, Barrios E, Muchane MN, Midega C, Jonsson M (2015) Effects of agroforestry on pest, disease and weed control: a meta-analysis. Basic Appl Ecol 16(7):573–582. https://doi.org/10.1016/j.baae.2015.08.006

Qiang W, Niu S, Liu A, Kastner T, Bie Q, Wang X, Cheng S (2020) Trends in global virtual land trade in relation to agricultural products. Land Use Policy 92:104439. https://doi.org/10.1016/j.landusepol.2019.104439

Quintarelli V, Radicetti E, Allevato E, Stazi SR, Haider G, Abideen Z, Bibi S, Jamal A, Mancinelli R (2022) Cover crops for sustainable cropping systems: a review. Agriculture 12(12). https://doi.org/10.3390/agriculture12122076

Ranasinghe CS, Thimothias KSH (2012) Estimation of carbon sequestration potential in coconut plantations under different agro-ecological regions and land suitability classes. J Natl Sci Found Sri Lanka 40(1):77. https://doi.org/10.4038/jnsfsr.v40i1.4171

Rathore SS, Babu S, El-Sappah AH, Shekhawat K, Singh VK, Singh RK, Upadhyay PK, Singh R (2022) Integrated agroforestry systems improve soil carbon storage, water productivity, and economic returns in the marginal land of the semi-arid region. Saudi J Biol Sci. https://doi.org/10.1016/j.sjbs.2022.103427

Romijn E, Coppus R, De Sy V, Herold M, Roman-Cuesta RM, Verchot L (2019) Land restoration in Latin America and the Caribbean: an overview of recent, ongoing and planned restoration initiatives and their potential for climate change mitigation. Forests 10(6):510. https://doi.org/10.3390/f10060510

Rosenfeld M, Marom G, Bitan A (2010) Numerical simulation of the airflow across trees in a windbreak. Bound-Layer Meteorol 135(1):89–107. https://doi.org/10.1007/s10546-009-9461-8

Sagastuy M, Krause T (2019) Agroforestry as a biodiversity conservation tool in the Atlantic forest? Motivations and limitations for small-scale farmers to implement agroforestry systems in North-Eastern Brazil. Sustainability 11(24):6932. https://doi.org/10.3390/su11246932

Senarathne SHS, Atapattu AJ, Raveendra T, Mensah S, Dassanayake KB (2019) Biomass allocation and growth performance of Tithonia diversifolia (Hemsl.) A. gray in coconut plantations in Sri Lanka. Agrofor Syst 93(5):1865–1875. https://doi.org/10.1007/s10457-018-0290-y

Shah S (2023) Policy interventions for scaling up agroforestry in Fiji. N Z Econ Pap 57(2):105–113. https://doi.org/10.1080/00779954.2023.2171904

Shamila SK, Udumann SS, Dissanayaka NS, Rajaratnam K, Atapattu AJ (2024) Assessing the impact of king coconut husk ash and biochar, combined with chemical fertilizer application, on enhancing soil fertility in coconut plantations. Crops 4(2):227–241. https://doi.org/10.3390/crops4020017

Sharma MP, Samanta A, Sharma KR (2018) Degraded lands and their management for rehabilitation and improving crop productivity. J Soil Water Conserv 17(1):25. https://doi.org/10.5958/2455-7145.2018.00004.8

Sharma R, Mina U, Kumar BM (2022) Homegarden agroforestry systems in achievement of Sustainable Development Goals. A review. Agron Sustain Dev 42(3):44. https://doi.org/10.1007/s13593-022-00781-9

Sileshi GW, Mafongoya PL, Akinnifesi FK, Phiri E, Chirwa P, Beedy T, Makumba W, Nyamadzawo G, Njoloma J, Wuta M, Nyamugafata P, Jiri O (2014) Agroforestry: fertilizer trees. In: Encyclopedia of agriculture and food systems, pp 222–234. https://doi.org/10.1016/B978-0-444-52512-3.00022-X

Sims NC, England JR, Newnham GJ, Alexander S, Green C, Minelli S, Held A (2019) Developing good practice guidance for estimating land degradation in the context of the United Nations Sustainable Development Goals. Environ Sci Pol 92:349–355. https://doi.org/10.1016/j.envsci.2018.10.014

Singh R, Babu S, Avasthe RK, Meena RS, Yadav GS, Das A, Mohapatra KP, Rathore SS, Kumar A, Singh C (2021) Conservation tillage and organic nutrients management improve soil properties, productivity, and economics of the maize-vegetable pea system in the Eastern Himalayas. Land Degrad Dev. https://doi.org/10.1002/ldr.4066

Siril S, Singh M, Shahina NN, Bhat JA, Shukla G, Panwar P, Chakravarty S (2022) Role of REDD+ in reducing land degradation and achieving SDGs. In: Panwar P, Shukla G, Bhat JA, Chakravarty S (eds) Land degradation neutrality: achieving SDG 15 by forest management. Springer Nature, Singapore, pp 303–313. https://doi.org/10.1007/978-981-19-5478-8_16

Sollen-Norrlin M, Ghaley BB, Rintoul NLJ (2020) Agroforestry benefits and challenges for adoption in Europe and beyond. Sustainability 12(17):7001. https://doi.org/10.3390/su12177001

Song C, Kim W, Kim J, Gebru BM, Adane GB, Choi YE, Lee W (2022) Spatial assessment of land degradation using MEDALUS focusing on potential afforestation and reforestation areas in Ethiopia. Land Degrad Dev 33(1):79–93. https://doi.org/10.1002/ldr.4130

Soumare A, Diedhiou AG, Thuita M, Hafidi M, Ouhdouch Y, Gopalakrishnan S, Kouisni L (2020) Exploiting biological nitrogen fixation: a route towards a sustainable agriculture. Plants 9(8):1011. https://doi.org/10.3390/plants9081011

Stavi I, Lal R (2015) Achieving zero net land degradation: challenges and opportunities. J Arid Environ 112:44–51. https://doi.org/10.1016/j.jaridenv.2014.01.016

Stocking MA (2001) Land degradation. In: International encyclopedia of the social & behavioral sciences, pp 8242–8247. https://doi.org/10.1016/B0-08-043076-7/04184-X

Stutter MI, Chardon WJ, Kronvang B (2012) Riparian buffer strips as a multifunctional management tool in agricultural landscapes: introduction. J Environ Qual 41(2):297–303. https://doi.org/10.2134/jeq2011.0439

Talukder B, Ganguli N, Matthew R, vanLoon GW, Hipel KW, Orbinski J (2021) Climate change-triggered land degradation and planetary health: a review. Land Degrad Dev 32(16):4509–4522. https://doi.org/10.1002/ldr.4056

Taylor M, Bhasme S (2018) Model farmers, extension networks and the politics of agricultural knowledge transfer. J Rural Stud 64:1–10. https://doi.org/10.1016/j.jrurstud.2018.09.015

Torres CMME, Jacovine LAG, Nolasco de Olivera Neto S, Fraisse CW, Soares CPB, de Castro Neto F, Ferreira LR, Zanuncio JC, Lemes PG (2017) Greenhouse gas emissions and carbon sequestration by agroforestry systems in southeastern Brazil. Sci Rep 7(1):16738. https://doi.org/10.1038/s41598-017-16821-4

Udawatta RP, Rankoth LM, Jose S (2019) Agroforestry and biodiversity. Sustainability 11(10):2879. https://doi.org/10.3390/su11102879

Udumann SS, Dissanayaka DMNS, Nuwarapaksha TD, Dissanayake DKRPL, Atapattu AJ (2023) *Megathyrsus maximus* as a raw material for organic fertilizer production: A feasibility study. Technol Hortic 3(1). https://doi.org/10.48130/TIH-2023-0009

Udumann SS, Dissanayaka NS, Atapattu AJ (2024a) Women's roles in building resilience to climate change in agriculture. In: Impact of women in food and agricultural development. IGI Global, pp 58–78. https://doi.org/10.4018/979-8-3693-3037-1.ch004

Udumann SS, Ranasinghe CS, Karunarathna LKNG, Kaliyadasa PE, Nuwarapaksha TD, Premathilaka UGAT, Atapattu AJ (2024b) Optimizing intercropping selection for coconut plantations based on PAR and agro-climatic zones. Agrofor Syst. https://doi.org/10.1007/s10457-024-00977-w

Ulain Q, Ali SM, Shah AA, Iqbal KMJ, Ullah W, Tariq MAUR (2022) Identification of soil erosion-based degraded land areas by employing a geographic information system—a case study of Pakistan for 1990–2020. Sustainability 14(19):11888. https://doi.org/10.3390/su141911888

van Ramshorst JGV, Siebicke L, Baumeister M, Moyano FE, Knohl A, Markwitz C (2022) Reducing wind erosion through agroforestry: a case study using large eddy simulations. Sustainability 14(20):13372. https://doi.org/10.3390/su142013372

Wang Y, Li C, Tu C, Hoyt GD, DeForest JL, Hu S (2017) Long-term no-tillage and organic input management enhanced the diversity and stability of soil microbial community. Sci Total Environ 609:341–347. https://doi.org/10.1016/j.scitotenv.2017.07.053

Wani OA, Kumar S, Hussain N, Wani AA, Babu S, Alam P, Rashid M, Popescu SM, Mansoor S (2022) Multi-scale processes influencing global carbon storage and land-carbon-climate nexus: a critical review. Pedosphere. https://doi.org/10.1016/j.pedsph.2022.07.002

Wani OA, Sharma V, Kumar SS, Babu S, Sharma KR, Rathore SS, Marwaha S, Ganai NA, Dar SR, Yeasin M, Raghavendra S, Tomar J (2023) Climate plays a dominant role over land management in governing soil carbon dynamics in North Western Himalayas. J Environ Manag. https://doi.org/10.1016/j.jenvman.2023.117740

Weeraratna S (2022) Factors causing land degradation. In: Weeraratna S (ed) Understanding land degradation. SpringerBriefs in geography. Springer, Cham, pp 5–22. https://doi.org/10.1007/978-3-031-12138-8_2

Yadav GS, Kandpal BK, Das A, Babu S, Mohapatra KP, Devi AG, Devi HL, Chandra P, Singh R, Barman KK (2021a) Impact of 28-year-old agroforestry systems on soil carbon dynamics in Eastern Himalayas. J Environ Manag. https://doi.org/10.1016/j.jenvman.2021.111978

Yadav GS, Babu S, Das A, Datta M, Mohapatra KP, Singh R, Singh VK, Rathore SS, Chakraborty M (2021b) Productivity, soil health, and carbon management index of Indian Himalayan intensified maize-based cropping systems under live mulch-based conservation tillage practices. Field Crop Res. https://doi.org/10.1016/j.fcr.2021.108080

Yadav D, Singh D, Babu S, Madegowda M, Singh DV, Mandal D, Rathore AC, Sharma VK, Singhal V, Kumawat A, Yadav DK, Yadav RK, Kumar S (2023) Intensified cropping reduces soil erosion and improves rainfall partitioning and soil properties in the marginal land of the Indian Himalayas. Int Soil Water Conserv Res. https://doi.org/10.1016/j.iswcr.2023.10.002

Yadav D, Singh D, Madhu M, Singh DV, Islam S, Kumar G, Gupta AK, Mandal D, Babu S, Barh A, Kumawat A, Dubey R, Pati K, Bhargavi B, Yadav DK (2024) Interplanted crops and rainfall variability regulate soil erosion in rainfed maize of the Indian Himalayas. Land Degrad Dev. https://doi.org/10.1002/ldr.5221

Yan T, Wang Z, Liao C, Xu W, Wan L (2021) Effects of the morphological characteristics of plants on rainfall interception and kinetic energy. J Hydrol 592:125807. https://doi.org/10.1016/j.jhydrol.2020.125807

Zhu X, Chen C, Wu J, Yang J, Zhang W, Zou X, Liu W, Jiang X (2019) Can intercrops improve soil water infiltrability and preferential flow in rubber-based agroforestry system? Soil Tillage Res 191:327–339. https://doi.org/10.1016/j.still.2019.04.017

Conservation Agriculture for Land Restoration and Food Security

4

Gunturi Alekhya, Kadapa Sreenivasa Reddy, Priyanka Saha, T. K. Das, Rishi Raj, and Subhash Babu

Abstract

Land degradation presents a substantial and widespread peril, influencing ~23% of surface area of the earth, with an annual expansion of 5–10 million ha (Mha). Conservation agriculture (CA) plays a pivotal part in arresting and expediting land restoration through its sustainable agricultural practices by minimizing soil disturbance, employing methodologies like minimal tillage, and introducing cover crops which also aids in preserving soil structure, preventing erosion, and improving water retention. These practices stimulate soil health, foster biodiversity, and promote vegetation growth. Furthermore, CA encourages judicious use of water resource utilization. Through these concerted actions, CA averts further land degradation and rejuvenates depleted lands by reestablishing soil health, mitigating erosion, and reinforcing overall ecosystem resilience. CA not only lessens agriculture's negative effects on the environment but also makes a substantial contribution to human welfare and global food security. A central focus is the Zero Net Land Degradation (ZNLD) policy, which aims to ensure that global degraded land either diminishes or stabilizes without surpassing the pace of land restoration. By 2030, this is to be accomplished, according to the United Nations Convention to Combat Desertification (UNCCD), accentuating the necessity for substantial land restoration across diverse landscapes, even those lacking immediate economic advantages. Financial backing for restoration initiatives is feasible by ecosystem services payments and economic management systems. Effective administration of assessment procedures, formulation of sup-

G. Alekhya · K. S. Reddy · T. K. Das · R. Raj (✉) · S. Babu
Division of Agronomy, ICAR-Indian Agricultural Research Institute, New Delhi, India

P. Saha
ICAR-Krishi Vigyan Kendra, West Garo Hills, ICAR-Research Complex for NEH Region, Umiam, Meghalaya, India

© The Author(s), under exclusive license to Springer Nature Singapore Pte Ltd. 2025

S. Babu et al. (eds.), *Ecological Solutions to Agricultural Land Degradation*,
https://doi.org/10.1007/978-981-96-3392-0_4

portive policies, and securing financing through innovative mechanisms are crucial steps towards realizing the ambitious yet vital objective of Zero Net Land Degradation

Keywords

Conservation agriculture · Land degradation · Soil fertility · Soil organic carbon

Abbreviations

CA	Conservation Agriculture
CEC	Cation Exchange Capacity
CT	Conventional tillage
CO_2	Carbon dioxide
FAO	Food and Agriculture Organization
FIRB	Furrow Irrigated Raised Bed
GHGs	Greenhouse gases
GLASOD	Global Assessment of Soil Degradation
ISRO	Indian Space Research Organization
M ha	Million hectares
MWD	Mean Weight Diameter
NT	No Tillage
SDGs	Sustainable Development Goals
SOC	Soil Organic Carbon
SOM	Soil Organic Matter
TGA	Total Geographical Area
UNCCD	United Nations Convention to Combat Desertification
USLE	Universal Soil Loss Equation
ZNLD	Zero Net Land Degradation
ZT	Zero Tillage

4.1 Introduction

The world faces a concerning rise in food demand, posing a significant challenge because of the expanding agriculture impacts on environment (Panwar et al. 2019). The Global demand for crops is projected to surge by an unprecedented 100–110% by 2050 (Tillman et al. 2011), highlighting a crucial balance between provision of food to the increasing population and safeguarding planet health. Unchecked agricultural development over the past five decades resulted in elevated CO_2 levels (Ainsworth and Long 2021) higher temperatures and widespread soil degradation. The escalating pressure on land resources from various uses, including agriculture, forestry (Wani et al. 2023) and energy production, poses a significant challenge highlighted (UNCCD 2012). Soil is a vital element of ecosystems, essential for terrestrial net primary productivity, carbon sequestration, and nutrient retention (Yadav et al. 2021a). Soil degradation can result from natural processes, human activities, or both. Agricultural practices, in particular, wield a significant influence on land productivity, water cycles, greenhouse gas (GHG) concentration, and biodiversity (Stavi and Lal 2013). Soil degradation caused by erosion, which deteriorates the soil's physical, chemical, and biological properties, negatively impacts vegetation growth and disrupts ecosystem processes (Wani et al. 2023).

Land degradation is a multifaceted concept with varied definitions presented across numerous scholarly literatures. However, the preeminent and universally acknowledged definition is articulated by UNCCD as "the decline in biological or economic productivity and complexity of pasture, irrigated and rainfed arable land, grassland, forest, and woodland due to a variety of factors, comprising land use & management practices".

Land degradation threatens carbon sequestration, endangers GHG reduction goals, and compromises global food security (Smith et al. 2020). Ongoing trends of intensified agriculture and increased clearing pose risks to ecosystems, biodiversity, and climate (Babu et al. 2023a). However, an alternative approach emphasizes moderate intensification, technology transfer, and global advancements (Tillman et al. 2011). The alternatives emphasize that effective conservation and management practices could serve as linchpins, significantly fortifying food security while concurrently mitigating the deleterious environmental impacts associated with expansive agricultural practices (Kumar et al. 2022a; Wani et al. 2023). A staggering 60% of the global land is now classified as the degraded land, posing a substantial objection to the achievement of sustainable land use. The critical issue of land degradation, fueled by ecological deterioration, escalating food demands, rapid urbanization, industrialization, and unsustainable land use, has become a focal point in the global environmental discourse. Addressing these issues requires producing an extra billion tons of food by 2050 and committing to soil restoration, reduced emissions, and overall environmental improvement (Lal 2015). A staggering 1.5 billion people worldwide are directly impacted by this phenomenon, affecting soils, vegetation, and water. Rapid land degradation, primarily due to water and wind erosion, has led to the annual loss of a colossal 24,000 million tonnes of fertile soil globally (Rhodes 2014). This degradation translates into an estimated loss of a staggering US $ 400 billion per year, according to the report of Global Soil Partnership, 2017. Global land degradation remains a critical issue, particularly in humid and subhumid regions, with anticipated escalation due to climate change (Kumar et al. 2022b; Kar et al. 2023). Carbon emissions, erosion, and runoff are outcomes of the loss of natural biodiversity. Intensive conventional cultivation practices degrade soil health, endangering the agricultural sustainability in the long run (Singh et al. 2021a; Das et al. 2022a). Dryland regions are vulnerable due to arid conditions and limited water resources, bear a significant brunt. With 38% of the world's population living in these regions, they are prime support of livestock and agriculture. (Farooq and Siddique 2016). In essence, navigating this complex landscape demands a harmonious blend of technological innovation, conservation practices, and global cooperation that guarantee a resilient yet sustainable future for both humanity and the planet.

Adopting sustainable land management practices is crucial in addressing the issues listed above. Here, conservation agriculture (CA) may be a cost-effective and climate-adaptive method of managing land to rebuild soil health. It involves minimizing soil disturbance, incorporating crop wastes, and rotating crops (Pittelkow et al. 2015). CA aims to achieve sustainable, profitable farming by enhancing natural biological processes in the soil and minimizing mechanical tillage (Yadav et al. 2019, 2020a; Babu et al. 2023b). Currently practiced on around 157 million hectares, mainly in the USA, Brazil, and Canada (Sharma and Singh 2014), CA improves soil organic matter, sequesters carbon, and reduces runoff and erosion compared to conventional methods (Yadav et al. 2023). Addressing these colossal challenges necessitates a comprehensive strategy involving the rehabilitation of degraded lands through the integration of agronomical, mechanical, and biological measures. This involves cultivation of the trees, shrubs, and crops/herbs to enhance productivity, provide ecological and economic security, and offer various ecosystem services that leads to direct or indirect land restoration. Conservation and sustainable resource management are crucial for yield enhancement and livelihood maintenance amid changing climates (Babu et al. 2023a; Kar et al. 2022). Implementing agronomical measures like mulching, residue retention and engineering measures like land levelling, CA remains effective in long-term management of land (Singh et al. 2023; Kumawat et al. 2023).

4.2 What Is Zero Net Land Degradation?

Zero Net Land Degradation (ZNLD) concept introduced in 2011, aiming to halt and offset land degradation, to ensure a balance between the loss and restoration of fertile lands. The strategy involves sound management to prevent degradation in non-degraded lands and simultaneous restoration of degraded lands. If the restoration matches the ongoing degradation rate, achieving ZNLD stabilizes the global area of fertile land. The Conference of Parties established various progress indicators to define Land Degradation Neutrality (LDN). Among the six indicators, three land-based metrics were selected: land cover and land cover change, land productivity status, and carbon stock value (Assennato et al. 2020). These indicators are utilized to assess and report progress towards LDN and are consistent with Sustainable Development Goal (SDG) 15, specifically indicator 15.3.1, which measures the proportion of degraded land relative to the total land area. All three indicators adhere to the "One Out, All Out" principle, which stipulates that if any sub-indicator within a specific land unit is negative (either during the baseline or reporting period), the entire unit is classified as degraded and must undergo ground validation by national authorities (UNCCD 2018). The UNCCD targets ZNLD by 2030, emphasizing restoration within the same landscape and ecosystem (Stavi and Lal 2015). However, the ZNLD framework lacks clarity, presenting challenges for implementation. It broadly outlines goals but lacks specific details and a clear roadmap. The types of land-use practices under ZNLD were also remain undefined. Although the restoration efforts focus on croplands, rangelands, and woodlands, they do not specifically address the restoration of non-economically valuable natural or semi-natural lands. Persistent soil degradation threatens global food security and millions of livelihoods, impacting 25–30% of Earth's surface and over 40% of agricultural lands (Singh and Tewari 2022). It contributes to biodiversity loss, climate change, droughts, food and water shortages, and various socioeconomic and environmental issues.

4.3 Conservation Agriculture: Concept, Principles, and Practices

Conservation agriculture is an ecosystem-based approach to agricultural land management, emphasizing three interconnected principles (Richards et al. 2014). It is crucial to adopt CA as a powerful tool to ensure sustainability in agriculture. These three integral principles of CA, should be considered together for effective design and implementation as shown (Fig. 4.1).

Minimum Soil Disturbance: Achieved through no-tillage or reduced tillage, where ideally, zero-till is preferred, but controlled tillage is acceptable as long as it disturbs no more than 20–25% of the soil.

Permanent Soil Mulch: Involves retaining 30% of crop residues/cover crops on the field, to maintain a permanent and organic soil cover according to CA definitions.

Crop Diversification: Encourages proper crop rotation and intercropping, particularly with legumes.

Conservation agriculture boosts biological processes by reducing tillage (Raj et al., 2023) and optimizing input use without disrupting ecosystems. It involves direct seeding into minimally disturbed soil, often after slashing or rolling weeds and crop residues. This method retains crop residues to provide soil cover and nutrients for

Fig. 4.1 Role of conservation agriculture in achieving land restoration

subsequent crops (Suraj and Behara 2014). There are several challenges in adoption of these practices at field level and are as follows:

4.3.1 Agronomic Challenges

In CA systems, managing weeds, pests, and diseases without tillage presents challenges, as does addressing nutrient stratification, poor plant establishment due to inadequate seeding technology, reduced nitrogen availability, and soil issues like surface crusting and compaction (Dang et al. 2015). Additionally, in regions with low yields, boosting soil organic carbon (SOC) and residue cover to improve fertility, prevent erosion, and control weeds remains difficult. These agronomic issues may be addressed by better tailoring CA practices to local conditions.

4.3.2 Social Challenges

Implementing Conservation Agriculture (CA) requires addressing technical, agronomic, social, cultural, and institutional factors (Lal 2015). In many farming communities, CA practices, particularly no-till farming, may conflict with traditional land management practices that are deeply rooted in cultural values and rural customs (Tekle 2016).

4.4 Relevance of CA in Land Restoration

Conservation Agriculture provides a sustainable approach to meeting global food needs, supporting SDGs, and revitalizing agricultural systems. It regenerates topsoil, promotes biodiversity, improves water management, enhances ecosystem services, boosts climate resilience, and strengthens overall soil health and vitality (Kumawat et al. 2023). Under CA, crop residue plays a crucial role in augmenting SOM, fostering nutrient cycling, conserving soil moisture and enhancing soil aggregation (Singh et al. 2021b), promoting biological activity, mitigating soil temperature fluctuations, ultimately mitigating land degradation (Das et al. 2020).

4.4.1 Improvement of Soil Physical Properties with CA Practices

Conservation agriculture techniques are increasingly recognized for their potential to reduce crop water footprints (Yadav et al. 2020b). These techniques enhance water infiltration, minimize soil erosion, runoff, and surface evaporation (Singh et al. 2023; Babu et al. 2023b), leading to improved water conservation and higher soil water content. Assessing the impact of CA on aggregate stability is crucial for understanding its effects on water savings and soil water storage as detailed (Table 4.1). Unstable soil aggregates are prone to slaking during irrigation or

Table 4.1 Irrigation water productivity of different cropping system under CA

Cropping system	IWP-CA[a] (kg grain m^{-3})	IWP-CT[b] (kg grain m^{-3})	References
Rice-wheat	0.60	0.41	Gathala et al. (2013)
Basmati rice-wheat-mungbean	0.67	0.59	Jat et al. (2019)
Rice-wheat	0.65	0.42	Jat et al. (2020)
Rice-wheat-mungbean	0.66	0.41	Jat et al. (2020)
Maize-wheat	1.35	0.99	Jat et al. (2020)
Maize-wheat-mungbean	1.28	0.91	Jat et al. (2020)
Winter wheat-summer maize	1.37	0.90	Wang et al. (2021)
Barley-maize	1.42	1.06	Fonteyne et al. (2021)
Wheat-maize	1.64	1.46	Bougari et al. (2020)
Common beans	1.76	0.55	Fatumah et al. (2021)
Potato-maize	1.60	1.20	Alam et al. (2015)
Wheat-maize	2.01	1.80	Shao et al. (2016)

[a]IWP-CA: Irrigation water productivity under CA, [b]IWP-CT: Irrigation water productivity under CT

rainfall, reducing water infiltration and increasing runoff and erosion (Raj et al. 2023; Yadav et al. 2023). Mulching further supports water conservation by decreasing runoff and enhancing infiltration rates (Raj et al. 2023). Surface mulching of crop residues has been demonstrated to decrease evaporation losses while increasing infiltration rates, leading to elevated soil water content within the CA framework (Singh et al. 2021b). Measures of aggregate stability are helpful in evaluating soil structural stability because they reflect an aggregate capacity to withstand various pressures. Zero tillage (ZT) with residue retention significantly improves dry aggregate size distribution and water stability compared to traditional tillage (CT), as evidenced by increased mean weight diameter (MWD) across various soil types and conditions (Chan et al. 2002; Pinheiro et al. 2004; Lichter et al. 2008). ZT with residue retention enhances soil stability, making it less prone to structural deterioration and water slaking, whereas CT, even with good structural distribution, lacks protection and is more susceptible to erosion (Das et al. 2021). Residues on the soil surface in CA shield the soil from rainfall effects, unlike CT, which offers no such protection (Yadav et al. 2020a; Singh et al. 2021a). Higher organic matter content in the topsoil improves macro porosity and resistance to deformation, with reduced tillage potentially leading to a new steady state of soil porosity (Das et al. 2022a). On the other hand, initial modifications can be too slight to differentiate from spontaneous variation (Kay and Van den Bygaart 2002). In Minnesota, a study on a silt loam with a maize-soybean rotation found higher soil bulk densities in the zero-tillage surface layer after 23 years compared to conventional tillage, though bulk densities were lower below 30 cm. This is attributed to the compacting effects of tillage implements and the surface rupturing caused by tillage (Dolan et al. 2006). Soil pores, especially those with diameters of 0.20–0.30 cm, are crucial for water storage and plant growth. Although zero tillage often results in larger micro and meso-porosity compared to CT, tillage effects can vary. Soil management that increases organic matter can enhance soil water retention (Ngangom et al. 2020; Das et al. 2022b). Higher SOM levels is linked to enhanced water availability across diverse soil types (Raj et al. 2017). Adopting CA can enhance water retention and help agriculture adapt to climate variability and extreme weather events. CA is recognized for its ability to offer erosion protection, lower soil temperatures, and improve infiltration rates, effectively addressing challenges such as heavy rainstorms, fluctuating daily temperatures, and frequent droughts (Kassam et al. 2009; Ansari et al. 2022). CA, particularly its residue retention and improved aggregate stability, has a substantial positive impact on soil water storage (Li et al. 2019). This positive effect is attributed to increased infiltration rates and reduced soil water evaporation, facilitated by improved aggregate stability in the soil profile surface where SOC improvements are prominent (Li et al. 2019; Yadav et al. 2019; Das et al. 2022b). The presence of crop residues plays a protective role by preventing raindrop impact and surface seal formation, subsequently promoting increased infiltration (Rajanna et al. 2022). Bulk density, indicating soil compaction and influencing seedling emergence and root growth, varies under CA systems. While studies report differing effects on bulk density under CA (Blanco-Canqui and Ruis 2018; Somasundaram et al. 2019), a recent global meta-analysis found an average increase of 1.4% with NT and residue retention (Li et al. 2019).

4.4.2 Improvement Soil Chemical Properties with CA Practices

Cation Exchange Capacity (CEC) is essential for soil fertility, reflecting the soil's ability to retain nutrients for plant uptake. High CEC indicates better nutrient availability and supports plant growth. No-till (NT) and reduced-till (RT) systems show significantly higher CEC than conventional tillage (CT), with CEC increasing with depth in NT and RT but remaining uniform in CT (Sithole and Magwaza 2019). In NT systems, surface mineralization of soil organic matter (SOM) and nitrification of nitrogen fertilizer can significantly reduce surface soil pH (Obour et al. 2017). CA systems often result in greater acidity due to the accumulation of plant residues and organic acids, affecting nitrogen mineralization and nitrate leaching (Sithole and Magwaza 2019; Das et al. 2022b). There is a close linear relationship between organic carbon and mean weight diameter (MWD) (Carter 1992). SOC variations impact macro-aggregate stability; NT systems show higher SOC at shallow depths compared to conventional tillage. Zibilske et al. (2002) reported 57.8% higher SOC at 0–4 cm and 15.1% higher at 4–8 cm under NT after 9 years, while Tabaglio et al. (2009) and Gal et al. (2007) found higher SOC in the 0–5 cm depth under NT, with variable results at greater depths. In CA systems, increased SOC improves plant nutrient availability by enhancing both nutrient quantity and distribution in the soil profile (Yadav et al. 2019). Effective CA practices, such as adding more residue and nutrient-rich organic materials, raise concentrations of essential nutrients like nitrogen, phosphorus, and potassium. Total nitrogen content is positively correlated with residue amounts, with no-till and permanent raised beds showing significantly higher levels compared to conventional tillage. For instance, (Thomas et al. 2007) reported 21% higher total nitrogen in the top 10 cm under no-till than conventional tillage. While higher nitrogen immobilization in the initial years can reduce crop yield and nitrogen fertilizer recovery, it ultimately improves yield and reduces nitrogen losses through leaching, runoff, and denitrification (Randall and Iragavarapu 1995). However, the lack of soil mixing in CA, particularly in no-till systems, can cause nutrient stratification, which may limit root access in arid regions (Dang et al. 2015). Although increases in soil nutrient reserves generally enhance plant nutrient availability (Pheap et al. 2019), this is not universally applicable. Specifically, with nitrogen, total N levels may rise under CA, yet plant-available N might decline, especially shortly after CA adoption. This reduction can necessitate additional N fertilizer applications to sustain yields (Sithole and Magwaza 2019). Reduced nitrogen (N) availability in Conservation Agriculture systems may arise from slower N mineralization due to less soil-stubble mixing or increased N immobilization linked to crop residues with high carbon-to-nitrogen ratios. Over time, N supply may improve as equilibrium between carbon (C) and N is re-established (Soane et al. 2012). In CA systems, available phosphorus (P) is generally higher under NT and RT compared to CT. NT systems show the highest P concentration at the surface, with a decrease from 0 to 10 cm depth and an increase from 20 to 30 cm depth, whereas CT and RT systems lack this gradient (Sithole and Magwaza 2019). NT with residue retention also improves K availability in surface soils, where root density is higher (Franzluebbers and Hons 1996). Permanent raised beds with residue retention have K concentrations 1.65 and 1.43 times higher in the 0–5 cm and 5–20 cm layers, respectively, compared to conventionally tilled raised beds (Govaerts et al. 2007). Although K concentration is typically higher in the 0–5 cm layer than in the 5–20 cm layer across tillage systems, some studies suggest that surface K accumulation may be independent of tillage practices (Duiker and Beegle 2006). In minimum tillage systems, P and K levels are 3.5 times higher in the 0–5 cm layer compared to the 5–15 cm layer. The lack of mechanical fertilizer incorporation in NT/MT systems concentrates P and K in the upper 5 cm of soil. Surface-placed residues decompose more slowly, reducing nutrient leaching and enhancing nutrient availability as roots decompose (Ismail et al. 1994; Balota et al. 2004; Qin et al. 2004).

4.4.3 Improvement of Soil Biological Properties with CA Practices

Tillage absence reduces soil disturbance, maintaining microbial habitats and preserving organic matter. Retaining crop residues supplies carbon and energy to microorganisms, thereby enhancing microbial biomass. This larger microbial community facilitates the breakdown of organic matter, releasing nutrients and organic compounds that boost enzyme activity. Moreover, residue cover moderates soil temperature and moisture, creating favorable conditions for microbial growth and enzyme function. These synergistic interactions under ZT and residue retention promote a robust microbial community and enzyme activity, enhancing nutrient cycling, organic matter turnover, and overall soil health (Kandeler et al. 2006). CA enhances earthworm populations compared to CT due to reduced soil disturbance and improved habitat conditions. The presence of crop residues on the soil surface in CA provides organic matter that earthworms feed on, improving their habitat and food availability (Briones and Schmidt 2017). Additionally, the mulch layer in CA aids in regulating moisture and temperature of soil, thereby establishing a more consistent environment for earthworms, whereas CT exposes the soil, leading to moisture loss and temperature fluctuations detrimental to earthworm survival (Kladivko 2001). CA systems often result in higher SOM content due to the decomposition of crop residues, serving as a food source for earthworms and promoting their growth and reproduction (Curry and Good 1992). Arthropods, including insects, spiders, and mites, also benefit from CA, as practices support higher arthropod diversity and abundance compared to CT due to the diverse plant cover and crop rotations that provide various niches and food resources (Altieri 1999). The reduced pesticide uses in CA's integrated pest management benefits non-target arthropods and fosters a more balanced ecosystem where natural predators can thrive (Birkhofer et al. 2008). The minimal soil disturbance in CA helps maintain soil structure, which supports arthropod habitats by allowing better water infiltration and aeration. Soil enzymes, indicators of microbial activity and soil health, are enhanced in CA due to improved soil conditions and microbial diversity (Dick 1994). The reduced disturbance and higher organic matter content in CA promote microbial biomass and diversity, leading to higher enzymatic activities necessary for breaking down organic matter and cycling nutrients (Wardle 1995). Furthermore, CA provides a more stable environment with consistent moisture and temperature levels, favoring enzyme activity, whereas CT can lead to fluctuating conditions that stress microbial communities and reduce enzyme production (Holland 2004). The continuous cover and root growth in CA contribute to nutrient cycling and availability, with enzymes such as phosphatases, dehydrogenases, and cellulases being more active in CA systems, facilitating nutrient transformations and improving soil fertility (Blanco-Canqui and Lal 2008). Studies have shown that CA systems can lead to a 50–100% increase in earthworm populations compared to CT, attributed to improved habitat conditions and organic matter availability (Briones and Schmidt 2017). Similarly, CA practices can enhance arthropod diversity and abundance by 30–70% compared to CT, linked to better habitat provision and reduced pesticide use (Crittenden et al. 2014). Enzymatic activities in CA systems are typically 20–50% higher than in CT, reflecting improved microbial activity and soil health, crucial for nutrient cycling and sustainable crop production (Lal 2004). Thus, CA offers significant ecological benefits over CT by improving earthworm and arthropod populations and enhancing enzymatic activities, contributing to better soil health, increased biodiversity, and more sustainable agricultural practices as the agricultural sector faces challenges related to soil degradation and biodiversity loss (Holland 2004). Adopting CA practices provides a viable solution to enhance ecosystem functions and ensure long-term agricultural productivity.

4.4.4 CA for Soil Organic Carbon Restoration

Soil carbon sequestration refers to the method of capturing carbon from the atmosphere and storing it in the soil through alterations in land use or management practices (West and Post 2002; Powlson et al. 2011). CA is now acknowledged as a practical solution for reducing GHGs, even though it was not initially intended to capture soil carbon. As a result, research aimed at comprehending and improving soil carbon sequestration has made CA a major focus mentioned in Table 4.2. Compared to conventional methods, CA techniques, such as crop rotations and surface residue retention, aim to boost carbon inputs compared to conventional methods (Rathore et al. 2022; Das et al. 2020). No-till (NT) practices enhance soil aggregation and protect soil carbon from decomposition, slowing down its decay (Das et al., 2022b). The interaction of these processes and their impact on soil carbon levels is influenced by factors like soil type, climate conditions, and additional management practices (Palm et al., 2014).

Conservation agriculture is vital for enhancing soil organic carbon (SOC) and maintaining soil health while supporting environmental functions (Yadav et al. 2019). It effectively sequesters carbon through the restoration of degraded lands and sustainable management practices (Das et al. 2022a). Key CA practices like crop rotation, reduced tillage, and organic residue incorporation significantly boost SOC levels (Wani et al. 2022). Improved soil structure, driven by increased microbial and fungal activity, helps stabilize SOC. CA is widely recognized for combating

Table 4.2 Enhancement of soil SOC through CA in different cropping system

Cropping system	Years	SOC-CA[a] (%)	SOC-CT[b] (%)	Soil depth (cm)	References[a]
Maize	6	0.64	0.49	0–15	Parihar et al. (2018)
Soybean-Pigeon pea	8	0.73	0.68	0–10	Naorem et al. (2023)
Soybean-Wheat	8	0.78	0.60	0–10	Naorem et al. (2023)
Maize-Gram	8	0.76	0.70	0–10	Naorem et al. (2023)
Rice-Maize	5	0.59	0.43	0–15	Singh et al. (2015)
Rice-Wheat	10	0.99	0.63	0–5	Roy et al. (2022)
Wheat-Fababean	7	0.85	0.75	0–10	Stagnari et al. (2020)
Maize	8	1.51	1.30	0–10	Swanepoel et al. (2018)
Wheat-Maize	11	1.30	0.75	0–20	Xu et al. (2019)
Peas-Wheat-Sunflower	3	0.97	0.86	0–5	Carbonell-Bojollo et al. (2015)
Peas-Wheat-Sunflower	3	0.88	0.71	5–10	Carbonell-Bojollo et al. (2015)

[a]SOC-CA: Soil organic carbon under CA; [b]SOC-CT: Soil organic carbon under CT

resource degradation and capturing atmospheric carbon. Important CA technologies include no-till, minimum/reduced tillage, Furrow Irrigated Raised Bed (FIRB), laser land leveling, and the retention of crop residues and mulches. Other practices such as cover crops, agroforestry, and biochar application also play roles in carbon sequestration (Tanveer et al. 2013). While agricultural soils can release CO_2, optimal soil and crop management practices enhance their carbon sequestration capacity, mitigating greenhouse gas emissions. The transition from conventional tillage to CA, along with effective nitrogen management, reduces CO_2, CH_4, and N_2O emissions, promotes carbon sequestration in the soil profile, and aids in restoring degraded agro-ecosystems. Retaining plant residues as mulch and employing no-till practices improve SOC concentration, soil moisture, and temperature, thus enhancing soil quality (Stavi et al. 2012). To boost SOC, agricultural practices should increase organic matter inputs, reduce decomposition, or combine both approaches (Paustian et al. 2000). Conservation tillage, including reduced tillage and residue retention, accumulates soil carbon across various climates (West and Post 2002). By minimizing soil disturbance and slowing stubble decomposition, these practices protect SOC through the formation of stable aggregates that act as barriers against microbial decomposition (Tripathi et al. 2014).

4.4.5 Conservation Agriculture for Enhancing Soil Fertility and Health

Soil health refers to a soil's ability to function and support ecosystem services, while soil quality denotes its inherent capacity to do so (Bonfante et al. 2020). Both concepts are used to assess soil management's impact on agricultural sustainability and guide decision-making to prevent degradation. CA aims to restore or maintain essential soil functions, such as carbon cycling, nutrient cycling, and soil structure maintenance, thereby improving crop yield, economic return, greenhouse gas mitigation, biodiversity, and soil health. CA is recommended as an alternative to conventional practices to enhance soil characteristics and processes (Caron et al. 2014; Palm et al. 2014). However, the impact of CA on soil health varies depending on local soil and climate conditions, the duration of implementation, specific practices used, and the farmer's expertise (Kassam et al. 2014; Indoria et al. 2017).

CA stands as a pragmatic approach to enhancing soil health, supporting both environmental and economic sustainability in agriculture (Ngangom et al. 2020; Saha et al. 2021). Recognizing the substantial impacts of CA, global endeavors are initiated to promote its adoption and implementation. SOM plays a crucial role in aggregate stability, influencing overall soil structure. In soils with low organic matter, aggregates break down into smaller units when wet, making the soil more prone to erosion and reducing water infiltration and seedling emergence (Blanco-Canqui and Ruis 2018). CA, by elevating SOC, typically leads to enhanced soil aggregate stability, facilitated by increased fungal populations and sustained root networks due to no tillage (Wang et al. 2010; Spurgeon et al. 2013).

A recent global meta-analysis found a 31% average increase in water-stable aggregates in no-till (NT) systems compared to conventional tillage without residue retention, which enhances soil water infiltration, reduces erosion (especially with residue retention), protects organic matter, and supports diverse microbial habitats (Li et al. 2019; Blanco-Canqui and Ruis 2018). However, despite higher total nitrogen (N) in CA, available N can decrease shortly after implementation due to slower mineralization and increased immobilization from high C: N ratio residues, potentially requiring additional N fertilizer to maintain yields (Rusinamhodzi et al. 2011; Sithole and Magwaza 2019). NT systems may also cause nutrient stratification at the soil surface (Dang et al., 2015). Additionally, CA, particularly NT, enhances the diversity of fungal and bacterial populations, with fungi thriving at the soil surface due to the absence of tillage, which benefits crop production by supporting plant growth and disease suppression (Spurgeon et al. 2013; Zhang et al. 2018; Gonzalez-Chávez et al. 2010; Wang et al. 2016; Yang et al. 2012).

4.4.6 Conservation Agriculture for Reducing GHG Emission

Conservation Agriculture influences greenhouse gas emissions by affecting soil carbon storage and health. Its principles—minimal soil disturbance, permanent cover, and crop rotation—can either reduce or increase emissions depending on conditions. CA boosts soil organic carbon sequestration and offsets CO_2 emissions and increasing organic matter (Lal 2004; Avasthe et al. 2023) as described in Table 4.3. In CA, using crop residue enhances soil quality by reducing erosion (Raj et al., 2022) and mitigating pollutant impacts (Singh et al. 2022a, b). Sharma et al. (2013) illustrated that immediate mulching post-maize harvest significantly reduced soil water loss through evaporation between the maize and wheat crops. Zero Tillage in conservation tillage improves soil structure (Babu et al. 2023b), preserves organic matter, boosts water retention, and reduces erosion and greenhouse gas emissions (Sharma et al. 2013; Yadav et al. 2020a). RT/NT practices decrease the oxidation of organic matter, resulting in lower CO_2 emissions from the soil (Six et al. 2002). In rice systems, CA can influence methane (CH_4) emissions, with practices like alternate wetting and drying reducing CH_4 emissions compared to continuous flooding. CA impact on nitrous oxide (N_2O) emissions is complex and varies depending on nitrogen management, soil conditions, and crop residue presence (Linquist et al. 2012). While RT can lower emissions by improving soil structure and reducing denitrification, it may increase emissions if residues with high nitrogen content remain on the surface (Van Kessel et al. 2013). Largely, CA mitigate GHG emissions by enhancing carbon sequestration and reducing CO_2 and CH_4 emissions, although its effects on N_2O emissions are variable and require careful management (Powlson et al. 2016). Adapting CA practices to local conditions and integrating them with other sustainable agriculture practices can enhance their effectiveness in mitigating climate change (Biswakarma et al. 2023).

Table 4.3 GHGs emission potential of CA under different regions of the globe

Cropping system	Soil type	ECE (Equivalent carbon emissions) (kg/ha)		Experimental site	References
		Conservation agriculture	Conventional tillage		
Oat and Maize-fava and Vetch	Andosol (Typic Ustivitrands)	856	958	Mexico, USA	Fuentes-Ponce et al. (2022)
Rice-Wheat	Sandy clay loam soil	2212	2691	Delhi, India	Ghosh et al. (2022a)
Wheat	Silt clay loam	1982	2027	Jamalpur, Bangladesh	Rahman et al. (2021)
Rice-Wheat	Clay loam	1497	2607	Haryana, India	Kakraliya et al. (2021)
Maize-Wheat	Sandy loam soil	386	842	Punjab, India	Nisar et al. (2021)
Wheat	Inceptisol, Sandy loam	266	321	Delhi, India	Nath et al. (2017)
Wheat monoculture	Sandy-loam Calcic Haploxeralf	116	1546	Madrid, Spain	Tellez-Rio et al. (2017)
Pigeon pea- Castor	Alfisol (Typic Haplustalf)	1042	1427	Telangana, India	Pratibha et al. (2016)
Rice-Wheat	Sandy loam to clay loam	771	951	Haryana, India	Tirol-Padre et al. (2016)
Rice-Wheat	Sandy clay loam	2316	2581	Haryana, India	Sapkota et al. (2015)
Pigeonpea-Castor	Typic Haplustalf	673.6	803:4	Telangana, India	Pratibha et al. (2015)
Maize-Wheat	Thermic Cumulic Halplustoll	1440	1872	Mexico, USA	Fuentes et al. (2012)
Maize-Wheat	Haplic Phaeozem	901	1156	Mexico, USA	Dendooven et al. (2012)

4.5 Conservation Agriculture for Achieving Food Security

An input-intensive production system boosts productivity but threatens natural resources and environmental balance. Coupled with rising food demand and climate change impacts on agriculture, this underscores the critical importance of CA. Recognizing the potential effects of CA, substantial initiatives have been undertaken for its widespread adoption and implementation globally (Kumawat et al. 2023). Attempting to produce food on degraded soils without proper management practices exacerbates environmental problems, rather than necessarily decreasing food security.

4.5.1 Crop Diversification and Cropping Systems in Conservation Agriculture

The conservation agriculture system, when combined with legume rotations and mineral nitrogen fertilization, tends to lower soil pH more effectively than conventional tillage systems lacking legume integration (Vieira et al. 2009). In arid regions, CA methods reduce soil temperature through residue retention, which can enhance crop growth and yields (Zhang et al. 2014; Steward et al. 2018). Including legumes and diverse crops in rotations enhances soil fertility and reduces pests and diseases, which can boost yields (Kaye and Quemada 2017). Creating locally suited seeding equipment and selecting appropriate crop rotations and cover crops are key to managing plant establishment, weeds, pests, and diseases without tillage (Thierfelder et al. 2018). Effective fertilizer uses or better legume integration boosts nitrogen availability and plant biomass, in turn increase residue cover and soil fertility (Verhulst et al. 2010). Various studies conducted across Asia, Africa and Latin America showed positive effects on soil quality parameters, nutrient recycling and improvement in environmental health through CA-based practices like residue retention and crop rotation with no tillage operations (Turmel et al. 2015). Gura and Nkeni (2019) reported that the Soil Management Assessment Framework, which was utilized to evaluate soil quality, indicated an enhancement in soil quality after 3 years of implementing conservation agriculture practices. The incorporation of soybean and its residues further contributed to a greater improvement in soil quality in Eastern Cape of South Africa. The long-term (>20 years) impacts of no-till versus plow-till (PT) on soil and corn yields was studied by Nunes et al. (2018) in New York. Long-term NT, enhanced by crop rotations and cover crops, improved soil health and corn yields, particularly in loamy sand and silt loam soils, with notable benefits from cover crops. Cover crops maintain soil cover during off-seasons and add organic material when crop residue is low. Identifying effective, region-specific practices requires thorough research such as strategic tillage, or occasional cultivation of no-till soils, helps manage weeds, pests, diseases, nutrient stratification, and soil compaction in long-term no-till systems.

4.5.2 Role of CA in Improving Crop Productivity and Achieving Food Security

Conservation agriculture plays a critical role in improving food security by increasing the productivity of cereals, pulses, and oilseeds. This farming approach, has demonstrated significant benefits in boosting crop yields across various regions. Recent studies have shown that CA can lead to substantial productivity gains as detailed in Table 4.4. For instance, in Sub-Saharan Africa, CA practices have increased maize yields by 15–25% and beans by up to 20% (Galani et al. 2022). In South Asia, CA has improved rice and wheat productivity by 10–20% (Jat et al. 2018) in various IGP regions of India. Similarly, in North America, conservation tillage has enhanced soybean and corn yields by 10–20% (Vetsch et al. 2007). CA

Table 4.4 Enhancement of crop productivity through utilization of CA-based practices across global regions

Cropping system	% increase in the productivity by CA over CT	Location	References
Cotton-Wheat	16.2–19.4%	Delhi, India	Alekhya et al. (2024)
Pigeonpea-Wheat	11.6–14.9%	Delhi, India	Sharma et al. (2023)
Rice-Wheat	6.3–22.7%	Delhi, India	Raj et al. (2022)
Maize-Wheat-Green gram	~56%	Delhi, India	Ghosh et al. (2022b)
Rice-Wheat	30.5–34.4%	Bihar, India	Rao et al. (2021)
Maize-Wheat	~6.6%	Central Pannonian subregion, Croatia	Jug et al. (2019)
Maize-Wheat	~9.2%	Shaanxi, China	Sun et al. (2018)
Wheat	~7.4%	Haryana, India	Choudhary et al. (2018)
Maize-Wheat	~11.54%	Delhi, India	Parihar et al. (2016a, b)
Maize-Wheat	~44.1%	Uttarakhand, India	Ghosh et al. (2015)
Rice-Wheat	~32.4%	Uttar Pradesh, India	Bohra and Kumar (2015)
Rice-Wheat	~19.6%	Bihar, India	Jat et al. (2014)
Cotton	~18.8%	Delhi, India	Das et al. (2014)
Rice-Maize	2.7–9.4%	Rangpur, Bangladesh	Gathala et al. (2012)
Rice-Wheat	~5.5%	Uttar Pradesh, India	Tomar et al. (2009)
Rice-Wheat	11–26%	IGP region of India	Laxmi et al. (2007) and Dhiman et al. (2003)

benefits are also evident in a study conducted in Europe, where wheat and barley yields have increased by 7–15% due to conservation practices (Basch et al. 2015). A study conducted in Australia also reported CA has led to a 12–18% rise in productivity for wheat and sorghum (Hussein et al. 2021). Likewise, study of Latin America has also inferred the impressive results, CA has achieved a 15–25% boost in soybean and maize yields (Jat et al. 2012). In Southern Africa, CA has resulted in a 20–30% increase in maize and sorghum productivity, enhancing regional food security (Thierfelder et al. 2015). Thus, CA's ability to significantly increase the productivity of cereals, pulses, and oilseeds makes it a valuable strategy in addressing food security challenges.

4.6 Conclusion

Over 25% of Earth's land is degraded, impacting the livelihood of 1.5 billion people. Land restoration is crucial to align with SDGs, focusing on degraded agricultural, forest, and wasteland areas. Urgent action is needed to combat soil and land degradation from natural processes and human activities. CA offers a sustainable way to meet food demands, support SDGs, and boost soil health and climate resilience. CA techniques prevents degradation by preserve soil, stabilizing carbon, and conserve water. As they provide diverse strategies to counteract degradation, boost productivity, resilience, and reduce costs. Despite scientific knowledge, a lack of consensus on effective strategies persists. Regular assessments and standardized methodologies are crucial for understanding degradation types, severity, and causes. Achieving global Land Degradation Neutrality by 2030 requires a universally accepted mechanism, disseminating practical knowledge and securing government support. International collaboration is key to realizing the 2030 target of Zero Net Land Degradation. Systematic addressing of challenges is vital. In conclusion, our study emphasizes CA-based practices in minimizing land degradation, aiming for Land Degradation Neutrality by 2030 and progressing towards SDGs.

References

Ainsworth EA, Long SP (2021) 30 years of free-air carbon dioxide enrichment (FACE): what have we learned about future crop productivity and its potential for adaptation? Glob Chang Biol 27(1):27–49

Alam MM, Ladha JK, Faisal MW, Sharma S, Saha A, Noor S, Rahman MA (2015) Improvement of cereal-based cropping systems following the principles of conservation agriculture under changing agricultural scenarios in Bangladesh. Field Crop Res 175:1–5. https://doi.org/10.1016/j.fcr.2014.12.015

Alekhya G, Das T, Kaur R, Raj R, Sudhishri S, Bhatia A, Biswas S, Sen S, Saha P, Sudarshan S, Rajpoot B (2024) Productivity, nutrient uptake and nitrogen economy of wheat (Triticum aestivum) in a conservation agriculture-based cotton (Gossypium herbaceum)–wheat system. Indian J Agric Sci 94(7):780–785

Altieri MA (1999) The ecological role of biodiversity in agroecosystems. In: Invertebrate biodiversity as bioindicators of sustainable landscapes. Elsevier

Ansari MA, Babu S, Choudhary J, Ravisankar N, Panwar AS (2022) Soil quality restoration and yield stabilization in acidic soils of northeastern Himalayas: five years impact of green manuring and crop residue management. Front Environ Sci 10:940349. https://doi.org/10.3389/fenvs.2022.940349

Assennato F, Di Leginio M, d'Antona M, Marinosci I, Congedo L, Riitano N, Luise A, Munafò M (2020) Land degradation assessment for sustainable soil management. Ital J Agron 15(4):299–305

Avasthe RK, Devi SH, Bhupenchandra I, Kumar A, Chongtham SK, Babu S, Singh R, Das A, Gudade BA, Bora SS (2023) Agro-tactics for reducing carbon footprint in agricultural production systems: a review. Indian J Agron 68(2):115–125. https://doi.org/10.59797/ija.v68i2.332

Babu S, Singh R, Avasthe RK, Yadav GS, Mohapatra KP, Selvan T, Das A, Singh VK, Valente D, Petrosillo I (2020a) Soil carbon dynamics in Indian Himalayan intensified organic rice-based cropping sequences. Ecol Indic 114:106292. https://doi.org/10.1016/j.ecolind.2020.106292

Babu S, Mohapatra KP, Das A, Yadav GS, Tahasildar M, Singh R, Panwar AS, Yadav V, Chandra P (2020b) Designing energy-efficient, economically sustainable and environmentally safe cropping system for the rainfed maize–fallow land of the Eastern Himalayas. Sci Total Environ 722:137874. https://doi.org/10.1016/j.scitotenv.2020.137874

Babu S, Singh R, Avasthe RK, Kumar S, Rathore SS, Singh VK, Ansari MA, Valente D, Petrosillo I (2023a) Soil carbon dynamics under organic farming: impact of tillage and cropping diversity. Ecol Indic 147:109940. https://doi.org/10.1016/j.ecolind.2023.109940

Babu S, Singh R, Avasthe R, Rathore SS, Kumar S, Das A, Layek J, Sharma V, Wani OA, Singh VK (2023b) Conservation tillage and diversified cropping enhance system productivity and eco-efficiency and reduce greenhouse gas intensity in organic farming. Front Sustain Food Syst 7:1114617. https://doi.org/10.3389/fsufs.2023.1114617

Balota EL, Colozzi A, Andrade DS, Dick RP (2004) Longterm tillage and crop rotation effecton microbial biomass and C and N mineralization in a Brazilian Oxisol. Soil Tillage Res 77:137–145

Basch G, Friedrich T, Kassam A, Gonzalez-Sanchez E (2015) Conservation agriculture in Europe. In: Conservation agriculture, pp 357–389

Birkhofer K, Bezemer TM, Bloem J, Bonkowski M, Christensen S, Dubois D, Ekelund F, Fließbach A, Gunst L, Hedlund K, Mäder P (2008) Long-term organic farming fosters below and aboveground biota: implications for soil quality, biological control and productivity. Soil Biol Biochem 40(9):2297–2308

Biswakarma N, Pooniya V, Zhiipao RR, Kumar D, Shivay YS, Das TK, Roy D, Das B, Choudhary AK, Swarnalakshmi K, Govindasamy P, Lakhena KK, Das K, Lama A, Jat RD, Babu S, Khan SA, Behara B (2023) Identification of a resource-efficient integrated crop management practice for the rice-wheat rotations in south Asian Indo-Gangetic Plains. Agric Ecosyst Environ 357:108675. https://doi.org/10.1016/j.agee.2023.108675

Blanco-Canqui H, Lal R (2008) Principles of soil conservation and management. Springer Science & Business Media

Blanco-Canqui H, Ruis SJ (2018) No-tillage and soil physical environment. Geoderma 326:164–200

Bohra JS, Kumar R (2015) Effect of crop establishment methods on productivity, profitability and energetics of rice (Oryza sativa)-wheat (Triticum aestivum) system. Indian J Agric Sci 85:217–223

Bonfante A, Basile A, Bouma J (2020) Targeting the soil quality and soil health concepts when aiming for the United Nations Sustainable Development Goals and the EU Green Deal. Soil 6(2):453–466

Bougari E, Asoodar MA, Marzban A, Kazemi N (2020) Investigating water use efficiency, energy productivity, economic and yield under different wheat-maize cropping system in the north of Khuzestan province. J Agric Sci Sustain Produc 30(4):295–310. https://doi.org/10.22034/saps.2020.12318

Briones MJ, Schmidt O (2017) Conventional tillage decreases the abundance and biomass of earthworms and alters their community structure in a global meta-analysis. Glob Chang Biol 23(10):4396–4419

Carbonell-Bojollo R, González-Sánchez EJ, de Torres MR, Ordóñez-Fernández R, Domínguez-Gimenez J, Basch G (2015) Soil organic carbon fractions under conventional and no-till management in a long-term study in southern Spain. Soil Res 53(2):113–124. https://doi.org/10.1071/SR13369

Caron P, Biénabe E, Hainzelin E (2014) Making transition towards ecological intensification of agriculture a reality: the gaps in and the role of scientific knowledge. Curr Opin Environ Sustain 8:44–52

Carter MR (1992) Influence of reduced tillage systems on organic matter, microbial biomass, macro-aggregate distribution and structural stability of the surface soil in a humid climate. Soil Tillage Res 23(4):361–372

Chan KY, Heenan DP, Oates A (2002) Soil carbon fractions and relationship to soil quality under different tillage and stubble management. Soil Tillage Res 63(3–4):133–139

Choudhary M, Sharma PC, Jat HS, McDonald A, Jat ML, Choudhary S, Garg N (2018) Soil biological properties and fungal diversity under conservation agriculture in Indo-Gangetic Plains of India. J Soil Sci Plant Nutr 18:1142–1156

Crittenden SJ, Eswaramurthy T, De Goede RG, Brussaard L, Pulleman MM (2014) Effect of tillage on earthworms over short-and medium-term in conventional and organic farming. Appl Soil Ecol 83:140–148

Curry JP, Good JA (1992) Soil faunal degradation and restoration. Soil Restoration 17:171–215

Dang YP, Seymour NP, Walker SR, Bell MJ, Freebairn DM (2015) Strategic tillage in no-till farming systems in Australia's northern grainsgrowing regions: I. Drivers and implementation. Soil Tillage Res 152:104–114. https://doi.org/10.1016/j.still.2015.03.0091

Das TK, Bhattacharyya R, Sudhishri S, Sharma AR, Saharawat YS, Bandyopadhyay KK, Sepat S, Bana RS, Aggarwal P, Sharma RK, Bhatia A, Singh G, Datta SP, Kar A, Singh B, Singh P, Pathak H, Vyas AK, Jat ML (2014) Conservation agriculture in an irrigated cotton-wheat system of the western Indo Gangetic Plains: crop & water productivity and economic profitability. Field Crop Res 158:24–33. https://doi.org/10.1016/j.fcr.2013.12.017

Das A, Layek J, Ramkrushna GI, Basavaraj S, Lal R, Krishnappa R, Yadav GS, Babu S, Ngachan SV (2020) Conservation tillage and residue management improves soil properties under upland rice–rapeseed system in subtropical Eastern Himalayas. Land Degrad Dev 31(14):775–1791. https://doi.org/10.1002/ldr.3568

Das TK, Bandyopadhyay KK, Ghosh PK (2021) Impact of conservation agriculture on soil health and crop productivity under irrigated ecosystems. Conservation agriculture: a sustainable approach for soil health and food security: Conserv Agric Sustain Agric:139–163

Das A, Babu S, Datta M, Kumar S, Singh R, Avasthe RK, Rathore SS, Yadav SK, Singh VK (2022a) Restoring soil carbon in marginal land of Indian Himalayas: impact of crop intensification and conservation tillage. J Environ Manag 318:115603. https://doi.org/10.1016/j.jenvman.2022.115603

Das A, Babu S, Singh R, Kumar S, Rathore SS, Singh VK, Datta M, Yadav SK, Wani OA, Yadav D (2022b) Impact of live mulch- based conservation tillage on soil properties and productivity of summer maize in Indian Himalayas. Sustainability 14:12078. https://doi.org/10.3390/su141912078

Dendooven L, Gutiérrez-Oliva VF, Patiño-Zúñiga L, Ramírez-Villanueva DA, Verhulst N, Luna-Guido M, Marsch R, Montes-Molina J, Gutiérrez-Miceli FA, Vásquez-Murrieta S, Govaerts B (2012) Greenhouse gas emissions under conservation agriculture compared to traditional cultivation of maize in the central highlands of Mexico. Sci Total Environ 431:237–244. https://doi.org/10.1016/j.scitotenv.2012.05.029

Dhiman SD, Kumar S, Om H (2003) Shallow tillage and drill technology for wheat. Indian Farm 53:10–13

Dick RP (1994) Soil enzyme activities as indicators of soil quality. In: Defining soil quality for a sustainable environment, vol 35, pp 107–124

Dolan MS, Clapp CE, Allmaras RR, Baker JM, Molina JAE (2006) Soil organic carbon and nitrogen in a Minnesota soil as related to tillage, residue and nitrogen management. Soil Tillage Res 89(2):221–231. https://doi.org/10.1016/j.still.2005.07.015

Duiker SW, Beegle DB (2006) Soil fertility distributions in long-term no-till, chisel/disk and mold board plow/disk systems. Soil Tillage Res 88:30–41. https://doi.org/10.1016/j.still.2005.04.004

Farooq M, Siddique KHM (2016) Research and developmental issues in dryland agriculture. In: Farooq M, Siddique K (eds) Innovations in dryland agriculture. Springer, Cham. https://doi.org/10.1007/978-3-319-47928-6_2

Fatumah N, Tilahun SA, Mohammed S (2021) Water use efficiency, grain yield, and economic benefits of common beans (Phaseolus vulgaris L.) under four soil tillage systems in Mukono District, Uganda. Heliyon 7(2):e06308. https://doi.org/10.1016/j.heliyon.2021.e06308

Fonteyne S, Flores García Á, Verhulst N (2021) Reduced water use in barley and maize production through conservation agriculture and drip irrigation. Front Sust Food Syst 5:734681. https://doi.org/10.3389/fsufs.2021.734681

Franzluebbers AJ, Hons FM (1996) Soil-profile distribution of primary and secondary plant available nutrients under conventional and no tillage. Soil Tillage Res 39:229–239. https://doi.org/10.1016/S0167-1987(96)01056-2

Fuentes M, Hidalgo C, Etchevers J, De León F, Guerrero A, Dendooven L, Verhulst N, Govaerts B (2012) Conservation agriculture, increased organic carbon in the top-soil macro-aggregates and reduced soil CO_2 emissions. Plant Soil 355(1):183–197. https://doi.org/10.1007/s11104-011-1092-4

Fuentes-Ponce MH, Gutiérrez-Díaz J, Flores-Macías A, González-Ortega E, Mendoza AP, Sánchez LMR, Novotny I, Espíndola IPM (2022) Direct and indirect greenhouse gas emissions under conventional, organic, and conservation agriculture. Agric Ecosyst Environ 340:1081–1048. https://doi.org/10.1016/j.agee.2022.108148

Gal A, Vyn TJ, Micheli E, Kladivko EJ, McFee WW (2007) Soil carbon and nitrogen accumulation with long-term no-till versus moldboard plowing overestimated with tilled-zone sampling depths. Soil Tillage Res 96(1–2):42–51. https://doi.org/10.1016/j.still.2007.02.007
Galani YJH, Ligowe IS, Kieffer M, Kamalongo D, Kambwiri AM, Kuwali P, Thierfelder C, Dougill AJ, Gong YY, Orfila C (2022) Conservation agriculture affects grain and nutrient yields of maize (Zea mays L.) and can impact food and nutrition security in Sub-Saharan Africa. Front Nutr 8:804663. https://doi.org/10.3389/fnut.2021.804663
Gathala MK, Timsina J, Islam S, Islam N, Rahman M, Ar H (2012) Effects of tillage and crop establishment methods on the system productivity profitability of rice-maize system in Bangladesh. In: 3rd international agronomy congress, 26–30 November 2012, New Delhi, vol I, pp 84–85
Gathala MK, Kumar V, Sharma PC, Saharawat YS, Jat HS, Singh M, Ladha JK (2013) Optimizing intensive cereal-based cropping systems addressing current and future drivers of agricultural change in the northwestern Indo-Gangetic Plains of India. Agric Ecosyst Environ 177:85–97. https://doi.org/10.1016/j.agee.2013.06.002
Ghosh BN, Dogra P, Sharma NK, Bhattacharyya R, Mishra PK (2015) Conservation agriculture impact for soil conservation in maize–wheat cropping system in the Indian sub-Himalayas. Int Soil Water Conserv Res 3:112–118. https://doi.org/10.1016/j.iswcr.2015.05.001
Ghosh S, Das TK, Rana KS, Biswas DR, Das DK, Singh G, Bhattacharyya R, Datta D, Rathi N, Bhatia A (2022a) Energy budgeting and carbon footprint of contrasting tillage and residue management scenarios in rice-wheat cropping system. Soil Tillage Res 223:105–445. https://doi.org/10.1016/j.still.2022.105445
Ghosh S, Das TK, Shivay YS, Bandyopadhyay KK, Sudhishri S, Bhatia A, Biswas DR, Yeasin M, Ghosh S (2022b) Weeds response and control efficiency, greengram productivity and resource-use efficiency under a conservation agriculture-based maize-wheat-greengram system. Indian J Weed Sci 3(1)
Gonzalez-Chávez MD, Aitkenhead-Peterson JA, Gentry TJ, Zuberer D, Hons F, Loeppert R (2010) Soil microbial community, C, N, and P responses to long-term tillage and crop rotation. Soil Tillage Res 106(2):285–293. https://doi.org/10.1016/j.still.2009.11.008
Govaerts B, Mezzalama M, Unno Y, Sayre KD, Luna-Guido M, Vanherck K, Dendooven L, Deckers J (2007) Influence of tillage, residue management, and crop rotation on soil microbial biomass and catabolic diversity. Appl Soil Ecol 37:18–30. https://doi.org/10.1016/j.apsoil.2007.03.006
Gura I, Mnkeni PNS (2019) Crop rotation and residue management effects under no till on the soil quality of a Haplic Cambisol in Alice, Eastern Cape, South Africa. Geoderma 337:927–934. https://doi.org/10.1016/j.geoderma.2018.10.042
Holland JM (2004) The environmental consequences of adopting conservation tillage in Europe: reviewing the evidence. Agric Ecosyst Environ 103(1):1–25. https://krishi.icar.gov.in/PDF/ICAR_Data_Use_Licence.pdf
Hussein MA, Antille DL, Kodur S, Chen G, Tullberg JN (2021) Controlled traffic farming effects on productivity of grain sorghum, rainfall and fertiliser nitrogen use efficiency. J Agric Food Res 3:100–111. https://doi.org/10.1016/j.jafr.2021.100111
Indoria AK, Rao CS, Sharma KL, Reddy KS (2017) Conservation agriculture—a panacea to improve soil physical health. Curr Sci:52–61. https://www.jstor.org/stable/24911616
Ismail I, Blevins RL, Frye WW (1994) Long-term no-tillage effects on soil properties and continuous corn yields. Soil Sci Soc Am J 58:193–198
Jat RA, Wani SP, Sahrawat KL (2012) Conservation agriculture in the semi-arid tropics: prospects and problems. Adv Agron 117:191–273. https://doi.org/10.1016/B978-0-12-394278-4.00004-0
Jat RK, Sapkota TB, Singh RG, Jat ML, Kumar M, Gupta RK (2014) Seven years of conservation agriculture in a rice–wheat rotation of Eastern Gangetic Plains of South Asia: yield trends and economic profit ability. Field Crop Res 164:199–210. https://doi.org/10.1016/j.fcr.2014.04.015
Jat RD, Jat HS, Nanwal RK, Yadav AK, Bana A, Choudhary KM, Kakraliya SK, Sutaliya JM, Sapkota TB, Jat ML (2018) Conservation agriculture and precision nutrient management practices in maize-wheat system: effects on crop and water productivity and economic profitability. Field Crop Res 222:111–120. https://doi.org/10.1016/j.fcr.2018.03.025

Jat HS, Kumar P, Sutaliya JM, Kumar S, Choudhary M, Singh Y, Jat ML (2019) Conservation agriculture based sustainable intensification of basmati rice-wheat system in North-West India. Arch Agron Soil Sci 65(10):1370–1386. https://doi.org/10.1080/03650340.2019.1566708

Jat ML, Chakraborty D, Ladha JK, Rana DS, Gathala MK, McDonald A, Gerard B (2020) Conservation agriculture for sustainable intensification in South Asia. Nat Sustain 3(4):336–343. https://doi.org/10.1038/s41893-020-0500-2

Jug D, Đurđević B, Birkás M, Brozović B, Lipiec J, Vukadinović V, Jug I (2019) Effect of conservation tillage on crop productivity and nitrogen use efficiency. Soil Tillage Res 194:104327. https://doi.org/10.1016/j.still.2019.104327

Kakraliya SK, Jat HS, Sapkota TB, Singh I, Kakraliya M, Gora MK, Sharma PC, Jat ML (2021) Effect of climate-smart agriculture practices on climate change adaptation, greenhouse gas mitigation and economic efficiency of rice-wheat system in India. Agriculture 11(12):12–69

Kandeler E, Deiglmayr K, Tscherko D, Bru D, Philippot L (2006) Abundance of narG, nirS, nirK, and nosZ genes of denitrifying bacteria during primary successions of a glacier foreland. Appl Environ Microbiol 72(9):5957–5962. https://doi.org/10.1128/AEM.00439-06

Kar SK, Kumar S, Sankar M, Patra S, Singh RM, Shrimali SS, Ojasvi PR (2022) Process-based modelling of soil erosion: scope and limitation in the Indian context. Curr Sci 122:533. https://doi.org/10.18520/cs/v122/i5/533-541

Kar SK, Patra S, Singh RM, Sankar M, Kumar S, Singh D, Singla S (2023) Impact of land use reformation on soil hydraulic properties and recovery potential of conservation tillage in India's North-West Himalayan region. Ecohydrol Hydrobiol

Kassam A, Friedrich T, Shaxson F, Pretty J (2009) The spread of conservation agriculture: justification, sustainability and uptake. Int J Agric Sustain 7(4):292–320. https://doi.org/10.3763/ijas.2009.0477

Kassam A, Derpsch R, Friedrich T (2014) Global achievements in soil and water conservation: the case of conservation agriculture. Int Soil Water Conserv Res 2(1):5–13. https://doi.org/10.1016/S2095-6339(15)30009-5

Kay BD, Van den Bygaart AJ (2002) Conservation tillage and depth stratification of porosity and soil organic matter. Soil Tillage Res 66(2):107–118. https://doi.org/10.1016/S0167-1987(02)00019-3

Kaye JP, Quemada M (2017) Using cover crops to mitigate and adapt to climate change. A review. Agron Sustain Dev 37(4):1–17. https://doi.org/10.1007/s13593-016-0410-x

Kladivko EJ (2001) Tillage systems and soil ecology. Soil Tillage Res 61(1-2):61–76. https://doi.org/10.1016/S0167-1987(01)00179-9

Kumar SS, Mir SA, Wani OA, Babu S, Yeasin M, Bhat MA, Hussain N, Wani AA, Kumar R, Devideen Y, Dar SR (2022a) Land-use systems regulate carbon geochemistry in the temperate Himalayas, India. J Environ Manag. https://doi.org/10.1016/j.jenvman.2022.115811

Kumar SS, Wani OA, Mir SA, Babu S, Sharma V, Chisti MH, Baba ZA, Sofi PA, Wani F, Dar SR, Singh R, Devideen Y (2022b) Soil carbon dynamics in the temperate Himalayas: impact of land use management. Front Environ Sci. https://doi.org/10.3389/fenvs.2022.1009660

Kumawat A, Yadav D, Srivastava P, Babu S, Kumar D, Singh D, Vishwakarma DK, Sharma VK, Madhu M (2023) Restoration of agroecosystems with conservation agriculture for food security to achieve sustainable development goals. Land Degrad Dev:1–19. https://doi.org/10.1002/ldr.4677

Lal R (2004) Soil carbon sequestration to mitigate climate change. Geoderma 123:1–22. https://doi.org/10.1016/j.geoderma.2004.01.032

Lal R (2015) The soil–peace nexus: our common future. Soil Sci Plant Nutr 61(4):566–578. https://doi.org/10.1080/00380768.2015.1065166

Laxmi V, Erenstein O, Gupta RK (2007) Assessing the impact of natural resource management research: the case of zero tillage in India's rice–wheat systems. In: The impact of natural resource management research: studies from the CGIAR. CAB International, Wallingford, pp 68–90

Li Y, Li Z, Cui S, Jagadamma S, Zhang Q (2019) Residue retention and minimum tillage improve physical environment of the soil in croplands: a global meta-analysis. Soil Tillage Res 194:104292. https://doi.org/10.1016/j.still.2019.06.009

Lichter K, Govaerts B, Six J, Sayre KD, Deckers J, Dendooven L (2008) Aggregation and C and N contents of soil organic matter fractions in a permanent raised-bed planting system in the Highlands of Central Mexico. Plant Soil 305:237–252. https://doi.org/10.1007/s11104-008-9557-9

Linquist BA, Van Groenigen KJ, Adviento-Borbe MA, Pittelkow C, Van Kessel C (2012) An agronomic assessment of greenhouse gas emissions from major cereal crops. Glob Chang Biol 18(1):194–209. https://doi.org/10.1111/j.1365-2486.2011.02502.x

Naorem A, Jayaraman S, Sinha NK, Mohanty M, Chaudhary RS, Hati KM, Lal R (2023) Eight-year impacts of conservation agriculture on soil quality, carbon storage, and carbon emission footprint. Soil Tillage Res 232:105748. https://doi.org/10.1016/j.still.2023.105748

Nath CP, Das TK, Rana KS, Bhattacharyya R, Pathak H, Paul S, Meena MC, Singh SB (2017) Greenhouse gases emission, soil organic carbon and wheat yield as affected by tillage systems and nitrogen management practices. Arch Agron Soil Sci 63(12):1644–1660. https://doi.org/10.1080/03650340.2017.1300657

Ngangom B, Das A, Lal R, Ramkrushna GI, Layek J, Basavaraj S, Subhash B, Yadav GS, Ghosh PK (2020) Double mulching improves soil properties and productivity of maize-based cropping system in eastern Indian Himalayas. Int Soil Water Conserv Res 8:308–320. https://doi.org/10.1016/j.iswcr.2020.07.001

Nisar S, Benbi DK, Toor AS (2021) Energy budgeting and carbon footprints of three tillage systems in maize-wheat sequence of north-western Indo-Gangetic Plains. Energy 229:120–661. https://doi.org/10.1016/j.energy.2021.120661

Nunes MR, van Es HM, Schindelbeck R, Ristow AJ, Ryan M (2018) No-till and cropping system diversification improve soil health and crop yield. Geoderma 328:30–43. https://doi.org/10.1016/j.geoderma.2018.04.031

Obour AK, Mikha MM, Holman JD, Stahlman PW (2017) Changes in soil surface chemistry after fifty years of tillage and nitrogen fertilization. Geoderma 308:46–53. https://doi.org/10.1016/j.geoderma.2017.08.020

Palm C, Blanco-Canqui H, DeClerck F, Gatere L, Grace P (2014) Conservation agriculture and ecosystem services: an overview. Agric Ecosyst Environ 187:87–105. https://doi.org/10.1016/j.agee.2013.10.010

Panwar AS, Shamim M, Babu S, Ravishankar N, Prusty AK, Alam NM, Singh DK, Bindhu JS, Kaur J, Dashora LN, Pasha LMD, Chaterjee S, Sanjay MT, Desai LJ (2019) Enhancement in productivity, nutrients use efficiency, and economics of rice-wheat cropping systems in India through Farmer's participatory approach. Sustainability 11:122. https://doi.org/10.3390/su11010122www.mdpi.com/journal/sustainability

Parihar CM, Jat SL, Singh AK, Kumar B, Pradhan S, Pooniya V, Dhauja A, Chaudhary V, Jat ML, Jat RK, Yadav OP (2016a) Conservation agriculture in irrigated intensive maize-based systems of north-west ern India: effects on crop yields, water productivity and economic profitability. Field Crop Res 193:104–116. https://doi.org/10.1016/j.fcr.2016.03.013

Parihar CM, Yadav MR, Jat SL, Singh AK, Kumar B, Pradhan S, Chakraborty D, Jat ML, Jat RK, Saharawat YS, Yadav OP (2016b) Long term effect of conservation agriculture in maize rotations on total organic carbon, physical and biological properties of a sandy loam soil in north-western Indo-Gangetic Plains. Soil Tillage Res 161:116–128. https://doi.org/10.1016/j.still.2016.04.001

Parihar CM, Yadav MR, Singh AK, Kumar B, Pooniya V, Pradhan S, Saharawat YS (2018) Long-term conservation agriculture and intensified cropping systems: effects on growth, yield, water, and energy-use efficiency of maize in northwestern India. Pedosphere 28(6):952–963. https://doi.org/10.1016/S1002-0160(17)60468-5

Paustian K, Six J, Elliott ET, Hunt HW (2000) Management options for reducing CO_2 emissions from agricultural soils. Biogeochemistry 48:147–163. https://doi.org/10.1023/A:1006271331703

Pheap S, Lefevre C, ThoumazeauA LV, BoulakiaS KR (2019) Multi-functional assessment of soil health under Conservation Agriculture in Cambodia. Soil Tillage Res 194:104349. https://doi.org/10.1016/j.still.2019.104349

Pinheiro EFM, Pereira MG, Anjos LHC (2004) Aggregate distribution and soil organic matter under different tillage systems for vegetable crops in a Red Latosol from Brazil. Soil Tillage Res 77(1):79–84. https://doi.org/10.1016/j.still.2003.11.005

Pittelkow CM, Liang X, Linquist BA, Van Groenigen KJ, Lee J, Lundy ME, Van Kessel C (2015) Productivity limits and potentials of the principles of conservation agriculture. Nature 517(7534):365–368. https://doi.org/10.1038/nature13809

Powlson DS, Gregory PJ, Whalley WR, Quinton JN, Hopkins DW, Whitmore AP, Goulding KW (2011) Soil management in relation to sustainable agriculture and ecosystem services. Food Policy 36:S72–S87

Powlson DS, Stirling CM, Thierfelder C, White RP, Jat ML (2016) Does conservation agriculture deliver climate change mitigation through soil carbon sequestration in tropical agro-ecosystems? Agric Ecosyst Environ 220:164–174. https://doi.org/10.1016/j.agee.2016.01.005

Pratibha G, Srinivas I, Rao KA, Raju BMK, Thyagaraj CR, Korwar GR, Venkateswarlu B, Shanker AK, Choudhary DK, Rao KS, Srinivasarao C (2015) Impact of conservation agriculture practices on energy use efficiency and global warming potential in rainfed pigeonpea–castor systems. Eur J Agron 66:30–40. https://doi.org/10.1016/j.eja.2015.02.001

Pratibha G, Srinivas I, Rao KV, Shanker AK, Raju BMK, Choudhary DK, Rao KS, Srinivasarao C, Maheswari M (2016) Net global warming potential and greenhouse gas intensity of conventional and conservation agriculture system in rainfed semi-arid tropics of India. Atmos Environ 145:239–250. https://doi.org/10.1016/j.atmosenv.2016.09.039

Qin RJ, Stamp P, Richner W (2004) Impact of tillage on root systems of winter wheat. Agric J 96:1523–1530

Rahman MM, Aravindakshan S, Hoque MA, Rahman MA, Gulandaz MA, Rahman J, Islam MT (2021) Conservation tillage (CT) for climate-smart sustainable intensification: assessing the impact of CT on soil organic carbon accumulation, greenhouse gas emission and water footprint of wheat cultivation in Bangladesh. Environ Sustain Ind 10:100–106. https://doi.org/10.1016/j.indic.2021.100106

Raj R, Kumar A, Solanki IS, Dhar S, Dass A, Gupta AK, Kumar V, Singh CB, Jat RK, Pandey UC (2017) Influence of crop establishment methods on yield, economics, and water productivity of rice cultivars under upland and lowland production ecologies of Eastern Indo-Gangetic Plains. Paddy Water Environ 15:861–877. https://doi.org/10.1007/s10333-017-0598-7

Raj R, Das TK, Banerjee T, Ghosh A, Bhattacharyya R, Chakraborty D, Prasad S, Babu S, Kumar V, Sen S, Ghosh S (2022) Co-implementation of conservation tillage and herbicides reduces weed and nematode infestation and enhances the productivity of direct-seeded rice in North-western Indo-Gangetic Plains. Front Sustain Food Syst 6:1017013. https://doi.org/10.3389/fsufs.2022.1017013

Raj R, Das TK, Chakraborty D, Bhattacharyya R, Babu S, Govindasamy P, Kumar V, Ekka U, Sen S, Ghosh S, Roy A (2023) Soil physical environment and active carbon pool in rice–wheat system of South Asia: impact of long-term conservation agriculture practices. Environ Technol Innov 29:102966. https://doi.org/10.1016/j.eti.2022.102966

Rajanna GA, Dass A, Suman A, Babu S, Venkatesh P, Singh VK, Upadhyay PK, Sudhishri S (2022) Co-implementation of tillage, irrigation, and fertilizers in soybean: impact on crop productivity, soil moisture, and soil microbial dynamics. Field Crop Res 288:108672. https://doi.org/10.1016/j.fcr.2022.108672

Randall G, Iragavarapu T (1995) Impact of long-term tillage systems for continuous corn on nitrate leaching to tile drainage. J Environ Qual 24:360–366. https://doi.org/10.2134/jeq1995.00472425002400020020x

Rao KK, Samal SK, Poonia SP, Kumar R, Mishra JS, Bhatt BP, Dwivedi SK, Mondal S, Choubey AK, Kumar S, Kumar M (2021) Conservation agriculture improves soil physical properties and crop productivity: a long-term study in middle Indo-Gangetic Plains of India. Soil Res 60(6):442–454. https://doi.org/10.1071/SR20329

Rathore SS, Babu S, Shekhawat K, Singh R, Yadav SK, Singh VK, Singh C (2022) Designing energy cum carbon-efficient environmentally clean production system for achieving green economy in agriculture. Sustain Energy Technol Assess. https://doi.org/10.1016/j.seta.2022.102190

Rhodes CJ (2014) Soil erosion, climate change and global food security: challenges and strategies. Sci Prog 97(2):97–153

Richards M, Sapkota T, Stirling C, Thierfelder C, Verhulst N, Friedrich T, Kienzle J (2014) Practice brief—climate smart agriculture. https://doi.org/10.1016/S2095-6339(15):30053-8

Roy D, Datta A, Jat HS, Choudhary M, Sharma PC, Singh PK, Jat ML (2022) Impact of long term conservation agriculture on soil quality under cereal based systems of North West India. Geoderma 405:115391. https://doi.org/10.1016/j.geoderma.2021.115391

Rusinamhodzi L, Corbeels M, van Wijk MT, Rufino MC, Nyaymangara J, Giller KE (2011) A meta-analysis of long-term effects of conservation agriculture on maize grain yield under rain-fed conditions. Agron Sustain Dev 31:657–673. https://doi.org/10.1007/s13593-011-0040-2

Saha S, Yadav D, Layak J, Babu S, Ansari MA, Das B, Choudhary S, Singh R, Rajkhowa DJ, Ghosh PK, Das A (2021) Potential of conservation agriculture in hill ecosystems: impact on resource sustainability. Indian J Agron 66:128–141

Sapkota TB, Jat ML, Shankar V, Singh LK, Rai M, Grewal MS, Stirling CM (2015) Tillage, residue and nitrogen management effects on methane and nitrous oxide emission from rice–wheat system of Indian Northwest Indo-Gangetic Plains. J Integr Environ Sci 12(1):31–46. https://doi.org/10.1080/1943815X.2015.1110181

Shao Y, Xie Y, Wang C, Yue J, Yao Y, Li X, Liu W, Zhu Y, Guo T (2016) Effects of different soil conservation tillage approaches on soil nutrients, water use and wheat-maize yield in rainfed dry-land regions of North China. Eur J Agron 81:37–45. https://doi.org/10.1016/j.eja.2016.08.014

Sharma AR, Singh VP (2014) Integrated weed management in conservation agriculture systems. Indian J Weed Sci 46(1):23–30

Sharma NK, Ghosh BN, Khola OPS, Dubey RK (2013) Residue and tillage management for soil moisture conservation in post maize harvesting period under rainfed conditions of North–West Himalayas. Indian J Soil Conserv 41:287–292

Sharma T, Das TK, Maity PP, Biswas S, Sudhishri S, Govindasamy P, Raj R, Sen S, Singh T, Paul AK, Roy A (2023) Long-term conservation agriculture influences weed diversity, water productivity, grain yield, and energy budgeting of wheat in north-western Indo-Gangetic plains. Sustain 15(9):7290. https://doi.org/10.3390/su15097290

Singh K, Tewari SK (2022) Does the road to land degradation neutrality in India is paved with restoration science? Restor Ecol 30(5):e13585

Singh R, Babu JN, Kumar R, Srivastava P, Singh P, Raghubanshi AS (2015) Multifaceted application of crop residue biochar as a tool for sustainable agriculture: an ecological perspective. Ecol Eng 77:324–347. https://doi.org/10.1016/j.ecoleng.2015.01.011

Singh R, Babu S, Avasthe RK, Meena RS, Yadav GS, Das A, Mohapatra KP, Rathore SS, Kumar A, Singh C (2021a) Conservation tillage and organic nutrients management improve soil properties, productivity, and economics of the maize-vegetable pea system in the Eastern Himalayas. Land Degrad Dev. https://doi.org/10.1002/ldr.4066

Singh R, Babu S, Avasthe RK, Yadav GS, Das A, Mohapatra KP, Kumar A, Singh VK, Chandra P (2021b) Crop productivity, soil health, and energy dynamics of Indian Himalayan intensified organic maize-based systems. Int Soil Water Conserv Res 9(2):260–270. https://doi.org/10.1016/j.iswcr.2020.11.003

Singh D, Mishra AK, Patra S, Mariappan S, Singh N, Kar SK (2022a) Spatial variability of soil hydraulic and physical properties in erosive sloping agricultural fields. Water Sci Eng 16:57–66. https://doi.org/10.1016/j.wse.2022.10.001

Singh D, Patra S, Mishra AK, Mariappan S, Singh N (2022b) Temporal variation of saturated and near-saturated soil hydraulic conductivity and water-conducting macroporosity in a maize-wheat rotation under conventional and conservation tillage practices. Land Degrad Dev 33(13):2208–2219. https://doi.org/10.1002/ldr.4251

Singh R, Kumar A, Babu S, Avasthe R, Das A, Rathore SS, Kumar S, Singh C, Sharma V, Bhupenchandra I (2023) Development of organic nutrients management system for profitable and soil-supportive French bean (Phaseolus vulgaris L.) farming in North Eastern Himalayas, India. Front Sustain Food Syst 7:1115521. https://doi.org/10.3389/fsufs.2023.1115521

Sithole NJ, Magwaza LS (2019) Long-term changes of soil chemical characteristics and maize yield in no-till conservation agriculture in a semi-arid environment of South Africa. Soil Tillage Res 194:147–156. https://doi.org/10.1016/j.still.2019.104317

Six J, Conant RT, Paul EA, Paustian K (2002) Stabilization mechanisms of soil organic matter: implications for C-saturation of soils. Plant Soil 241:155–176. https://doi.org/10.1023/A:1016125726789

Smith P, Calvin K, Nkem J, Campbell D, Cherubini F, Grassi G, Korotkov V, Le Hoang A, Lwasa S, McElwee P, Nkonya E (2020) Which practices co-deliver food security, climate change mitigation and adaptation, and combat land degradation and desertification? Glob Chang Biol 26(3):1532–1575. https://doi.org/10.1111/gcb.14878

Soane BD, Ball BC, Arvidsson J, Basch G, Moreno F, Roger-Estrade J (2012) No-till in northern, western and south-western Europe: a review of problems and opportunities for crop production and the environment. Soil Tillage Res 118:66–87. https://doi.org/10.1016/j.still.2011.10.015

Somasundaram J, Salikram M, Sinha NK, Mohanty M, Chaudhary RS, Dalal RC, Mitra NG, Blaise D, Coumar MV, Hati KM, Thakur JK (2019) Corrigendum to: conservation agriculture effects on soil properties and crop productivity in a semiarid region of India. Soil Res 57(2):200. https://doi.org/10.1071/SR18145_CO

Spurgeon DJ, Keith AM, Schmidt O, Lammertsma DR, Faber JH (2013) Land-use and land-management change: relationships with earthworm and fungi communities and soil structural properties. BMC Ecol 13:1–3. https://doi.org/10.1186/1472-6785-13-46

Stagnari F, Pagnani G, Galieni A, D'Egidio S, Matteucci F, Pisante M (2020) Effects of conservation agriculture practices on soil quality indicators: a case-study in a wheat-based cropping systems of Mediterranean areas. Soil Sci Plant Nutr 66(4):624–635. https://doi.org/10.1080/00380768.2020.1779571

Stavi I, Lal R (2013) Agriculture and greenhouse gases, a common tragedy. A review. Agron Sustain Dev 33:275–289. https://doi.org/10.1007/s13593-012-0110-0

Stavi I, Lal R (2015) Achieving zero net land degradation: challenges and opportunities. J Arid Environ 112:44–51

Stavi I, Lal R, Jones S, Reeder RC (2012) Implications of cover crops for soil quality and geo diversity in a humid-temperate region in the mid-Western USA. Land Degrad Dev 23:322–330. https://doi.org/10.1002/ldr.2148

Steward PR, Dougill AJ, Thierfelder C, Pittelkow CM, Stringer LC, Kudzala M, Shackelford GE (2018) The adaptive capacity of maize-based conservation agriculture systems to climate stress in tropical and subtropical environments: a meta-regression of yields. Agric Ecosyst Environ 251:194–202. https://doi.org/10.1016/j.agee.2017.09.019

Sun L, Wang S, Zhang Y, Li J, Wang X, Wang R, Lyu W, Chen N, Wang Q (2018) Conservation agriculture based on crop rotation and tillage in the semi-arid Loess Plateau, China: effects on crop yield and soil water use. Agric Ecosyst Environ 251:67–77. https://doi.org/10.1016/j.agee.2017.09.011

Suraj, Behera UK (2014) Conservation agriculture in India—problems, prospects and policy issues. Int Soil Water Conserv Res 2(4):1–12. https://doi.org/10.1016/S2095-6339(15)30053-8. Elsevier Masson SAS

Swanepoel CM, Rötter RP, Van der Laan M, Annandale JG, Beukes DJ, du Preez CC, Swanepoel LH, van der Merwe A, Hoffmann MP (2018) The benefits of conservation agriculture on soil organic carbon and yield in southern Africa are site-specific. Soil Tillage Res 183:72–82. https://doi.org/10.1016/j.still.2018.05.016

Tabaglio V, Gavazzi C, Menta C (2009) Physico-chemical indicators and microarthropod communities as influenced by no-till, conventional tillage and nitrogen fertilisation after four years of continuous maize. Soil Tillage Res 105(1):135–142. https://doi.org/10.1016/j.still.2009.06.006

Tanveer SK, Wen X, Lu XL, Zhang J, Liao Y (2013) Tillage, Mulch and N fertilizer affect emissions of CO_2 under the rain fed condition. PLoS ONE 8(9):e72140. https://doi.org/10.1371/journal.pone.0072140

Tekle AT (2016) Adaptation and constraints of conservation agriculture. J Biol Agric Healthc 6(1):1–14

Tellez-Rio A, Vallejo A, Garcia-Marco S, Martin-Lammerding D, Tenorio JL, Rees RM, Guardia G (2017) Conservation agriculture practices reduce the global warming potential of rainfed low N input semi-arid agriculture. Eur J Agron 84:95–104. https://doi.org/10.1016/j.eja.2016.12.013
Thierfelder C, Matemba-Mutasa R, Rusinamhodzi L (2015) Yield response of maize (Zea mays L.) to conservation agriculture cropping system in Southern Africa. Soil Tillage Res 146:230–242. https://doi.org/10.1016/j.still.2014.10.015
Thierfelder C, Baudron F, Setimela P, Nyagumbo I, Mupangwa W, Mhlanga B, Lee N, Gérard B (2018) Complementary practices supporting conservation agriculture in southern Africa-a review. Agron Sustain Dev 38(2):16. https://doi.org/10.1007/s13593-018-0492-8
Thomas GA, Dalal RC, Standley J (2007) No-till effects on organic matter, pH, cation exchange capacity and nutrient distribution in a Luvisol in the semi-arid subtropics. Soil Tillage Res 94(2):295–304. https://doi.org/10.1016/j.still.2006.08.005
Tillman D, Balzer C, Hill J, Befort BL (2011) Global food demand and the sustainable intensification of agriculture. PNAS 108(50):20260–20265
Tirol-Padre A, Rai M, Kumar V, Gathala M, Sharma PC, Sharma S, Nagar RK, Deshwal S, Singh LK, Jat HS, Sharma DK (2016) Quantifying changes to the global warming potential of rice wheat systems with the adoption of conservation agriculture in northwestern India. Agric Ecosyst Environ 219:125–137. https://doi.org/10.1016/j.agee.2015.12.020
Tomar RK, Singh D, Gangwar KS, Garg RN, Chakraborty GVK, Sahoo D, Rajeev RNR, Chakravarty NVK (2009) Effect of tillage systems and irrigation schedules on soil cracking pattern, water requirement and performance of rice-wheat cropping system in inceptisols in semi-arid regions. J Soil Water Conserv 8:26–33
Tripathi R, Nayak AK, Bhattacharyya P, Shukla AK, Shahid M, Raja R, Panda BB, Mohanty S, Kumar A, Thilagam AK (2014) Soil aggregation and distribution of carbon and nitrogen in different fractions after 41 years long term fertilizer experiment in tropical rice–rice system. Geoderma 213:280–286. https://doi.org/10.1016/j.geoderma.2013.08.031
Turmel MS, Speratti A, Baudron F, Verhulst N, Govaerts B (2015) Crop residue management and soil health: a systems analysis. Agric Syst 134:6–16. https://doi.org/10.1016/j.agsy.2014.05.009
UNCCD (2012) Zero net land degradation, a Sustainable Development Goal for Rioþ20. http://www.unccd.int/Lists/SiteDocumentLibrary/Rioþ20/UNCCDPolicyBrief_pdf
UNCCD (2018) Default data: methods and interpretation. A guidance document for the 2018 UNCCD reporting. https://prais.unccd.int/sites/default/files/helper_documents/pdf
Van Kessel C, Venterea R, Six J, Adviento-Borbe MA, Linquist B, Groenigen KJ (2013) Climate, duration, and N placement determine N_2O emissions in reduced tillage systems: a meta-analysis. Glob Chang Biol 19(1):33–44
Verhulst N, Govaerts B, Verachtert E, Castellanos-Navarrete A, Mezzalama M, Wall PC, Chocobar A, Deckers J, Sayre KD (2010) Conservation agriculture, improving soil quality for sustainable production systems? In: Lal R, Stewart BA (eds) Food security and soil quality. Advances in soil science. CRC Press, Boca Raton
Vetsch JA, Randall GW, Lamb JA (2007) Corn and soybean production as affected by tillage systems. Agronomy 99(4):952–959
Vieira FCB, Bayer C, Zanatta J, Ernani PR (2009) Organic matter kept Al toxicity low in a subtropical no-tillage soil under long-term (21-year) legume-based crop systems and N fertilisation. Aust J Soil Res 47:707–714. https://doi.org/10.1071/SR08273
Wang Y, Xu J, Shen J, Luo Y, Scheu S, Ke X (2010) Tillage, residue burning and crop rotation alter soil fungal community and water-stable aggregation in arable fields. Soil Tillage Res 107(2):71–79. https://doi.org/10.1016/j.still.2010.02.008
Wang Z, Chen Q, Liu L, Wen X, Liao Y (2016) Responses of soil fungi to 5-year conservation tillage treatments in the drylands of northern China. Appl Soil Ecol 101:132–140. https://doi.org/10.1016/j.apsoil.2016.02.002
Wang C, Zhao J, Feng Y, Shang M, Bo X, Gao Z, Chen F, Chu Q (2021) Optimizing tillage method and irrigation schedule for greenhouse gas mitigation, yield improvement, and water conservation in wheat–maize cropping systems. Agric Water Manag 248:106762. https://doi.org/10.1016/j.agwat.2021.106762

Wani OA, Kumar S, Hussain N, Wani AA, Babu S, Alam P, Rashid M, Popescu SM, Mansoor S (2022) Multi-scale processes influencing global carbon storage and land-carbon-climate nexus: a critical review. Pedosphere. https://doi.org/10.1016/j.pedsph.2022.07.002

Wani OA, Sharma V, Kumar SS, Babu S, Sharma KR, Rathore SS, Marwaha S, Ganai NA, Dar SR, Yeasin M, Raghavendra S, Tomar J (2023) Climate plays a dominant role over land management in governing soil carbon dynamics in North Western Himalayas. J Environ Manag. https://doi.org/10.1016/j.jenvman.2023.117740

Wardle DA (1995) Impacts of disturbance on detritus food webs in agro-ecosystems of contrasting tillage and weed management practices. In: Advances in ecological research, vol 26. Academic, pp 105–185

West TO, Post WM (2002) Soil organic carbon sequestration rates by tillage and crop rotation. Soil Sci Soc Am J 66:1930–1946

Xu J, Han H, Ning T, Li Z, Lal R (2019) Long-term effects of tillage and straw management on soil organic carbon, crop yield, and yield stability in a wheat-maize system. Field Crop Res 233:33–40. https://doi.org/10.1016/j.fcr.2018.12.016

Yadav GS, Das A, Lal R, Babu S, Datta M, Meena RS, Patil SB, Singh R (2019) Impact of no-till and mulching on soil carbon sequestration under rice (*Oryza Sativa* L.)—rapeseed (*Brassica campestris* L. Var. rapeseed) cropping system in hilly agro-ecosystem of the eastern Himalayas, India. Agric Ecosyst Environ 275:81–92. https://doi.org/10.1016/j.agee.2019.02.001

Yadav GS, Das A, Babu S, Mohapatra KP, Lal R, Rajkhowa DJ (2020a) Potential of conservation tillage and altered land configuration to improve soil properties, carbon sequestration and productivity of maize based cropping system in eastern Himalayas, India. Int Soil Water Conserv Res. https://doi.org/10.1016/j.iswcr.2020.12.003

Yadav GS, Babu S, Das A, Mohapatra KP, Singh R, Avasthe RK, Roy S (2020b) No-till and mulching enhance energy use efficiency and reduce carbon footprint of a direct-seeded upland rice production system. J Clean Prod 271:122700. https://doi.org/10.1016/j.jclepro.2020.122700

Yadav GS, Kandpal BK, Das A, Babu S, Mohapatra KP, Devi AG, Devi HL, Chandra P, Singh R, Barman KK (2021a) Impact of 28-year-old agroforestry systems on soil carbon dynamics in Eastern Himalayas. J Environ Manag. https://doi.org/10.1016/j.jenvman.2021.111978

Yadav GS, Babu S, Das A, Datta M, Mohapatra KP, Singh R, Singh VK, Rathore SS, Chakraborty M (2021b) Productivity, soil health, and carbon management index of Indian Himalayan intensified maize-based cropping systems under live mulch-based conservation tillage practices. Field Crop Res. https://doi.org/10.1016/j.fcr.2021.108080

Yadav D, Singh D, Babu S, Madegowda M, Singh D, Mandal D, Rathore AC, Sharma VK, Singhal V, Kumawat A, Yadav DK (2023) Intensified cropping reduces soil erosion and improves rainfall partitioning and soil properties in the marginal land of the Indian Himalayas. Int Soil Water Conserv Res. https://doi.org/10.1016/j.iswcr.2023.10.002

Yang A, Hu J, Lin X, Zhu A, Wang J, Dai J, Wong MH (2012) Arbuscular mycorrhizal fungal community structure and diversity in response to 3-year conservation tillage management in a sandy loam soil in North China. J Soils Sediments 12:835–843. https://doi.org/10.1007/s11368-012-0518-9

Zhang HL, Lal R, Zhao X (2014) Opportunities and challenges of soil carbon sequestration by conservation agriculture in China. Adv Agron 124:1–36. https://doi.org/10.1016/B978-0-12-800138-7.00001-2

Zhang X, Xin X, Zhu A, Yang W, Zhang J, Ding S, Mu L, Shao L (2018) Linking macroaggregation to soil microbial community and organic carbon accumulation under different tillage and residue managements. Soil Tillage Res 178:99–107. https://doi.org/10.1016/j.still.2017.12.020

Zibilske LM, Bradford JM, Smart JR (2002) Conservation tillage induced changes in organic carbon, total nitrogen and available phosphorus in a semi-arid alkaline subtropical soil. Soil Tillage Res 66(2):153–163. https://doi.org/10.1016/S0167-1987(02)00023-5

Regenerative Agricultural Practices for Sustainable Soil Health and Food Production

5

Shashi S. Udumann, Nuwandhya S. Dissanayaka, Tharindu D. Nuwarapaksha, and Anjana J. Atapattu

Abstract

Regenerative agriculture (RA) is a combined management of the soil and its health for human food production. RA practices minimizing the soil disturbance, maximizing crop diversity, maintaining soil cover, integrating livestock, and enhancing biodiversity. The best management practices are no-till farming, cover cropping, and holistic grazing management alongside their proper application and precise effectiveness for enhancing sustainability of the ecosystems. Many innovative practices of RA resulting in improved soil health, enhanced carbon sequestration, nutrient cycling, and greater resilience to climate stresses. RA related associated problems that users have to face, namely, initial costs, knowledge requirements, and lack of policy support as the critical factors in the adoption. The future directions and research needs, emphasizing the importance of scaling up practices, developing supportive policies, advancing technologies, and improving education in the field. Thus, the RA can reshape the global food systems by making them productive, resilient, and sustainable in the face of increasing challenges.

Keywords

Carbon sequestration · Cover cropping · Soil health · Sustainable agriculture

S. S. Udumann · N. S. Dissanayaka · T. D. Nuwarapaksha · A. J. Atapattu (✉)
Agronomy Division, Coconut Research Institute, Lunuwila, Sri Lanka

© The Author(s), under exclusive license to Springer Nature Singapore Pte Ltd. 2025

S. Babu et al. (eds.), *Ecological Solutions to Agricultural Land Degradation*,
https://doi.org/10.1007/978-981-96-3392-0_5

5.1 Introduction

Regenerative agriculture has come up as one of the feasible solutions that can reverse some of the effects sparked by soil degradation, climate change, and food insecurity. As a comprehensive farming approach, this system seeks to restore and indeed lift the total environment of the farm with an emphasis on soil health (Muhie 2022; Udumann et al. 2024a, b, c). Based on the given criteria, this is described as a farming system that restores soil health, enhances micro-climate, provides food security, and restores ecosystem services. It transcends sustainability to a level where it tries to enhance resources while using them instead of merely preserving them (Rhodes 2017). It is therefore, imperative to have some of the best policies, practices, and strategies in place that will ensure healthy soil for food production in the future. Soil health is fundamental to maintain efficient and productive food production systems, as soil serves as the primary source of nutrients, water, and microbial ecosystems (Lehman et al. 2015). The increasing global population and climate change are exerting unprecedented pressure on our limited natural resources (Chew et al. 2024). In this context, sustaining and enhancing soil fertility is crucial for ensuring long-term food security and agricultural sustainability (Atapattu et al. 2024a, b; Lal 2020).

Various issues and factors to be taken into consideration once there is a shift or plan to move toward the practice of regenerative agriculture. The emerging challenges in modern agriculture, including climate change impacts, water scarcity, and biodiversity loss, and how regenerative practices can help to mitigate these issues. For instance, cover crops can enhance soil water retention in drought-prone areas, and diverse crop rotations can improve pest resilience in the face of changing climate patterns. Furthermore, the various issues and factors to be considered when shifting towards regenerative agriculture, including economic considerations, knowledge gaps, and policy support needs. An exploration of future research directions, potential policy initiatives, and the potential of regenerative agriculture practices to enhance global food system sustainability will give the answer to feed the burgeoning population. A comprehensive understanding of regenerative agricultural practices, their benefits for soil health and food production, and how they address current and future challenges in agriculture is being given. This knowledge will be crucial for developing resilient and sustainable agri-food systems in the face of increasing environmental and socio-economic pressures.

5.2 Principles of Regenerative Agriculture

Regenerative agriculture encompasses various sustainable farming approaches, including organic farming, natural farming, and integrated farming systems. These systems share common principles aimed at improving soil health, enhancing biodiversity, and promoting sustainable food production (Lal 2020). Organic farming focuses on producing food without synthetic inputs, emphasizing soil health and ecological balance (Reganold and Wachter 2016). Natural farming, developed by

Masanobu Fukuoka, promotes minimal intervention and working with natural processes (Fukuda 2018). Integrated farming systems combine different farm enterprises for optimal resource utilization and sustainability (Bhagat et al. 2024). The principles discussed in this section (Fig. 5.1) reflect the core aspects of these systems, demonstrating how regenerative agriculture integrates and builds upon established sustainable farming practices.

5.2.1 Minimizing Soil Disturbance

Soil disturbance as much as possible has to be reduced in regenerative agriculture as this will improve the properties of the soil. Conventional tillage methods disintegrate the soil structure, decrease organic matter, and negatively affect microorganisms, reducing fertility and increasing erosion rates (Nunes et al. 2020; Rempelos et al. 2023). It does not involve mechanical inversion of the soil; it checks the disintegration of soil aggregates; has a way of improving diversification; and helps to increase carbon stocks. Out of the two, crop residues help prevent soil erosion, and they also offer homes for soil microorganisms (Somasundaram et al. 2020). No-till methods can increase soil organic carbon (SOC) by 9.9%, and together with crop residue return and rotation, can raise SOC to 12.77%, improving soil quality and addressing food insecurity and adverse climatic conditions (Cui et al. 2024). No-till systems also increase microbial biomass nitrogen and microbial biomass carbon, leading to improved nutrient cycling (Nunes et al. 2020).

Regenerative agriculture practices significantly reduce soil losses compared to conventional tillage systems. A meta-analysis found reductions in soil erosion due to no-till practices compared to conventional tillage, with reductions ranging from

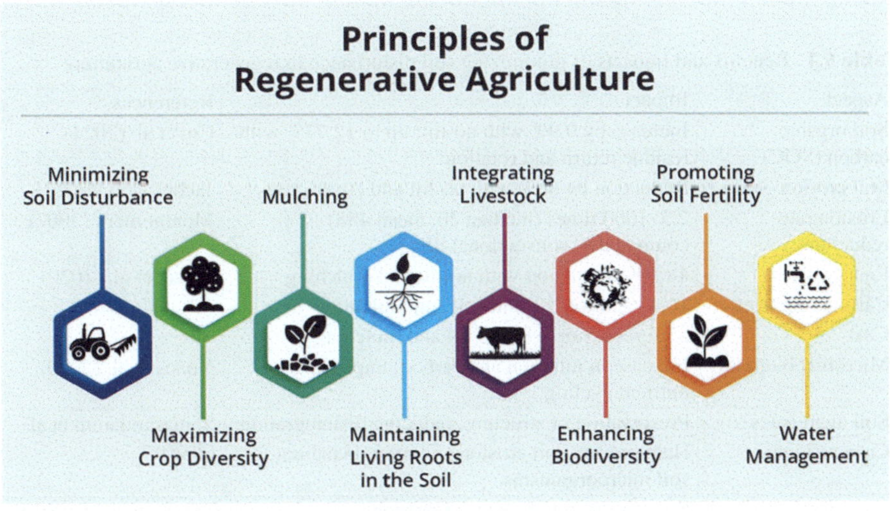

Fig. 5.1 Principles of regenerative agriculture

2.5 to around 1000 times, with median and mean values of 20 and 488 times, respectively. These reductions are significant and bring agricultural erosion rates more in line with rates of soil production (Montgomery 2007). In a long-term study reported that no-till practices reduced soil erosion and no-till practices are crucial for sustaining long-term crop productivity by reducing erosion and are commonly adopted to achieve agronomic and environmental goals (Pittelkow et al. 2015). Another study found that soil loss decreased by 86% in the conservation system with no-till and cover crops compared to the conventional system, highlighting its potential to protect against accelerated soil erosion, particularly under climate change-induced weather extremes (Jacobs et al. 2022). A 4-year study on no-till with mulching for upland rice showed a 48.50% reduction in energy use, a 16.48% lower carbon footprint, and a 35% cost reduction while improving efficiency and profitability (Yadav et al. 2020). Changing from conventional tillage to no-till could imply new practices of weed management and decrease the production temporarily (Pittelkow et al. 2015). However, despite these challenges, there are many long-term advantages surrounding the improvement of soil health and sustainability as well as general farm economics. The benefits and impacts of minimizing soil disturbance in regenerative agriculture are summarized in Table 5.1.

5.2.2 Maximizing Crop Diversity

Crop diversification is one of the principal aspects of regenerative agriculture that works in favour of diversification to develop strong systems for operations in farming. Ecosystem, the state of the soil quality is enhanced, the functional capabilities of an ecosystem are increased, as well as crop and food production rates are increased by crop diversification even in areas of monoculture (Dissanayaka et al.

Table 5.1 Benefits and impacts of minimizing soil disturbance in regenerative agriculture

Aspect	Impact	References
Soil organic carbon (SOC)	Increase by 9.9% with no-till; up to 12.77% with residue return and rotation	Cui et al. (2024)
Soil erosion	Reduction by 86% with no-till and cover crops	Jacobs et al. (2022)
Erosion rate reduction	2.5–1000 times (median 20, mean 488) compared to conventional tillage	Montgomery (2007)
Energy use	48.50% reduction with no-till and mulching	Yadav et al. (2020)
Carbon footprint	16.48% lower with no-till and mulching	
Cost	35% reduction with no-till and mulching	
Microbial biomass	Increase in nitrogen and carbon, improving nutrient cycling	Nunes et al. (2020)
Soil aggregates	Preservation of structure, reducing disintegration	Somasundaram et al. (2020)
Crop residues	Help prevent soil erosion and provide habitat for soil microorganisms	
Long-term productivity	Sustained through reduced erosion	Pittelkow et al. (2015)

2024a, b; Nuwarapaksha et al. 2024). Crop diversification helps in improving the structure of soil, organic content, and quantity and quality of microorganisms hence cycling nutrients (Bender et al. 2016). Diversified crops also minimize pests and diseases as they interrupt pest life cycles and support the places for beneficial insects (Atapattu and Udumann 2024). Studies reveal that diversified cropping systems are less affected by pests compared to monoculture systems (Lin 2011). Some of the best practices to diversify the crops include intercropping, cover cropping, crop rotation, and integrating perennials. Intercropping, which is growing more than one crop at the same time on the same land, enhances productivity and land utilization (Udumann et al. 2024a, b, c; Yu et al. 2015). A meta-analysis found that intercropping systems increased land equivalent ratios by 22% on average, indicating higher productivity compared to monocultures (Yu et al. 2015). The use of cover crops during the fallow maintains the spread of soil and increases the content of organic carbon (Blanco-Canqui et al. 2015). Another study reported that cover crops increased subsequent cash crop yields by an average of 3.4%, with corn yields increasing by 4% (Marcillo and Miguez 2017). Often, planting and managing different crops in a single area in a successive pattern is termed crop rotation, and this type of farming is much better as compared to monoculture (Bowles et al. 2020). A long-term study found that diversified crop rotations increased maize yields by 7% and soybean yields by 22% compared to simple maize-soybean rotations (Smith et al. 2008). Integrating annuals in perennial systems improves the characteristics of the soil, the levels of carbon, and the organisms in the rhizosphere (Atapattu et al. 2017; Crews et al. 2016)

In terms of system resilience, diversified cropping systems have shown significant advantages. A study reported that more diverse rotations increased corn and soybean yields by 7% and 22% respectively during drought years (Gaudin et al. 2015). Additionally, a meta-analysis found that diversification practices increased crop production by 14% on average while also enhancing ecosystem services provision by 41% (Tamburini et al. 2020). Agroforestry systems, which integrate trees with crops or livestock, reported yield improvements of up to 56% compared to monoculture systems, along with enhanced biodiversity and climate resilience (Torralba et al. 2016). On the economic side, operational risks and market dependence are mitigated, and costs can be economized through the gain of net revenues by increasing the degree of operating as a scope economy (Davis et al. 2012; Duru et al. 2015). However, there is the problem of poor management when it comes to implementing crop diversification especially because of the market constraints. Nevertheless, crop rotation and crop diversification remain important aspects of sustainable farming practices that will in the long run improve and sustain soil and ecosystems' health as well as the profitability of farming (Atapattu et al. 2017).

5.2.3 Mulching

This practice involves maintaining the soil cover by different means like cover crop, residue etc. all through the year, it offers many advantages. Soil cover prevents the soil from being washed away or blown away by the wind because it acts as a barrier (Blanco-Canqui et al. 2015). It also helps regulate the soil temperature, benefits the soil creatures and root systems of the crops (Kader et al. 2017). Also, covered soil reduces water evaporation compared to exposed soils; this is important for areas that are usually faced with water shortages (Ranaivoson et al. 2017). Recent studies by Indian researchers have further reinforced these benefits. In the semi-arid regions of India, conservation agriculture with residue retention improved soil moisture by 17–30% compared to conventional tillage (Jat et al. 2020). Additionally, a long-term study in northwestern India assessed the impact of tillage practices and maize-based crop rotations on soil health. Conservation agriculture (CA) techniques like zero tillage (ZT) and permanent raised beds (PB) enhanced soil properties compared to conventional tillage (CT). ZT and PB improved water-stable aggregates (16.1–32.5%), reduced bulk density (11.0–14.3%), and increased soil organic carbon by 34.6–5.3% (0–15 cm depth) and 23.6–26.5% (15–30 cm). Microbial biomass rose by 45–48.9%, promoting sustainable intensification for South Asian agro-ecologies (Parihar et al. 2016). Planting of other crops during the intercropping season known as cover crops helps in the protection of the soil as well as improving the organic matter and the nutrient content of the soil as it indexes (Dissanayaka et al. 2024a, b; Lamichhane and Alletto 2022). Cutting crop residues and retaining them on the field following harvest enhances the rate of water infiltration (Palm et al. 2014). Application of mulches on the soil surface reduces the temperature and holds moisture for the crops, the compost does the same by regulating the soil temperature and moisture (Prem et al. 2020). They also contribute to the creation of diversity in a certain ecosystem since soil cover contains food and shelter for soil inhabitants. Organic mulches especially assist in vegetation regeneration exercise, which ensures that ecological connectivity and nutrient cycling are promoted (Leclercq-Dransart et al. 2020). In addition, cover crops play other roles as outlined herein One of the essential roles is that it helps in carbon sequestration, therefore helping to increase the level of soil organic carbon stocks (Poeplau and Don 2015). There are disadvantages associated with maintaining the soil cover among them being a delay in the soil's warming up in the spring however, it is a crucial practice in regenerative agriculture.

5.2.4 Maintaining Living Roots in the Soil

Living roots in the soil all year are part of the regenerative agricultural practices that have a significant impact on soil health and the ecosystem. This means getting as many actively growing plants through the year to spend as much time in the soil as possible. Living roots are very useful in enhancing the quality of the soil. It should be noted that these roots release organic compounds to the rhizosphere for the support of the microbial populace and the escalation of microbial functioning, which increases nutrient cycling and soil structure (Pausch and Kuzyakov 2018). Roots physically and chemically bind soil particles, reducing erosion and improving soil aggregation (Bodner et al. 2021; Galloway et al. 2020). Roots also help hold the soil particles together and consequently reduce soil erosion and enhance soil aggregation (Bettermann et al. 2021). Also, roots can sustain soil microbial community especially the mycorrhizal fungi that play a role in the exchanges of nutrients and water (Barto et al. 2012). The outcomes of the studies also demonstrate that successful management of mycorrhizal fungi is boosted as long as living roots are preserved rather than letting fields remain fallow (Cavagnaro et al. 2015). In the case of living roots being in the soil, cover cropping, perennial cropping, crop rotation, and intercropping can be employed by farmers. The use of cover crops leads to the enhancement of the organic carbon stock in the soil thus fighting climate change (Blanco-Canqui et al. 2015; Dissanayaka et al. 2022). Crop rotations when well-timed increase the amounts of soil organic matter than in monocultures (McDaniel et al. 2014). This helps increase the opportunities for nutrient cycling, and intercropping increases the soil microbial biomass and activity (Nuwarapaksha et al. 2022; Wang et al. 2015). These activities must be planned and involve changes in the management of farms but offer numerous advantages regarding the health and functions of the soil and ecosystem along with environmental conservation.

5.2.5 Integrating Livestock in Farming System

Crop-livestock linkages are widely practiced in regenerative systems which has benefits to the ecosystem, soil and overall productivity of the farms (Dissanayaka et al. 2023a, b, c; Nuwarapaksha et al. 2023a, b, c). This system of farming is called mixed farming or integrated crop-animal production, which has advantages in terms of sustainability and/or resilience. There is enhanced nutrient cycling which is a big boost to the plants. Ruminant animals generate manure thus in the case of application, grazed crops have the same yields as crops produced through the use of mineral fertilizers (Lemaire et al. 2014). This of course reduces costs but also reduces the use of synthetic fertilizers hence reducing effects on the environment (Udumann et al. 2023a, b). Extension grazing practices such as intensity, duration and mob/stocking density, discrete grazing, adaptive rotational or multiple-paddock grazing improve the health of the soil by promoting plant growth, soil organic matter and water infiltration (Teague et al. 2016). Besides, livestock hooves contribute to soil crushing and the smoothening of its structure and aeration (Shah et al. 2017).

Integration of livestock increases farm biological diversity of the plants, animals, insects, and microorganisms in the field and the sub-soil (Derner et al. 2018). Integrated systems have higher plant species diversity and more arthropods that help in pest suppression and pollination. Economically, these systems are more sustainable, because the farms have diverse ways of generating income, and therefore no risks from the market and the climate (Ryschawy et al. 2012). Optimum grazing enhances the organic carbon content of the soil (Stanley et al. 2018) and boosts the water's capacity to infiltrate and then to be retained (Weber and Gokhale (2011). Animals also assist in weed management and interrupt pest's life cycles (Liebman and Davis 2000). There are numerous advantages to livestock integration but disadvantages are also there such as; the complexity involved in animal management and the possibility of soil compaction to occur. However, all these challenges can be met if there is proper planning and management of integrated crop-livestock Systems (Nuwarapaksha et al. 2023a, b, c).

5.2.6 Enhancing Biodiversity

Biodiversity ought to be improved because it plays an important function in regenerative agriculture which brings out the sustainability of the ecosystem. This includes the encouragement of diversity at the genetic, species, and even at the above as well as below ground levels.

Studies have established that greater levels of species in production landscapes result in higher provision, stability, and resistance to different emergence challenges of ecosystem goods and services (Tscharntke et al. 2012). A few of the major components of biodiverse regenerative agriculture include crop, livestock, soil, habitat, and genetic points. Crop diversification through crop rotation, intercropping, and polyculture increase yields and control of pests. Diversity in production can improve the production of crops and animals (Tamburini et al. 2020). Integrated production of animals and crops promotes the utilization of the available resources hence improving the provision of ecosystem services, vegetation cover, and quality soil (Wang et al. 2019). Preservation of a range of indigenous populations in the soil through zero and reduced tillage, cover crops, and organic matter is vital. It is possible to explain the effect as the different crops improve the quantity and quality of nutrients in the soil, which may increase the number of microbes (Venter et al. 2016). This approach also increases the number of beneficial insects in and around farms while not diminishing yields by also developing various habitats on or near farms (Pywell et al. 2015). Diversifying the crops planted and the animals produced ensures that even when disease attacks, not all crops or animals will be affected, and it's emphasized that heritage and locally adapted crops and animals are more resilient to diseases (Reiss and Drinkwater 2018).

Thus, greater availability of bio-diversity in agriculture brings in the following advantages (Fig. 5.2). With advanced pest control, it is effective in enhancing the diet of predators and parasitoids, hence lowering pest abundance (LaCanne and Lundgren 2018). Higher plant diversity increases pollination services and wild bees and other pollinators, improving the fruit set globally (Garibaldi et al. 2013). Plant and microbial assemblages enhance nutrient cycling (Atapattu et al. 2020) and hence enhance the carbon and nitrogen stocks in the soil (Isbell et al. 2013). Biodiversity can also reduce vulnerability of the agriculture to stress factors hence improving agricultural production. Crop diversification enhances yield stability and drought tolerance of crops due to the differences in the rotations which enhances soil health (Gaudin et al. 2015). Also, it was noted that different plant species are more productive than a single species in the aspect of biomass output (Cardinale et al. 2011).

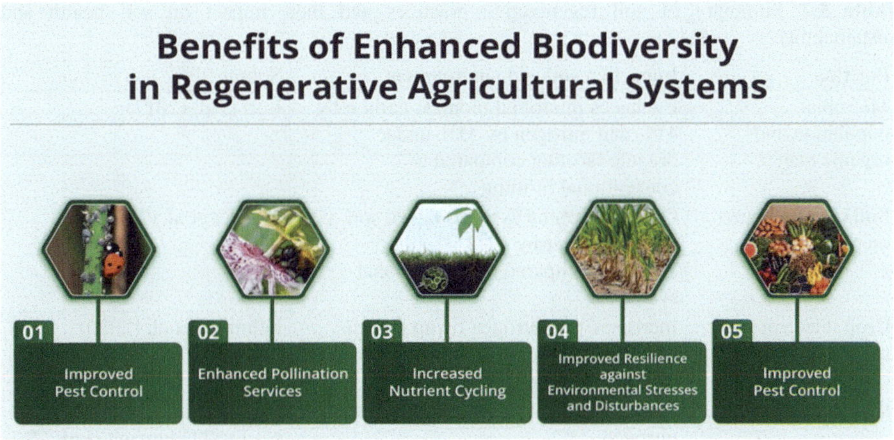

Fig. 5.2 Benefits of enhanced biodiversity in regenerative agricultural systems

However, many tensions come with the increase in the scale of species and functional types; including the complexity of management and potential yield penalties, certain aspects of increasing the scope of species and functional types are critical for regenerative agriculture because of their benefits for ecosystem health, production system resilience, and sustainable food production.

5.2.7 Promoting Soil Fertility

The principle of regenerative agriculture widely addresses the enhancement of the soil's ability to support plant growth through biological means rather than through chemical means (Table 5.2). This approach recognizes the soil as a living biology, which attempts to work for the growth of good biomass, improve conversion of nutrients, and improve the organic matter. Soil regenerative practices involve improving the microbial population, addition of organic matter, crop rotations, cover crops, composting, and bio-fertilizers (Dissanayake et al. 2023a, b). Thus, microbial biomass carbon can be enhanced by 41% and nitrogen by 51% under organic farming in contrast to conventional farming (Lori et al. 2017).

A step in enhancing the properties of soil includes the promotion of the formation of structure, water capacity and nutrients in the presence of organic matter. After 21 years, the research found that organic farming systems raised the soil organic matter by 12–15% more than conventional systems (Maeder et al. 2002). Intercropping as well as crop rotation techniques that include planting legumes has tremendously increased soil nitrogen by up to 50% as well as phosphorus by about 30% (Lehmann et al. 2020).

Cover crops protect the soil, add organic matter, and fix nitrogen (Nuwarapaksha et al. 2023a, b, c; Udumann et al. 2023a, b, 2024a, b, c). Compost and organic amendments enhance the nutrient content and organic matter (Dissanayaka et al.

Table 5.2 Summary of soil regenerative practices and their impact on soil health and sustainability

Practice	Impact on soil and environment	Source
Microbial population and organic matter	Enhances microbial biomass carbon by 41% and nitrogen by 51% under organic farming compared to conventional farming.	Lori et al. (2017)
Soil organic matter increase	Organic farming systems raised soil organic matter by 12–15% after 21 years compared to conventional systems.	Maeder et al. (2002)
Crop intercropping and rotation	Increases soil nitrogen by up to 50% and phosphorus by 30% with legume-based intercropping and rotations.	Lehmann et al. (2020)
Cover crops	Protect soil, add organic matter, and fix nitrogen.	Nuwarapaksha et al. (2023a, b, c) and Udumann et al. (2023a, b, 2024a, b, c)
Compost and organic amendments	Enhance nutrient content and organic matter; increase soil organic carbon and total nitrogen.	Martínez-Blanco et al. (2013)
Biofertilizers	Improve nutrient uptake and can raise yields by up to 17%.	Martínez-Blanco et al. (2013)
Organic manure and fertilizers	Efficient nutrient utilization, better water retention, less chemical fertilizer use, improved soil status.	Wu and Ma (2015)
Legume rotation in organic systems	Reduces nitrogen losses compared to conventional systems.	Drinkwater et al. (1998)
Soil structure and organic matter	Improved soil structure and organic matter content enhance water-holding capacity.	Libohova et al. (2018)
Energy consumption in organic systems	Organic farming uses 45-97% less energy than conventional systems.	Reganold and Wachter (2016)
Carbon storage	Organic farming stores 450 kg C ha^{-1} yr^{-1} more carbon compared to conventional methods.	Gattinger et al. (2012)
Environmental and soil health benefits	Reduces dependency on synthetic inputs, improves soil fertility, and contributes to long-term farm stability.	Pimentel and Burgess (2014)

2023a, b, c; Dissanayake et al. 2023a, b). Martínez-Blanco et al. (2013) reported that the application of compost increases soil organic carbon and total nitrogen compared to unamended soils.

Some biofertilizers include mycorrhizal fungi and nitrogen-fixing bacteria that improve the uptake of nutrients and raise the yields by up to 17% (Martínez-Blanco et al. 2013). The above concepts of nutrient management could be incorporated by using both organic manure and fertilizers to increase yields and fertility of the soil (Wu and Ma 2015). Some of the advantages include efficiency in nutrient utilization, water retention, less reliance on chemical fertilizers, and improved soil status.

Organic systems-based legume rotation can reduce nitrogen losses compared to the conventional system (Drinkwater et al. 1998). When the soil structure is enhanced and organic matter content heightened the water shifting capacity improves (Libohova et al. 2018).

Regenerative practices also reduce dependency on external inputs by fostering natural soil fertility, significantly decreasing the need for synthetic fertilizers. The organic system consumes 45–97% less energy than the conventional system (Reganold and Wachter 2016) and organic soils are highly organic matter and biologically active (Pimentel and Burgess 2014). These practices also help in the reduction of carbon emissions as they allow the natural processes of carbon storage. Soil organic matter in organic farming can be stored between 450 kg C ha^{-1} $year^{-1}$ than conventional methods (Gattinger et al. 2012).

Though transition may cause short-term problems for the farmers, the long-term advantage of the health of the soil, adaptability of the environment, and the stability of the farm make regenerative practices key to sustainable agriculture.

5.2.8 Water Management

Water management is an important aspect of regenerative agriculture, especially looking at climate change hence variability in rainfall patterns and at times water scarcity. The regenerative practices will focus on increasing the infiltration, storage and utilization efficiency of water resources for crop production and to improve the water use and health of ecosystems. Key aspects of water management in regenerative agriculture are shown in Table 5.3.

Constructions can help in the collection of rainwater to help supply water during dry seasons besides increasing agricultural production for the smallholder farmers. Hence, small-scale water harvesting strategy is feasible in assisting communities in developing countries to cope with the future adverse effects of climate change on water resources (Lasage and Verburg 2015). It should be recognized here that regenerative approaches give importance to water use. Scheduling irrigation with the help of incorporation of the soil moisture condition as well as water demand by crops can save water by 20–40% without loss of production (Qiao et al. 2019). The use of cover crops decreases water evaporation and increases water infiltration and water retention in the soil thus leading to enhanced soil water content in the view that it increases by an average of 7% of the total soil water content (Basche et al. 2016). Water management is quite crucial in the implementation of regenerative agriculture practices. This is because; it enhances water conservation, which in turn enhances topsoil water retention and facilitates crops to withstand dry conditions. No-till systems produce higher dry matter than conventional and ridge-till systems during the years of drought. Improved water retention in the soil combined with the efficient use of water minimizes water use and consequently the cost of irrigation (Pimentel and Burgess 2014). It also reduces flood impact, since watering infiltration and water retention in the soil layers enhance soil water holding capacity when there is a heavy downpour there will be less surface runoff (Zhang et al. 2007).

Table 5.3 Key aspects of water management in regenerative agriculture

Aspect of water management	Description	References
Enhancing soil water infiltration	No-till practices can increase water infiltration rates compared to conventional tillage, reducing runoff and soil erosion while improving water availability for plants	Franzluebbers (2002)
Increasing soil water retention	Improved SOM content significantly enhances soil water-holding capacity, reducing irrigation needs and improving crop resilience during dry periods	Libohova et al. (2018)
Implementing water harvesting techniques	Small-scale water harvesting techniques collect and store surface runoff for domestic or agricultural uses, helping communities adapt to climate change impacts on water resources	Lasage and Verburg (2015)
Optimizing irrigation efficiency	Improved irrigation scheduling based on soil moisture monitoring and crop water demand can reduce water use without compromising yield	Qiao et al. (2019)
Utilizing cover crops	Cover crops reduce evaporation, improve infiltration, and enhance soil water retention. A meta-analysis found that cover crops increased soil water content	Basche et al. (2016)
Implementing agroforestry practices	Agroforestry systems can reduce surface runoff and increase water use efficiency compared to conventional monocultures	Jose (2009)

Besides, water quality is enhanced through the reduction of runoff and erosion which is pleasing for the environment and health department (Reganold and Wachter 2016).

Enhanced water management helps maintain ecosystem services including a recharged aquifer and stable streamflow, which create measurable economic advances and ecosystem values (Power 2010). Despite these costs, it is critical since it contributes to enhancing farms' sustainability, effective use of water resources, and protection of the environment.

5.3 Benefits of Regenerative Agriculture

Regenerative agriculture offers numerous benefits that address many of the challenges conventional farming systems face (Fig. 5.3). By focusing on soil health, biodiversity, and ecosystem restoration, regenerative practices can improve agricultural productivity while enhancing environmental sustainability.

5.3.1 Improved Soil Structure and Water Retention

A clear description of regenerative practices reveals that it sharply improve the physical attributes of soil, a factor that leads to better water-holding capacity. A study revealed that there is a higher soil organic matter on regenerative farms than the conventional ones (LaCanne and Lundgren 2018). These lead to better aggregation of the soil particles thus providing a better-structured soil which one that allows high infiltration and water retention. This also boosts the water retention capacity of the soil hence improving water availability to the crops. It was also revealed that regenerative farms' water availability was higher than that of the fields managed under conventional practices (Basche et al. 2016). This increased water retention is particularly important in areas where the incidence of drought is high, this will make the crops survive when water is scarce. Another study showed that there is a way in which regenerative activities are proven to help enhance the plant-available water capacity alongside a rise in soil's organic carbon (Lal 2020). It is thus for these reasons that this advancement in the ability to retain water not only aids the processes of crop growth but also makes the crops more durable in conditions of drought.

5.3.2 Increased Soil Organic Matter and Carbon Sequestration

Regenerative practices like cover crops, minimum tillage, and diversified cropping enhances the organic material levels in the soil which can potentially helps in reversing the climate change. A meta-analysis conducted revealed that regenerative agriculture practices could return up to 3–6 Gt CO_2 per annum globally (Griscom et al. 2017). The increase in organic matter within the soil helps in combating adverse climatic conditions while enhancing the condition of the soil. Array of studies noted that by improving soil organic matter, one could boost crop production. This clearly shows that carbon sequestration is also being achieved simultaneously with the enhancement of agricultural capability (Lal 2020). Studies also

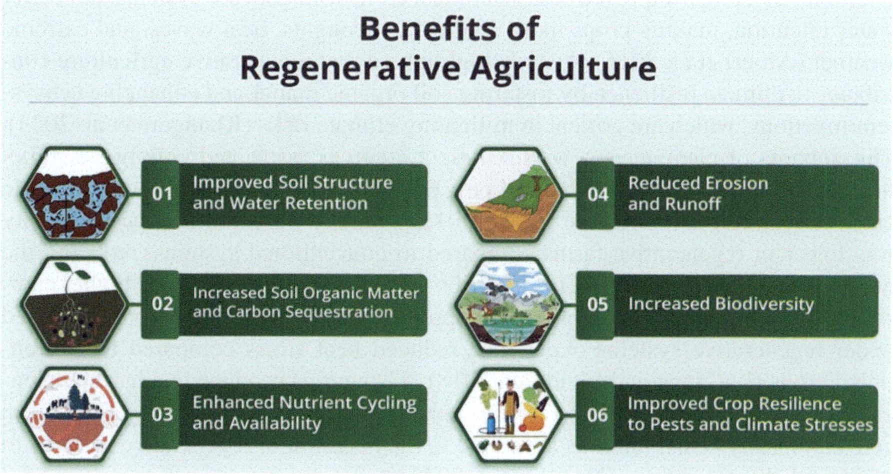

Fig. 5.3 Benefits of regenerative agriculture

indicated that the measures designed to increase the share of regenerative agroecosystems can neutralize the overall anthropogenic emissions (Paustian et al. 2019). This goes to show the relevance that regenerative agriculture calls for in the conception of climate change mitigation measures.

5.3.3 Enhanced Nutrient Cycling and Availability

Techniques for regeneration agriculture enhance nutrient cycling in the soil ecology as compared to ordinary practices. Organic farming improves plant nutrient uptake because of more diverse microorganisms in the soil and such techniques as rotation and multi-cropping. Another study revealed that regenerative fuels majority an amazingly higher level of potentially mineralizable nitrogen than conventional fuels (Bowles et al. 2014). Higher nutrient availability as a result of these practices may lower the dependency on artificial fertilizers hence cutting on costs of inputs for the farmers as well as eradicating on negative effects of fertilizer leaching into water bodies (Atapattu et al. 2024a, b). Also, a current study established that for total nitrogen, potassium, phosphorus, and phosphorus availability regenerative practices are higher than conventional systems hence reducing reliance on synthetic fertilizers (Luján Soto et al. 2021).

5.3.4 Improved Crop Resilience to Pests and Climate Stresses

Regenerative agriculture practices significantly enhance crop resilience against pests and climate instabilities by improving soil health, biodiversity, and agroecosystem stability. These systems emphasize holistic management, including crop rotation, cover cropping, and minimal tillage, which bolster soil structure and microbial diversity. Improved soil health results in stronger root systems and better water retention, making crops more resilient to droughts, heat waves, and extreme weather (Altieri et al. 2015). A study highlighted that regenerative agriculture contributes to climate resilience by fostering soil organic matter and enhancing ecosystem functions, which are critical in mitigating climate risks (Khangura et al. 2023). The richness of plant species within these systems is associated with higher functional diversity, leading to enhanced pest control and better adaptation to variable environmental conditions (Duru et al. 2015). A study observed that pest intensity was lower in regenerative farms compared to conventional systems, reducing the need for chemical pesticides. This is attributed to the healthier soils and higher natural enemy populations in diversified cropping systems. Similarly, crops produced under regenerative systems experience reduced heat stress compared to conventional crops (LaCanne and Lundgren 2018). Also, crops produced under the regenerative systems experienced a lower heat stress compared to the crops from conventional systems; this was attributed to the potential of regenerative systems to withstand climate change (Nwosu et al. 2024).

Furthermore, regenerative practices, by integrating crop-livestock systems, reduce input costs, improve nutrient cycling, and support a more robust agroecosystem (Kronberg et al. 2021). These factors contribute to more stable yields even under challenging conditions, ensuring food security. Diversified and resilient production systems are critical for future-proofing agriculture against environmental and socio-economic challenges, particularly in the face of climate change (Stetkiewicz et al. 2023). Moreover, regenerative agriculture provides numerous benefits by enhancing soil quality, promoting biodiversity, and increasing the resilience of agroecosystems. These strategies collectively make agricultural production more sustainable and adaptable to future climate challenges, thereby supporting long-term food security.

5.3.5 Reduced Erosion and Runoff

Regenerative farming reduces the effect of soil erosion and water runoff on a large scale through regenerative agriculture. Contour farming and growing plants on the surface of the soil all year round reduce the vulnerability of the surface to the forces of wind and water. A review stated that regenerative agriculture reduces soil erosion by adopting cover cropping, no-tilling, and agroforestry which also improves the soil's structure, fertility and sustainability (Dissanayaka et al. 2023a, b, c; Kodaparthi et al. 2024). This has the added effect of preserving topsoil as well as eliminating the loss of nutrients and polluting of water from agricultural runoff. Some literature has suggested that regenerative farms can perform better than conventional systems reducing rates of soil erosion (Khangura et al. 2023). Such a sharp decline in the erosion rates is beneficial in terms of the long-term preservation of soil surfaces coupled with enhanced water quality in bodies of water adjacent to the sites.

5.3.6 Increased Biodiversity

A sustainability model that falls under regenerative agriculture includes enhancement of the biotic some of the soil as well as on ground cover. Such practices tend to diminish the application of chemicals and promote the practice of crop diversification hence providing habitation to different forms of life. An analysis revealed that regenerative farms were more beneficial to bird species' diversity when compared with conventional farms (Lundgren and Fausti 2015). Down below the soil, the variety of microorganisms is also somewhat greater. Regenerative practices improved the soil bacterial and fungal community more than the conventional practices (Lori et al. 2017). Data obtained in the course of the last few years have revealed that regenerative farms harbor superior pollinator diversity to conventional farms, and thereby a better ecosystem service and crop pollination (Kremen and Miles 2012).

5.4 Challenges and Barriers

Despite the advantages that regenerative agriculture provides on soil health, biodiversity, and climate change, the adoption of this practice encounters several serious difficulties. This section discusses some of the main challenges that farmers, policymakers and other actors need to address next to see more widespread adoption of regenerative practices.

5.4.1 Initial Costs and Transition Period

The first challenge in embracing regenerative agriculture solutions is the cost and potential revenue loss while transitioning to the new system. This is because farmers are usually reluctant to reverse their trends and patterns because of initial expenses that come with the changes in practices and with purchasing different tools (LaCanne and Lundgren 2018). For example, shift from a conventional tillage system to no-tillage entails the need to acquire new seeders, an investment that may be expensive, especially for small and medium-sized farming ventures. Also, cover crop seeds and the amount of manpower needed to sow them are extra costs that may not return an income quickly (Carlisle 2016). This process can take several years, and the crop yield may be lower during the transition period because of shifts in the soil biota responding to the new management regime. This yield dip can cause the farmers to have various forms of financial hardships in the course of fulfilling their economic requirements through farming practices. According to research, it may take 3–5 years of using organic systems to get the same yield as conventional farming; any regenerative practices may take the same duration (Crowder and Reganold 2015).

5.4.2 Knowledge and Skill Requirements

Regenerative agriculture is a concept that focuses on the provision of the profound knowledge and skills needed to practice integrated farming systems that are grounded on ecological principles. This is because many farmers have been trained using conventional practices of farming and may not gather sufficient information to implement regenerative practices (Gosnell et al. 2020). For instance, successful cover crops need one to understand the plant species, when to plant the crop when to remove it and how it interacts with the cash crops. Likewise, practising integrated grazing management requires an appreciation of factors such as rates of grass regrowth, animal activities, and the broader biophysical environment context. The learning process may take time, and the farmers may be required to make considerable efforts to acquire the necessary knowledge. As much as there are increasing resources available, with key categories of workshops, online courses and farmers' networks, it is still a challenge to access such information and support sometimes (Carlisle et al. 2019).

5.4.3 Market and Policy Support

Another barrier to the adoption of regenerative agriculture is the absence of concrete market returns and friendly policies. The existing agricultural and subsidy systems are yet another form of subsidizing conventional industrial agriculture and commodity production, which makes it financially difficult for regenerative farmers(Schulte et al. 2019). For example, crop insurance programs have been developed based on monoculture systems and might not sufficiently address diversified regenerative farms. Also, due to the lack of universally recognized standards or certification for regenerative products, farmers are missing out on the market premium they might require to make up for the initial transition costs and incentivize the shift toward sustainable agricultural practices (Elevitch et al. 2018). There is insufficient policy provision for research and education from a regenerative point of view as well as financial incentives. Although some countries have already started adopting policies promoting the health of the soil and carbon storage, they are not yet popular, let alone complete enough to globally push through massive changes to the kind of farming that is being practiced, or regenerative agriculture.

5.4.4 Addressing Misconceptions and Resistance to Change

Various psychological and cultural barriers threaten the adoption of regenerative agriculture: There is a lack of understanding about what it entails and primary resistance to change in the farming community as well as the wider agricultural community at large. Farmers and agricultural professionals have put years of practice into the current practices and may not readily accept any approach that goes against the norm (Gosnell et al. 2020). Misconceptions are often as follows: regenerative agriculture is only applicable to small or organic farming; regenerative agriculture always causes a decrease in yield and profitability. These misconceptions therefore need to be corrected through various awareness creation and examples of effective regenerative systems across scales. Additionally, most industries involved in input provision, processing, and retailing of agricultural produce are dominated by conventional farming systems. While the conventional structures have a long-term and extensive setup, they are rigid and take time to respond to the needs of regenerative farmers exacerbating the challenges of the transition period (Schulte et al. 2019). Change management resistance involves not only an educational aspect but also the resultant cultural changes within farming societies. Supporting other farmers, presenting successful farmers as examples and fostering risk-taking and experimentation are the key measures necessary to overcome this obstacle. All in all, it is imperative to acknowledge that the problems regarding the implementation of regenerative agriculture at a large scale are considerable, yet solvable. Overcoming these barriers will entail multi-stakeholder engagement from farmers, researchers, policy-makers and market players to foster an enabling environment that will drive change and sustain transition in regenerative farming systems.

5.5 Future Directions and Research Needs

As the practice of regenerative agriculture begins to take root, several promising fields need refinement, so that more and more farmers can embrace and sustain the practice. Possible future development and the research requirements that will define the advancement of the regenerative practice in the coming years are discussed hereunder.

5.5.1 Scaling Up Regenerative Practices

However, transitioning to regenerative agriculture at the regional or even a global level is a challenge that has been observed despite the good performances that have been realized at the farm level. Therefore, more detailed studies should be conducted to define the strategies for the adoption of regenerative practices within the different agroecological regions and farming typologies. It has been observed that most prime success stories of regeneration attempted by farmers have been reported from small to medium-sized, diversified farms. Further studies are required to understand how practices such as coverage cropping, reduced tillage, and integrated livestock farming can be integrated into large-scale monoculture systems without compromising yield and income (Newton et al. 2020). Additionally, studies on the landscape-level impacts of widespread regenerative adoption are crucial. Also, research into the degree of regenerative adoption at the landscape level is important. This includes analyzing how the application of regenerative practices influences the health and productivity of watersheds, species distinctiveness, and climate change vulnerability within agricultural regions (DeLonge et al. 2016).

5.5.2 Policy Support and Incentives

Policies and incentives to successfully implement regenerative agriculture are going to be among the research agendas in the future. Future work that aims at selecting policies, which can foster technology adoption at an optimal rate with desirable impacts on the farmers, communities, and ecosystems should be conducted. The other promising avenue reflects the further launch of payment for ecosystem services (PES) for regenerative practices. Further investigation should be directed toward estimating the contribution of ecosystem services that are offered by regenerative farms (Kremen and Miles 2012). This data can be used to design the PES programmes where farmers are paid in proportion to the stewardship services they undertake. Also worthy of mention is the third area of focus, on changing the incentives provided through agricultural subsidies and insurance to encourage regenerative practices. Research works should be carried out to determine the possible effects of transferring subsidies towards enhancing the quality of soil, implementing crop diversification and crop-animal production systems. (Reganold and Wachter 2016).

5.5.3 Technology and Innovation in Regenerative Agriculture

Innovations and technological adoptions will be vital in the future of regenerative agriculture. There's a demand for more studies that seek to inform how regenerative practices can be applied at large and how this process can be facilitated and audited. One area is precision agriculture technology applied to regenerative systems where platforms for variable rate technology and environmental sensors are emphasized. This includes soil moisture and nutrient monitoring sensors, emergence, growth, and vigour sensing systems, as well as species and biotic diversity monitoring tools. Combining these technologies with artificial intelligence and machine learning algorithms may assist farmers in determining when and where to rotate crops, how to manage cover crop, and how to integrate livestock (Wolfert et al. 2017). Another promising avenue to address the problem under consideration is the development of plant varieties and animal breeds that are suitable for regenerative practices. Efforts are warranted to create plants with higher root biomass and root/shoot ratio for sequestering carbon, better nutrient uptake ability and enhanced resistance to diseases and pests in various crop establishment methods (Gaudin et al. 2015).

5.5.4 Education and Training Programmes

Since regenerative agriculture is based on innovative practices that require significant levels of knowledge, the identification of efficient training and education models is paramount for the advancement of the concept. Further studies have to be conducted to figure out the most effective ways of sharing regenerative knowledge and skills with farmers, agricultural specialists, policy-makers, and other stakeholders. Another key concern is the identification of intervention strategies that amalgamate research information with farmers' practical experience to enhance learning. Further studies should be aimed at active promotion and research concerning increased networking of farmer-to-farmer education, on-farm demonstrations, and farmer-centered research ventures (Krebs and Bach 2018). Furthermore, in the further research agenda, it is crucial to highlight the need for teaching regenerative agriculture principles as part of agricultural education. Researches of education efficiency are required to compare the effectiveness of applied strategies and standard curricula adaptable for different cultural settings and agricultural systems (Carlisle et al. 2019). Therefore, to enhance the progress of regenerative agriculture, the following research needs and development areas must be pursued in the future. Overall, if the agricultural community aims to enhance the quality and sustainability of current systems, then efforts towards scaling up such practices, supportive policies for similar practices, and advancements in technology and education can play a significant role in transitioning towards a more regenerative system. As this field develops, influencing work with farmers, researchers, policymakers, and other field stakeholders will be crucial to achieving the full potential of regenerative agriculture.

5.6 Conclusion

Regenerative agriculture is a revolutionary concept towards a sustainable form of farming that focuses on maintaining and improving the health of the entire farm system. Regenerative practices present an avenue to form more sustainable and resilient agricultural systems owing to key principles like soil disturbance reduction, crop diversification, maintaining living roots, integrating livestock, and improving the overall biological diversity. The advantages of regenerative agriculture are numerous and cut across different aspects of the community. Soil health becomes better, carbon stock rises higher, water holding capacity increases, and last but not least, the bio-diversity also increases which not only benefits the farmer productivity-wise but also environment and climate change-wise. Given that regeneration practices can potentially tackle climate change, food insecurity, and ecosystem degradation at once, it can be posited as a solution to challenges affecting global agriculture. Yet, the switch to regenerative agriculture comes with its shortcomings as well. Farmers are also hampered by costs, paucity of information, and inadequate policies and markets. Addressing these challenges will, therefore, need collaborative efforts from various players in the agricultural sector to undertake appropriate research, policy changes, and educational strategies. As to prospects, several problems require a solution. Primarie's to integration, including how to upscale regenerative practices to larger farming operations, how to develop technologies for regenerative systems, and what forms of policy incentives will be genuinely effective, are essential tasks. Moreover, enhancing the quality of educative and training programmes will also be crucial to ensuring farmers are knowledgeable and capable of undertaking these technically demanding ecosystem-based practices on their farms.

In light of the current food crises and the further increasing global population on one hand and the effects of climate change on the other, regenerative agriculture can presents an opportunity for the future. Therefore, these techniques are noble since they are pro-nature and may revolutionize our current food production systems to be productive, sustainable, and stable for upcoming generations. The transition to large-scale implementation of regenerative farming practices will require sustained research and development efforts, cooperation, as well as engagement of farmers, scientists, government, and consumers.

References

Altieri MA, Nicholls CI, Henao A, Lana MA (2015) Agroecology and the design of climate change-resilient farming systems. Agron Sustain Dev 35(3):869–890. https://doi.org/10.1007/s13593-015-0285-2

Atapattu AJ, Udumann SS (2024) Leveraging agroforestry principles for nature-based climate-smart solutions for coconut cultivation. In: Handbook of nature-based solutions to mitigation and adaptation to climate change. Springer, pp 1–28. https://doi.org/10.1007/978-3-030-98067-2_166-1

Atapattu AAAJ, Senarathne SHS, Raveendra SAST, Egodawatte WCP, Mensah S (2017) Effect of short term agroforestry systems on soil quality in marginal coconut lands in Sri Lanka. Agric Res J 54(3):324–328. https://doi.org/10.5958/2395-146X.2017.00060.6

Atapattu AJ, Xia S, Cao M, Zhang W, Mishra S, Yang X (2020) Can dominant canopy species leaf litter determine soil nutrient heterogeneity? A case study in a tropical rainforest in southwest China. J Soil Sci Plant Nutr 20(4):2479–2489. https://doi.org/10.1007/s42729-020-00314-x

Atapattu AJ, Ranasinghe CS, Nuwarapaksha TD, Udumann SS, Dissanayaka NS (2024a) Sustainable agriculture and Sustainable Development Goals (SDGs), pp 1–27. https://doi.org/10.4018/979-8-3693-4864-2.ch001

Atapattu AJ, Udumann SS, Nuwarapaksha TD, Dissanayaka NS (2024b) Upcycling coconut husk by-products: transitioning from traditional applications to emerging high-value usages. In: Agricultural waste to value-added products. Springer Nature, Singapore, pp 249–273. https://doi.org/10.1007/978-981-97-2535-9_12

Barto EK, Weidenhamer JD, Cipollini D, Rillig MC (2012) Fungal superhighways: do common mycorrhizal networks enhance below ground communication? Trends Plant Sci 17(11):633–637. https://doi.org/10.1016/j.tplants.2012.06.007

Basche AD, Archontoulis SV, Kaspar TC, Jaynes DB, Parkin TB, Miguez FE (2016) Simulating long-term impacts of cover crops and climate change on crop production and environmental outcomes in the Midwestern United States. Agric Ecosyst Environ 218:95–106. https://doi.org/10.1016/j.agee.2015.11.011

Bender SF, Wagg C, van der Heijden MGA (2016) An underground revolution: biodiversity and soil ecological engineering for agricultural sustainability. Trends Ecol Evol 31(6):440–452. https://doi.org/10.1016/j.tree.2016.02.016

Bettermann A, Zethof JHT, Babin D, Cammeraat ELH, Solé-Benet A, Lázaro R, Luna L, Nesme J, Sørensen SJ, Kalbitz K, Smalla K, Vogel C (2021) Importance of microbial communities at the root-soil interface for extracellular polymeric substances and soil aggregation in semiarid grasslands. Soil Biol Biochem 159:108301. https://doi.org/10.1016/j.soilbio.2021.108301

Bhagat R, Walia SS, Sharma K, Singh R, Singh G, Hossain A (2024) The integrated farming system is an environmentally friendly and cost-effective approach to the sustainability of agri-food systems in the modern era of the changing climate: a comprehensive review. Food Energy Secur 13(1). https://doi.org/10.1002/fes3.534

Blanco-Canqui H, Shaver TM, Lindquist JL, Shapiro CA, Elmore RW, Francis CA, Hergert GW (2015) Cover crops and ecosystem services: insights from studies in temperate soils. Agron J 107(6):2449–2474. https://doi.org/10.2134/agronj15.0086

Bodner G, Mentler A, Keiblinger K (2021) Plant roots for sustainable soil structure management in cropping systems. In: The root systems in sustainable agricultural intensification. Wiley, pp 45–90. https://doi.org/10.1002/9781119525417.ch3

Bowles TM, Acosta-Martínez V, Calderón F, Jackson LE (2014) Soil enzyme activities, microbial communities, and carbon and nitrogen availability in organic agroecosystems across an intensively-managed agricultural landscape. Soil Biol Biochem 68:252–262. https://doi.org/10.1016/j.soilbio.2013.10.004

Bowles TM, Mooshammer M, Socolar Y, Calderón F, Cavigelli MA, Culman SW, Deen W, Drury CF, Garcia y Garcia A, Gaudin ACM, Harkcom WS, Lehman RM, Osborne SL, Robertson GP, Salerno J, Schmer MR, Strock J, Grandy AS (2020) Long-term evidence shows that crop-rotation diversification increases agricultural resilience to adverse growing conditions in North America. One Earth 2(3):284–293. https://doi.org/10.1016/j.oneear.2020.02.007

Cardinale BJ, Matulich KL, Hooper DU, Byrnes JE, Duffy E, Gamfeldt L, Balvanera P, O'Connor MI, Gonzalez A (2011) The functional role of producer diversity in ecosystems. Am J Bot 98(3):572–592. https://doi.org/10.3732/ajb.1000364

Carlisle L (2016) Factors influencing farmer adoption of soil health practices in the United States: a narrative review. Agroecol Sustain Food Syst 40(6):583–613. https://doi.org/10.1080/21683565.2016.1156596

Carlisle L, de Wit MM, DeLonge MS, Calo A, Getz C, Ory J, Munden-Dixon K, Galt R, Melone B, Knox R, Iles A, Press D (2019) Securing the future of US agriculture: the case for invest-

ing in new entry sustainable farmers. Elementa Sci Anthropocene 7. https://doi.org/10.1525/elementa.356
Cavagnaro TR, Bender SF, Asghari HR, Heijden MGAVD (2015) The role of arbuscular mycorrhizas in reducing soil nutrient loss. Trends Plant Sci 20(5):283–290. https://doi.org/10.1016/j.tplants.2015.03.004
Chew B-H, Maxwell L, Anyiam FE, Menouni A, Kurniawan TA, Dimobe K, Sharma TPP, Ali GAM, Shah RDT, Saleem R, Mashroofa MM, Nasr M, Abbas B, Atapattu AJ, Mahmoud M, Singh N, Sarker MR (2024) Statements on open science for sustainable development goals. Data Sci J 23:49. https://doi.org/10.5334/dsj-2024-049
Crews TE, Blesh J, Culman SW, Hayes RC, Jensen ES, Mack MC, Peoples MB, Schipanski ME (2016) Going where no grains have gone before: from early to mid-succession. Agric Ecosyst Environ 223:223–238. https://doi.org/10.1016/j.agee.2016.03.012
Crowder DW, Reganold JP (2015) Financial competitiveness of organic agriculture on a global scale. Proc Natl Acad Sci 112(24):7611–7616. https://doi.org/10.1073/pnas.1423674112
Cui Y, Zhang W, Zhang Y, Liu X, Zhang Y, Zheng X, Luo J, Zou J (2024) Effects of no-till on upland crop yield and soil organic carbon: a global meta-analysis. Plant Soil 499(1–2):363–377. https://doi.org/10.1007/s11104-022-05854-y
Davis AS, Hill JD, Chase CA, Johanns AM, Liebman M (2012) Increasing cropping system diversity balances productivity, profitability and environmental health. PLoS ONE 7(10):e47149. https://doi.org/10.1371/journal.pone.0047149
DeLonge MS, Miles A, Carlisle L (2016) Investing in the transition to sustainable agriculture. Environ Sci Pol 55:266–273. https://doi.org/10.1016/j.envsci.2015.09.013
Derner JD, Smart AJ, Toombs TP, Larsen D, McCulley RL, Goodwin J, Sims S, Roche LM (2018) Soil health as a transformational change agent for US grazing lands management. Rangel Ecol Manag 71(4):403–408. https://doi.org/10.1016/j.rama.2018.03.007
Dissanayaka D, Nuwarapaksha T, Udumann S, Dissanayake D, Atapattu AJ (2022) A sustainable way of increasing productivity of coconut cultivation using cover crops: a review. Circular Agric Syst 2(1):1–9. https://doi.org/10.48130/CAS-2022-0007
Dissanayaka DMNS, Dissanayake DKRPL, Udumann SS, Nuwarapaksha TD, Atapattu AJ (2023a) Agroforestry—a key tool in the climate-smart agriculture context: a review on coconut cultivation in Sri Lanka. Front Agron 5. https://doi.org/10.3389/fagro.2023.1162750
Dissanayaka DMNS, Udumann SS, Dissanayake DKRPL, Nuwarapaksha TD, Atapattu AJ (2023b) Review on aquatic weeds as potential source for compost production to meet sustainable plant nutrient management needs. Waste 1(1):264–280. https://doi.org/10.3390/waste1010017
Dissanayaka NS, Udumann SS, Nuwarapaksha TD, Atapattu AJ (2023c) Agroforestry: an avenue for resilient and productive farming through integrated crops and livestock production. In: Transitioning to zero hunger. MDPI. https://doi.org/10.3390/books978-3-03897-863-3-5
Dissanayaka DMNS, Udumann SS, Atapattu AJ (2024a) Synergies between tree crops and ecosystems in tropical agroforestry. In: Agroforestry. Wiley, pp 49–87. https://doi.org/10.1002/9781394231164.ch3
Dissanayaka DMNS, Udumann SS, Nuwarapaksha TD, Atapattu AJ (2024b) Harnessing the potential of Mucuna cover cropping: a comprehensive review of its agronomic and environmental benefits. Circular Agric Syst:1–10. https://doi.org/10.48130/cas-0024-0001
Dissanayake DKRPL, Dissanayaka DMNS, Udumann SS, Nuwarapaksha TD, Atapattu AJ (2023a) Is biochar a promising soil amendment to enhance perennial crop yield and soil quality in the tropics? Technol Agron 3(1). https://doi.org/10.48130/TIA-2023-0004
Dissanayake DKRPL, Udumann SS, Dissanayaka DMNS, Nuwarapaksha TD, Atapattu AJ (2023b) Effect of biochar application rate on macronutrient retention and leaching in two coconut growing soils. Technol Agron 3(1). https://doi.org/10.48130/TIA-2023-0005
Drinkwater LE, Wagoner P, Sarrantonio M (1998) Legume-based cropping systems have reduced carbon and nitrogen losses. Nature 396(6708):262–265. https://doi.org/10.1038/24376
Duru M, Therond O, Fares M (2015) Designing agroecological transitions; A review. Agron Sustain Dev 35(4):1237–1257. https://doi.org/10.1007/s13593-015-0318-x

Elevitch CR, Mazaroli DN, Ragone D (2018) Agroforestry standards for regenerative agriculture. Sustainability 10(9):3337. https://doi.org/10.3390/su10093337

Franzluebbers AJ (2002) Water infiltration and soil structure related to organic matter and its stratification with depth. Soil Tillage Res 66(2):197–205. https://doi.org/10.1016/S0167-1987(02)00027-2

Fukuda K (2018) The advantage of natural farming as an eco-friendly way of living: practice and discourse on the "learners' fields" in Fukuoka, Japan. Culture Agric Food Environ 40(1):15–23. https://doi.org/10.1111/cuag.12101

Galloway AF, Akhtar J, Marcus SE, Fletcher N, Field K, Knox P (2020) Cereal root exudates contain highly structurally complex polysaccharides with soil-binding properties. Plant J 103(5):1666–1678. https://doi.org/10.1111/tpj.14852

Garibaldi LA, Steffan-Dewenter I, Winfree R, Aizen MA, Bommarco R, Cunningham SA, Kremen C, Carvalheiro LG, Harder LD, Afik O, Bartomeus I, Benjamin F, Boreux V, Cariveau D, Chacoff NP, Dudenhöffer JH, Freitas BM, Ghazoul J, Greenleaf S et al (2013) Wild pollinators enhance fruit set of crops regardless of honey bee abundance. Science 339(6127):1608–1611. https://doi.org/10.1126/science.1230200

Gattinger A, Muller A, Haeni M, Skinner C, Fliessbach A, Buchmann N, Mäder P, Stolze M, Smith P, Scialabba NE-H, Niggli U (2012) Enhanced top soil carbon stocks under organic farming. Proc Natl Acad Sci 109(44):18226–18231. https://doi.org/10.1073/pnas.1209429109

Gaudin ACM, Tolhurst TN, Ker AP, Janovicek K, Tortora C, Martin RC, Deen W (2015) Increasing crop diversity mitigates weather variations and improves yield stability. PLoS ONE 10(2):e0113261. https://doi.org/10.1371/journal.pone.0113261

Gosnell H, Charnley S, Stanley P (2020) Climate change mitigation as a co-benefit of regenerative ranching: insights from Australia and the United States. Interface Focus 10(5):20200027. https://doi.org/10.1098/rsfs.2020.0027

Griscom BW, Adams J, Ellis PW, Houghton RA, Lomax G, Miteva DA, Schlesinger WH, Shoch D, Siikamäki JV, Smith P, Woodbury P, Zganjar C, Blackman A, Campari J, Conant RT, Delgado C, Elias P, Gopalakrishna T, Hamsik MR et al (2017) Natural climate solutions. Proc Natl Acad Sci 114(44):11645–11650. https://doi.org/10.1073/pnas.1710465114

Isbell F, Reich PB, Tilman D, Hobbie SE, Polasky S, Binder S (2013) Nutrient enrichment, biodiversity loss, and consequent declines in ecosystem productivity. Proc Natl Acad Sci 110(29):11911–11916. https://doi.org/10.1073/pnas.1310880110

Jacobs AA, Evans RS, Allison JK, Garner ER, Kingery WL, McCulley RL (2022) Cover crops and no-tillage reduce crop production costs and soil loss, compensating for lack of short-term soil quality improvement in a maize and soybean production system. Soil Tillage Res 218:105310. https://doi.org/10.1016/j.still.2021.105310

Jat ML, Chakraborty D, Ladha JK, Rana DS, Gathala MK, McDonald A, Gerard B (2020) Conservation agriculture for sustainable intensification in South Asia. Nature Sustain 3(4):336–343. https://doi.org/10.1038/s41893-020-0500-2

Jose S (2009) Agroforestry for ecosystem services and environmental benefits: an overview. Agrofor Syst 76(1):1–10. https://doi.org/10.1007/s10457-009-9229-7

Kader MA, Senge M, Mojid MA, Ito K (2017) Recent advances in mulching materials and methods for modifying soil environment. Soil Tillage Res 168:155–166. https://doi.org/10.1016/j.still.2017.01.001

Khangura R, Ferris D, Wagg C, Bowyer J (2023) Regenerative agriculture—a literature review on the practices and mechanisms used to improve soil health. Sustainability 15(3):2338. https://doi.org/10.3390/su15032338

Kodaparthi A, Ranjit P, Deepu PG, Kaushik DM, Valli LA, Ashrutha P, Harihara J, Chepuri K (2024) Impact of regenerative agriculture on soil erosion. In: Prospects for soil regeneration and its impact on environmental protection. Springer, pp 269–295. https://doi.org/10.1007/978-3-031-53270-2_12

Krebs J, Bach S (2018) Permaculture—scientific evidence of principles for the agroecological design of farming systems. Sustainability 10(9):3218. https://doi.org/10.3390/su10093218

Kremen C, Miles A (2012) Ecosystem services in biologically diversified versus conventional farming systems: benefits, externalities, and trade-offs. Ecol Soc 17(4):40. https://doi.org/10.5751/ES-05035-170440

Kronberg SL, Provenza FD, van Vliet S, Young SN (2021) Review: Closing nutrient cycles for animal production—current and future agroecological and socio-economic issues. Animal 15:100285. https://doi.org/10.1016/j.animal.2021.100285

LaCanne CE, Lundgren JG (2018) Regenerative agriculture: merging farming and natural resource conservation profitably. PeerJ 6:e4428. https://doi.org/10.7717/peerj.4428

Lal R (2020) Regenerative agriculture for food and climate. J Soil Water Conserv 75(5):123A–124A. https://doi.org/10.2489/jswc.2020.0620A

Lamichhane JR, Alletto L (2022) Ecosystem services of cover crops: a research roadmap. Trends Plant Sci 27(8):758–768. https://doi.org/10.1016/j.tplants.2022.03.014

Lasage R, Verburg PH (2015) Evaluation of small scale water harvesting techniques for semi-arid environments. J Arid Environ 118:48–57. https://doi.org/10.1016/j.jaridenv.2015.02.019

Leclercq-Dransart J, Demuynck S, Douay F, Grumiaux F, Pernin C, Leprêtre A (2020) Comparison of the interest of four types of organic mulches to reclaim degraded areas: a field study based on their relative attractiveness for soil macrofauna. Ecol Eng 158:106066. https://doi.org/10.1016/j.ecoleng.2020.106066

Lehman R, Cambardella C, Stott D, Acosta-Martinez V, Manter D, Buyer J, Maul J, Smith J, Collins H, Halvorson J, Kremer R, Lundgren J, Ducey T, Jin V, Karlen D (2015) Understanding and enhancing soil biological health: the solution for reversing soil degradation. Sustainability 7(1):988–1027. https://doi.org/10.3390/su7010988

Lehmann J, Bossio DA, Kögel-Knabner I, Rillig MC (2020) The concept and future prospects of soil health. Nature Rev Earth Environ 1(10):544–553. https://doi.org/10.1038/s43017-020-0080-8

Lemaire G, Franzluebbers A, Carvalho PCDF, Dedieu B (2014) Integrated crop–livestock systems: strategies to achieve synergy between agricultural production and environmental quality. Agric Ecosyst Environ 190:4–8. https://doi.org/10.1016/j.agee.2013.08.009

Libohova Z, Seybold C, Wysocki D, Wills S, Schoeneberger P, Williams C, Lindbo D, Stott D, Owens PR (2018) Reevaluating the effects of soil organic matter and other properties on available water-holding capacity using the National Cooperative Soil Survey Characterization Database. J Soil Water Conserv 73(4):411–421. https://doi.org/10.2489/jswc.73.4.411

Liebman M, Davis AS (2000) Integration of soil, crop and weed management in low-external-input farming systems. Weed Res 40(1):27–47. https://doi.org/10.1046/j.1365-3180.2000.00164.x

Lin BB (2011) Resilience in agriculture through crop diversification: adaptive management for environmental change. Bioscience 61(3):183–193. https://doi.org/10.1525/bio.2011.61.3.4

Lori M, Symnaczik S, Mäder P, De Deyn G, Gattinger A (2017) Organic farming enhances soil microbial abundance and activity—a meta-analysis and meta-regression. PLoS ONE 12(7):e0180442. https://doi.org/10.1371/journal.pone.0180442

Luján Soto R, Martínez-Mena M, Cuéllar Padilla M, de Vente J (2021) Restoring soil quality of woody agroecosystems in Mediterranean drylands through regenerative agriculture. Agric Ecosyst Environ 306:107191. https://doi.org/10.1016/j.agee.2020.107191

Lundgren JG, Fausti SW (2015) Trading biodiversity for pest problems. Sci Adv 1(6). https://doi.org/10.1126/sciadv.1500558

Maeder P, Fliessbach A, Dubois D, Gunst L, Fried P, Niggli U (2002) Soil fertility and biodiversity in organic farming. Science 296(5573):1694–1697. https://doi.org/10.1126/science.1071148

Marcillo GS, Miguez FE (2017) Corn yield response to winter cover crops: an updated meta-analysis. J Soil Water Conserv 72(3):226–239. https://doi.org/10.2489/jswc.72.3.226

Martínez-Blanco J, Lazcano C, Christensen TH, Muñoz P, Rieradevall J, Møller J, Antón A, Boldrin A (2013) Compost benefits for agriculture evaluated by life cycle assessment. A review. Agron Sustain Dev 33(4):721–732. https://doi.org/10.1007/s13593-013-0148-7

McDaniel MD, Tiemann LK, Grandy AS (2014) Does agricultural crop diversity enhance soil microbial biomass and organic matter dynamics? A meta-analysis. Ecol Appl 24(3):560–570. https://doi.org/10.1890/13-0616.1

Montgomery DR (2007) Soil erosion and agricultural sustainability. Proc Natl Acad Sci 104(33):13268–13272. https://doi.org/10.1073/pnas.0611508104
Muhie SH (2022) Novel approaches and practices to sustainable agriculture. J Agric Food Res 10:100446. https://doi.org/10.1016/j.jafr.2022.100446
Newton P, Civita N, Frankel-Goldwater L, Bartel K, Johns C (2020) What is regenerative agriculture? A review of scholar and practitioner definitions based on processes and outcomes. Front Sustain Food Syst 4. https://doi.org/10.3389/fsufs.2020.577723
Nunes MR, Karlen DL, Veum KS, Moorman TB, Cambardella CA (2020) Biological soil health indicators respond to tillage intensity: an US meta-analysis. Geoderma 369:114335. https://doi.org/10.1016/j.geoderma.2020.114335
Nuwarapaksha T, Udumann S, Dissanayaka D, Dissanayake D, Atapattu AJ (2022) Coconut based multiple cropping systems: an analytical review in Sri Lankan coconut cultivations. Circular Agric Syst 2(1):1–7. https://doi.org/10.48130/CAS-2022-0008
Nuwarapaksha TD, Dissanayake WK, Gunathilaka WS, Udumann SS, Dissanayaka NS, Atapattu AJ (2023a) Assessing the optimum harvesting stage of Tithonia diversifolia as climate smart soil amendment for coconut plantations. In: The 2nd international online conference on agriculture, p 1. https://doi.org/10.3390/IOCAG2023-15967
Nuwarapaksha TD, Rajapaksha UN, Ekanayake J, Weerasooriya SA, Atapattu AJ (2023b) Exploring the economic viability of integrating Jamnapari goat into underutilized pastures under coconut cultivations in Coconut Research Institute, Sri Lanka. IECAG 2023:27. https://doi.org/10.3390/IECAG2023-14990
Nuwarapaksha TD, Udumann SS, Dissanayaka NS, Atapattu AJ (2023c) Coconut-based livestock farming: a sustainable approach to enhancing food security in Sri Lanka. In: Transitioning to zero hunger. MDPI. https://doi.org/10.3390/books978-3-03897-863-3-8
Nuwarapaksha TD, Udumann SS, Atapattu AJ (2024) Fostering food and nutritional security through agroforestry practices. In: Agroforestry. Wiley, pp 285–318. https://doi.org/10.1002/9781394231164.ch10
Nwosu NJ, Oshunsanya SO, Okon PB, Isong IA, Adetuga AT, Singh H (2024) Sustainable regenerative practices for land restoration and climate change management in the tropics. In: Handbook of nature-based solutions to mitigation and adaptation to climate change. Springer, Cham, pp 1–25. https://doi.org/10.1007/978-3-030-98067-2_119-1
Palm C, Blanco-Canqui H, DeClerck F, Gatere L, Grace P (2014) Conservation agriculture and ecosystem services: an overview. Agric Ecosyst Environ 187:87–105. https://doi.org/10.1016/j.agee.2013.10.010
Parihar CM, Yadav MR, Jat SL, Singh AK, Kumar B, Pradhan S, Chakraborty D, Jat ML, Jat RK, Saharawat YS, Yadav OP (2016) Long term effect of conservation agriculture in maize rotations on total organic carbon, physical and biological properties of a sandy loam soil in north-western Indo-Gangetic Plains. Soil Tillage Res 161:116–128. https://doi.org/10.1016/j.still.2016.04.001
Pausch J, Kuzyakov Y (2018) Carbon input by roots into the soil: quantification of rhizodeposition from root to ecosystem scale. Glob Chang Biol 24(1):1–12. https://doi.org/10.1111/gcb.13850
Paustian K, Larson E, Kent J, Marx E, Swan A (2019) Soil C sequestration as a biological negative emission strategy. Front Clim 1. https://doi.org/10.3389/fclim.2019.00008
Pimentel D, Burgess M (2014) An environmental, energetic and economic comparison of organic and conventional farming systems. In: Integrated pest management. Springer, Dordrecht, pp 141–166. https://doi.org/10.1007/978-94-007-7796-5_6
Pittelkow CM, Linquist BA, Lundy ME, Liang X, van Groenigen KJ, Lee J, van Gestel N, Six J, Venterea RT, van Kessel C (2015) When does no-till yield more? A global meta-analysis. Field Crop Res 183:156–168. https://doi.org/10.1016/j.fcr.2015.07.020
Poeplau C, Don A (2015) Carbon sequestration in agricultural soils via cultivation of cover crops—a meta-analysis. Agric Ecosyst Environ 200:33–41. https://doi.org/10.1016/j.agee.2014.10.024
Power AG (2010) Ecosystem services and agriculture: tradeoffs and synergies. Philos Trans R Soc B Biol Sci 365(1554):2959–2971. https://doi.org/10.1098/rstb.2010.0143

Prem M, Ranjan P, Seth N, Patle GT (2020) Mulching techniques to conserve the soil water and advance the crop production—a review. Curr World Environ Special Issue 1:10–30. https://doi.org/10.12944/CWE.15.Special-Issue1.02
Pywell RF, Heard MS, Woodcock BA, Hinsley S, Ridding L, Nowakowski M, Bullock JM (2015) Wildlife-friendly farming increases crop yield: evidence for ecological intensification. Proc R Soc B Biol Sci 282(1816):20151740. https://doi.org/10.1098/rspb.2015.1740
Qiao X, Sai L, Chen X, Xue L, Lei J (2019) Impact of fruit-tree shade intensity on the growth, yield, and quality of intercropped wheat. PLoS ONE 14(4):e0203238. https://doi.org/10.1371/journal.pone.0203238
Ranaivoson L, Naudin K, Ripoche A, Affholder F, Rabeharisoa L, Corbeels M (2017) Agro-ecological functions of crop residues under conservation agriculture. A review. Agron Sustain Dev 37(4):26. https://doi.org/10.1007/s13593-017-0432-z
Reganold JP, Wachter JM (2016) Organic agriculture in the twenty-first century. Nature Plants 2(2):15221. https://doi.org/10.1038/nplants.2015.221
Reiss ER, Drinkwater LE (2018) Cultivar mixtures: a meta-analysis of the effect of intraspecific diversity on crop yield. Ecol Appl 28(1):62–77. https://doi.org/10.1002/eap.1629
Rempelos L, Kabourakis E, Leifert C (2023) Innovative organic and regenerative agricultural production. Agronomy 13(5):1344. https://doi.org/10.3390/agronomy13051344
Rhodes CJ (2017) The imperative for regenerative agriculture. Sci Prog 100(1):80–129. https://doi.org/10.3184/003685017X14876775256165
Ryschawy J, Choisis N, Choisis JP, Joannon A, Gibon A (2012) Mixed crop-livestock systems: an economic and environmental-friendly way of farming? Animal 6(10):1722–1730. https://doi.org/10.1017/S1751731112000675
Schulte RPO, O'Sullivan L, Vrebos D, Bampa F, Jones A, Staes J (2019) Demands on land: Mapping competing societal expectations for the functionality of agricultural soils in Europe. Environ Sci Pol 100:113–125. https://doi.org/10.1016/j.envsci.2019.06.011
Shah AN, Tanveer M, Shahzad B, Yang G, Fahad S, Ali S, Bukhari MA, Tung SA, Hafeez A, Souliyanonh B (2017) Soil compaction effects on soil health and crop productivity: an overview. Environ Sci Pollut Res 24(11):10056–10067. https://doi.org/10.1007/s11356-017-8421-y
Smith RG, Gross KL, Robertson GP (2008) Effects of crop diversity on agroecosystem function: crop yield response. Ecosystems 11(3):355–366. https://doi.org/10.1007/s10021-008-9124-5
Somasundaram J, Sinha NK, Dalal RC, Lal R, Mohanty M, Naorem AK, Hati KM, Chaudhary RS, Biswas AK, Patra AK, Chaudhari SK (2020) No-till farming and conservation agriculture in South Asia—issues, challenges, prospects and benefits. Crit Rev Plant Sci 39(3):236–279. https://doi.org/10.1080/07352689.2020.1782069
Stanley PL, Rowntree JE, Beede DK, DeLonge MS, Hamm MW (2018) Impacts of soil carbon sequestration on life cycle greenhouse gas emissions in Midwestern USA beef finishing systems. Agric Syst 162:249–258. https://doi.org/10.1016/j.agsy.2018.02.003
Stetkiewicz S, Menary J, Nair A, Rufino MC, Fischer ARH, Cornelissen M, Duchesne R, Guichaoua A, Jorasch P, Lemarié S, Nanda AK, Wilhelm R, Davies JAC (2023) Crop improvements for future-proofing European food systems: a focus-group-driven analysis of agricultural production stakeholder priorities and viewpoints. Food Energy Secur 12(1). https://doi.org/10.1002/fes3.362
Tamburini G, Bommarco R, Wanger TC, Kremen C, van der Heijden MGA, Liebman M, Hallin S (2020) Agricultural diversification promotes multiple ecosystem services without compromising yield. Sci Adv 6(45). https://doi.org/10.1126/sciadv.aba1715
Teague WR, Apfelbaum S, Lal R, Kreuter UP, Rowntree J, Davies CA, Conser R, Rasmussen M, Hatfield J, Wang T, Wang F, Byck P (2016) The role of ruminants in reducing agriculture's carbon footprint in North America. J Soil Water Conserv 71(2):156–164. https://doi.org/10.2489/jswc.71.2.156
Torralba M, Fagerholm N, Burgess PJ, Moreno G, Plieninger T (2016) Do European agroforestry systems enhance biodiversity and ecosystem services? A meta-analysis. Agric Ecosyst Environ 230:150–161. https://doi.org/10.1016/j.agee.2016.06.002

Tscharntke T, Clough Y, Wanger TC, Jackson L, Motzke I, Perfecto I, Vandermeer J, Whitbread A (2012) Global food security, biodiversity conservation and the future of agricultural intensification. Biol Conserv 151(1):53–59. https://doi.org/10.1016/j.biocon.2012.01.068

Udumann SS, Dissanayaka DMNS, Nuwarapaksha TD, Dissanayake DKRPL, Atapattu AJ (2023a) Megathyrsus maximus as a raw material for organic fertilizer production: A feasibility study. Technol Horticult 3(1). https://doi.org/10.48130/TIH-2023-0009

Udumann SS, Dissanayaka NS, Nuwarapaksha TD, Thelwadana EP, Atapattu AJ (2023b) Assessing the growth potential of Sunn hemp (Crotalaria juncea L.) as a cover crop for major coconut-growing soils. Trends Horticult 6(2):3579. https://doi.org/10.24294/th.v6i2.3579

Udumann SS, Ranasinghe CS, Karunarathna LKNG, Kaliyadasa PE, Nuwarapaksha TD, Premathilaka UGAT, Atapattu AJ (2024a) Optimizing intercropping selection for coconut plantations based on PAR and agro-climatic zones. Agrofor Syst. https://doi.org/10.1007/s10457-024-00977-w

Udumann SS, Dissanayaka NS, Atapattu AJ (2024b) Women's roles in building resilience to climate change in agriculture. In: Impact of women in food and agricultural development, pp 58–78. https://doi.org/10.4018/979-8-3693-3037-1.ch004

Udumann SS, Sanduni Darshika Bandara KG, Ranawana SRWMCJK, Premathilake UGAT, Atapattu AJ (2024c) Identification of the optimal lopping stage of Gliricidia sepium used as manure for coconut plantations. Contemp Agric 73(1–2):104–113. https://doi.org/10.2478/contagri-2024-0013

Venter ZS, Jacobs K, Hawkins H-J (2016) The impact of crop rotation on soil microbial diversity: a meta-analysis. Pedobiologia 59(4):215–223. https://doi.org/10.1016/j.pedobi.2016.04.001

Wang Z, Bao X, Li X, Jin X, Zhao J, Sun J, Christie P, Li L (2015) Intercropping maintains soil fertility in terms of chemical properties and enzyme activities on a timescale of one decade. Plant Soil 391(1–2):265–282. https://doi.org/10.1007/s11104-015-2428-2

Wang L, Delgado-Baquerizo M, Wang D, Isbell F, Liu J, Feng C, Liu J, Zhong Z, Zhu H, Yuan X, Chang Q, Liu C (2019) Diversifying livestock promotes multidiversity and multifunctionality in managed grasslands. Proc Natl Acad Sci 116(13):6187–6192. https://doi.org/10.1073/pnas.1807354116

Weber KT, Gokhale BS (2011) Effect of grazing on soil-water content in semiarid rangelands of southeast Idaho. J Arid Environ 75(5):464–470. https://doi.org/10.1016/j.jaridenv.2010.12.009

Wolfert S, Ge L, Verdouw C, Bogaardt M-J (2017) Big data in smart farming—a review. Agric Syst 153:69–80. https://doi.org/10.1016/j.agsy.2017.01.023

Wu W, Ma B (2015) Integrated nutrient management (INM) for sustaining crop productivity and reducing environmental impact: a review. Sci Total Environ 512–513:415–427. https://doi.org/10.1016/j.scitotenv.2014.12.101

Yadav GS, Babu S, Das A, Mohapatra KP, Singh R, Avasthe RK, Roy S (2020) No-till and mulching enhance energy use efficiency and reduce carbon footprint of a direct-seeded upland rice production system. J Clean Prod 271:122700. https://doi.org/10.1016/j.jclepro.2020.122700

Yu Y, Stomph T-J, Makowski D, van der Werf W (2015) Temporal niche differentiation increases the land equivalent ratio of annual intercrops: a meta-analysis. Field Crop Res 184:133–144. https://doi.org/10.1016/j.fcr.2015.09.010

Zhang G, Chan K, Oates A, Heenan D, Huang G (2007) Relationship between soil structure and runoff/soil loss after 24 years of conservation tillage. Soil Tillage Res 92(1–2, 122):–128. https://doi.org/10.1016/j.still.2006.01.006

Cereal-Legume Integration: Association for Food Security, Soil and Environmental Sustainability

6

Sunil Kumar Prajapati, Megha Kumari, Twinkle Jena, Sk Asraful Ali, Deo Narayan Singh, and Sarla Yadav

Abstract

Cereal-legume integration is crucial for maintaining the long-term productivity and sustainability. Cereal-legume integration can potentially stabilize ecosystems, mitigate global climate change, and enhance food-security by improving soil organic carbon, land management, erosion prevention, and reducing the use of harmful chemicals. Cereal-legume intercropping utilizes the available resources efficiently, improving crop yield by ~10–25%, and promoting biodiversity. Legumes fix atmospheric nitrogen, which benefits cereal crops by improving nitrogen availability, reducing reliance on synthetic nitrogen fertilizers and improving soil health. Legumes, through their symbiotic relationship with rhizobia, convert atmospheric nitrogen into ammonia and provide up to 73% of the nitrogen requirements of other crops. This integration also reduces negative environmental outcomes, by improving carbon retention, nitrogen use-efficiency (~25%), while reducing NO_3-N leaching (~10–16%), GHGs emission (~10–25% CO_2 and ~15–35% N_2O). It also improves soil microbial diversity, which is essential for nutrient cycling and plant growth. The complementary root systems of cereals and legumes demnify nutrient competition, encouraging greater nitrogen fixation and improving soil fertility. Cereal-legume systems enhance phosphorus acquisition due to increased root surface area, improving nutrient exploration. Overall, cereal-legume integration offers a promising sus-

S. K. Prajapati (✉) · M. Kumari · T. Jena · S. A. Ali
Division of Agronomy, ICAR-Indian Agricultural Research Institute, New Delhi, India

D. N. Singh
Department of Agronomy, Udai Pratap College, Varanasi, Uttar Pradesh, India

S. Yadav
ICAR-Central Potato Research Station, Patna, Bihar, India

© The Author(s), under exclusive license to Springer Nature Singapore Pte Ltd. 2025

S. Babu et al. (eds.), *Ecological Solutions to Agricultural Land Degradation*,
https://doi.org/10.1007/978-981-96-3392-0_6

tainable approach to agriculture by boosting productivity, reducing environmental impacts, and enhancing soil health.

Keywords

Biodiversity · Cereals · Legume · N_2-fixation · Soil health · Sustainability

Abbreviations

BNF	Biological Nitrogen Fixation
CEC	Cation Exchange Capacity
GHG	Greenhouse Gas
LER	Land equivalent ratio
MBC	Microbial biomass carbon
MBN	Microbial Biomass Nitrogen
SDGs	Sustainable Development Goals
SOC	Soil Organic Carbon
SOM	Soil Organic Matter
SWHC	Soil Water Holding Capacity

6.1 Introduction

Accelerating soil-erosion, diminishing soil health and, organic carbon storage, dwindling biodiversity and environmental sustainability are the major challenges agriculture faces globally (Ahmad et al. 2022; Babu et al. 2023). Hence, designing and development of environmentally robust and soil supportive production technologies are crucial for ensuring the nutritional-food and livelihood security of ensuing generations. Agricultural practices, such as clearing land, crop production, irrigation water, livestock management, fisheries and other related activities, influence GHG emissions and contribute to climate fluctuations. Sustainable soil health management involves addressing current needs while ensuring that future generations can also rely on the soil to meet their requirements. Soil preservation has been essential for centuries, but the advancements of technology and population expansion has had both positive and negative impacts. Over-intensive farming, urban sprawl and climate shifts have negative impact on health of soil. It is crucial to focus on soil-conservation, soil quality and soil vitality to safeguard soils for future sustainability (Yadav et al. 2023). Sustainable soil utilization stabilizes the climate and strengthens resilient supply chains, particularly in terms of food security. Key aspects of soil preservation include increasing soil organic carbon, preventing erosion, and minimizing chemicals use. Soil preservation is vital for healthy crop production. Soil system acts as a reservoir of nutrients, water and air, creating optimal conditions for plant growth. Sustainable soil management practices can store carbon in the ground and mitigate GHG emissions (Singh et al. 2021). Soil can also become compacted, saline, or chemically contaminated by injudicious fertilizer use, excess, insufficient, or poor-quality water. Excessive irrigation can cause leaching, diseases, and other negative environmental and agricultural outcome. Soil compaction can affect the energy efficiency of field equipment, greenhouse gas emissions. Declining biodiversity can potentially affect crop pollination.

The integrated cropping system aims to advance sustainable agriculture and rural development by making efficient use of resources, encouraging biodiversity and minimizing harmful environmental effects. Legume embedded intensified cropping produce more food with a lesser environmental impact, besides increasing biodiversity and soil-health (Singh et al. 2019). Integrated systems can efficiently utilize resources like water, sunlight and nutrients both time and spatial dimensions, enhance crop yields and land-use efficiency; reduce weed, diseases and pests incidence. Legumes serve multiple purposes, like improving human nutrition, fodder availability and boosting crop production by restoring the soil fertility through biological N_2 fixation. The cereal-legume integration (Fig. 6.1) system extensively recommended in food crop production to gain both agronomic and ecological advantages. In cereal-legume integration systems, legumes are capable of efficiently fixing atmospheric nitrogen, which can subsequently be utilized by the cereal crops and supply up to 73% of the nitrogen requirement of cereal crop, depending on age stand, background soil available-nitrogen (N), cropping system and crop species (Thilakarathna et al. 2016). The efficiency of symbiotic biological nitrogen fixation

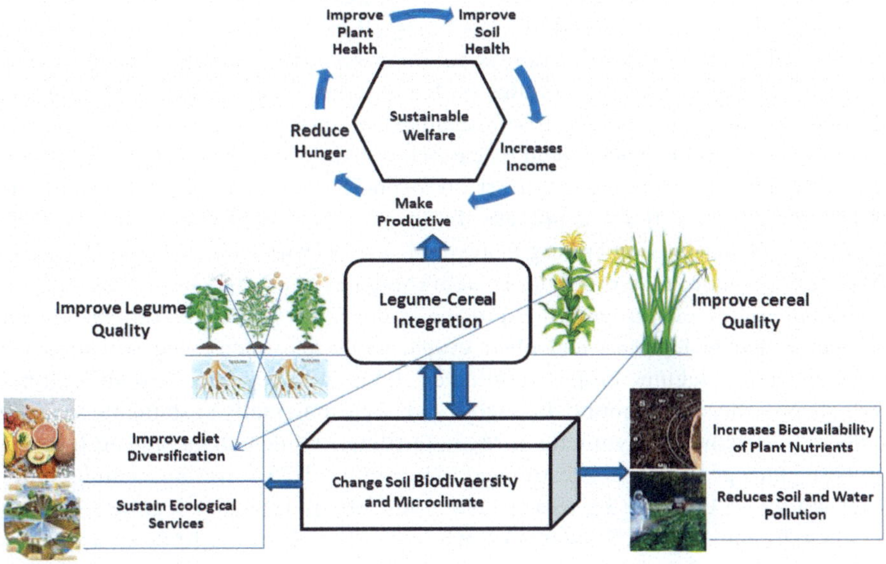

Fig. 6.1 Legume cereal integration

(BNF) in legume crops is enhanced by reducing the inhibitory effects of soil nitrogen on nitrogenase activity, which is achieved by the crops competitive absorption of soil nitrogen. Legume and non-legume integration minimized the harmful effects linked to fertilizer use, and cereal dominating cropping systems (Ashworth et al. 2015) Cereal-legume integration offer several environmental benefits, including enhanced soil carbon-retention and improved nitrogen use-efficiency. These advantages are mainly due to the way nitrogen is shared and utilized between the legumes and cereals, which can significantly cut down on the need for chemical nitrogen fertilizers. This reduction in fertilizer use is critical for mitigating the environmental issues associated with excessive nitrogen applications. To advance sustainable agriculture, it is essential to examine the nitrogen-fixation rates and the nitrogen dynamics of transfer and distribution in cereal-legume integration. Moreover, in such systems, microorganisms can move between the roots of different plants, leading to improved soil conditions. Cereal-legume integration boosts soil health by increasing the biological-diversity and functionality of the soil microbial community like MBC, MBN, dehydrogenase, sulfatase and phosphatase. Soil microbial diversity is crucial for various soil processes, including nutrient-cycling, organic-matter decomposition, mineralization, nitrification and improve soil structure, all of which support healthy plant growth.

6.2 Biological Mechanism of Cereal-Legume Integration

Heavy load of inorganic fertilizers in agriculture has frequently led to suboptimal results, characterized by low nitrogen use efficiency and significant adverse effects on soil, air, water and other ecological environments impacts. Symbiotic biological nitrogen fixation (BNF) has emerged as a viable solution to enhance agricultural production while addressing environmental concerns (Fig. 6.2). The discovery of nitrogen-fixing nodules on leguminous roots marked the beginning of understanding the potential of legumes in nitrogen assimilation (Glaze-Corcoran et al. 2020). Subsequent research has demonstrated that rhizobia associated with legume crops can convert atmospheric nitrogen (N_2) into ammonia under ambient conditions, a process that requires minimal energy inputs. This conversion makes the fixed nitrogen available not only to the host-legume but also to other crops within the cropping-system (Stagnari et al. 2017). By providing an additional source of nitrogen, this process is expected to reduce competition for nitrogen between legumes and other crops, while also increasing soil nitrogen content after the legume crop is terminated. Integrating legumes into agricultural systems presents an economically viable and environmentally sustainable strategy to decrease reliance on synthetic nitrogen fertilizers while enhancing soil fertility. Globally, nitrogen fixation is estimated to be around 1.75×10^{11} kg, with legumes contributing ~8.0×10^{10} kg (Srivastava et al. 2023). Typically legumes, can fix between 20 and 200 kg N/ha/annum, while nearly half of total nitrogen fixation occurs through industrial processes that produce nitrogen fertilizers.

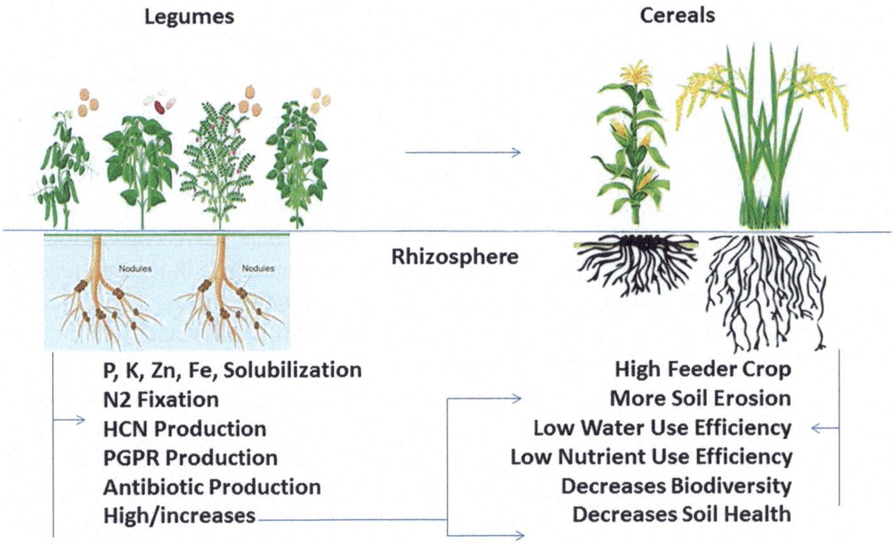

Fig. 6.2 Beneficial impacts of legume-cereal integration

Intercropping of wheatgrass with perennial legumes effectively reduces nitrogen inhibition and improve the overall nitrogen content in the soil, facilitating long-term nitrogen release while minimizing the adverse effects typically associated with fertilizers application (Ashworth et al. 2015). Additionally, legumes-cereals combinations improve nitrogen use efficiency, by decreasing nitrate leaching. These benefits likely stem from the complementary nitrogen utilization and transfer. Maize + beans intercropping offers clear benefits over monoculture, primarily because the roots of both crops intertwine during their common growth period, reducing competition for root nutrients, especially nitrogen (Singh et al. 2021; Yadav et al. 2021). This competition enhances the ability of cereal crops to access soil nitrogen effectively and reduces the nitrogen repression experienced by legume root nodules, resulted in yield advantages for both cereal and legume crops (Rathore et al. 2022; Avasthe et al. 2023). Nitrogen application significantly affects both the percentage of nitrogen sourced from the atmosphere and the amount fixed by intercropped alfalfa. When 200 mg of nitrogen per kg of soil is applied, atmospheric nitrogen contribution and nitrogen fixation in intercropped alfalfa drop by 41% and 46%, respectively (Wang et al. 2020). An integration of legume-pasture enhances symbiotic biological nitrogen fixation (BNF) efficiency, increasing fixed nitrogen by 18% compared to sole legume cropping. In soybean-sugarcane integration, nitrogen application rates notably impact the quantity and biomass of soybean root nodules.

Hei et al. (2021) reported that water Mimosa achieved the highest efficiency in biological nitrogen fixation under the reduced-nitrogen (RN, 140 kg N ha^{-1}) than conventional-nitrogen (CN, 180 kg N ha^{-1}). The ability for nitrogen fixation, nitrogenase activity, and the transformation and distribution of nitrogen in intercropping systems can vary depending on the region and the particular crop species used. To enhance nitrogen utilization in cereal-legume integration, it is essential to explore strategies that maximize symbiotic biological nitrogen fixation in legumes and accurately assess nitrogen-distribution within these systems, that the integration of cereal & legumes can improve soil characteristics by enhancing the functional biodiversity of the soil microbial community, which is vital for plant growth. In intercropping systems, microorganisms can move between the roots of different plants, aiding in the cycling of mineral nutrients, the turnover of organic matter, nitrification, mineralization, and the formation of soil structure. A positive-correlation has been noted between legume development, nitrogen accumulation in legume tissues, and up to a 30% increase in the yield of subsequent maize crops. The potential advantages of legume-cereal integration in enhancing soil fertility and boosting the yields of subsequent crops is given in Table 6.1.

Intercropping of faba bean with maize notably improve above ground biomass and grain yield compared to a wheat-maize system, particularly when phosphorus levels are low (75 kg P_2O_5 ha^{-1}), but this advantage diminishes at higher phosphorus rates (>112.5 kg P_2O_5 ha^{-1}) (Li et al. 2007). This yield benefit arises from the different growth seasons and rooting depths of faba bean and maize, making intercropping particularly effective on soils with low phosphorus availability. Legumes can transfer nitrogen to cereal crops mainly through the breakdown of legume litter and the addition of root to the soil (Rathore et al. 2024). The maize-legume-wheat

Table 6.1 Effects of legumes-cereals on crop yields, nutrient availability, and nutrient uptake

Study area	Cropping system	Findings	References
South Punjab (Pakistan)	Chickpea, wheat, soybean sole systems, and wheat-chickpea, wheat-soybean intercropping systems	Higher nitrogen and phosphorus and crop yield under cereal-legume integration over sole cropping	Raza et al. (2023)
Haidian (China)	Monocropping and intercropping systems with legumes (soybean, alfalfa and mung bean) and cereals (maize and sorghum)	Improve cereals biomass production and soil sustainability	Zhao et al. (2022)
Karnataka (India)	Oats, maize, pearl millet, sorghum and bajra napier hybrid (cereal crops) lucerne and cowpea, desmanthus and sesbania (legume crops)	Lowers fiber fractions; enhanced crude protein, dry matter intake and digestibility	Manoj et al. (2021)
Dowa (Malawi)	Sole systems (pigeon pea, cowpea and maize), and legume–legume and legume–cereal intercrops	Higher yields and net returns in all intercrops than sole crops	Njira et al. (2021)
Kanpur (India)	Lowland rice-based system: rice–wheat-rice–chickpea, rice–wheat–mung bean and rice–wheat, rice–chickpea	In addition of green gram in rice -wheat increased rice yield (grain) by 10–14% Substituting wheat with chickpea in the rice-wheat system boosted rice grain yields by 5–8%	Ghosh et al. (2020)

cropping system has been found to be more advantageous than monoculture, allowing for a 25% reduction in nitrogen usage for subsequent wheat crops while maintaining sufficient residual fertility to significantly boost wheat yields without additional fertilization (Lai et al. 2022). In system perspectives, legume intercropping not only conserves nitrogen but also improve soil health indicators (Yadav et al. 2021). Cereal crops tend to rapidly absorb nitrogen from the soil, and their root exudates can stimulate nodulation in legume roots, thereby increasing the rate of symbiotic nitrogen fixation and ultimately enriching the soil nitrogen pool (Singh et al. 2021).

Cereal-legume integration systems (Fig. 6.2) have been demonstrated to increases the phosphorus (P) acquisition however; rhizosphere dynamics play a crucial role in determining the amount of phosphorus (P) available to crops (Hinsinger et al. 2011). Integration of legumes with other crops (cereals) increased root-surface area enhances soil exploration for nutrients and legume enhances the P acquisition. The complementary growth periods and root distributions of legumes and cereals allow for more adequate use of resources like light and nutrients, leading to yield advantages compared to monoculture systems (Fig. 6.2).

6.3 Agronomic Practices for Cereal-Legume Integration

Integrating legumes in cereal-based system holds a prominent role in enhancing soil sustainability. Agronomic practices for integration involve the judicious combination of different agricultural practices and crops to increases the long term sustainability and resilience of the cropping-system. Legumes can be integrated, by crop rotation, intercropping, and covers crop in the systems.

6.3.1 Crop Rotation

The principle of crop rotation (Fig. 6.3) aims at the strategic integration of various crops in the same field at different cropping seasons (Dadhich et al. 2015). Exhaustive crops (cereals) should be followed by restorative crops (pulses) which improve the soil conditions. Deep rooted crops such as cowpea and beans should be followed by shallow-rooted crops like cereals minimizing nutrient exhaustion from the soil systems (Watson and Stoddard 2017).

Alternating crops with varied root structures, different crop durations as well as having different peak requirements of water, fertilizers, labor, etc. leads to the development of effective crop rotation. Various factors affect crop rotation via climatic conditions, availability of labor, fertilizers, seeds, and irrigation, as well as their profitability, marketing, and processing aspects (Tanveer et al. 2019). Integration of legumes in cereal-based cropping-system offers a multitude of benefits that account for the breaking of insect, weed, and disease cycle, reducing the fertilizer input for the succeeding crop (Stagnari et al. 2017), increasing biodiversity, improving fertility, controlling erosion (Ratnadass et al. 2013), balanced uptake of nutrients from

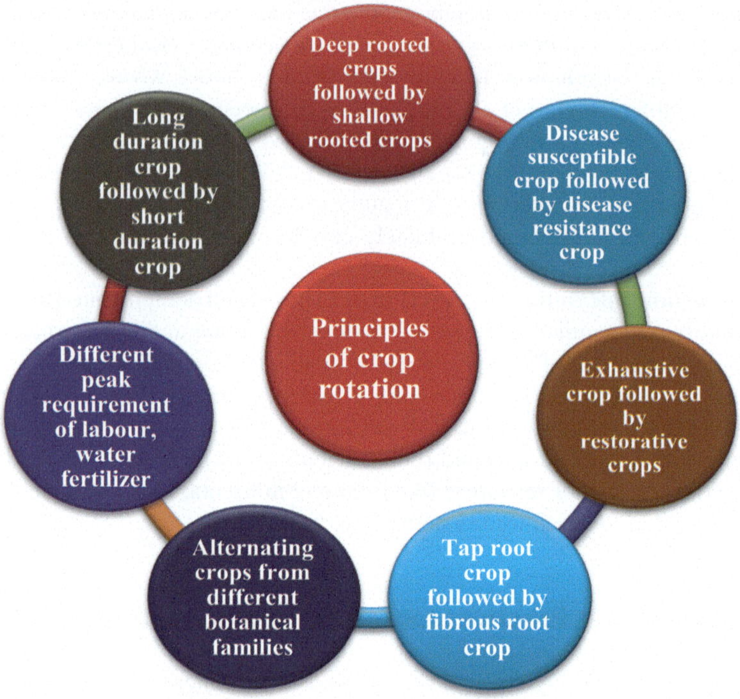

Fig. 6.3 Principles of crop rotation

different layers of soil, producing higher biomass, thereby increasing SOC stocks and water retention in the soil. Legume growth stage while incorporated decides to a large extent the magnitude of N fixed and available to the succeeding crops. Improved crop yield, addition of nitrogen in soil and enhances carbon pool due to legumes over the non-legume systems under various regions of the earth (Franke et al. 2018; Kumar et al. 2018). Benefits of legume integration are listed in Table 6.2.

6.3.2 Intercropping

Two or more then two crops are cultivated on the same-piece of land in a definite row arrangement in time and spatial dimension. Its main aim is to enhances production per unit area and efficiently use the available resources (Maitra et al. 2020). In an intercropping system, the ecological principles of facilitation and complementarity operate simultaneously. In facilitation, one species modifies the biotic and abiotic environment in a positive aspect. While complementarity refers to the improved sharing of resources thereby reducing interspecific competition, and enhancing better resource utilization (Li et al. 2020). Row, relay, mixed, and strip are the four types of crop integration usually practiced. The component crops exhibit varying

Table 6.2 Crop rotation involving legumes and their associated benefits

Cropping system	Location	Remarks	References
Lowland Rice-wheat-mungbean Upland Maize-wheat-mungbean	Kanpur	Grain yield, system productivity, net return, soil organic carbon, available nitrogen and phosphorus↑	Ghosh et al. (2020)
Maize-chickpea-sesbania Maize-mustard-mungbean	Delhi	Total soil nitrogen, soil organic carbon↑	Yadav et al. (2017)
Rice-potato-green gram	Varanasi	System productivity, net returns ↑	Banjara et al. (2022)
Groundnut–wheat	Delhi	System productivity, net returns, SOC, available nitrogen and phosphorus ↑	Kumar et al. (2016)
Maize-wheat-mungbean Pigeonpea-wheat	Kanpur	SOC, enzymes activity (acid/alkaline phosphatase, β-glucosidase and arylsulfatase) ↑	Borase et al. (2020)
Rice–wheat–cowpea	Patna	Bulk density, penetration resistance, micro-aggregate, ↓ macro-aggregates, ↑ mean weight diameter, aggregate ratio, geometric mean diameter, energy use efficiency, profitability ↑	Rao et al. (2021)

competitive abilities in utilizing available resources such as nutrients, water, space and sunlight. Therefore, the selection of a suitable crop and seedling ratio is of utmost importance to effectively meet the demand for resources (Lithourgidis et al. 2011). Therefore, tall and short plants, legume and cereal, C_3 and C_4 plants, and plants with different rooting, growth patterns, crop duration, water and nutrient requirements should be integrated (Brooker et al. 2015). Replacement series and additive series is another form of classification of intercropping on plant population percentage. In the additive series, the base crop population is 100% and an intercrop is added while in the replacement series, no component crop has a population of 100%. Legumes are weak weed suppressors due to slow development in early stages but when grown as an intercrop with cereals suppression ability against weed increases, also the allelopathic compounds released by the component crop add to the benefit. Intercropping increases, the population of parasites and predators of the pests and the changed environment making it difficult for the spread of pests (Luo et al. 2022). Intercropping provides an effective strategy for removing heavy metals from the soil by releasing root exudates, modifying microbial diversity and enhancing biomass production *viz*., microbial biomass carbon (MBC) (Wiche et al. 2016; Bian et al. 2021).

6.4 Effect of Cereal-Legume Integration on Soil Health and Ecosystem Services

Cereal-legume integration plays a significant role in improving soil-health and providing valuable ecosystem services. Legumes enhances soil structure by the addition of biomass and fertility by promoting BNF also reducing GHG emissions, increasing resource use efficiency; reduce the energy and fertilizers consumption (Yadav et al. 2021; Rathore et al. 2024). Legume based integration can reduce the carbon footprint of a system making it more sustainable. Legume residues enhance soil quality and health by positively affecting its physical, chemical and biological properties (Mousavi et al. 2009). Intercropping with legumes was found to improve invertase, urease, alkaline phosphatase, and catalase activity. Integration of legumes in cropping system reduces the activity of plant parasitic nematodes and increases the microbial biodiversity in rhizospheric zone (Zhang et al. 2010). Improvement in soil quality due to legumes attributes to their low C:N ratio, extensive root system, increased microbial activity and nutrient-cycling (Dhakal et al. 2015). Legume inclusion binds pesticide with soil particles for longer time thereby reducing contamination (Fester et al. 2014). Further, Rhizobia can help in reducing soil contamination by degrading soil pollutants (Jin et al. 2013), or increasing the activity of various microbial enzymatic activity, adsorption as well as accumulation of heavy-metals and volatilization by activity of microbes facilitating bioremediation (Hao et al. 2012). Hydrogen, a by-product of the nitrogen fixation process, has recently been found to enhance plant tolerance to heavy metal toxicity (Cui et al. 2013). By integrating legumes with cereals, farmers can create a more balanced and productive agro-ecosystem that supports long-term soil health and delivers essential ecosystem services.

6.4.1 Soil Physical Properties

The fibrous root system of cereals binds the soil upper layer and legumes by their deep root system breaks the hardpan in deeper layer of soil thereby facilitating utilization as well as circulation of water and nutrients (Fig. 6.4). Soil physical-properties such as soil structure porosity and aggregate stability are influenced by the soil organic carbon, which playing an important role in regulating the soil water-holding capacity (SWHC), infiltration, bulk density, and penetration resistance (Singh et al. 2021; Yadav et al. 2024). Leguminous crops add more organic matter to the soil after decomposition which improves water retention, chelation thereby improving soil aggregation. Legumes as green manure crops improve soil physical characteristics and soil structure by improving soil aggregation ability as well as increasing water-infiltration (Mousavi et al. 2009). The macro and micro aggregates have increased in legume based cropping-system (Garland et al. 2016). Inclusion of legumes in the cropping systems increases aggregate stability which influences water retention, water infiltration, and movement of air in addition to reduces bulk density and penetration resistance (Wilhelm et al. 2004). Legumes had deeper tap

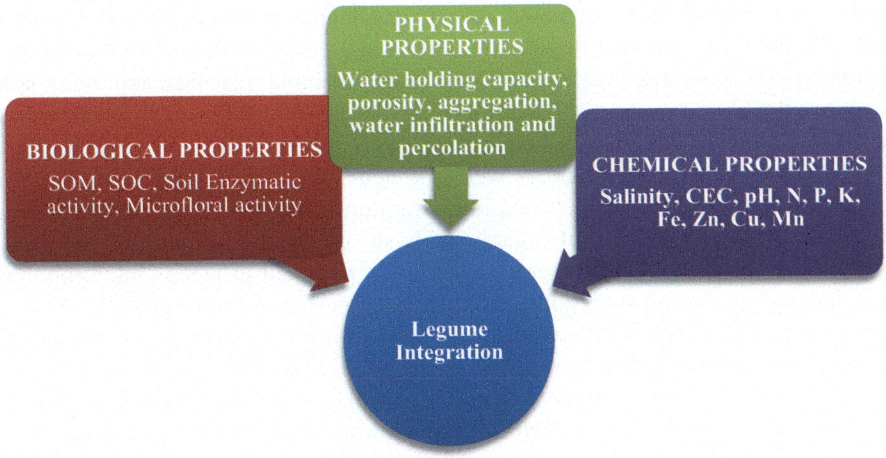

Fig. 6.4 Effects of legume integration on soil health

root system, penetrate soil hardpan thereby preventing soil compaction and provides residual root channel (Peoples et al. 2015).

6.4.2 Soil Chemical Properties

Legumes, with their deep root systems, ability to fix symbiotic biological nitrogen and complementary association with cereals and other non-legume crops, can help restore the declining crop productivity of cereal-based systems. Total nitrogen, SOC, phosphorus, and potassium was reported to be increased with legume based cropping system whereas exchangeable cation and C: N ratio was decreased (Adesoji et al. 2014). Legumes integration improves various soil chemical indicators such as pH, EC, cation exchange capacity (CEC), nutrient content and SOC concentration that are important to assess soil sustainability (Yuvaraj et al. 2020). Legumes aids in soil buffering by maintaining soil pH by increasing it by reduction of H^+ to H_2O and CO_2, simultaneously decrease it by production of CO_2 and organic acids (Mitran et al. 2018). When incorporated as green manure increases soil nitrogen as well as phosphorus (Adesoji et al. 2014). In upper layer of soil, the total nitrogen (N) content has shown increment by 4 times in pigeon pea-based cropping system as compared to cropping system without legumes (Kumawat et al. 2022). Grain legumes with their deep root system, higher root cation exchange capacity, higher root exudates, particularly carbon-rich rhizo-deposition, aid in phosphorus cycling and enhance the bioavailability of non-labile phosphorus by acidifying the rhizosphere soil, increasing the activity of vesicular-arbuscular mycorrhizae, promoting greater root length, and boosting phosphorus transporter activity. Owing to the phosphatase enzymes as well as organic acids, lupin and faba bean can recover phosphorus from non-available form (Nuruzzaman et al. 2005). The extensive deep root system including microbial diversity in integration of cereal and legumes increases uptake and availability various nutrients. Legumes influence the pH of the soil in the root zone due to release of acids from roots and increased activity of soil microbes thereby impacting soil acidification. The ex-situ and in-situ incorporation of legume biomass has positive-impact on soil chemistry as well as nutrient availability. Soil organic carbon and N were 4% and 20% higher respectively on top 20 cm soil surface in legume based intercropping as compared to monocropping (Cong et al. 2015). Release of organic acids such as carboxylic acids from roots solubilizes phosphorus and increases availability to cereal grown as intercropping with legumes including succeeding crops in rotation (Nuruzzaman et al. 2005).

The BNF, in addition to benefitting legumes, improves yield in crops flowing legumes through interplant transfer of nitrogen (Varma et al. 2017). Addition of nitrogen-fixing legumes to increase supply of nitrogen to succeeding crop as well as thorough understanding of the factors affecting transfer is needed to manage nitrogen in sustainable manner. The biologically fixed N annually by legumes varies 25–200 kg N per ha depending on several factors (Carlsson and Huss-Danell 2014). Legumes can improve nutrient availability due to biological nitrogen fixation, nutrient recycling, enhancement of soil organic carbon and nitrogen (Keeler et al. 2009). Most of the nitrogen though harvested as grain yield, significant portion of nitrogen is deposited in soil thereby adding N stock to soil (Dhakal et al. 2015). The N is added in soil through root exudates, fallen leaves and N leachate from leaves (Addo-Quaye et al. 2011) The grain legume will meet the requirement of soil N by fixing atmospheric N_2 (Naudin et al. 2010). Phosphorus mobilization was reported to be higher in intercropping with legumes due to secretion of organic acids by the legume as reported by several studies such as in cowpea and maize (Latati et al. 2016) and faba bean and lupin (Wang et al. 2012).

6.4.3 Soil Biological Properties

The altered root distribution as well as structure with inclusion of legume in cropping system along with symbiosis with rhizobia helps in secretion of organic acids in the rhizosphere which shows positive correlation with various soil microorganism and their interaction with crops (Swarnalakshmi et al. 2020). Legume biomass provides nutrients, amino acids, and organic acids enhancing growth and activity of microorganisms such as fungi, bacteria and, other soil biota that help in the organic matter decomposition and nutrient cycling (Fig. 6.4). Integration of cereals-legumes, enhanced soil MBC and MBN to the tune of 8.1–19.7% and 52.4–75.8%, respectively over non-legume integration (Kumawat et al. 2022). Decomposition of nitrogen rich legume residues increase soil biopores and earthworm population in rhizospere thus facilitating movement of air and water in soil. Increased population of soil microbes like fungi, bacteria, actinomycetes, and diazotrophs, in plant rhizosphere and endosphere including activity of various enzymes such as dehydrogenase, nitrogenase, urease, and acid phosphatase was reported with inclusion of legumes in the cropping system (Li et al. 2016; Rathore et al. 2024). Legumes inclusion increases the microbial activity along with diversity of microbial population thereby facilitating decomposition of OM, nutrient cycling, N fixation (Ludwig et al. 2015). Lectins produced by legumes influence plant growth-promoting rhizobacteria (PGPR) mobility, activity and colonization (Schelud'ko et al. 2009). Maize-soybean integration increased actinomycetes, halophile, firmicutes, cysts, and corynebacter population (Lin et al. 2022). Pivato et al. (2021) reported an abundance of *Chlorobacteria, Proteobacteria, Actinobacteria*, and bacteroidetes in wheat-pea intercropping and an abundance of N-fixing microorganisms was reported in groundnut + maize intercropping (Chen et al. 2018). Through microbial activity improvement, legumes contribute to a more dynamic, holistic, and functional soil ecosystem. Inclusion of legumes increases microbial diversity as well as activity (Alvey et al. 2003) improve inherent fertility of soil as microbial carbon and nitrogen constitutes about up to 7% and 5% of total soil carbon and nitrogen respectively. The SMB, soil microbial community structure as well as soil health is improved by including legumes (Chintala et al. 2014; Singh et al. 2021). SMB also facilitates nutrient uptake in crops besides improving soil structure, chemistry, degradation of various agrochemicals and suppression of harmful pathogens (Liang et al. 2010). Inclusion of pulses after maize enhanced the soil microbial biomass carbon (SMBC), SOC, dehydrogenase (DHA) to the tune of 5.7–6.9%, 4.7–8.4% and 11.0–32.3%, respectively over non-legume inclusion (Ngangom et al. 2017). Effect of legume- cereal integration on microbial population given in Table 6.3.

6.4.4 Ecosystem Services

Legumes inclusion provides a multitude of ecosystem services (Fig. 6.5) that has the potential to enhance soil-health, mitigate GHG emissions thus addressing the issues of land degradation and climate-change (Mitran et al. 2018). Legumes inclusion in the cropping-system provides multitude of benefits to agro-ecosystem and society which constitutes reduction of GHGs emission particularly carbon dioxide

Table 6.3 Effect of legume- cereal integration on microbial population

Cropping system	Microorganisms proportion	Beneficial diazotrophs	References	Place
Soybean–sugarcane	Increases Arthrobacter, rhizobia, Chinese rhizobi Streptomyces, Bacillus, Pantomyces, Enterobacter	Enrichment of diazotrophs in the both rootzone including endosphere Effective bio-inoculants for sugarcane by improving plant growth	Solanki et al. (2019)	China
Soybean–maize	Actinomycetes, Firmicutes, Corynebacter, Cysts, Halophile bacillus, Phagocytes		Lin et al. (2022)	China

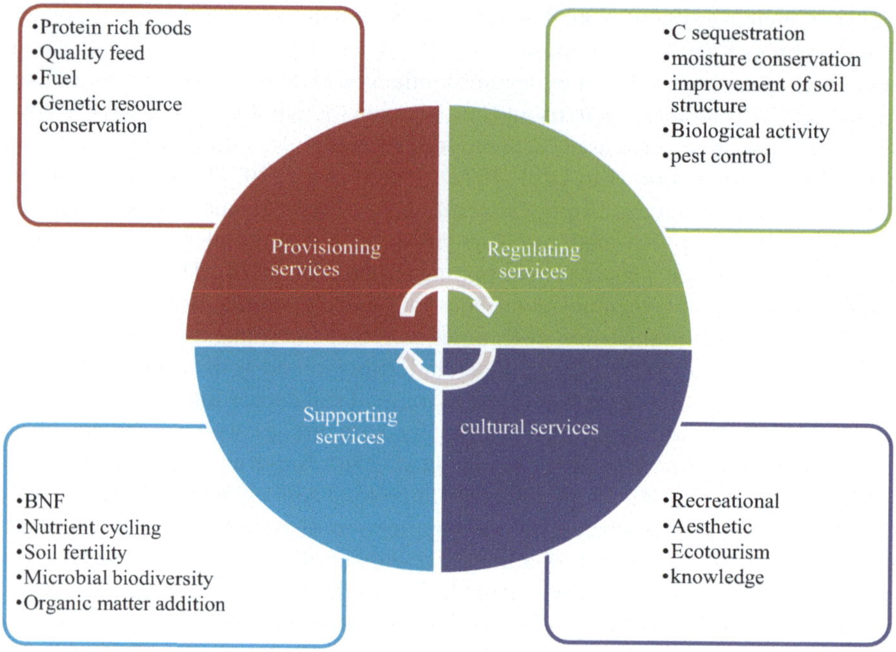

Fig. 6.5 Multiple ecosystem services by legume based cropping system

(CO_2) and nitrous oxide (N_2O) by 5–7 times per unit area (Jeuffroy et al. 2013) and restoration of soil-health, control of insect-pests and diseases, natural resources-conservation such as soil and water (Lithourgidis et al. 2011) besides decrease in use of fertilizers and energy consumption (Watson and Stoddard 2017). Further it provides various additional ecosystem services such as food and nutritional-security (Ness et al. 2010), safeguarding health of human being, ecological balance, nutrient-cycling and improvement of soil-biota (Chen et al. 2003).

6.5 Beneficial Impact of Cereal-Legume Interaction

Beneficial impact of legumes is summarised in Fig. 6.6.

6.5.1 Soil Conservation

Xtractive farming practices results in soil erosion and disintegration of soil structure. Legume can play role in protecting soil by way of inclusion in cropping system (Srinivasarao et al. 2012). Legumes have dense foliage which protects soil from the impact of raindrop. Legumes roots bind soil particles near the surface and improve water infiltration. Intercropping can reduce soil erosion than that of sole cropping

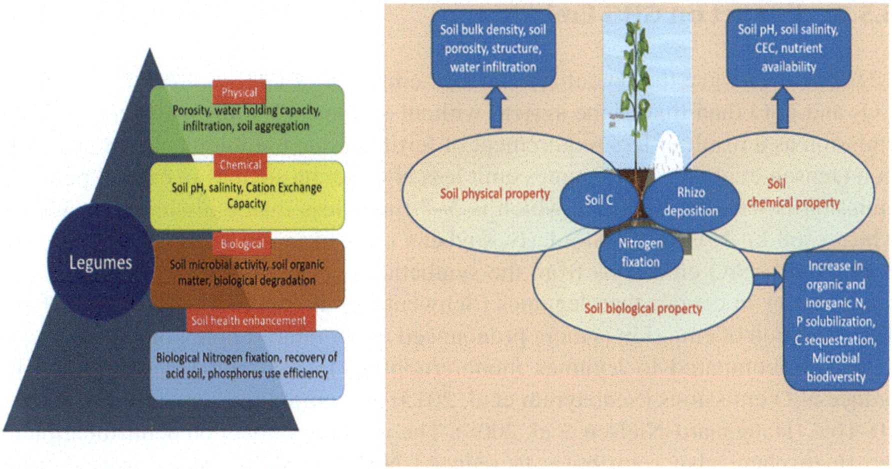

Fig. 6.6 Beneficial impact of legumes

(Wen et al. 2022) and intercropping cereals with legumes particularly more effective in minimizing the soil loss due to erosion (Hauggaard-Nielsen et al. 2008). Further rotation of crops with different rooting pattern such as cereal-legume can increase macro and micro aggregate thereby increasing infiltration of water and reduces soil erosion as a result of decreased surface runoff. The release of root exudates particularly organic acids acts as chelating agent thereby reducing soil erodibility as well as improving aggregation. Strip cropping of legumes in cropping system reduces soil loss due to erosion (Lithourgidis et al. 2011) as reported by various studies such as maize and cowpea, sorghum and cow pea to the tune of 5–10% (Dass and Sudhishri 2010). Legumes can also be planted as hedgerows. Cover crops such as annual legumes provide faster cover for a short span of 6 weeks and control erosion faster. Plaza-Bonilla et al. (2016) reported inclusion of cover-crops in conventionally tilled legume-based rotation reduces organic carbon and nitrogen loss of soil, thereby improving N use efficiency. Legume integration with cereal enhances water use efficiency (WUE) due to a reduction in evaporation and conservation of soil moisture as compared to maize mono-cropping. P-solubilizing/mobilizing microorganism activity is increased by legume integration, reducing external P requirements. Soil erosion, which has been hastened by anthropogenic activities, can be minimized by integrating legumes as cover crop (Yadav et al. 2024) which has the capacity to act as soil conditioner by providing substrate to the micro-flora and enhancing residue decomposition. Soil physical properties can be enhanced by incorporating legumes as green manure-crops. In addition, intercropping with legumes can protect the soil from the impact of raindrop by providing a dense-cover (Maitra et al. 2020). Soybean + maize and maize + pigeon pea system reduced the soil loss by 34.5% and 24.4% over sole maize cropping owing to their dense canopy (Singh et al. 2014).

6.5.2 Impact on GHG Emission

Legume crops have the potential to reduce emission of GHGs (Table 6.4) such as CO_2 and N_2O than that of the system without legumes. The legumes reduce GHG emission as a result of less requirement of fertilizer due to biological nitrogen fixation (Jensen et al. 2020). Legumes emit less N_2O per ha such as in case of peas the emission is −69 kg N_2O ha^{-1} which is 5–7 times less than emission from winter wheat (368 kg N_2O ha^{-1}). Similarly, soybean emitted less GHG than cereals (Yue et al. 2017). N_2O emissions from the synthetic fertilizers were reported to be 2.5 times higher in cereals than legumes (Schwenke et al. 2015; Guardia et al. 2016). The reduction of emission is more pronounced under intercropping of legumes with cereals as compared to legumes mono-cropping. Intercropping was reported to reduce N_2O emissions (Senbayram et al. 2015) and nitrate leaching to the extent of 10–16% (Hauggaard-Nielsen et al. 2008). The inhibitory effect on denitrification in the rhizosphere also contributes to reduced N_2O emissions. Feeds derived from legume have the potential to reduce CH_4 emission due to less fibre content, presence of tannin and saponin as well as faster rate of passage ultimately leading to modified rumen methanogenesis (Eckard et al. 2010). Legumes emit lesser CO_2 than cereals due to greater root biomass including rhizo-deposition in the cereal crops (Sehy et al. 2003). Further, legumes can tolerate elevated CO_2 by translocation of above-ground photosynthates to the nodules which is a carbon as well as energy source for the nitrogenase enzyme (Irigoyen et al. 2014).

6.5.3 Yield Advantage

Legume inclusion enhances biomass and grain yield of cereal based systems. Higher yield as compared to monoculture with increased efficiency was reported with introduction of legumes into intercropping system (Inal et al. 2007). The yield advantage in intercropping might be as a result of improvement of soil-functions through the symbiotic effect of legume-rhizobia (Fustec et al. 2010). Legume crops

Table 6.4 Effect of cereals-legume on GHG emission in cropping system

S. No.	Cropping system	GHG emission	References
1.	Maize-soybean rotation	Reduce CO_2 emission by 16%	Omonode et al. (2007)
2.	Maize-soybean (1:2) intercropping	Reduce CO_2 and N_2O flux by 21.31% and 18.1–20.4%, respectively	Dyer et al. (2012)
3.	Maize-peanut and maize-soybean intercropping	Reduce CO_2 and N_2O flux by 24.73% and 25.55%, respectively	Huang et al. (2014)
4.	Maize-soybean intercropping	Reduce N_2O emission by 32%	Chen et al. (2021)
5.	Sorghum- cowpea intercropping	Reduce CO_2 emission by 11–19%	Mogale et al. (2023)

increase the yield of main crop due to addition of more N into the system (Ghaley et al. 2005). Sowing of barley and pea (50:50 and 25:75), barley and faba bean (70:30) showed 50% higher biomass yield (Soufan and Al-Suhaibani 2021) and 52% higher herbage yield (Akdeniz et al. 2019). Similar finding of enhanced biomass when sorghum mixed with lima bean in 80:20 ratio (Yadav et al. 2020) and sorghum + cowpea in 100% (Agegnehu et al. 2006) population increased biomass yield significantly. Higher yield is due to effective transfer of N to cereal under intercropping system as well as capacity to uptake and utilize resources during the cropping season. Inclusion of legumes can increase crop productivity due to their impact on disease control, suppression of weeds, diversification of cropping system including recovery of soil functions (Deutsch et al. 2006; Singh et al. 2021). Addo-Quaye et al. (2011) reported higher land equivalent ratio (LER) in the intercropping-system which indicates the significant utilization of land resource by introduction of legumes.

6.5.4 Reduction of Production Cost

Legumes provide nitrogen to the succeeding crop thereby saving 40–70 kg N ha^{-1} whereas need only starter dose for their cultivation saving about 170–220 kg^{-1} ha of N (Varma et al. 2017). Reduction in fertilizer requirement particularly nitrogen reduces the cost of fertilizers in addition to GHG mitigation (Jensen et al. 2020). About 35–108 kg fertilizer nitrogen can be saved in cereal crops by including legume in preceding year as a result of better utilization of resources during the crop growth. The yield benefit to the tune of 8–25% had been reported with reduction in nitrogenous fertilizer. Higher yield in succeeding maize crop grown after legumes such as *Tephrosia, Mucuna, and Crotalaria* has been reported without application of external fertilizer. Cereal-legume integration is a low-cost environment friendly agricultural practice which minimizes chemical fertilizer application by enriching soil nitrogen. The non-legume crops grown simultaneously or in sequence with legumes in cropping system meet their nitrogen requirement from the biological nitrogen fixation or from exudates from root and leachates from leaf, decomposition of residue (Addo-Quaye et al. 2011) thereby reducing the fertilizer requirement and input cost. The increasing cost of nitrogen fertilizer and energy cost of natural gas-based nitrogen fertilizer can be addressed by relying more on BNF. Reduction in tillage decreases maintenance cost, cost of fuel including depreciation of machinery and various implements. Combinations of reduced tillage and legume rotations have shown significant impact in Germany, leading to 21% savings on cost (Preissel et al. 2015). Intercropping with legumes also reduces cost due to saving in labour requirement (Hauggaard-Nielsen et al. 2008; Singh et al. 2021). Positive economic impact of inclusion of cereal-legumes in cropping system given in Table 6.5.

Table 6.5 Economic impact of inclusion of legumes in cropping system

Cropping systems	Rice/maize equivalent yield (Mg ha^{-1})	B:C ratio	Employment (man-days ha^{-1} year^{-1})
Rice-pea	10.72	5.78	230
Rice-lentil	6.42	2.24	210
Rice-fallow	3.26	2.17	100
Maize-French bean	7.07	3.88	150
Maize-rapeseed	3.84	2.30	110
Maize-fallow	2.64	2.39	60

Source: Das et al. (2014) modified

6.6 Challenges and solutions

Cereal-legume integration, frequently utilized in intercropping, mixed cropping, or crop rotation systems, plays a significant role in sustainable-agriculture and offers numerous benefits, including enhance soil-fertility, improved nutrient-cycling and sustainable crop yields. However, this integration also poses several challenges, including variations in growth patterns, competition for resources, allelopathic interactions, pest and disease management, and market limitations. For example, when fertilizer was not applied to maize, strong competition can lead to a 33% decrease in pigeonpea productivity (Kiwia et al. 2019). Additionally, an important consideration is that scaling up pigeonpea intercropping may expose crops to diseases, such as *Fusarium* wilt, and pest pressures that could jeopardize production and productivity. A study by Attallah et al. (2024) reported a decrease in chickpea yield when intercropped compared to its yield in sole farming. This decrease is attributed to intense competition with durum wheat for nitrogen resources. The most notable yield differences between intercropped and mono cropped systems were observed in specific combinations, such as fox millet-groundnut (yield reductions ranging from 20% to 53%), maize-dry bean (10–35%), and in wheat-chickpea (12–32%). These lower yields in intercropped cereals and legumes, when compared to their mono-crop counterparts, are primarily due to increased competition for resources among the intercropped species (Chimonyo et al. 2016). Besides, incorporating intercrops into monoculture systems complicates various aspects of crop production, such as planting, management, and harvesting. This added complexity can lead to increased labour costs (Hong et al. 2017) and may not be well-suited for large-scale mechanized farming operations. Introducing an additional crop into a field can greatly complicate the schedule for crop rotation. Furthermore, when intercropped species have varying maturity times, multiple harvests is necessary. Harvesting crops with similar maturity at the same time can be advantageous; however, it requires specialized equipment, which may not be readily available to farmers with limited resources (Bedoussac and Justes 2010). Additionally, intercropping systems complicate seed-saving efforts of farmers who aim to reduce external-inputs. A possible solution to some harvesting challenges is strip intercropping, where crops are sown in wider bands spanning multiple rows. This method allows for the use of existing machinery; however, the advantages of complementary intercropping diminish as the distance between the crops increases within these wider-strips (Van Oort et al. 2020). Although intercropping may be more compatible with non-mechanized harvesting systems, it can significantly increase labour demands due to the technical requirements for weed control, fertility management, and overall crop management when multiple crops are cultivated together in the same field (Guo et al. 2006). In some instances, these complexities can render intercropping an unprofitable option. Such management challenges create barriers that hinder the widespread adoption of inter-cropping practices among-farmers (Himanen et al. 2016).

The adoption of cereal-legume integration is often limited by concerns that non-crop plants may act as a 'green bridge,' harbouring pests-pathogens that can migrate to cash-crops. Living plants and the residues from intercrops can provide food and shelter for over-wintering pests and pathogens, facilitating their infestation of newly planted crops. This 'green-bridge' effect can enable the movement of pests between growing seasons or neighbouring fields (Ocimati et al. 2018). Additionally, Sarkodie-Addo and Abdul-Rahaman (2012) noted that legumes can become problematic in intercropping systems by overshadowing the main crop(s) and consequently diminishing yields.

Economic constraints and market dynamics also pose challenges to cereal-legume integration. Farmers may face difficulties accessing markets for both types of crops, particularly if one crop has a more established market than the other. Furthermore, the prices of legumes and cereals can fluctuate differently, which can lead to economic instability for farmers relying on both crops for their livelihood. This economic uncertainty can deter farmers from adopting legume-cereal integration practices, even if they offer agronomic benefits. A lack of technical knowledge and barriers to adoption are additional challenges. Farmers may not be fully aware of the best practices for integrating legumes and cereals, such as optimal planting times, crop spacing, and rotation schedules. This knowledge gap can result in sub-optimal integration, reducing the potential benefits of these systems. Moreover, traditional farming practices and resistance to change can hinder the adoption of new practices, with some farmers hesitant to move away from familiar mono-cropping systems.

Enhancements of water use efficiency in cereal legume systems necessitate the adoption of efficient irrigation techniques and the use of drought-resilient crop varieties. Implementing drip or sprinkler irrigation systems can deliver water more precisely to crops, thereby reducing competition and improving overall water use efficiency (Evans and Sadler 2008). Furthermore, breeding and selecting legume and cereal varieties that exhibit greater drought tolerance can help mitigate the risks associated with water scarcity and bolster the resilience of these integrated systems. Integrated pest and disease management (IPM) offers solutions to the complex pest dynamics present in cereal-legume systems. IPM strategies can include biological control methods, such as introducing natural predators or utilizing bio-pesticides, to manage pests in an environmentally sustainable manner. Economic and market interventions are crucial for supporting legume-cereal integration. Strengthening value chains for both crop types, including improvements in storage, processing, and marketing infrastructure, can enhance market access and provide farmers with more stable and diversified income sources. Building farmers' capacity through education and extension services is vital for the successful integration of legumes and cereals. Investment in research and development (R&D) is also necessary to create new technologies, crop varieties, and farming practices that address the specific challenges of legume-cereal integration.

6.7 Future Directions and Research Needs

While cereal-legume interaction offers multiple benefits for sustainable agriculture, however several areas require further studies for harnessing the full potential of this system.

(a) Optimizing crop combinations: Further study is required to determine the most effective cereal-legume combinations for different agro-ecological zones to harness full yield potential and ecosystem services.
(b) Improving mechanization: Creating specialized machinery and methodologies for the planting, management and harvesting of intercropping systems to reduce labour demand and facilitate adoption in large-scale agriculture.
(c) Improving nutrient dynamics: Additional research on nutrient transfer mechanisms between legumes and cereals will pave the way for effective nutrient management and more demand tailored application of fertilizers.
(d) Crop improvement: Investigation into the selection and breeding of cereal and legume varieties that exhibit superior adaptation to the intercropping systems under the changing climatic conditions will help in wider adaptability of the system among farmers.
(e) Economic analysis: In-depth economic assessments to measure the long-term advantages of cereal-legume integration encompassing ecosystem services and reduced input cost will help in comprehensive understanding of the profitability of the system.
(f) Ecosystem based studies: Studies on the effect of cereal-legume integration on ecosystem services like biodiversity, carbon sequestration GHGs emission need to be done for wider acceptability of this system under changing climatic scenario.

6.8 Conclusion

The integration of cereals and legumes offers a sustainable solution to global agricultural challenges by improving soil health, increasing crop yields, and reducing the dependency on chemical fertilizers. This system harnesses the benefits of biological nitrogen fixation by legumes, which naturally enriches soil nitrogen levels, supporting the growth of neighbouring cereal crops. The reduction of chemical fertilizers, improvement of soil properties as well as mitigation of GHG emission makes legumes a viable option for cropping system diversification for long term sustainability. By improving nutrient cycling and resource utilization, intercropping can enhance environmental outcomes, such as soil carbon retention and reduced nitrate leaching. Additionally, this approach promotes biodiversity, improves water management, and mitigates climate change impacts. Legume-cereal intercropping represents a key strategy for promoting the sustainable agriculture, ensuring food and, nutrient security and protecting environmental quality.

References

Addo-Quaye AA, Darkwa AA, Ocloo GK (2011) Growth analysis of component crops in a maize-soybean intercropping system as affected by time of planting and spatial arrangement. ARPN J Agric Biol Sci 6:34–44

Adesoji AG, Abubakar IU, Labe DA (2014) Soil chemical properties as affected by incorporated legumes and nitrogen in soil with maize(Zea mays L.) in a semi-arid environment. Int J Agric Innov Res 3:888–894

Agegnehu G, Ghizaw A, Sinebo W (2006) Yield performance and land-use efficiency of barley and faba bean mixed cropping in Ethiopian highlands. Eur J Agron 25:202–207. https://doi.org/10.1016/J.EJA.2006.05.002

Ahmad F, Saeed Q, Shah SMU et al (2022) Environmental sustainability: challenges and approaches. In: Natural resources conservation and advances for sustainability, pp 243–270. https://doi.org/10.1016/B978-0-12-822976-7.00019-3

Akdeniz H, Hosaflioglu I, Koc A et al (2019) Evaluation of herbage yield and nutritive value. Appl Ecol Environ Res 17:5571–5581

Alvey S, Yang CH, Buerkert A, Crowley DE (2003) Cereal/legume rotation effects on rhizosphere bacterial community structure in West African soils. Biol Fertil Soils 37:73–82. https://doi.org/10.1007/S00374-002-0573-2

Ashworth AJ, West CP, Allen FL et al (2015) Biologically fixed nitrogen in legume intercropped systems: comparison of nitrogen-difference and nitrogen-15 enrichment techniques. Agron J 107:2419–2430. https://doi.org/10.2134/agronj14.0639

Attallah A, Hamdi W, Souid A et al (2024) Impact of cereal–legume intercropping on changes in soil nutrients contents under semi-arid conditions. Sustainability 16:2725. https://doi.org/10.3390/SU16072725

Avasthe RK, Devi SH, Bhupenchandra I, Kumar A, Chongtham SK, Babu S, Singh R, Das A, Gudade BA, Bora SS (2023) Agro-tactics for reducing carbon footprint in agricultural production systems: a review. Indian J Agron 68(2):115–125. https://doi.org/10.59797/ija.v68i2.332

Babu S, Das A, Singh R, Mohapatra KP, Kumar S, Rathore SS, Yadav SK, Yadav P, Ansari MA, Panwar AS, Wani OA, Singh M, Ravishankar N, Layek J, Chandra P, Singh VK (2023) Designing an energy efficient, economically feasible, and environmentally robust integrated farming system model for sustainable food production in the Indian Himalayas. Sustain Food Technol. https://doi.org/10.1039/D2FB00016D

Banjara TR, Bohra JS, Kumar S et al (2022) Sustainable alternative crop rotations to the irrigated rice-wheat cropping system of Indo-Gangetic plains of India. Arch Agron Soil Sci 68:1568–1585. https://doi.org/10.1080/03650340.2021.1912324

Bedoussac L, Justes E (2010) Dynamic analysis of competition and complementarity for light and N use to understand the yield and the protein content of a durum wheat-winter pea intercrop. Plant Soil 330:37–54. https://doi.org/10.1007/s11104-010-0303-8

Bian F, Zhong Z, Li C et al (2021) Intercropping improves heavy metal phytoremediation efficiency through changing properties of rhizosphere soil in bamboo plantation. J Hazard Mater 416:125898. https://doi.org/10.1016/J.JHAZMAT.2021.125898

Borase DN, Nath CP, Hazra KK et al (2020) Long-term impact of diversified crop rotations and nutrient management practices on soil microbial functions and soil enzymes activity. Ecol Indic 114:106322. https://doi.org/10.1016/J.ECOLIND.2020.106322

Brooker RW, Bennett AE, Cong WF et al (2015) Improving intercropping: a synthesis of research in agronomy, plant physiology and ecology. New Phytol 206:107–117. https://doi.org/10.1111/NPH.13132

Carlsson G, Huss-Danell K (2014) Does nitrogen transfer between plants confound 15N-based quantifications of N2 fixation? Plant Soil 374:345–358. https://doi.org/10.1007/S11104-013-1802-1/METRICS

Chen G, Zhu H, Zhang Y (2003) Soil microbial activities and carbon and nitrogen fixation. Res Microbiol 154:393–398. https://doi.org/10.1016/S0923-2508(03)00082-2

Chen J, Arafat Y, Wu L et al (2018) Shifts in soil microbial community, soil enzymes and crop yield under peanut/maize intercropping with reduced nitrogen levels. Appl Soil Ecol 124:327–334. https://doi.org/10.1016/J.APSOIL.2017.11.010

Chen JS, Amin AS, Hamani AKM et al (2021) Effects of maize (zea mays l.) Intercropping with legumes on nitrous oxide (N_2O) emissions. Appl Ecol Environ Res 19:3393–3407. https://doi.org/10.15666/aeer/1905_33933407

Chimonyo VGP, Modi AT, Mabhaudhi T (2016) Assessment of sorghum–cowpea intercrop system under water limited conditions using a decision support tool. Water SA 42:316–327. https://doi.org/10.4314/WSA.V42I2.15

Chintala R, Mollinedo J, Schumacher TE et al (2014) Effect of biochar on chemical properties of acidic soil. Arch Agron Soil Sci 60:393–404. https://doi.org/10.1080/03650340.2013.789870

Cong WF, Hoffland E, Li L et al (2015) Intercropping enhances soil carbon and nitrogen. Glob Chang Biol. 21:1715–1726. https://doi.org/10.1111/GCB.12738

Cui W, Gao C, Fang P et al (2013) Alleviation of cadmium toxicity in Medicago sativa by hydrogen-rich water. J Hazard Mater 260:715–724. https://doi.org/10.1016/J.JHAZMAT.2013.06.032

Dadhich RK, Meena RS, Reager ML, Kansotia BC (2015) Response of bio-regulators to yield and quality of Indian mustard (Brassica juncea L. Czernj. and Cosson) under different irrigation environments. J Appl Natural Sci 7:52–57. https://doi.org/10.31018/jans.v7i1.562

Das A, Ramkrushna GI, Choudhury BU et al (2014) Conservation agriculture in rice and maize based cropping systems for enhancing crop and water productivity - participatory technology demonstration in north east India. Indian J Soil Conserv 42(1):196–203 http://dx.doi.org/10.1007/978-981-13-0253-4_2

Das A, Babu S, Datta M, Kumar S, Singh R, Avasthe RK, Rathore SS, Yadav SK, Singh VK (2022) Restoring soil carbon in marginal land of Indian Himalayas: impact of crop intensification and conservation tillage. J Environ Manag. https://doi.org/10.1016/j.jenvman.2022.115603

Dass A, Sudhishri S (2010) Intercropping in finger millet (*Eleusine coracana*) with pulses for enhanced productivity, resource conservation and soil fertility in uplands of Southern Orissa. Indian J Agron 55:89–94

Deutsch B, Kahle P, Voss M (2006) Assessing the source of nitrate pollution in water using stable N and O isotopes. Agron Sustain Dev 26:263–267. https://doi.org/10.1051/agro:2006025

Dhakal Y, Meena RS, De N et al (2015) Growth, yield and nutrient content of mungbean (Vigna radiata L.) in response to INM in eastern Uttar Pradesh, India. Bangladesh J Bot 44:479–482. https://doi.org/10.3329/bjb.v44i3.38559

Dyer L, Oelbermann M, Echarte L (2012) Soil carbon dioxide and nitrous oxide emissions during the growing season from temperate maize-soybean intercrops. J Plant Nutr Soil Sci 175:394–400. https://doi.org/10.1002/JPLN.201100167

Eckard RJ, Grainger C, de Klein CAM (2010) Options for the abatement of methane and nitrous oxide from ruminant production: a review. Livest Sci 130:47–56. https://doi.org/10.1016/J.LIVSCI.2010.02.010

Evans RG, Sadler EJ (2008) Methods and technologies to improve efficiency of water use. Water Resour Res 44. https://doi.org/10.1029/2007WR006200

Fester T, Giebler J, Wick LY et al (2014) Plant–microbe interactions as drivers of ecosystem functions relevant for the biodegradation of organic contaminants. Curr Opin Biotechnol 27:168–175. https://doi.org/10.1016/J.COPBIO.2014.01.017

Franke AC, van den Brand GJ, Vanlauwe B, Giller KE (2018) Sustainable intensification through rotations with grain legumes in Sub-Saharan Africa: a review. Agric Ecosyst Environ 261:172–185. https://doi.org/10.1016/j.agee.2017.09.029

Fustec J, Lesuffleur F, Mahieu S, Cliquet JB (2010) Nitrogen rhizodeposition of legumes. A review. Agron Sustain Dev 30:57–66. https://doi.org/10.1051/AGRO/2009003

Garland G, Bünemann EK, Oberson A et al (2016) Plant-mediated rhizospheric interactions in maize-pigeon pea intercropping enhance soil aggregation and organic phosphorus storage. Plant Soil 415:37–55. https://doi.org/10.1007/S11104-016-3145-1

Ghaley BB, Hauggaard-Nielsen H, Høgh-Jensen H, Jensen ES (2005) Intercropping of wheat and pea as influenced by nitrogen fertilization. Nutr Cycl Agroecosyst 73:201–212. https://doi.org/10.1007/S10705-005-2475-9/METRICS

Ghosh PK, Hazra KK, Venkatesh MS et al (2020) Grain legume inclusion in cereal–cereal rotation increased base crop productivity in the long run. Exp Agric 56:142–158. https://doi.org/10.1017/S0014479719000243

Glaze-Corcoran S, Hashemi M, Sadeghpour A et al (2020) Understanding intercropping to improve agricultural resiliency and environmental sustainability. In: Advances in agronomy. Academic, pp 199–256

Guardia G, Tellez-Rio A, García-Marco S et al (2016) Effect of tillage and crop (cereal versus legume) on greenhouse gas emissions and Global Warming Potential in a non-irrigated Mediterranean field. Agric Ecosyst Environ 221:187–197. https://doi.org/10.1016/J.AGEE.2016.01.047

Guo Z, Zhang Y, Deegen P, Uibrig H (2006) Economic analyses of rubber and tea plantations and rubber-tea intercropping in Hainan, China. Agrofor Syst 66:117–127. https://doi.org/10.1007/S10457-005-4676-2/METRICS

Hao X, Lin Y, Johnstone L et al (2012) Draft genome sequence of plant growth-promoting Rhizobium Mesorhizobium amorphae, isolated from zinc-lead mine tailings. J Bacteriol 194:736–737. https://doi.org/10.1128/JB.06475-11

Hauggaard-Nielsen H, Jørnsgaard B, Kinane J, Jensen ES (2008) Grain legume—cereal intercropping: the practical application of diversity, competition and facilitation in arable and organic cropping systems. Renew Agric Food Syst 23:3–12. https://doi.org/10.1017/S1742170507002025

Hei Z, Xiang H, Zhang J et al (2021) Rice intercropping with water mimosa (Neptunia oleracea Lour.) can facilitate soil N utilization and alleviate apparent N loss. Agric Ecosyst Environ 313. https://doi.org/10.1016/j.agee.2021.107378

Himanen SJ, Mäkinen H, Rimhanen K, Savikko R (2016) Engaging farmers in climate change adaptation planning: assessing intercropping as a means to support farm adaptive capacity. Agriculture 6:34. https://doi.org/10.3390/AGRICULTURE6030034

Hinsinger P, Betencourt E, Bernard L et al (2011) P for two, sharing a scarce resource: soil phosphorus acquisition in the rhizosphere of intercropped species. Plant Physiol 156:1078–1086. https://doi.org/10.1104/pp.111.175331

Hong Y, Heerink N, Jin S et al (2017) Intercropping and agroforestry in China—current state and trends. Agric Ecosyst Environ 244:52–61. https://doi.org/10.1016/J.AGEE.2017.04.019

Huang JX, Chen YQ, Sui P et al (2014) Soil nitrous oxide emissions under maize-legume intercropping system in the North China Plain. J Integr Agric 13:1363–1372. https://doi.org/10.1016/S2095-3119(13)60509-2

Inal A, Gunes A, Zhang F, Cakmak I (2007) Peanut/maize intercropping induced changes in rhizosphere and nutrient concentrations in shoots. Plant Physiol Biochem 45:350–356. https://doi.org/10.1016/J.PLAPHY.2007.03.016

Irigoyen JJ, Goicoechea N, Antolín MC et al (2014) Growth, photosynthetic acclimation and yield quality in legumes under climate change simulations: an updated survey. Plant Sci 226:22–29. https://doi.org/10.1016/J.PLANTSCI.2014.05.008

Jensen ES, Carlsson G, Hauggaard-Nielsen H (2020) Intercropping of grain legumes and cereals improves the use of soil N resources and reduces the requirement for synthetic fertilizer N: a global-scale analysis. Agron Sustain Dev 40:1–9. https://doi.org/10.1007/S13593-020-0607-X/TABLES/3

Jeuffroy MH, Baranger E, Carrouée B et al (2013) Nitrous oxide emissions from crop rotations including wheat, oilseed rape and dry peas. Biogeosciences 10:1787–1797. https://doi.org/10.5194/BG-10-1787-2013

Jin Q, Zhu K, Cui W et al (2013) Hydrogen gas acts as a novel bioactive molecule in enhancing plant tolerance to paraquat-induced oxidative stress via the modulation of heme oxygenase-1 signalling system. Plant Cell Environ 36:956–969. https://doi.org/10.1111/pce.12029

Keeler BL, Hobbie SE, Kellogg LE (2009) Effects of long-term nitrogen addition on microbial enzyme activity in eight forested and grassland sites: implications for litter and soil organic matter decomposition. Ecosystems 12:1–15. https://doi.org/10.1007/S10021-008-9199-Z/METRICS

Kiwia A, Kimani D, Harawa R et al (2019) Sustainable intensification with cereal-legume intercropping in Eastern and Southern Africa. Sustainability 11:2891. https://doi.org/10.3390/SU11102891

Kumar S, Sheoran S, Kumar SK et al (2016) Drought: a challenge for Indian farmers in context to climate change and variability. Progress Res Int J 11:6243–6246

Kumar N, Hazra KK, Nath CP et al (2018) Grain legumes for resource conservation and agricultural sustainability in South Asia. In: Legumes for soil health and sustainable management. Springer, Singapore, pp 77–107

Kumawat A, Bamboriya SD, Meena RS, et al (2022) Legume-based inter-cropping to achieve the crop, soil, and environmental health security. In: Advances in legumes for sustainable intensification. Elsevier, pp 307–328

Lai H, Gao F, Su H, Zheng P, Li Y, Yao H (2022) Nitrogen distribution and soil microbial community characteristics in a legume–cereal intercropping system: a review. Agronomy 12(8):1900. https://doi.org/10.3390/agronomy12081900

Latati M, Bargaz A, Belarbi B et al (2016) The intercropping common bean with maize improves the rhizobial efficiency, resource use and grain yield under low phosphorus availability. Eur J Agron 72:80–90. https://doi.org/10.1016/J.EJA.2015.09.015

Li L, Li S-M, Sun J-H et al (2007) Diversity enhances agricultural productivity via rhizosphere phosphorus facilitation on phosphorus-deficient soils. Proc Natl Acad Sci 104:11192–11196. https://doi.org/10.1073/pnas.0704591104

Li QS, Wu LK, Chen J et al (2016) Biochemical and microbial properties of rhizospheres under maize/peanut intercropping. J Integr Agric 15:101–110. https://doi.org/10.1016/S2095-3119(15)61089-9

Li C, Hoffland E, Kuyper TW et al (2020) Yield gain, complementarity and competitive dominance in intercropping in China: a meta-analysis of drivers of yield gain using additive partitioning. Eur J Agron 113:125987. https://doi.org/10.1016/J.EJA.2019.125987

Liang B, Lehmann J, Sohi SP et al (2010) Black carbon affects the cycling of non-black carbon in soil. Org Geochem 41:206–213. https://doi.org/10.1016/J.ORGGEOCHEM.2009.09.007

Lin WW, Li N, Chen LS et al (2022) Effects of maize and sobean interspecific interactions on rhizospheric bacteria community structure and diversity. Chin J Eco-Agric 30:26–37

Lithourgidis AS, Dordas CA, Damalas CA, Vlachostergios DN (2011) Annual intercrops: an alternative pathway for sustainable agriculture. Aust J Crop Sci 5:396–410

Ludwig M, Achtenhagen J, Miltner A et al (2015) Microbial contribution to SOM quantity and quality in density fractions of temperate arable soils. Soil Biol Biochem 81:311–322. https://doi.org/10.1016/J.SOILBIO.2014.12.002

Luo C, Lv J, Guo Z, Dong Y (2022) Intercropping of Faba bean with wheat under different nitrogen levels reduces Faba bean rust and consequent yield loss. Plant Dis 106:2370–2379. https://doi.org/10.1094/PDIS-11-21-2451-RE/ASSET/IMAGES/LARGE/PDIS-11-21-2451-RET3.JPEG

Maitra S, Shankar T, Banerjee P (2020) Potential and advantages of maize-legume intercropping system. In: Maize—production and use, pp 1–14. https://doi.org/10.5772/intechopen.91722

Manoj KN, Shekara BG, Sridhara S et al (2021) Biomass quantity and quality from different year-round cereal–legume cropping systems as forage or fodder for livestock. Sustainability (Switzerland) 13. https://doi.org/10.3390/su13169414

Mitran T, Meena RS, Lal R, Layek J, Kumar S, Datta R (2018) Role of soil phosphorus on legume production. In: Legumes for soil health and sustainable management, pp 487–510. https://doi.org/10.1007/978-981-13-0253-4_15

Mogale TE, Ayisi KK, Munjonji L et al (2023) Understanding the impact of the intercropping system on carbon dioxide (CO_2) emissions and soil carbon stocks in Limpopo Province, South Africa. Int J Agron 2023:6307673. https://doi.org/10.1155/2023/6307673

Mousavi SF, Yousefi-Moghadam S, Mostafazadeh-Fard B et al (2009) Effect of puddling intensity on physical properties of a silty clay soil under laboratory and field conditions. Paddy Water Environ 7:45–54. https://doi.org/10.1007/S10333-008-0148-4

Naudin C, Corre-Hellou G, Pineau S et al (2010) The effect of various dynamics of N availability on winter pea–wheat intercrops: crop growth, N partitioning and symbiotic N2 fixation. Field Crop Res 119:2–11. https://doi.org/10.1016/J.FCR.2010.06.002

Ness B, Anderberg S, Olsson L (2010) Structuring problems in sustainability science: the multi-level DPSIR framework. Geoforum 41:479–488. https://doi.org/10.1016/J.GEOFORUM.2009.12.005

Ngangom B, Das AR, Krishnappa S (2017) Soil physical properties and productivity as influenced by soil moisture conservation measures under maize based cropping system in acid soils of North East India. Int J Curr Microbiol App Sci 6(3):428–436. https://doi.org/10.20546/ijcmas.2017.603.049

Njira KOW, Semu E, Mrema JP, Nalivata PC (2021) Productivity of pigeon pea, cowpea and maize under sole cropping, legume–legume and legume–cereal intercrops on Alfisols in Central Malawi. Agrofor Syst 95:279–291. https://doi.org/10.1007/s10457-020-00589-0

Nuruzzaman M, Lambers H, Bolland MDA et al (2005) Phosphorus benefits of different legume crops to subsequent wheat grown in different soils of Western Australia. Plant Soil 271:175–187. https://doi.org/10.1007/S11104-004-2386-6/METRICS

Ocimati W, Were E, Groot JCJ et al (2018) Risks posed by intercrops and weeds as alternative hosts to xanthomonas campestris pv. Musacearum in Banana Fields. Front Plant Sci 9:379805. https://doi.org/10.3389/FPLS.2018.01471/BIBTEX

Omonode RA, Vyn TJ, Smith DR et al (2007) Soil carbon dioxide and methane fluxes from long-term tillage systems in continuous corn and corn–soybean rotations. Soil Tillage Res 95:182–195. https://doi.org/10.1016/J.STILL.2006.12.004

Peoples MB, Hauggaard-Nielsen H, Jensen ES (2015) The potential environmental benefits and risks derived from legumes in rotations. Nitrogen Fix Crop Prod 52:349–385. https://doi.org/10.2134/AGRONMONOGR52.C13

Pivato B, Semblat A, Guégan T et al (2021) Rhizosphere bacterial networks, but not diversity, are impacted by pea-wheat intercropping. Front Microbiol 12:674556. https://doi.org/10.3389/FMICB.2021.674556/BIBTEX

Plaza-Bonilla D, Nolot JM, Passot S et al (2016) Grain legume-based rotations managed under conventional tillage need cover crops to mitigate soil organic matter losses. Soil Tillage Res 156:33–43. https://doi.org/10.1016/J.STILL.2015.09.021

Preissel S, Reckling M, Schläfke N, Zander P (2015) Magnitude and farm-economic value of grain legume pre-crop benefits in Europe: a review. Field Crop Res 175:64–79. https://doi.org/10.1016/J.FCR.2015.01.012

Rao KK, Samal SK, Poonia SP et al (2021) Conservation agriculture improves soil physical properties and crop productivity: a long-term study in middle Indo-Gangetic Plains of India. Soil Res 60:442–454. https://doi.org/10.1071/SR20329

Rathore SS, Babu S, Shekhawat K, Singh R, Yadav SK, Singh VK, Singh C (2022) Designing energy cum carbon-efficient environmentally clean production system for achieving green economy in agriculture. Sustain Energy Technol Assess. https://doi.org/10.1016/j.seta.2022.102190

Rathore SS, Babu S, Shekhawat K, Kumar V, Gairola A, Wani OA, Singh VK (2024) Exploring sustainable agricultural production models to coordinate the system productivity, soil biological health, and eco-efficiency in the semi-arid region. Environ Sustain Indic. https://doi.org/10.1016/j.indic.2024.100480

Ratnadass A, Blanchart É, Lecomte P (2013) Ecological interactions within the biodiversity of cultivated systems.In: Cultivating biodiversity to transform agriculture, pp 141–179. https://doi.org/10.1007/978-94-007-7984-6_5

Raza MA, Zhiqi W, Yasin HS et al (2023) Effect of crop combination on yield performance, nutrient uptake, and land use advantage of cereal/legume intercropping systems. Field Crop Res 304:109144. https://doi.org/10.1016/J.FCR.2023.109144

Sarkodie-Addo J, Abdul-Rahaman I (2012) Spatial arrangements and time of introduction an intercrop on the productivity of component crops in Maize–soybean intercropping systems. 2:103–107

Schelud'ko AV, Makrushin KV, Tugarova AV et al (2009) Changes in motility of the rhizobacterium Azospirillum brasilense in the presence of plant lectins. Microbiol Res 164:149–156. https://doi.org/10.1016/J.MICRES.2006.11.008

Schwenke GD, Herridge DF, Scheer C et al (2015) Soil N_2O emissions under N_2-fixing legumes and N-fertilised canola: a reappraisal of emissions factor calculations. Agric Ecosyst Environ 202:232–242. https://doi.org/10.1016/J.AGEE.2015.01.017

Sehy U, Ruser R, Munch JC (2003) Nitrous oxide fluxes from maize fields: relationship to yield, site-specific fertilization, and soil conditions. Agric Ecosyst Environ 99:97–111. https://doi.org/10.1016/S0167-8809(03)00139-7

Senbayram M, Wenthe C, Lingner A et al (2015) Legume-based mixed intercropping systems may lower agricultural born N_2O emissions. Energy Sustain Soc 6:1–9. https://doi.org/10.1186/S13705-015-0067-3/FIGURES/4

Singh R, Chaudhary R, Somasundaram J et al (2014) Impact of crop covers on soil properties, runoff, soil-nutrients losses and crop productivity in Vertisols of central India. Indian J Soil Conserv 42:268–275

Singh DN, Bohra JS, Banjara T (2019) Diversification of rice-wheat cropping system for sustainability and livelihood security. In: Crop diversification for resilience in agriculture and doubling farmers income. ICAR—Indian Agricultural Research Institute (IARI), New Delhi, pp 78–97

Singh R, Babu S, Avasthe RK, Yadav GS, Das A, Mohapatra KP, Kumar A, Singh VK, Chandra P (2021) Crop productivity, soil health, and energy dynamics of Indian Himalayan intensified organic maize-based systems. Int Soil Water Conserv Res 9(2):260–270. https://doi.org/10.1016/j.iswcr.2020.11.003

Solanki MK, Wang FY, Wang Z et al (2019) Rhizospheric and endospheric diazotrophs mediated soil fertility intensification in sugarcane-legume intercropping systems. J Soils Sediments 19:1911–1927. https://doi.org/10.1007/s11368-018-2156-3

Soufan W, Al-Suhaibani NA (2021) Optimizing yield and quality of silage and hay for pea–barley mixtures ratio under irrigated arid environments. Sustainability 13:13621. https://doi.org/10.3390/SU132413621

Srinivasarao C, Venkateswarlu B, Lal R et al (2012) Long-term effects of soil fertility management on carbon sequestration in a rice–lentil cropping system of the Indo-Gangetic Plains. Soil Sci Soc Am J 76:168–178. https://doi.org/10.2136/SSSAJ2011.0184

Srivastava S, Checcucci A. Maingi JM et al (2023) Frontiers REVIEW in Sustainable Food Systems published: 16 November 2021. Nitrogen use to improve sustainable yields in agricultural systems 16. https://doi.org/10.3389/978-2-8325-3800-5

Stagnari F, Maggio A, Galieni A, Pisante M (2017) Multiple benefits of legumes for agriculture sustainability: an overview. Chem Biol Technol Agric 4:1–13. https://doi.org/10.1186/s40538-016-0085-1

Swarnalakshmi K, Yadav V, Tyagi D et al (2020) Significance of plant growth promoting rhizobacteria in grain legumes: growth promotion and crop production. Plants 9:1596. https://doi.org/10.3390/PLANTS9111596

Tanveer A, Ikram RM, Ali HH (2019) Crop rotation: principles and practices. In: Agronomic crops: volume 2: management practices, pp 1–12. https://doi.org/10.1007/978-981-32-9783-8_1

Thilakarathna MS, McElroy MS, Chapagain T et al (2016) Belowground nitrogen transfer from legumes to non-legumes under managed herbaceous cropping systems. A review. Agron Sustain Dev 36. https://doi.org/10.1007/s13593-016-0396-4

Van Oort PAJ, Gou F, Stomph TJ, van der Werf W (2020) Effects of strip width on yields in relay-strip intercropping: a simulation study. Eur J Agron 112:125936. https://doi.org/10.1016/J.EJA.2019.125936

Varma D, Meena RS, Kumar S, Kumari E (2017) Response of mungbean to NPK and lime under the conditions of vindhyan region of Uttar Pradesh. Legum Res 40:542–545. https://doi.org/10.18805/LR.V0I0.7854

Wang Y, Marschner P, Zhang F (2012) Phosphorus pools and other soil properties in the rhizosphere of wheat and legumes growing in three soils in monoculture or as a mixture of wheat and legume. Plant Soil 354:283–298. https://doi.org/10.1007/S11104-011-1065-7/METRICS

Watson CA, Stoddard FL (2017) Introduction—perspectives on legume production and use in European agriculture. In: Legumes in cropping systems, pp 1–17. https://doi.org/10.1079/9781780644981.0001

Wen Z, Wu J, Yang Y et al (2022) Implementing intercropping maintains soil water balance while enhancing multiple ecosystem services. Catena (Amst) 217:106426. https://doi.org/10.1016/J.CATENA.2022.106426

Wiche O, Székely B, Kummer NA et al (2016) Effects of intercropping of oat (Avena sativa L.) with white lupin (Lupinus albus L.) on the mobility of target elements for phytoremediation and phytomining in soil solution. Int J Phytoremed 18:900–907. https://doi.org/10.1080/15226514.2016.1156635

Wilhelm WW, Johnson JMF, Hatfield JL et al (2004) Crop and soil productivity response to corn residue removal. Agron J 96:1–17. https://doi.org/10.2134/AGRONJ2004.1000A

Yadav GS, Babu S, Das A et al (2020) No-till and mulching enhance energy use efficiency and reduce carbon footprint of a direct-seeded upland rice production system. J Clean Prod 271:122700. https://doi.org/10.1016/J.JCLEPRO.2020.122700

Yadav GS, Das A, Kandpal BK, Babu S, Rattan L, Datta M, Das B, Singh R, Singh VK, Mohapatra KP, Chakraborty M (2021) The food-energy-water-carbon nexus in a maize-maize-mustard cropping sequence of the Indian Himalayas: an impact of tillage-cum-live mulching. Renew Sust Energ Rev. https://doi.org/10.1016/j.rser.2021.111602

Yadav D, Singh D, Babu S, Madegowda M, Singh DV, Mandal D, Rathore AC, Sharma VK, Singhal V, Kumawat A, Yadav DK, Yadav RK, Kumar S (2023) Intensified cropping reduces soil erosion and improves rainfall partitioning and soil properties in the marginal land of the Indian Himalayas. Int Soil Water Conserv Res. https://doi.org/10.1016/j.iswcr.2023.10.002

Yadav D, Singh D, Madhu M, Singh DV, Islam S, Kumar G, Gupta AK, Mandal D, Babu S, Barh A, Kumawat A, Dubey R, Pati K, Bhargavi B, Yadav DK (2024) Interplanted crops and rainfall variability regulate soil erosion in rainfed maize of the Indian Himalayas. Land Degrad Dev. https://doi.org/10.1002/ldr.5221

Yue Q, Xu X, Hillier J et al (2017) Mitigating greenhouse gas emissions in agriculture: from farm production to food consumption. J Clean Prod 149:1011–1019. https://doi.org/10.1016/J.JCLEPRO.2017.02.172

Yuvaraj M, Pandiyan M, Gayathri P (2020) Role of legumes in improving soil fertility status. In: Legume crops-prospects, production and uses, pp 16–27. https://doi.org/10.5772/intechopen.93247

Wang J, Zhang X, Yao H (2020) Optimizing ultracentrifugation conditions for DNA-based stable isotope probing (DNA-SIP). Journal of Microbiological Methods 173:105938. https://doi.org/10.1016/j.mimet.2020.105938

Zhang NN, Sun YM, Li L et al (2010) Effects of intercropping and Rhizobium inoculation on yield and rhizosphere bacterial community of faba bean (Vicia faba L.). Biol Fertil Soils 46:625–639. https://doi.org/10.1007/S00374-010-0469-5/METRICS

Zhao Y, Tian Y, Li X et al (2022) Nitrogen fixation and transfer between legumes and cereals under various cropping regimes. Rhizosphere 22:100546. https://doi.org/10.1016/J.RHISPH.2022.100546

System Approaches for Enhancing Soil Health and Food Security

7

Sougata Roy, Subrata Bag, Devendra Kumar Dadhich, Tanmay Das, and Suman Dutta

Abstract

Maintaining agricultural production, environmental sustainability, and human health worldwide all depend on healthy soil, which is the cornerstone of sustainable agro-ecosystems. A thorough knowledge of soil health is necessary because anthropogenic activities, such as intensive land use and farming techniques, have a major influence on the soil processes. In the past, non-biological characteristics like nutrients and soil structure were the main emphasis of soil health evaluations. Nonetheless, it is now acknowledged that soil microbes play a very critical role in preserving soil health. Given the complexity of agricultural systems and the need to accomplish numerous objectives, systems approaches and multidisciplinary assessments are necessary. Understanding the complete agronomic production system requires the use of systems techniques, which have been more popular in recent decades. These methods use scientific models and information from knowledgeable stakeholders to consider the temporal and geographical aspects of agronomic systems. With examples given in this work, the choice of

S. Roy (✉)
International Potato Centre (CGIAR-CIP), Patna, Bihar, India
e-mail: Sougata.Roy@cgiar.org

S. Bag
ICAR-Central Research Institute for Dryland Agriculture, Hyderabad, Telangana, India

D. K. Dadhich
ICAR-National Dairy Research Institute, Karnal, Haryana, India

T. Das
ICAR-Indian Agricultural Research Institute, New Delhi, India

S. Dutta
Ramkrishna Mission Vivekananda Educational and Research Institute,
Kolkata, West Bengal, India

© The Author(s), under exclusive license to Springer Nature Singapore Pte Ltd. 2025

S. Babu et al. (eds.), *Ecological Solutions to Agricultural Land Degradation*,
https://doi.org/10.1007/978-981-96-3392-0_7

suitable approaches, ranging from basic to sophisticated, is contingent upon particular study requirements. In order to improve research and implementation, systems methods are increasingly including stakeholders, such as farmers, residents, and policymakers. Developing creative, sustainable agro-ecosystems requires a thorough multidisciplinary examination, with soil health being a critical element for both ecosystem sustainability and future food security. This comprehensive strategy aids in the creation of laws and procedures that protect the world's food systems and maintain healthy soils.

Keywords

Soil degradation · Agroecology · Cropping systems · Biodiversity · Microbiome

Abbreviation

C: N	Carbon to Nitrogen Ratio
N	Nitrogen
P	Phosphorus
FSN	Food Security and Nutrition
SDG	Sustainable Development Goal
ISMP	Integrated Soil Management Practices
BNF	Biological Nitrogen Fixation
WSMA	Water Stable Macro-Aggregates
SOC	Soil Organic Carbon
CA	Conservation Agriculture
IFS	Integrated Farming System
IOBC	International Organization for Biological Control
ICAR	Indian Council of Agricultural Research
NEH	North-Eastern Hill
Tg N	Teragram of Nitrogen
Mg	Megagram
FYM	Farmyard Manure

7.1 Introduction

One of the main problems facing agriculture worldwide is the 1.93% annual population increase. 3.2 billion people's livelihoods are at risk due to the degradation of ~6 billion hectares of land worldwide. Strategies that successfully address these problems must be put into place in order to guarantee sustainable farming. Conventional crop management techniques often target specific crops while ignoring the crops' long-term impacts on following plantings. Crop-specific strategies often result in wasteful use of resources. Using a systems approach in agriculture improves production stability, increases net returns, and makes optimal use of resources without compromising soil health (Lal 2015). An intricate and priceless ecosystem, soil is essential from an ecological and human standpoint. It is essential for the production of sufficient food, fiber, and fuel as well as for the decomposition of organic matter, recycling of nutrients, detoxification of contaminants, sequestration of carbon, management of water, and conservation of biodiversity. However, the health of people and ecosystems is seriously threatened by the worldwide soil degradation caused by intensive agriculture and other human activities. This deterioration is especially concerning since soil is a non-renewable resource that is used throughout a person's lifetime (Acton and Gregorich 1995). Over time, the meaning of soil health has changed, especially since the 1990s, when the emphasis moved from crop output and soil quality to a more comprehensive understanding of soil's capacity to sustain vital ecological activities. According to Van Es and Karlen (2019), soil health is today understood to be the capacity of soil to operate within an ecosystem, providing essential functions without deteriorating or endangering the environment. While "soil health" highlights the dynamic, living aspect of soil, "soil quality" refers to quantifiable physical, chemical, and biological qualities. Despite this, the phrases are often used interchangeably. Recent research has included physicochemical and microbiological characteristics as important markers to our knowledge of soil health. These consist of the C:N ratio, enzyme activity, soil microbial composition, and other physical and chemical characteristics of the soil (Ozlu et al. 2019; VeVerka et al. 2019). Consistently defining and evaluating soil health remains difficult, even with improvements in soil health assessment. Establishing trustworthy reference soils for comparison is challenging due to the intrinsic variety of soil and its temporal changes. Accordingly, assessing soil functional sustainability requires tracking long-term developments in soil management techniques (Karlen et al. 2001). Furthermore, the sustainability and profitability of farming are threatened by elements including shrinking land holdings, nutrient-deficient and degraded soils, erratic weather patterns, and insufficient investments. Given that location-specific integrated farming systems improve resilience and flexibility to climatic variability, a systems approach is essential, particularly for the 126 million marginal and small agricultural holdings in India and the 570 million worldwide. This strategy also tackles problems including diminishing agricultural production and profitability, decreased resource-use efficiency, and resource degradation.

7.2 Concept of System Approach, Its Component and Principles

In agriculture, a system is a deliberate blending of two or more components, driven by the ideas of enhancing complementarity and avoiding rivalry. Utilizing cutting-edge agronomic management techniques to increase farm revenue, better family nutrition, and provide ecosystem services in a eco-friendly and sustainable manner is the aim. Maintaining biodiversity, varying crops and farming methods, and optimizing recycling are essential elements of this strategy (Singh and Ravisankar 2015; Fatima et al. 2023). In essence, a system is the beneficial interplay of many elements on a farm to increase output and profitability without endangering the land sustainability and the environment. Selecting businesses should prioritize strengthening complementarity and lowering competition. A system production method aims to improve farm production efficiency and farm household systems and rural communities in a sustainable manner.

- Ensuring nutritional security and meeting basic requirements.
- Recycling and resource efficiency inside the system to increase agricultural revenue.

In agriculture, cropping systems are an essential part of system study. It includes the farming practices used on a farm and how they interact with farm resources, technologies and other businesses. The distribution of different crops across a given region and time, reflecting the order and spatial arrangement of crops and fallow periods, is referred to as a cropping pattern. Although the terms crop sequence and crop rotation are sometimes used interchangeably, crop rotation refers explicitly to the yearly or long-term succession of various crops grown on the same piece of land. A crop system is a configuration of crop populations that produces useful biomass from inputs such as water, fertilizers, and sunlight. Within the larger cropping system, it functions as a subsystem. Agronomic research has traditionally focused on interdisciplinary studies in crop production, nutrition, growth physiology, and pesticide-assisted crop protection (Fargione et al. 2018). It is now clear, therefore, that concentrating on a single component often ignores the interplay between other manufacturing variables. This insight emphasizes how important it is to approach agriculture holistically. In order to create agro-ecosystems that satisfy many goals, contemporary agronomic research currently places a strong emphasis on examining complete production systems. Interdisciplinary studies are necessary to do this, combining expertise from several disciplines to get a thorough understanding of the cropping system and how it affects soil health (Fig. 7.1).

7.3 Soil Degradation from Global Cropping System

A serious environmental and social problem, soil degradation is caused by a variety of human activities. Soil deterioration is mostly caused by agricultural practices such as drainage, heavy equipment adaption, intense tillage, and excessive use of fossil fuels (Wani et al. 2024). Additional negative impacts include soil compaction, salinization, organic matter loss, water and wind erosion, and biodiversity loss, all of which endanger soil health (Yadav et al. 2021a; Stolte et al. 2015). Food security and safety are seriously threatened by this deterioration, especially in light of the world's rapidly expanding population and the escalating impacts of climate change on food production (Smith and Gregory 2013). Almost 95% of food and feed production depends on healthy soils, hence agricultural soil degradation is an urgent

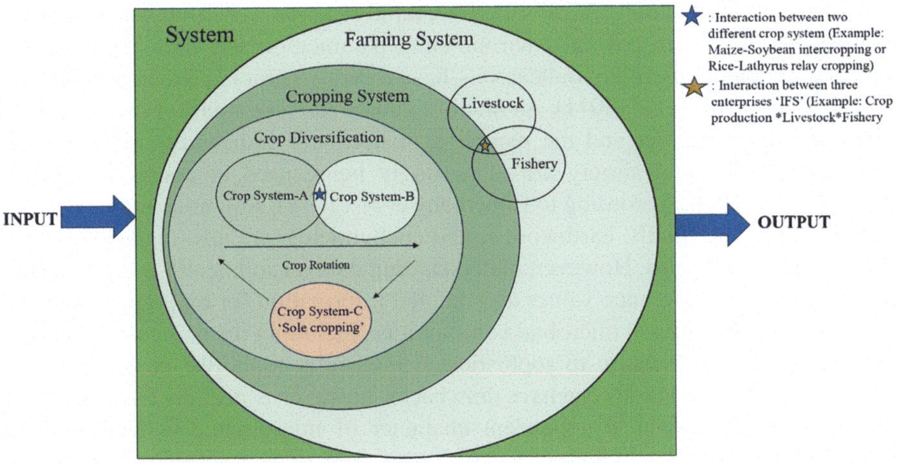

Fig. 7.1 A schematic diagram of system approach in agronomy

problem that requires attention. According to Bhattacharya (2019), the main cause of soil deterioration in agriculture is insufficient and unbalanced fertilization, since the supply-demand gap for minerals like phosphorus (P) and nitrogen (N) grows with time. In addition to raising food prices, mineral fertilization's poor effectiveness-In cereal crops, N efficiency ranges from 50% to 60%, while P efficiency is between 15% and 20%-also leads to soil erosion and other environmental problems. By changing soil structure, influencing pH, organic compounds, and nutrient availability, as well as making the soil more vulnerable to erosion, tillage-a common agronomic practice—further aggravates soil degradation (Grahmann et al. 2020; Carr 2017). Furthermore, nutritional imbalances, the buildup of soil-borne diseases, and decreased microbial diversity have resulted from the extensive use of monoculture systems, especially for cereal crops (Bai et al. 2019; Wang et al. 2018). In light of the rising need for food, these difficulties highlight the necessity for alternative agronomic techniques to preserve crop output and enhance soil health.

7.4 Indicators for Assessing Soil Health in Cropping Systems

Soil is a complex bio-matrix with a wide variety of organisms and a complicated architecture that may change depending on the cropping system, assessing soil health in cropping systems is challenging. For an appropriate assessment of soil health, physical, chemical, and biological properties must all be considered (Bünemann et al. 2018). The most effective markers of soil health ought to be linked to soil processes and responsive to modifications in the environment and management. Thus far, soil physicochemical properties such as texture, depth, ρb, water-holding capacity, porosity, pH, electrical conductivity, organic matter, cation exchange capacity, and nutrient content have been employed as indicators.

Because of their ecological significance, rapid reaction, sensitivity, and capacity to integrate data from several environmental parameters, soil biological properties—in particular, microbial indicators—have recently garnered attention (Galende et al. 2014; Barrutia et al. 2011). As microorganisms are essential for the decomposition of organic matter and the recycling of nutrients, indices of soil microbial biomass, activity, and variety are increasingly being used to assess soil health (Epelde et al. 2010). According to Bünemann et al. (2018), soil biota, includes nematodes, mites, springtails, earthworms, and enchytraeids, are also regarded as bio-indicators of soil health. However, understanding data on soil health is complicated by the absence of reference values and defined information for biological markers. Our understanding of soil microbial communities, including the non-culturable portion, has increased because to sophisticated molecular tools like next-generation sequencing. These methods can have drawbacks, however, especially when it comes to interpreting the context-dependent character of microbial data, which differ depending on variables including location, soil type, and sample period (Garbisu et al. 2011). More general characteristics including biodiversity, stability, resilience, and ecosystem services have been suggested by a number of writers as markers of soil health (Velásquez et al. 2007). For example, high biodiversity is very crucial component of soil health because it increases the capacity for resistance to perturbations. According to Klimkowicz-Pawlas et al. (2019), soil managers may utilize soil health indicators alone or in combination to make interpretation and decision-making easier. However, this approach may oversimplify complicated responses of soil ecosystems.

7.5 Agroecology Principles of System Approach

The global agriculture and food system is failing to meet key parameters such as environmental impact, decent agricultural livelihoods, and hunger and malnutrition rates. To achieve Sustainable Development Goal 2 (SDG2), which aims to eradicate hunger and malnutrition by 2030, a substantial shift is required. Understanding the fundamentals of agroecology and how they work is crucial before putting sustainable agricultural practices into practice. These methods, which are informed by agroecology, seek to increase soil health via crop rotation integration, biodiversity enhancement, and reduction of synthetic inputs (Fig. 7.2). Utilizing natural processes, agroecology is a transdisciplinary strategy that integrates research, practices, and social movements to create agricultural systems that are resilient and productive. Agroecology seeks to increase soil health while reducing synthetic inputs by promoting beneficial soil organisms, recycling organic matter, and boosting biodiversity. This strategy is seen to be a workable way to combat climate change, hunger, and the economic viability of rural populations (Pretty et al. 2018; Caron et al. 2018). Agroecological principles may have an influence on Food Security and Nutrition (FSN) in the Global South, especially by lowering small-scale farmers' food insecurity by decreasing their dependency on bought inputs. Agroecology-based economic diversification may improve resistance to price

Fig. 7.2 Research based on system approach of International Potato Centre (CIP) in CPRS, Patna station

volatility and financial independence (Kanmennang et al. 2017). Fair pay and working conditions are essential for FSN, while social values that support biodiversity in farming have a direct impact on nutrition. Local economies may be strengthened by improving the living conditions of agricultural laborers and strengthening the bonds between farmers and customers. However, local food programs may inadvertently increase gaps in access to fresh, locally produced food if they fail to address underlying structural challenges (Bezner Kerr et al. 2019). Principles of agroecology were summarised by the Wezel et al. (2020) (Table 7.1).

7.6 Sustainable Agriculture Strategies for Soil Health Improvement

7.6.1 Cropping Systems and Soil Health

One of the most prominent cropping systems for ensuring global food security is the rice (*Oryza sativa* L.)—wheat (*Triticum aestivum* L.) cropping system (RWCS), which supplies staple foods to a substantial section of the worldwide population (Banjara et al. 2021). With 13.5 million hectares (Mha) under cultivation, RWCS is extensively grown across Asia, with 57% of that area being in South Asia, namely in the Indo-Gangetic Plains (Ladha et al. 2009). With this approach, rice is grown as

Table 7.1 Agroecological principles and description

S. No.	Principle	Description
1.	Recycling	Use locally available renewable resources wherever feasible, and try to keep the cycles of biomass and nutrients
2.	Reduction of input	Decrease or do away with reliance on bought inputs and boost independence
3.	Soil health	Controlling soil organic matter and boosting soil biological activity
4.	Biodiversity	Maintaining and enhancing genetic diversity helps to sustain agroecosystem biodiversity over time and place at the field, farm, and landscape scales
5.	Economic diversification	Increase the diversity of on-farm earnings by giving small-scale farmers more possibilities for value addition and financial independence, allowing them to better meet consumer demand
6.	Animal health	Ensure animal health and welfare
7.	Synergy	Cooperation, integration, and complementarity among the earth, water, trees, crops, and animals
8.	Social values and diets	Create food systems that provide wholesome, varied, seasonally and culturally appropriate meals by drawing on the culture, identity, tradition, social justice, and gender equality of the surrounding communities
9.	Co-creation of knowledge	Encourage horizontal information exchange and co-creation, especially between farmers, incorporating scientific and local innovation
10.	Interaction	To maintain intimacy and trust between farmers and consumers, encourage fair and efficient distribution networks and reintegrate food systems into local economies
11.	Fairness	Encourage all participants in the food system to have sustainable, dignified livelihoods, and respect for intellectual property rights
12.	Participation	To support decentralized governance, local adaptive management of agricultural and food systems, social organization, and more participation in decision-making by food suppliers and producers
13.	Land and natural resource management	Recognize and support family farmers, smallholders, and rural food producers as important guardians of genetic resources and natural resources by improving institutional frameworks

Source: Wezel et al. (2020)

a lowland crop from June to October while wheat is cultivated as an upland crop from November to April. In rice fields, puddling weakens the soil's structure by reducing aeration and generating compaction, which hinders the establishment of roots for later wheat harvests (Pathak et al. 2011). As a result, compared to direct-seeded rice, yield of wheat decrease by 8% when planted after puddled rice (Kumar et al. 2008). The issue is made worse by groundwater depletion brought on by inadequate management techniques (Humphreys and Gaydon 2015). Furthermore, rice and wheat both are nutrient-exhaustive crops, and burning waste after mechanical harvesting causes serious air pollution and health problems, especially in northwest India (NAAS 2017). Therefore, water conservation, improving soil health, and environmental protection all depend on viable alternatives to extensive tillage. There is

Fig. 7.3 Maize -wheat-mung bean system as an alternative of rice-wheat system

an urgent need for a methodical strategy to increasing input usage efficiency and protecting natural resources. Cropping systems have a major influence on soil health and quality at both geographical and temporal scales. These systems include crop rotation, diversification, and intercropping (Vukicevich et al. 2016; Rathore et al. 2022). Maize-wheat-mungbean is a promising alternative to rice-wheat system in northwest India (Fig. 7.3). Environmental sustainability in cropping systems is currently a top focus in modern agriculture since it ensures soil health for long-term production through nutrient cycling, water management, biodiversity, and physical stability. (Hatfield et al. 2017).

7.6.2 Crop Diversification

Crop diversification, often referred to as "planned diversity" in cropping systems (Matson et al. 1997), is crucial for enhancing crop yield and soil health. Diversification has several advantages, including lowering soil-borne diseases, increasing nutrient usage efficiency, and balancing soil biodiversity (Barbieri et al. 2019). Better soil health and crop production result from the enhancement of soil nutrients, physical structures, and microbial activities (Maron et al. 2011). However, the effects of crop variety on plant and soil health might differ according on pathogen specificity and species redundancy (Zhu et al. 2000). The variety of crop species and plant functional categories, such as legumes, non-leguminous forbs, C_3 and C_4 grasses, and others, may contribute to improvements in soil health (Milcu et al. 2013). Because of their diverse ecological requirements, these groups get along well in communities, increasing the variety and functioning of soil microbes (Roscher et al. 2013). Soil health may be further enhanced by contemporary breeding initiatives that create crop types compatible with beneficial soil biota (Pan et al. 2017). According to Bull et al. (2018), genetically modified cultivars, such as a novel cassava variety with a lower amylose concentration, might influence food industry preferences but must be carefully considered in terms of the environment. Accordingly, improved cultivars and optimum crop diversity may increase soil productivity, nutrient efficiency, and environmental effects, resulting in sustainable agroecosystems and enhanced soil health (Ellouze et al. 2014).

7.6.3 Crop Rotation

By improving soil health, decreasing pests and disease outbreaks, and increasing yields, crop rotation is a successful method for maintaining agroecosystem biodiversity (Barbieri et al. 2019). Numerous variables, including crop kinds, rotation series, frequency, and length, regulates the efficiency of crop rotations (Tiemann et al. 2015; Das et al. 2022). These factors affect the health of the soil by encouraging good bacteria and disrupting the life cycles of soil-borne illnesses. For example, rotating crops with grain legumes may increase soil nitrogen levels, which will boost the productivity and protein content of subsequent crops (Gan et al. 2003). Legumes often increase microbial diversity and soil nutrient availability, whereas other crops have varying effects on soil microbial communities and nitrogen cycling (Yang et al. 2013). Canola and other non-mycorrhizal crops, on the other hand, could need more mineral fertilizers and might interfere with beneficial mycorrhizal associations, which might eventually lower soil health (Ellouze et al. 2014). The duration and frequency of rotations are crucial because longer rotations might improve soil health and yield stability, whereas shorter rotations may raise disease risks and lower yields (Lupwayi et al. 2018). For example, compared to a 3-year rotation, a 5-year rotation demonstrated increased soil microbial biomass, which enhanced soil health and wheat yields (Lupwayi et al. 2018). In comparison to continuous wheat, including many legume phases into rotations in Western Canada improved soil health by doubling soil nitrogen levels and changing rhizosphere bacterial populations (Hamel et al. 2018). However, increasing the frequency of the same crop could counteract the benefits of rotation by increasing the levels of pathogens (Bainard et al. 2017). The physical and chemical properties of the soil also affect rotation design, affecting the number and diversity of microorganisms (Allison and Martiny 2008). Inclusion of legumes and oilseeds in to the wheat-based rotations in semi-arid of Canada, improvs soil nutrient levels and preserving moisture (Gan et al. 2011). For controlling soil-borne illnesses, effective rotation design also entails choosing non-host plants; varied cropping increases pathogen resistance. In general, adding extra crops to rotations sustainably may improve agricultural productivity and soil health (Bennett et al. 2012; Rathore et al. 2024).

7.6.4 Intercropping System

Reducing chemical induced pollution (Lemaire et al. 2014), managing soil diseases (Vukicevich et al. 2016), improving root function (Bukovsky-Reyes et al. 2019), and increasing spatial and nutrient use efficiency (Hinsinger et al. 2011) are few advantages of intercropping. Compared to rotational cropping, intercropping systems with maize, wheat, and faba beans in a semi-arid area of Gansu, China, produced 23%, 4%, and 11% increased root biomass as well as organic carbon and nitrogen content in the top 20 cm of soil, respectively (Cong et al. 2015). Compared to monoculture, intercropping of cassava with pigeon pea and beans in Pernambuco, Brazil, reduced black root rot (*Scytalidium lignicola*) in cassava by up to 50% while improving soil organic carbon, nutritional levels, microbial biomass, and enzyme activity (de Medeiros et al. 2019). More plant variety in intercropping systems frequently leads to higher resource efficiency. For instance, legumes, with their nitrogen-fixing bacteria, help control soil nitrogen levels, maximizing nutrient usage and lowering carbon emissions, whereas grasses flourish in high-nitrogen soils (de Araújo Santos et al. 2019). Nevertheless, knowledge is less about the biological roles and ecological effects of intercropped species. Improved soil health does not necessarily translate into higher intercropping yields. For example, compared to sole cropping, intercropping of *Medicago sativa* and *Dactylis glomerata* increased total shoot biomass, but it also resulted in higher N_2O emissions. This suggests that further research is necessary to fully understand the ecological and environmental implications of intercropping systems (Jungers et al. 2019; Graf et al. 2019).

7.7 Interactions Between Soil-Microbes- Plants and Its Impact on Soil Health

For more than 400 million years, microbes and plants have coevolved to build intricate soil-microbe-plant systems that are essential for carbon storage, nutrient cycling, soil fertility, and ecosystem resilience (Fierer 2017). The use of well-chosen crop species in agriculture intensifies these interactions by enhancing the "host effects" on soil microorganisms. In order to efficiently regulate their root microbiomes, plants have evolved a variety of chemical communication mechanisms (Fierer 2017).

Because it depends on root exudates for carbon and energy, the root microbiome-which is influenced by plant species and soil characteristics (Ellouze et al. 2014)-significantly changes the rhizosphere's soil characteristics (Ji et al. 2015). Additionally, these exudates have a great impact on soil microorganism eco-functions and microhabitats (Maltais-Landry et al. 2014). Hormones generated by plants, such as strigolactones, jasmonic acid, gibberellic acid, ethylene, salicylic acid, auxin, and cytokinin, are essential for plant-microbe interactions because they control defense and recognition mechanisms (Maruri-López et al. 2019). For example, strigolactones induce hyphal branching in mycorrhizal symbioses and initiate pathogen responses, while salicylic acid and jasmonic acid activate the pathogen defense mechanisms (Kretzschmar et al. 2012). Plant peptides may act as hormones or enzymes to control environmental stressors, and flavonoids start a symbiotic relationship with diazotrophs (Miransari et al. 2013; Fritig et al. 1998; Turrini et al. 2004). Furthermore, plant roots contain volatile chemical substances that might stop the development of pathogens like Fusarium spp. (Cruz et al. 2012).

7.7.1 The Symbiotic Microbiome and Its Impact on Soil Health

By improving plant health, nutrient absorption, and tolerance to environmental challenges, the symbiosis of various soil microbes in cropping systems provides substantial advantages for crops (Fierer 2017). Important symbiotic partners include mycorrhizal fungi (Bolan 1991), plant growth-promoting bacteria (PGPR) (Peoples and Craswell 1992), and beneficial endophytic fungi (Rodriguez and Redman 2008). To comprehend and make use of these microbial populations in sustainable agriculture, the idea of a positive microbiome is essential. Vandenkoornhuyse et al. (2015) define three types of microbiomes: the "pan-microbiome," which consists of microorganisms associated with a specific plant species; the "eco-microbiome," which consists of microorganisms associated with a plant population in a specific environment; and the "core-microbiome," which consists of microorganisms that are consistently associated with a plant species. Because core-microbiomes are stable throughout time and place and may be controlled by plant selection, they are essential for agricultural output. According to Lay et al. (2018), four bacteria and one fungus had a positive correlation with canola production, but only 14 bacterial and one fungal operational taxonomic unit (OTU) were found to be core-microbiomes in canola roots from the Western Canadian Prairie. Although it is still difficult to connect taxonomic information to functional effects, understanding the taxonomic and phylogenetic structure of these symbiotic microbes is crucial for sustainable agriculture (Fierer 2017). Shotgun metagenomics and tools like PICRUST2 (Douglas et al. 2019) and Fun Guild (Nguyen et al. 2016) help connect the functional functions of microbial communities to their data.

7.7.2 Free-Living Microorganisms and Their Impact on Soil Health

In contrast to their symbiotic relatives, free-living soil microorganisms are also essential for improving plant development and soil health in cropping systems (Müller et al. 2016). The interactions between plants and microbes in the rhizosphere are greatly influenced by these microorganisms, which live outside of plant

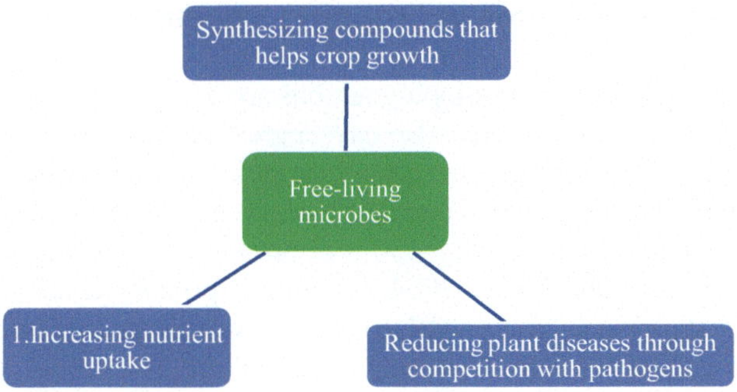

Fig. 7.4 Role of free-living microbes in cropping system

cells. Free living microbes has the potential to improve nutreint uptake and crop growth (Fig. 7.4). To boost plant output, for instance, strains of the biological nitrogen-fixing bacteria *Azotobacter*, *Azospirillum*, *Bacillus*, and *Klebsiella* have been used globally (Lynch 1983). In a similar vein, P-solubilizing microbes such as *Bacillus* and *Paenibacillus* increase plants' access to phosphorus (Brown 1974). Free-living microorganisms from *Actinobacteria* or *Firmicutes* that are present throughout wheat rotations have been shown to increase wheat yields and head counts (Yang et al. 2012). They can create growth-regulating hormones, bio-control substances to combat infections, and enzymes to reduce ethylene levels, which promotes root formation and growth. They also improve tolerance to salinity and drought by solubilizing soil minerals (Hayat et al. 2010). Free-living microorganisms may also help with soil restoration (Zhuang et al. 2007). Making the most of these advantageous microbes is essential for both healthy soil systems and sustainable agriculture. Various new-generation chemicals are now available in the market that have minimal impact on microbial biomass carbon, a key indicator of soil microbial activity. One such herbicide, tembotrione (34.4%), has gained popularity in zero-tillage maize cultivation due to its effectiveness, while demonstrating no adverse effects on other crops within the cropping system (Roy et al. 2024a, b).

7.8 Integrated Approaches for Soil Management

Integrated Soil Management Practices (ISMP) guarantee sustainable farming methods by improving crop productivity and soil health. ISMP significantly contributes to improving food security. Soil is the foundation of all terrestrial life. Degradation of soil health is a key problem in the present generation. Several ISMP, such as maximizing the utilization of soil resources, decreasing soil degradation, and boosting cropping systems' resilience, play a vital role in improving soil health and raising food security. Below is a detailed analysis of ISMP's contribution to food security:

7.8.1 Inclusion of Legume in Cropping System

Food poverty and nutrition in developing nations are being made worse by soil degradation and climate change, which also have a great on food production and nutritional quality (Lal 2017). Pulses are edible legumes that are harvested as dry seeds and are essential to preserving food and nutritional security. They are the best source of proteins, minerals, and fiber. 180–430 g of protein is found in one kilogram of grain legumes (Crozat and Fustec 2004). Legumes' primary characteristic is their natural biological nitrogen fixation (BNF), which is facilitated by microorganisms. Therefore, legumes play a significant role in agriculture and may offset the significant energy consumption and emissions associated with the Haber-Bosch process of producing anthropogenically inorganic N fertilizer. The worldwide natural BNF is now 110 Tg N year on land and 140 Tg N annually in the water (Meena and Lal 2018). Legumes' enhanced root systems may improve the physical structure and aggregation of the soil and aid in the solubilization of insoluble phosphorus (P) in the soil (Gogoi et al. 2018). Legume crops improve water control and nutrient cycling. Additionally, legume leftovers may be used into mulch (Sarkar et al. 2020). In nations like India, where lowland rice or cereal-cereal (rice-wheat) is the prevalent cropping pattern, legumes may improve soil aggregation, aggregate carbon concentration, and soil carbon pool. The rice-chickpea cycle, for instance, had the lowest water stable macro-aggregates (WSMA) content (57%) in an Indian study on the rice-wheat cropping system, whereas the rice-wheat rotation had the highest (65%). Grain legume inclusive rotations also increased the carbon management index (5–7%), SOC (6–17%), and active carbon pool (9–18%) in comparison to rice–wheat rotation (Nath et al. 2019).

7.8.2 Green and Brown Manurings

One of the practical ways to improve the soil health is to combine legumes with green or brown manure. Higher C:N ratio slows down the breakdown process and results in a temporary immobilization of nutrients, hence to manage the C:N ratio, green manure must be added prior to blooming. Although non-legume may be used for green manuring, legumes are the favored choice due to their ability to fix nitrogen (Lin et al. 2019). *Sunhemp* and *Sesbania* are often employed in-situ green manure crops, whereas *Subabul*, neem, *Gliricidia*, cassia, and other ex-situ crops are also frequently used. By lowering ρb and increasing the soil's porosity and organic carbon content, green manure may help increase crop output (Wani et al. 2003). A great source of *ex-situ* green manure for improving the physical, chemical, and biological health of soil in rainfed locations is *Gliricidia* grown on agricultural bunds or fallow ground. One tonne of *Gliricidia* leaf manure per acre can yield 21 kg N, 2.5 kg P, 18 kg K, 85 g Zn, 164 g Mn, 365 g Cu, and 728 g Fe (Srinivasa Rao 2011). One kind of green manure is brown manure, which kills crops using specific herbicides rather than cultivating them (Fig. 7.5). In rice fields, brown manuring with *Sesbania* is a typical technique. In brown manuring, 2,4 D and

Fig. 7.5 Brown manuring of Sesbania knocked down by 2,4-D in kharif maize crop

bispyribac sodium are often employed herbicides. In rice systems, brown manuring combined with sesbania may enhance soil health and system production. For instance, after 2 years of experimentation, Sesbania brown manuring in direct-seeded rice produced the lowest soil ρb (17.2%) and the maximum soil porosity (19.3%) in research including brown manuring in a rice-wheat combination. Co-cultivating Sesbania with zero tillage in wheat in the rice-growing plot resulted in a 13–22% increase in total organic carbon in both years (Farooq et al. 2021). Another study at ICAR-Indian Agricultural Research Insititue, New Delhi by Roy et al. (2024a, b) on brown manuring, the results indicated that brown manuring in zero-tillage maize plots reduced total weed population by 28.4% at 60 days after sowing, compared to conventional tillage with sole maize.

7.8.3 Sustainable Greening of Rice Fallow

The term "rice fallow area" refers to the sizeable portion of South Asia's rice-growing region that is left uncultivated after rice is harvested. The major section (88.3%) of its 22.3 Mha in South Asia is in India, with the remaining portions being in Bangladesh (8.7%), Nepal (1.4%), Sri Lanka (1.1), Pakistan (0.5%), and Bhutan (0.02%), (Gumma et al. 2016). In India, >84% of the land is left fallow for rice in the eastern and northeastern regions (Singh et al. 2017). The rice-fallow system is unsustainable because to factors such poor diversification, declining soil quality, and income (Lal et al. 2016). There are several reasons for remaining the land fallow after rice harvest, like the cultivation of long-duration rice varieties postpones the adoption of a second crop; monsoon rains predominate but winter rains do not; low soil moisture reserves, lack of irrigation facilities and infrastructure. Growing short- or medium-duration rice and then planting pulse crops in the post-kharif season might boost the productivity of rice fallow land. The ability of the rice fallow system to become more productive and sustainable can be enhanced by various resource conservation techniques, such as rainwater management, conservation agriculture, and the preservation of kharif rice residue. Legumes may enhance soil health and boost cropping intensity and output when added to a rice fallow system. When lentil added during the winter season, the seed output was greater under mulching and standing stubbles than with no mulching. The average production of lentil (1.53 Mg ha^{-1}) was higher than that of the native long duration rice varieties after the medium duration rice varieties (Layek et al. 2021). After rice harvested, residue retention techniques may increase soil moisture content, which can facilitate the adoption of short-duration crops. Kumar et al. (2020) suggested that the zero-tillage direct seeded rice with rice residue retention (ZTDSRR+) and zero-tillage direct seeded rice (ZTDSR) outperformed the puddled transplanted rice (PTR) t in terms of grain output for winter crops.

7.8.4 Cover Cropping

The crops sown to cover the earth are referred to as "cover crops" in this context. In order to reduce soil erosion, leaching, and surface movement of nutrients from deep layers, cover crops should be planted with care (Yadav et al. 2021a; Kaye and Quemada 2017). Cover crops (CCs) improve the physical and chemical properties of soil while also lowering production costs and offering biological advantages. Furthermore, CCs help in better insect control, increased soil aggregate stability, and greater carbon (C) sequestration (O'Reilly et al. 2011). Through biological N fixation, CCs also increase nitrogen (N) availability and aid in storing extra soil N for the succeeding crops. By sequestering atmospheric CO_2 and N_2O, CCs reduce the negative effects of global warming (Fageria et al. 2005). CCs reduces water erosion and halts the sediment and dissolved nutrients losses (Blanco-Canqui et al. 2011). Rye-enriched soils showed a 54% reduction in rill erosion after 3 years, while oat-enriched soils showed an 89% reduction (Kaspar et al. 2001). Interseeding

Fig. 7.6 Smothering effect of Sesbania as a cover crop in maize-wheat-mungbean system at ICAR-IARI, New Delhi

of red clover, rye (Kleinman et al. 2005), and alfalfa (Osterholz et al. 2019) into silage corn dramatically decreased the quantity of phosphorus (P) lost in runoff over without cover crops. Owing the capacity to cover the soil's surface and enhance its physical, chemical, and biological qualities, legumes are regarded as the primary component of cover cropping systems (Fig. 7.6). Rapid germination and emergence, resilience to severe weather conditions, nitrogen fixation, lack of rivalry with the main crop, resistance to disease and insect pests, and affordability are the important characteristics of the ideal cover crop (Reddy 2016). Cover crops provide a favorable microclimate that boosts soil biological activity and moisture content (Gallaher 1977). Inoculating the roots of cover crops with microbial fungus aids in nutrient mobilization and influences the development of primary crops. However, the length of the cover crop is crucial in this situation. When the wheat cover crop was stopped at 4–6 weeks of age, Elmore et al. (1992) recorded no impact on the growth and yield of succeeding soybean crops. Cover crops decrease NO_3-leaching, the decrease in top soil loss due to runoff is also a prominent advantages of cover crops. Besides the lower incidence of pest and diseases the combination of sorghum + pigeonpea-hedge lucerne and sorghum + pigeonpea- guinea grass reduced runoff by 51–52% and sediment loss by 52% (Visha Kumari et al. 2023).

7.8.5 Conservation Tillage

Conventional farming practices like recurrent tilling, monoculture, and residue exclusion cause ~24% of land degradation globally (Bai et al. 2008). The best management strategies for preserving soil and water include minimizing soil disturbance and maintaining residue cover, which are promoted worldwide to boost sequestration (Lal et al. 2018). Conservation agriculture (CA), is based on three principles—zero tillage, residue retention, and crop rotation—is a significant approach that is now in the spotlight. Cultivation of maize, black gram, and buckwheat with zero tillage is a more effective way to manage land resources in terms of enhancing the SOC and its various pools and raising farm productivity in the Indian Himalaya region (Babu et al. 2023b). The rice-wheat cropping pattern that predominates in India's Indo-Gangetic Plain (IGP) is unsustainable under traditional agricultural methods over the conservation tilling (Fig. 7.7). Enhancing crop production, soil quality, nutrient cycling, and C-sequestration potential may be achieved by the use of CA in conjunction with site-specific nutrient management (Parihar et al. 2020). Growing two crops in a year is very difficult in rainfed regions with Alfisol soil because of a shortage of water and poor soil quality, but CA has a significant potential to increase crop yield in semiarid zones (Lal 1995). Compared to conventional (0.29–0.42%), CA produced a slightly higher soil OC (–.31% to –.45%) and a greater crop yield (Kundu et al. 2013).

7.8.6 Integrated Farming System (IFS)

IFS is one of the effective climate-resilient techniques that may raise farmers' standard of living and preserve sustainability by strengthening their resource base (Kumar et al. 2018). In order to achieve maximum replacement of off-farm inputs, guarantee high-quality food production, and maintain sustainability by lowering the current environmental pollution from agriculture, IFS is a biologically integrated system that combines natural resources and regulatory systems (Babu et al. 2023a). IFS's two primary characteristics are residue recycling, which turns wastes from one component into an input for another, and increased land use efficiency via the division of land among several businesses. IFS components vary by area based on agroclimatic conditions, such as soil type and water availability,

Fig. 7.7 Zero Tillage practice in direct seeded rice in research plot of International Potato Centre, CPRS, Patna

market demand, and farmers' socioeconomic standing (Singh et al. 2008). The Indian government launched a number of programs to support livestock and agricultural output while also giving small and marginal farmers a stable source of income (Fatima et al. 2023). The IFS practices includes livestock and fisheries with crops have been shown to improve soil microbial activity, nitrogen cycling, and nutrient utilization efficiency (Sujatha and Bhat 2015). IFS has enormous potential to improve soil health, increase economy, conserve biodiversity, provide food and nutritional security, and increase climate resilience. By lowering the amount of external fertilizer used and enhancing the nutrient cycling inside the IFS system, IFS has a lot of promise (Paramesh et al. 2022). However, in order for IFS to be sustainable in a given environment, its components need be carefully chosen after both on- and off-farm study. The elements of an integrated agricultural system should work together to support sustainability. The crops + livestock + fisheries system, a popular IFS approach in India. With the help of pond water, crops may be cultivated with guaranteed irrigation, increasing crop yields and supplying the household with adequate food. The availability of forage for cattle production on the farm can provide the farm family with milk and protein. FYM improves soil health and uses less chemical fertilizers. The generation of biogas may lower the energy consumption of traditional fossil fuels. Fish in the pond have the ability to lessen water pollution since they consume the plants and animals. On the farm, ecological variety is preserved. Thus, this approach enhances food production, lowers farm family poverty and hunger, and IFS aids in resource conservation and efficient use, lowering environmental pollution, and eventually contributing to sustainability.

7.8.7 Agroforestry

Agroforestry systems has a great potential to enhance soil quality and environmental sustainability (Dollinger and Jose 2018; Kumar et al. 2022). Agroforestry is an integrated system of crops, forages, woody perennials, palms, and animals that maximizes output and makes effective use of available resources. Array of interactions between trees and crops in agroforestry contribute to the sustainability of agriculture (Pretty 2018; Wani et al. 2023). Agroforestry has the ability to restore the degraded soil, lowering GHG emissions, and enhancing global food security in degraded landscapes, particularly in rainfed or dryland areas. Crop type, soil, and climate all affect how agroforestry affects soil quality, either directly or indirectly (Yadav et al. 2021b; Kumar et al. 2022). Nutrient cycling depends on the dynamics of tree root systems, which act as a "safety net" to stop nutrient loss through leaching. Trees contribute significantly to the cycling of nutrients because their deep roots serve as a "safety net" against nutrient losses from the nutrient cycle (Fahad et al. 2022). Because tree roots, liters, and understorey plants slow down runoff flow velocity and promote sedimentation, agroforestry systems minimize runoff water and soil loss (Udawatta et al. 2010). Agrosilvopastoral systems based on *Hardiwickia binata* + finger millet + Stylos and *Dalbergia latifolia* + finger millet + Stylos increase soil N (36% and 33%), P (65% and 63%), and organic carbon (36% and 35%) while decreasing the ρb by 6% over initial status (Venkatesh et al. 2024). In Bangladesh, agroforestry plots had considerably greater concentrations of SOM (4.75%), accessible P (12.17 mg kg^{-1}), and exchangeable K (0.39 mg kg^{-1}) than afforestation, reforestation, or slash-and-burn plots (sandy loam and clayey soil) (Chowdhury et al. 2022).

7.8.8 Residue Recycling

India produces around 686 million tonnes of agricultural residues per year from 26 crops (Hiloidhari et al. 2014). Of these, cereals—rice, wheat, maize, pearl millet, barley, small millets, and sorghum—contribute the most, making up 368 million tonnes (54%) of the total residue. After grain harvesting, farmers in northwest India and other regions of eastern and southern India often burn extra crop remains in the fields, especially before to sowing the next crop. In Punjab, India alone, ~20 million tons of rice and wheat wastes are burned in situ annually. This results in the loss of ~8 million tonnes of carbon, or a carbon dioxide load of nearly 29 million tonnes yearly. Along with the loss of sulfur and beneficial soil microflora, this also results in the loss of around 100,000 tons of nitrogen (Yadvinder-Singh et al. 2010). There is 5–8 kg of nitrogen in one tonne of rice straw (Dobermann and Fairhurst 2002). In order to promote soil health, increase crop yield, and lessen environmental effects, a systems approach to residue recycling incorporates crop residues into agricultural operations. This method encourages the utilization of agricultural wastes in a closed-loop system that recycles nutrients back into the soil by seeing them as useful resources rather than trash. Around the globe, a variety of agricultural residue management approaches are used in real-world field settings, such as conservation tillage, nutrient cycling, soil conservation methods, zero tillage, residue mulching, usage as animal feed, and vermicomposting (Raza et al. 2019). By controlling the biochemical characteristics of the soil, efficient crop residue management enhances input efficiency (Sarkar et al. 2020). Gupta et al. (2002) found that adding agricultural wastes rather than burning straw increases phosphorus release, decreases phosphorus sorption, and improves the availability of both organic and inorganic phosphorus. By incorporating residue, it is possible to recycle between 50% and 80% of the micronutrients (zinc, iron, copper, and manganese) that are absorbed by rice and wheat crops. Rice's availability of micronutrients like iron and zinc is also influenced by crop wastes (Singh et al. 2005; Gupta et al. 2002). Microbial activity in the topsoil layer was enhanced by applying crop residue mulching (Samui et al. 2020). This is most likely due to the fact that it adjusted the microclimate between plants and soil, enhanced nutrient and water availability, and enhanced soil temperature regulation. Applying wheat crop residue considerably raised soil microbial biomass carbon (MBC) (Chatterjee et al. 2018).

7.9 Conclusion

To improve food security and guarantee a sustainable future, a systems approach is essential. Sustainable system approaches enhance the soil ecosystem's general resilience and health. This strategy greatly enhances soil health by emphasizing varied cropping systems, such as crop diversification, rotation, and intercropping, and encouraging advantageous interactions between soil, microorganisms, and plants. It has been shown that important techniques including incorporating legumes into cropping systems, greening rice fallows, and using conservation tillage improve soil organic matter, nitrogen cycling, and soil structure, all of which boost crop yields and environmental sustainability. The significance of these behavior is supported by statistical data. For example, cover crops incorporation may increase soil organic carbon by 8–12% over 5 years, whereas conservation tillage can minimize soil erosion by up to 60%. Additionally, research has shown that crop variety may boost soil microbial biomass by as much as 20%, resulting in a more robust soil ecosystem.

Agroecological concepts are also in line with the systems approach, which emphasizes the value of biodiversity, the use of few synthetic inputs, and the blending of old knowledge with contemporary methods. By using these ideas, farmers may develop more productive and sustainable farming systems that address more general issues like food security and climate change in addition to enhancing soil health. Moreover, a systems approach to soil health actively improves the soil's capacity to sustain agricultural output in the face of escalating global problems rather than just preserving the status quo.

References

Acton DF, Gregorich LJ (1995) The health of our soils: toward sustainable agriculture in Canada

Allison SD, Martiny JBH (2008) Resistance, resilience, and redundancy in microbial communities. Proc Natl Acad Sci USA 105:11512e11519. https://doi.org/10.1073/pnas.0801925105

Babu S, Das A, Singh R, Mohapatra KP, Kumar S, Rathore SS, Yadav SK, Yadav P, Ansari MA, Panwar AS, Wani OA, Singh M, Ravishankar N, Layek J, Chandra P, Singh VK (2023a) Designing an energy efficient, economically feasible, and environmentally robust integrated farming system model for sustainable food production in the Indian Himalayas. Sustain Food Technol. https://doi.org/10.1039/D2FB00016D

Babu S, Singh R, Avasthe R, Kumar S, Rathore SS, Singh VK et al (2023b) Soil carbon dynamics under organic farming: Impact of tillage and cropping diversity. Ecol Indic 147:109940. https://doi.org/10.1016/j.ecolind.2023.109940

Bai T, Xu S, Rupp F, Fan H, Yin K, Guo Z, Zhang L, Yang B, Huang Y, Li Y, Li X, Zeng L, Zheng S-J (2019) Temporal variations of Fusarium oxysporum f. sp. cubense tropical race 4 population in a heavily infected banana field in Southwest China. Acta Agric Scand Sect B Soil Plant Sci 69:641e648. https://doi.org/10.1080/09064710.2019.1635200

Bai ZG, Dent DL, Olsson L, Schaepman ME (2008) Proxy global assessment of land degradation. Soil use and management 24(3):223–234

Bainard LD, Navarro-Borrell A, Hamel C, Braun K, Hanson K, Gan Y (2017) Increasing the frequency of pulses in crop rotations reduces soil fungal diversity and increases the proportion of fungal pathotrophs in a semiarid agroecosystem. Agric Ecosyst Environ 240:206e214. https://doi.org/10.1016/j.agee.2017.02.020

Banjara TR, Bohra JS, Kumar S, Singh T, Shori A, Prajapat K (2021) Sustainable alternative crop rotations to the irrigated rice-wheat cropping system of Indo-Gangetic Plains of India. Arch Agron Soil Sci 67:1–18. https://doi.org/10.1080/03650340.2021.1912324

Barbieri P, Pellerin S, Seufert V, Nesme T (2019) Changes in crop rotations would impact food production in an organically farmed world. Nat Sustain 2:378e385. https://doi.org/10.1038/s41893-019-0259-5

Barrutia O, Garbisu C, Epelde L, Sampedro M, Goicolea M, Becerril J (2011) Plant tolerance to diesel minimizes its impact on soil microbial characteristics during rhizoremediation of diesel-contaminated soils. Sci Total Environ 409:4087e4093. https://doi.org/10.1016/j.scitotenv.2011.06.025

Bennett AJ, Bending GD, Chandler D, Hilton S, Mills P (2012) Meeting the demand for crop production: the challenge of yield decline in crops grown in short rotations. Biol Rev 87:52e71. https://doi.org/10.1111/j.1469-185X.2011.00184.x

Bezner Kerr R, Hickey C, Lupafya E, Dakishoni L (2019) Repairing rifts or reproducing inequalities? Agroecology, food sovereignty, and gender justice in Malawi. J Peasant Stud 46(7):1499–1518. https://doi.org/10.1080/03066150.2018.1547897

Bhattacharya A (2019) Chapter 1—global climate change and its impact on agriculture. In: Bhattacharya A (ed) Changing climate and resource use efficiency in plants. Academic, p 1e50. https://doi.org/10.1016/B978-0-12-816209-5.00001

Blanco-Canqui H, Mikha MM, Presley DR, Claassen MM (2011) Addition of cover crops enhances no-till potential for improving soil physical properties. Soil Sci Soc Am J 75(4):1471–1482. https://doi.org/10.2136/sssaj2010.0430

Bolan NS (1991) A critical review on the role of mycorrhizal fungi in the uptake of phosphorus by plants. Plant Soil 134:189e207. https://doi.org/10.1007/bf00012037

Brown ME (1974) Seed and root bacterization. Annu Rev Phytopathol 12:181e197. https://doi.org/10.1146/annurev.py.12.090174.001145

Bukovsky-Reyes S, Isaac ME, Blesh J (2019) Effects of intercropping and soil properties on root functional traits of cover crops. Agric Ecosyst Environ 285. https://doi.org/10.1016/j.agee.2019.106614

Bull SE, Seung D, Chanez C, Mehta D, Kuon J-E, Truernit E, Hochmuth A, Zurkirchen I, Zeeman SC, Gruissem W, Vanderschuren H (2018) Accelerated ex situ breeding of GBSS- and PTST1-edited cassava for modified starch. Sci Adv 4:eaat6086. https://doi.org/10.1126/sciadv.aat6086

Bünemann EK, Bongiorno G, Bai Z, Creamer RE, De Deyn G, de Goede R, Fleskens L, Geissen V, Kuyper TW, Mader P, Pulleman M, Sukkel W, van Groenigen JW, Brussaard L (2018) Soil quality e a critical review. Soil Biol Biochem 120:105e125. https://doi.org/10.1016/j.soilbio.2018.01.030

Caron P, Ferrero y de Loma-Osorio G, Nabarro D, Hainzelin E, Guillou M, Andersen I et al (2018) Food systems for sustainable development: proposals for a profound four-part transformation. Agron Sustain Dev 38:1–12. https://doi.org/10.1007/s13593-018-0519-1

Carr P (2017) Guest editorial: conservation tillage for organic farming. Multidisciplinary Digital Publishing Institute. https://doi.org/10.3390/agriculture7030019

Chowdhury FI, Barua I, Chowdhury AI, Resco de Dios V, Alam MS (2022) Agroforestry shows higher potential than reforestation for soil restoration after slash-and-burn: a case study from Bangladesh. Geol Ecol Landsc 6(1):48–54. https://doi.org/10.1080/24749508.2020.1743528

Cong WF, Hoffland E, Li L, Six J, Sun JH, Bao XG, Zhang FS, Van Der Werf W (2015) Intercropping enhances soil carbon and nitrogen. Glob Chang Biol 21:1715e1726. https://doi.org/10.1111/gcb.12738

Crozat Y, Fustec J (2004) Assessing the role of grain legumes in crop rotation: some agronomical concepts that can help. In: Grain legumes and the environment: how to assess benefits and impacts. Proceedings of AEP workshop, pp 18–19

Cruz AF, Hamel C, Yang C, Matsubara T, Gan Y, Singh AK, Kuwada K, Ishii T (2012) Phytochemicals to suppress Fusarium head blight in wheat chickpea rotation. Phytochemistry 78:72e80. https://doi.org/10.1016/j.phytochem.2012.03.003

Das A, Babu S, Datta M, Kumar S, Singh R, Avasthe RK, Rathore SS, Yadav SK, Singh VK (2022) Restoring soil carbon in marginal land of Indian Himalayas: impact of crop intensification and conservation tillage. J Environ Manag. https://doi.org/10.1016/j.jenvman.2022.115603

de Araújo Santos GA, Moitinho MR, de Oliveira Silva B, Xavier CV, Teixeira DDB, Cora JE, Júnior NLS (2019) Effects of long-term no-tillage systems with different succession cropping strategies on the variation of soil CO_2 emission. Sci Total Environ 686:413e424. https://doi.org/10.1016/j.scitotenv.2019.05.398

de Medeiros EV, Notaro KDA, de Barros JA, Duda GP, Moraes MDCHDS, Ambrosio MMDQ, Negreiros AMP, Sales Júnior R (2019) Soils from intercropped fields have a higher capacity to suppress black root rot in cassava, caused by Scytalidium lignicola. J Phytopathol 167:209e217. https://doi.org/10.1111/jph.12788

Dobermann ATHF, Fairhurst TH (2002) Rice straw management. Better Crops Int 16(1):7–11

Dollinger J, Jose S (2018) Agroforestry for soil health. Agrofor Syst 92:213–219. https://doi.org/10.1007/s10457-018-0223-9

Douglas GM, Maffei VJ, Zaneveld J, Yurgel SN, Brown JR, Taylor CM, Huttenhower C, Langille MGI (2019) PICRUSt2: an improved and extensible approach for metagenome inference. bioRxiv 672295. https://doi.org/10.1101/672295
Ellouze W, Esmaeili Taheri A, Bainard LD, Yang C, Bazghaleh N, Navarro-Borrell A, Hanson K, Hamel C (2014) Soil fungal resources in annual cropping systems and their potential for management. Biomed Res Int. https://doi.org/10.1155/2014/531824
Elmore CD, Wesley RA, Heatherly LG (1992) Stale seedbed production of soybeans with a wheat cover crop. J Soil Water Conserv 47(2):187–190
Epelde L, Becerril JM, Kowalchuk GA, Deng Y, Zhou J, Garbisu C (2010) Impact of metal pollution and Thlaspi caerulescens growth on soil microbial communities. Appl Environ Microbiol 76:7843e7853. https://doi.org/10.1128/AEM.01045-10
Fageria NK, Baligar VC, Bailey BA (2005) Role of cover crops in improving soil and row crop productivity. Commun Soil Sci Plant Anal 36(19–20):2733–2757. https://doi.org/10.1080/00103620500303939
Fahad S, Chavan SB, Chichaghare AR, Uthappa AR, Kumar M, Kakade V et al (2022) Agroforestry systems for soil health improvement and maintenance. Sustainability 14(22):14877. https://doi.org/10.3390/su142214877
Fargione JE, Bassett S, Boucher T, Bridgham SD, Conant RT, Cook-Patton SC et al (2018) Natural climate solutions for the United States. Sci Adv 4(11):eaat1869. https://doi.org/10.1126/sciadv.aat1869
Farooq M, Ullah N, Nadeem F, Nawaz A, Siddique KH (2021) Sesbania brown manuring improves soil health, productivity, and profitability of post-rice bread wheat and chickpea. Exp Agric 57(3):145–162. https://doi.org/10.1017/S0014479721000090
Fatima A, Singh VK, Babu S, Singh RK, Upadhyay PK, Rathore SS, Kumar B, Hasanain M, Parween H (2023) Food production potential and environmental sustainability of different farming system models in northwest India. Front Sustain Food Syst. https://doi.org/10.3389/fsufs.2023.95946
Fierer N (2017) Embracing the unknown: disentangling the complexities of the soil microbiome. Nature Reviews Microbiology 15(10):579–590
Fritig B, Heitz T, Legrand M (1998) Antimicrobial proteins in induced plant defense. Current opinion in immunology 10(1):16–22
Galende MA, Becerril JM, Barrutia O, Artetxe U, Garbisu C, Hernandez A (2014) Field assessment of the effectiveness of organic amendments for aided. https://doi.org/10.1016/j.gexplo.2014.06.006
Gallaher RN (1977) Soil moisture conservation and yield of crops no-till planted in rye. Soil Sci Soc Am J 41(1):145–147. https://doi.org/10.2136/sssaj1977.03615995004100010040x
Gan YT, Miller PR, McConkey BG, Zentner RP, Stevenson FC, McDonald CL (2003) Influence of diverse cropping sequences on durum wheat yield and protein in the semiarid northern Great Plains. Agron J 95:245e252. https://doi.org/10.2134/agronj2003.2450
Gan Y, Liang C, Hamel C, Cutforth H, Wang H (2011) Strategies for reducing the carbon footprint of field crops for semiarid areas. A review. Agron Sustain Dev 31:643e656. https://doi.org/10.1007/s13593-011-0011-7
Garbisu C, Alkorta I, Epelde L (2011) Assessment of soil quality using microbial properties and attributes of ecological relevance. Appl Soil Ecol:1e4. https://doi.org/10.1016/j.apsoil.2011.04.018
Gogoi N, Baruah KK, Meena RS (2018) Grain legumes: impact on soil health and agroecosystem. In: Legumes for soil health and sustainable management, pp 511–539. https://doi.org/10.1007/978-981-13-0253-4_16
Government of India and WFP sign cooperation agreement for 2023–27 website. https://www.wfp.org/news/government-india-and-wfp-sign-cooperation-agreement-2023-27. Accessed 23 Aug 2024
Graf DRH, Saghaï A, Zhao M, Carlsson G, Jones CM, Hallin S (2019) Lucerne (Medicago sativa) alters N2O-reducing communities associated with cocksfoot (Dactylis glomerata) roots and

promotes N2O production in intercropping in a greenhouse experiment. Soil Biol Biochem 137. https://doi.org/10.1016/j.soilbio.2019.107547

Grahmann K, Rubio Dellepiane V, Terra JA, Quincke JA (2020) Long-term observations in contrasting crop-pasture rotations over half a century: statistical analysis of chemical soil properties and implications for soil sampling frequency. Agric Ecosyst Environ 287:106710. https://doi.org/10.1016/j.agee.2019.106710

Gumma MK, Thenkabail PS, Teluguntla P, Rao MN, Mohammed IA, Whitbread AM (2016) Mapping rice-fallow cropland areas for short-season grain legumes intensification in South Asia using MODIS 250 m time-series data. Int J Digit Earth 9(10):981–1003. https://doi.org/10.1080/17538947.2016.1168489

Gupta RK, Shukla AK, Ashraf M, Ahmed ZU, Sinha RKP, Hobbs PR (2002) Options for establishment of rice and issues constraining its productivity and sustainability in eastern Gangetic plains of Bihar, Nepal and Bangladesh. In Rice-wheat consortium travelling seminar report series, vol 4, p 36

Hamel C, Gan Y, Sokolski S, Bainard LD (2018) High frequency cropping of pulses modifies soil nitrogen level and the rhizosphere bacterial microbiome in 4-year rotation systems of the semiarid prairie. Appl Soil Ecol 126:47e56. https://doi.org/10.1016/j.apsoil.2018.01.003

Hatfield JL, Sauer TJ, Cruse RM (2017) Chapter one—soil: the forgotten piece of the water, food, energy Nexus. In: Sparks DL (ed) Advances in agronomy, vol 143. Academic, p 1e46. https://doi.org/10.1016/bs.agron.2017.02.001

Hayat R, Ali S, Amara U, Khalid R, Ahmed I (2010) Soil beneficial bacteria and their role in plant growth promotion: a review. Ann Microbiol 60:579e598. https://doi.org/10.1007/s13213-010-0117-1

Hiloidhari M, Das D, Baruah DC (2014) Bioenergy potential from crop residue biomass in India. Renew Sust Energ Rev 32:504–512. https://doi.org/10.1016/j.rser.2014.01.025

Hinsinger P, Betencourt E, Bernard L, Brauman A, Plassard C, Shen J, Tang X, Zhang F (2011) P for two, sharing a scarce resource: soil phosphorus acquisition in the rhizosphere of intercropped species. Plant Physiol 156:1078e1086. https://doi.org/10.1104/pp.111.175331

Humphreys E, Gaydon DS (2015) Options for increasing the productivity of the rice–wheat system of north-west India while reducing groundwater depletion. Part 2. Is conservation agriculture the answer? Field Crop Res 173:81–94. https://doi.org/10.1016/j.fcr.2014.11.019

Ji H, Ding Y, Liu X, Li L, Zhang D, Li Z, Sun J, Lashari MS, Joseph S, Meng Y, Kuzyakov Y, Pan G (2015) Root-derived short-chain suberin diacids from rice and rape seed in a paddy soil under rice cultivar treatments. PLoS One 10. https://doi.org/10.1371/journal.pone.0127474

Jungers JM, DeHaan LH, Mulla DJ, Sheaffer CC, Wyse DL (2019) Reduced nitrate leaching in a perennial grain crop compared to maize in the Upper Midwest, USA. Agric Ecosyst Environ 272:63e73. https://doi.org/10.1016/j.agee.2018.11.007

Kangmennaang J, Kerr RB, Lupafya E, Dakishoni L, Katundu M, Luginaah I (2017) Impact of a participatory agroecological development project on household wealth and food security in Malawi. Food Secur 9:561–576. https://doi.org/10.1007/s12571-017-0669-z

Karlen DL, Andrews SS, Doran JW (2001) Soil quality: current concepts and applications. https://doi.org/10.1016/S0065-2113(01)74029-1

Kaspar TC, Radke JK, Laflen JM (2001) Small grain cover crops and wheel traffic effects on infiltration, runoff, and erosion. J Soil Water Conserv 56(2):160–164

Kaye JP, Quemada M (2017) Using cover crops to mitigate and adapt to climate change. A review. Agron Sustain Dev 37:1–17. https://doi.org/10.1007/s13593-016-0410-x

Kleinman PJ, Salon P, Sharpley AN, Saporito LS (2005) Effect of cover crops established at time of corn planting on phosphorus runoff from soils before and after dairy manure application. J Soil Water Conserv 60(6):311–322. https://www.jswconline.org/content/60/6/311.short

Klimkowicz-Pawlas A, Ukalska-Jaruga A, Smreczak B (2019) Soil quality index for agricultural areas under different levels of anthropopressure. Int Agrophys 33:455e462. https://doi.org/10.31545/intagr/113349

Kretzschmar T, Kohlen W, Sasse J, Borghi L, Schlegel M, Bachelier JB, Reinhardt D, Bours R, Bouwmeester HJ, Martinoia E (2012) A petunia ABC protein controls strigolactone-dependent symbiotic signalling and branching. Nature 483:341e344. https://doi.org/10.1038/nature10873

Kumar V, Bellinder RR, Gupta RK, Malik RK, Brainard DC (2008) Role of herbicide-resistant rice in promoting resource conservation technologies in rice–wheat cropping systems of India: a review. Crop Prot 27:290–301. https://doi.org/10.1016/j.cropro.2007.05.016

Kumar S, Bhatt BP, Dey A, Shivani S, Kumar U, Idris MD, Mishra JS, Kumar S (2018) Integrated farming system in India: Current status, scope and future prospects in changing agricultural scenario. Indian J Agric Sci 88(11):1661–1675. https://doi.org/10.56093/ijas.v88i11.84880

Kumar R, Mishra JS, Rao KK, Mondal S, Hazra KK, Choudhary JS et al (2020) Crop rotation and tillage management options for sustainable intensification of rice-fallow agro-ecosystem in eastern India. Sci Rep 10(1):11146. https://doi.org/10.1038/s41598-020-67973-9

Kumar SS, Mir SA, Wani OA, Babu S, Yeasin M, Bhat MA, Hussain N, Wani AA, Kumar R, Devideen Y, Dar SR (2022) Land-use systems regulate carbon geochemistry in the temperate Himalayas, India. J Environ Manag. https://doi.org/10.1016/j.jenvman.2022.115811

Kumari VV, Balloli SS, Ramana DBV, Kumar M, Maruthi V, Prabhakar M et al (2023) Crop and livestock productivity, soil health improvement and insect dynamics: impact of different fodder-based cropping systems in a rainfed region of India. Agric Syst 208:103646. https://doi.org/10.1016/j.agsy.2023.103646

Kundu S, Srinivasarao C, Mallick RB, Satyanarayana T, Prakash Naik R, Johnston A, Venkateswarlu B (2013) Conservation agriculture in maize (Zea mays L.) e horsegram (Macrotyloma uniflorum L.) system in rainfed Alfisols for carbon sequestration and climate change mitigation. J Agrometeorol 15(1):144–149

Ladha JK, Kumar V, Alam MM, Sharma S, Gathala MK, Chandna P et al (2009) Integrating crop and resource management technologies for enhanced productivity, profitability and sustainability of the rice–wheat system in South Asia. In: Ladha JK, Singh Y, Erenstein O, Hardy B (eds) Integrated crop and resource management in the rice–wheat system of South Asia. International Rice Research Institute, Los Banos, pp 69–108

Lal R (1995) Tillage and mulching effects on maize yield for seventeen consecutive seasons on a tropical alfisol. J Sustain Agric 5(4):79–93. https://doi.org/10.1300/J064v05n04_07

Lal R (2015) A system approach to conservation agriculture. J Soil Water Conserv 70(4):82A–88A. https://doi.org/10.2489/jswc.70.4.82A

Lal R (2016) Soil health and carbon management. Food and energy security 5(4):212–222

Lal B, Gautam P, Panda BB, Raja R, Singh T, Tripathi R et al (2017) Crop and varietal diversification of rainfed rice based cropping systems for higher productivity and profitability in Eastern India. PLoS One 12(4):e0175709. https://doi.org/10.1371/journal.pone.0175709

Lal R, Smith P, Jungkunst HF, Mitsch WJ, Lehmann J, Nair PR, Ravindranath NH (2018) The carbon sequestration potential of terrestrial ecosystems. Journal of soil and water conservation 73(6):145A–152A

Lay CY, Bell TH, Hamel C, Harker KN, Mohr R, Greer CW, Yergeau E, St-Arnaud M (2018) Canola root-Associated microbiomes in the Canadian Prairies. Front Microbiol 9. https://doi.org/10.3389/fmicb.2018.01188

Layek J, Das A, Idapuganti R, Rangappa K, Ghosh P, Lal R, Choudhury B, Mohapatra KP, Babu S, Yadav G, Dey U (2021) Managing rice fallow lands of eastern Indian Himalayas: impacts of residue management and varietal interventions on soil properties, carbon stocks and productivity. Land Degrad Dev 32. https://doi.org/10.1002/ldr.4067

Lemaire G, Franzluebbers A, Carvalho PCDF, Dedieu B (2014) Integrated crop-livestock systems: strategies to achieve synergy between agricultural production and environmental quality. Agric Ecosyst Environ 190:4e8. https://doi.org/10.1016/j.agee.2013.08.009

Lin Y, Ye G, Kuzyakov Y, Liu D, Fan J, Ding W (2019) Long-term manure application increases soil organic matter and aggregation, and alters microbial community structure and keystone taxa. Soil Biol Biochem 134:187–196. https://doi.org/10.1016/j.soilbio.2019.03.030

Lupwayi NZ, Larney FJ, Blackshaw RE, Pearson DC, Eastman AH (2018) Soil microbial biomass and its relationship with yields of irrigated wheat under long-term conservation management. Soil Sci 183:179e187. https://doi.org/10.1097/SS.0000000000000242

Lynch JM (1983) Soil biotechnology: microbiological factors in crop productivity. Blackwell Scientific Publications

Maltais-Landry G, Scow K, Brennan E (2014) Soil phosphorus mobilization in the rhizosphere of cover crops has little effect on phosphorus cycling in California agricultural soils. Soil Biol Biochem 78:255e262. https://doi.org/10.1016/j.soilbio.2014.08.013

Maron JL, Marler M, Klironomos JN, Cleveland CC (2011) Soil fungal pathogens and the relationship between plant diversity and productivity. Ecol Lett 14:36e41. https://doi.org/10.1111/j.1461-0248.2010.01547.x

Maruri-Lopez I, Aviles-Baltazar NY, Buchala A, Serrano M (2019) Intra and extracellular journey of the phytohormone salicylic acid. Front Plant Sci 10. https://doi.org/10.3389/fpls.2019.00423

Matson PA, Parton WJ, Power AG, Swift MJ (1997) Agricultural intensification and ecosystem properties. Science 277:504e509. https://doi.org/10.1126/science.277.5325.504

Meena RS, Lal R (2018) Legumes and sustainable use of soils. Legumes for soil health and sustainable management. Springer, Singapore, pp 1–31. https://doi.org/10.1007/978-981-13-0253-4_1

Milcu A, Allan E, Roscher C, Jenkins T, Meyer ST, Flynn D, Bessler H, Buscot F, Engels C, Gubsch M, Konig S, Lipowsky A, Loranger J, Renker C, Scherber C, Schmid B, Thebault E, Wubet T, Weisser WW, Scheu S, Eisenhauer N (2013) Functionally and phylogenetically diverse plant communities key to soil biota. Ecology 94:1878e1885. https://doi.org/10.1890/12-1936.1

Miransari M, Riahi H, Eftekhar F, Minaie A, Smith DL (2013) Improving soybean (Glycine max L.) N2 fixation under stress. J Plant Growth Regul 32:909e921. https://doi.org/10.1007/s00344-013-9335-7

Müller DB, Vogel C, Bai Y, Vorholt JA (2016) The plant microbiota: systems-level insights and perspectives. Annu Rev Genet:211e234. https://doi.org/10.1146/annurev-genet-120215-034952

NAAS (2017) Innovative viable solution to rice residue burning in rice-wheat cropping system through concurrent use of super straw management system-fitted combines and turbo happy seeder. Policy Brief No. 2. National Academy of Agricultural Sciences, New Delhi

Nath CP, Hazra KK, Kumar N, Praharaj CS, Singh SS, Singh U, Singh NP (2019) Including grain legume in rice–wheat cropping system improves soil organic carbon pools over time. Ecol Eng 129:144–153. https://doi.org/10.1016/j.ecoleng.2019.02.004

Nguyen NH, Song Z, Bates ST, Branco S, Tedersoo L, Menke J, Schilling JS, Kennedy PG (2016) FUNGuild: an open annotation tool for parsing fungal community datasets by ecological guild. Fungal Ecol 20:241e248. https://doi.org/10.1016/j.funeco.2015.06.006

O'Reilly KA, Robinson DE, Vyn RJ, Van Eerd LL (2011) Weed populations, sweet corn yield, and economics following fall cover crops. Weed Technol 25(3):374–384. https://doi.org/10.1614/WT-D-10-00051.1

Osterholz WR, Renz MJ, Jokela WE, Grabber JH (2019) Interseeded alfalfa reduces soil and nutrient runoff losses during and after corn silage production. J Soil Water Conserv 74(1):85–90. https://doi.org/10.2489/jswc.74.1.85

Ozlu E, Sandhu SS, Kumar S, Arriaga FJ (2019) Soil health indicators impacted by long-term cattle manure and inorganic fertilizer application in a corn soybean rotation of South Dakota. Sci Rep 9. https://doi.org/10.1038/s41598-019-48207-z

Pan WL, Schillinger WF, Young FL, Kirby EM, Yorgey GG, Borrelli KA, Brooks ES, McCracken VA, Maaz TM, Machado S, Madsen IJ, Johnson- Maynard JL, Port LE, Painter K, Huggins DR, Esser AD, Collins HP, Stockle CO, Eigenbrode SD (2017) Integrating historic agronomic and policy lessons with new technologies to drive farmer decisions for farm and climate: the case of Inland Pacific Northwestern U.S. Front Environ Sci 5. https://doi.org/10.3389/fenvs.2017.00076

Parihar CM, Singh AK, Jat SL, Dey A, Nayak HS, Mandal BN et al (2020) Soil quality and carbon sequestration under conservation agriculture with balanced nutrition in intensive cereal-based system. Soil Tillage Res 202:104653. https://doi.org/10.1016/j.still.2020.104653

Paramesh V, Singh SK, Mohekar DS, Arunachalam V, Misra SD, Jat SL, Bhagat T (2022) Impact of sustainable land-use management practices on soil carbon storage and soil quality in Goa State, India. Land Degradation & Development 33(1):28–40

Pathak H, Tewari AN, Sankhyan S, Dubey DS, Mina U, Singh VK, Jain N (2011) Direct-seeded rice: potential, performance and problems—a review. Curr Adv Agric Sci (Int J) 3(2):77-88

Peoples MB, Craswell ET (1992) Biological nitrogen fixation: investments, expectations and actual contributions to agriculture. Plant Soil 141:13e39. https://doi.org/10.1007/978-94-017-0910-1_2

Pretty J (2018) Intensification for redesigned and sustainable agricultural systems. Science 362(6417):eaav0294. https://doi.org/10.1126/science.aav0294

Pretty J, Benton TG, Bharucha ZP, Dicks LV, Flora CB, Godfray HCJ et al (2018) Global assessment of agricultural system redesign for sustainable intensification. Nature Sustain 1(8):441–446. https://doi.org/10.1038/s41893-018-0114-0

Rathore SS, Babu S, Shekhawat K, Singh R, Yadav SK, Singh VK, Singh C (2022) Designing energy cum carbon-efficient environmentally clean production system for achieving green economy in agriculture. Sustain Energy Technol Assess. https://doi.org/10.1016/j.seta.2022.102190

Rathore SS, Babu S, Shekhawat K, Kumar V, Gairola A, Wani OA, Singh VK (2024) Exploring sustainable agricultural production models to coordinate the system productivity, soil biological health, and eco-efficiency in the semi-arid region. Environ Sustain Indic. https://doi.org/10.1016/j.indic.2024.100480

Raza MH, Abid M, Yan T, Naqvi SAA, Akhtar S, Faisal M (2019) Understanding farmers' intentions to adopt sustainable crop residue management practices: a structural equation modeling approach. J Clean Prod 227:613–623. https://doi.org/10.1016/j.jclepro.2019.04.244

Reddy PP (2016) Sustainable intensification of crop production. Springer, Singapore, pp 143–154. https://doi.org/10.1007/978-981-10-2702-4_1

Rodriguez R, Redman R (2008) More than 400 million years of evolution and some plants still can't make it on their own: plant stress tolerance via fungal symbiosis. J Exp Bot 59:1109e1114 10.1093/jxb/erm342

Roscher C, Schumacher J, Lipowsky A, Gubsch M, Weigelt A, Pompe S, Kolle O, Buchmann N, Schmid B, Schulze E-D (2013) A functional trait-based approach to understand community assembly and diversityeproductivity relationships over 7 years in experimental grasslands. Perspect Plant Ecol Evol System 15:139e149. https://doi.org/10.1016/j.ppees.2013.02.004

Roy S, Shekhawat K, Rathore SS, Sanketh GD, Kumar V (2024a) Brown manuring in conservation agriculture-based maize-wheat- green gram cropping system: effects on weed flora, crop yield, and profitability of maize (*Zea mays.*). Indian J Weed Sci 56(3):266–273. https://www.isws.org.in/IJWSn/Article.aspx?Article_id=1764

Roy S, Shekhawat K, Rathore SS, Sanketh GD, Kumar V (2024b) Selectivity of new generation herbicide 'Tembotrione' under conservation agriculture-based maize (*Zea mays*) in maize-wheat-mungbean cropping system. Indian J Agron 69(3):283–291. https://doi.org/10.59797/ija.v69i3.5523

Samui I, Skalicky M, Sarkar S, Brahmachari K, Sau S, Ray K, Hossain A, Ghosh A, Nanda MK, El Sabagh A (2020) Yield response, nutritional quality and water productivity of tomato (Solanum lycopersicum L.) are influenced by drip irrigation and straw mulch in the coastal saline ecosystem of Ganges delta, India. Sustainability 12(17):6779. https://doi.org/10.3390/su12176779

Sarkar S, Skalicky M, Hossain A, Brestic M, Saha S, Garai S, Dey K, Brahmachari K (2020) Management of crop residues for improving input use efficiency and agricultural sustainability. Sustainability 12(23):9808. https://doi.org/10.3390/su12239808

Singh JP, Ravisankar N (2015) Integrated farming system—towards organic agriculture. Modern concepts and practices of organic farming for safe secured and sustainable food production, p 50

Singh Y, Singh B, Timsina J (2005) Crop residue management for nutrient cycling and improving soil productivity in rice-based cropping systems in the tropics. Adv Agron 85:269–407. https://doi.org/10.1016/S0065-2113(04)85006-5

Singh SP, Gangwar B, Singh MP (2008) Economics of sugarcane-based farming system in western Uttar Pradesh. Agric Econ Res Rev 21(1):109–117

Singh RN, Praharaj CS, Kumar R, Singh SS, Kumar N, Singh U (2017) Influence of rice (Oryza sativa) habit groups and moisture conservation practices on soil physical and microbial properties in rice+ lathyrus relay cropping system under rice fallows in Eastern Plateau of India. Indian J Agric Sci 87(12):1633–1639. https://www.researchgate.net/publication/321835761

Smith P, Gregory PJ (2013) Climate change and sustainable food production. Proc Nutr Soc 72:21e28. https://doi.org/10.1017/S0029665112002832

Srinivasa Rao CH (2011) Soil health improvement with gliricidia green leaf manuring in rainfed agriculture: on farm experience, Central Research Institute for Dryland Agriculture, Hyderabad, p 16

Stolte J, Tesfai M, Øygarden L, Kværnø S, Keizer J, Verheijen F, Panagos P, Ballabio C, Hessel R (2015) Soil threats in Europe. Publications Office Luxembourg. https://doi.org/10.2788/828742

Sujatha S, Bhat R (2015) Resource use and benefits of mixed farming approach in arecanut ecosystem in India. Agric Syst 141:126–137. https://doi.org/10.1016/j.agsy.2015.10.005

Sumanta Chatterjee SC, Bandyopadhyay KK, Pradhan S, Singh R, Datta SP (2018) Effects of irrigation, crop residue mulch and nitrogen management in maize (Zea mays L.) on soil carbon pools in a sandy loam soil of Indo-gangetic plain region. https://doi.org/10.1016/j.catena.2018.02.005

Tiemann LK, Grandy AS, Atkinson EE, Marin-Spiotta E, McDaniel MD (2015) Crop rotational diversity enhances belowground communities and functions in an agroecosystem. Ecol Lett 18:761e771. https://doi.org/10.1111/ele.12453

Turrini A, Sbrana C, Pitto L, Ruffini Castiglione M, Giorgetti L, Briganti R, Bracci T, Evangelista M, Nuti MP, Giovannetti M (2004) The antifungal Dm-AMP1 protein from Dahlia merckii expressed in Solanum melongena is released in root exudates and differentially affects pathogenic fungi and mycorrhizal symbiosis. New Phytol 163:393e403. https://doi.org/10.1111/j.1469-8137.2004.01107.x

Udawatta RP, Garrett HE, Kallenbach RL (2010) Agroforestry and grass buffer effects on water quality in grazed pastures. Agrofor Syst 79:81–87. https://doi.org/10.1007/s10457-010-9288-9

van Es HM, Karlen DL (2019) Reanalysis validates soil health indicator sensitivity and correlation with long-term crop yields. Soil Sci Soc Am J 83(3):721–732. https://doi.org/10.2136/sssaj2018.09.0338

Vandenkoornhuyse P, Quaiser A, Duhamel M, Le Van A, Dufresne A (2015) The importance of the microbiome of the plant holobiont. New Phytol 206:1196e1206. https://doi.org/10.1111/nph.13312

Velásquez E, Lavelle P, Andrade M (2007) GISQ, a multifunctional indicator of soil quality. Soil biology and biochemistry 39(12):3066–3080

Venkatesh G, Gopinath KA, Ramana DBV, Kumari VV, Srinivas I, Shanker AK et al (2024) Agrosilvopastoral systems for improved crop and fodder productivity and soil health in the rainfed environments of South India. Agric Syst 214:103812. https://doi.org/10.1016/j.agsy.2023.103812

VeVerka JS, Udawatta RP, Kremer RJ (2019) Soil health indicator responses on Missouri claypan soils affected by landscape position, depth, and management practices. J Soil Water Conserv 74:126e137. https://doi.org/10.2489/jswc.74.2.126

Vukicevich E, Lowery T, Bowen P, Úrbez-Torres JR, Hart M (2016) Cover crops to increase soil microbial diversity and mitigate decline in perennial agriculture. A review. Agron Sustain Dev 36. https://doi.org/10.1007/s13593-016-0385-7

Wang F, Chen S, Wang Y, Zhang Y, Hu C, Liu B (2018) Long-term nitrogen fertilization elevates the activity and abundance of nitrifying and denitrifying microbial communities in an upland soil: implications for nitrogen loss from intensive agricultural systems. Front Microbiol 9:2424. https://doi.org/10.3389/fmicb.2018.02424

Wani SP, Pathak P, Jangawad LS, Eswaran H, Singh P (2003) Improved management of Vertisols in the semiarid tropics for increased productivity and soil carbon sequestration. Soil Use Manag 19(3):217–222. https://doi.org/10.1111/j.1475-2743.2003.tb00307.x

Wani OA, Sharma V, Kumar SS, Babu S, Sharma KR, Rathore SS, Marwaha S, Ganai NA, Dar SR, Yeasin M, Raghavendra S, Tomar J (2023) Climate plays a dominant role over land management in governing soil carbon dynamics in North Western Himalayas. J Environ Manag. https://doi.org/10.1016/j.jenvman.2023.117740

Wani OA, Sharma VK, Shasang S, Malik A, Raouf P, Aastika D, Khushboo K, Vipin G, Ananya Y, Devideen V, Donatella P, Irene, Babu S (2024) Geostatistical modelling of soil properties towards long-term ecological sustainability of agroecosystems. Ecol Indic. https://doi.org/10.1016/j.ecolind.2024.112540

Wezel A, Herren BG, Kerr RB, Barrios E, Gonçalves ALR, Sinclair F (2020) Agroecological principles and elements and their implications for transitioning to sustainable food systems. A review. Agron Sustain Dev 40:1–13. https://doi.org/10.1007/s13593-020-00646-z

Yadav GS, Babu S, Das A, Datta M, Mohapatra KP, Singh R, Singh VK, Rathore SS, Chakraborty M (2021a) Productivity, soil health, and carbon management index of Indian Himalayan intensified maize-based cropping systems under live mulch-based conservation tillage practices. Field Crop Res. https://doi.org/10.1016/j.fcr.2021.108080

Yadav GS, Kandpal BK, Das A, Babu S, Mohapatra KP, Devi AG, Devi HL, Chandra P, Singh R, Barman KK (2021b) Impact of 28-year-old agroforestry systems on soil carbon dynamics in Eastern Himalayas. J Environ Manag. https://doi.org/10.1016/j.jenvman.2021.111978

Yadvinder-Singh, Gupta RK, Singh J, Singh G, Gobinder-Singh, Ladha JK (2010) Placement effects on rice residue decomposition and nutrient dynamics on two soil types during wheat cropping in rice- wheat system in northwestern India. Nutr Cycl Agroecosyst 88:471–480. https://doi.org/10.1007/s10705-010-9370-8

Yang C, Hamel C, Gan Y, Vujanovic V (2012) Bacterial endophytes mediate positive feedback effects of early legume termination times on the yield of subsequent durum wheat crops. Can J Microbiol 58:1368e1377. https://doi.org/10.1139/w2012-114

Yang C, Hamel C, Gan Y, Vujanovic V (2013) Pyrosequencing reveals how pulses influence rhizobacterial communities with feedback on wheat growth in the semiarid Prairie. Plant Soil 367:493e505. https://doi.org/10.1007/s11104-012-1485-z

Zhu Y, Chen H, Fan J, Wang Y, Li Y, Chen J, Fan J, Yang S, Hu L, Leung H, Mew TW, Teng PS, Wang Z, Mundt CC (2000) Genetic diversity and disease control in rice. Nature 406:718e722. https://doi.org/10.1038/35021046

Zhuang X, Chen J, Shim H, Bai Z (2007) New advances in plant growth-promoting rhizobacteria for bioremediation. Environ Int 33:406e413. https://doi.org/10.1016/j.envint.2006.12.005

Biomass Recycling for Soil and Environmental Sustainability in the Indian Himalayas

8

Amit Kumar, Raman Jeet Singh, Trisha Roy,
Raghavendra Singh, Subhash Babu, and Gautam Kumar

Abstract

The Himalayan region of India, faces significant challenges related to soil degradation, environmental sustainability, and climate change. Biomass recycling is a critical strategy to enhance soil fertility, reduce environmental footprints, and sustain agricultural productivity in the region. This chapter explores various biomass recycling techniques, including mulching, cover crops, intercropping, composting, direct incorporation, biochar production, residue removal and utilization, agro-geotextiles, and the use of municipal solid wastes and industrial byproducts for their potential soil and environmental sustainability benefits. Mulching with organic materials like crop residues and forest litter reduces erosion in rainy season and conserves soil moisture in post-rainy season, and improves soil organic matter level of slopping arable fields of the Himalaya. Composting and direct incorporation of organic residues enrich the Himalayan soils with essential nutrients and enhance microbial activity, leading to improved soil structure and fertility. Biochar production, offers carbon sequestration, improved soil water retention, and increased nutrient availability for long-term soil health. Residue removal for livestock feed and utilization of their cow-dung as composts prevent the loss of valuable organic carbon as CO_2 in the air by residue-burning while providing feedstock for other recycling processes. Valorization of huge biomass

A. Kumar (✉) · G. Kumar
ICAR Research Complex for NEH Region, Sikkim Centre, Tadong, Gangtok, Sikkim, India

R. J. Singh · T. Roy
ICAR-Indian Institute of Soil and Water Conservation, Dehradun, Uttarakhand, India

R. Singh
ICAR-Indian Institute of Farming Systems Research, Meerut, Uttar Pradesh, India

S. Babu
Division of Agronomy, ICAR-Indian Agricultural Research Institute, New Delhi, India

© The Author(s), under exclusive license to Springer Nature Singapore Pte Ltd. 2025
S. Babu et al. (eds.), *Ecological Solutions to Agricultural Land Degradation*,
https://doi.org/10.1007/978-981-96-3392-0_8

of perennial grasses such as *Arundo donax* as agro-geotextiles stabilize fertile soil on steep slopes, reduce erosion, conserve moisture and promote crop growth. Recycling industrial byproducts and municipal solid waste may be a useful strategy to improve soil fertility and reduce the environmental problems related to waste disposal while also adding organic matter and vital nutrients. Field studies across different agro-climatic zones of Himalayas show that integrating these biomass recycling techniques leads to significant improvements in soil quality indicators, climate change alleviation, and soil fertility. Hence, biomass recycling presents viable and sustainable nature-based positive solutions for soil and environmental sustainability in the Himalayan region, supporting local livelihoods and global environmental goals.

Keywords

Biomass recycling · Environment quality · Soil health · Sustainability

Abbreviations

IHR	Indian Himalayan Region
AGT	Agro-geotextiles
AD	*Arundo donax*
NWHR	North-Western Himalayan Region
NEHR	North-Eastern Himalayan Region
GM	Green-manure
FYM	Farmyard manure
CO_2	Carbon-di-oxide
SOC	Soil organic carbon
OM	Organic matter

8.1 Introduction

Agri-residue burning is a pervasive issue in India, with significant environmental, health, and economic consequences. While the practice is most commonly associated with the northwestern states like Punjab and Haryana, it is also prevalent in other parts of the country, including the northeastern states. The burning of crop residues releases significant quantities of particulate matter (PM2.5 and PM10), which are linked to respiratory and cardiovascular diseases. A study conducted by the Central Pollution Control Board (CPCB) revealed that the contribution of crop residue burning to the overall PM2.5 levels in Delhi can range from 15% to 45% during the peak burning season (CPCB 2020). The health impacts of residue burning is profound, particularly in densely populated regions. The smoke from burning residues contains a complex mixture of harmful pollutants, including carbon monoxide (CO), volatile organic compounds (VOCs), and polycyclic aromatic hydrocarbons (PAHs), which are known carcinogens. Exposure to high levels of particulate matter from agricultural burning can lead to an increase in respiratory and cardiovascular diseases, including asthma, bronchitis, and heart attacks (WHO 2018). A study by the Indian Council of Agricultural Research (ICAR) found that burning one ton of paddy straw results in the loss of approximately 5.5 kg of nitrogen, 2.3 kg of phosphorus, and 25 kg of potassium, which are essential for crop growth (ICAR 2018). Additionally, the practice contributes to greenhouse gas emissions, with each ton of crop residue burned releasing approximately 1.46 kg of PM, 60 kg of CO, 1460 kg of CO_2, along with other harmful gases like methane and nitrous oxide (Gupta et al. 2019). These emissions play a role in global warming and the creation of ground-level ozone, which poses risks to human health and reduces crop productivity. The Himalayan region of India is among the most ecologically diverse and environmentally critical areas globally, stretching across northern India and forming a natural barrier between the Indian sub-continent and the Tibetan Plateau. This formidable mountain range, includes the world's highest peaks like the Mount Everest, is characterized by its rich ecosystems, varying from subtropical forests in the foothills to alpine meadows and glacial landscapes at higher altitudes. Notably, about 22% of the Earth's land surface is covered by mountain ecosystems, where agriculture forms the backbone of livelihoods for millions of people. In India, the Himalayan range is mostly contained within the Indian Himalayan Region (IHR), which includes twelve states such as Jammu & Kashmir, Ladakh, Himachal Pradesh, Uttarakhand (north-western Himalayan region-NWHR), Sikkim, Meghalaya, Mizoram, Nagaland, Arunachal Pradesh, Manipur, Tripura and Assam (north-eastern Himalayan region-NEHR). Around 70% of the IHR's population is rural, with states like Himachal Pradesh and Uttarakhand having particularly high rural populations (FAO 2019). The IHR is vital to India as it provides water to much of the subcontinent, is a reservoir of medicinal plants, and contains significant biodiversity. Tropical evergreen rainforests are found in the eastern and central Himalayas, whereas tropical deciduous forests are characteristic of the western Himalayas. The region's climate, ranging from tropical conditions at lower elevations to permanent ice at the highest peaks, supports various agricultural activities, including crop cultivation, horticulture, and flower farming.

Agriculture is the primary source of livelihood for a significant portion of the Himalayan population. However, the region's steep slopes, fragile soils, and harsh climate present challenges to sustainable farming. Traditional farming methods, adapted over centuries to these conditions, face growing pressure from modern agricultural demands, climate change, and population growth (Kumar et al. 2024). In the northeastern states of India, residue burning is also a common practice, particularly in areas where shifting cultivation (*Jhum* farming) is prevalent. About 150,000 hectares of land in the northeastern region is affected by shifting cultivation annually, with residue burning being a significant contributing factor to land degradation (NESAC 2018). Soil health is a critical factor for sustainable agriculture in the Himalayas, where soils are often fragile and prone to degradation (Bhattacharyya et al. 2023). The combination of steep terrain and heavy rainfall causes intense soil erosion, leading to topsoil loss, decreased soil fertility, and lower agricultural productivity. A substantial proportion of soils in the Himalayan region are acidic, with pH levels < 5.5. In the eastern Himalayas, more than 60% of soils are acidic, restricting the availability of key nutrients such as P, Ca, and Mg. (Singh et al. 2010). The acidic nature of these soils is further exacerbated by the high rainfall in the region, which leaches away vital nutrients, leaving the soil nutrient-poor and less productive. Nutrient deficiency is another major concern in Himalayan soils. A large proportion of soils lack essential nutrients, including nitrogen (N), phosphorus (P), and potassium (K). For instance, in the state of Uttarakhand, approximately 45% of soils are deficient in nitrogen, 25% in phosphorus, and 15% in potassium (ICAR 2015). Likewise, micronutrient deficiencies, particularly of zinc (Zn) and boron (B), are also widespread. Approximately 30% of soils in the central Himalayan region are estimated to be deficient in zinc (Zn) (Sharma and Bhatt 2017). Soil erosion is a pervasive problem in the Himalayan region, driven by the steep topography, intense rainfall, and deforestation. In the Himalayan region, soil erosion rates can vary between 20 and 40 tons per hectare annually (Bhattacharyya et al. 2015). In some of the most affected areas, soil erosion rates have been recorded as high as 80 tons per hectare per year, significantly higher than the global average of 12 tons per hectare per year (MoEFCC 2014). This alarming rate of soil erosion not only depletes the soil's fertility but also contributes to the siltation of rivers and reservoirs, leading to reduced water quality and increased flood risks downstream. In the Himachal Pradesh region, studies have indicated that nearly 50% of the agricultural soils are deficient in organic carbon, which is crucial for maintaining soil structure and fertility (Pathak et al. 2023). In response to these challenges, biomass recycling has emerged as a key strategy for promoting soil and environmental sustainability in the region. Biomass recycling involves collecting, converting, and reusing organic waste materials to improve soil fertility and support sustainable agricultural practices. Historically, Himalayan farmers have practiced biomass recycling, recognizing the importance of returning organic matter to the soil to maintain its fertility. Recently, this practice has gained renewed attention as an essential component of sustainable agriculture and environmental conservation (Singh et al. 2023a, b).

The Indian Himalayan states produce approximately 25–30 million tons of crop residues (Sharma et al. 2023). This includes residues from major crops such as rice, wheat, maize, and barley, as well as horticultural residues from apple, citrus, and other fruit orchards. In states like Uttarakhand and Himachal Pradesh, rice and wheat are the primary sources of agricultural residues, contributing to nearly 60% of the total residue production. For instance, in Himachal Pradesh alone, it is estimated that around 6 million tons of crop residues were generated during the 2022–2023, with wheat straw and rice husk constituting the bulk of this volume (ICAR 2023). Similarly, in the northeastern states, maize and paddy are the major contributors, with the region producing an estimated 8–10 million tons of residues annually (NESAC 2023). Recycling biomass boosts soil health by raising organic matter levels, thereby enhancing the soil's physical, chemical, and biological characteristics (Thakur et al. 2021). Improving soil structure, increasing water retention, and promoting the development of beneficial microbes are all achieved by adding organic matter via biomass recycling. By aiding in the cycling of nutrients, these microbes increase the availability of vital nutrients for plants. (Verma et al. 2024). Biomass recycling also offers a sustainable alternative to chemical inputs by providing natural sources of nutrients and pest control. Organic amendments like compost and manure, derived from biomass recycling, are rich in essential nutrients that are gradually released into the soil, reducing the need for synthetic fertilizers. Organic amendments improve soil structure, making soils more resilient to erosion and degradation, and enhance nutrient retention, minimizing nutrient leaching and runoff. Furthermore, by sequestering carbon, biomass recycling is essential to mitigating climate change (Zhang et al. 2020; Roy et al. 2023a, b). When organic matter is added to the soil, a portion of the carbon in the biomass is stored in the soil as organic carbon. This process enhances soil health while also lowering atmospheric CO_2 levels, thereby contributing to the mitigation of the greenhouse effect. Soils abundant in organic matter can eventually serve as substantial carbon sinks, aiding global initiatives to lower greenhouse gas emissions. In addition to carbon sequestration, biomass recycling helps reduce GHG emissions by decreasing the need for chemical fertilizers, which are often linked to nitrous oxide (N_2O) emissions (Singh et al. 2017; Babu et al. 2023a). The use of crop residues and other organic materials for mulching and soil cover further reduces soil erosion, conserves soil moisture, and buffers the impacts of extreme weather events, enhancing the resilience of agricultural systems to climate change (Singh et al. 2024). Biomass recycling also contributes to biodiversity conservation and profit maximization by promoting sustainable agricultural practices that are less disruptive to natural ecosystems (Singh et al. 2016a, b; Akram et al. 2019; Gupta et al. 2019). Hence, sustainable agricultural practices, particularly biomass recycling, is crucial for tackling climate change, maintaining soil health, and increasing agricultural output in this fragile region.

8.2 Biomass Recycling Practices and Their Effects on Natural Resources Management

A Working Paper published by NITI Aayog in July 2023 reports that India produces an average of about 650 million tonnes of agricultural residue each year. Biomass recycling refers to the practices employed to handle the remains of crops, including stems, leaves, stalks, and roots, left in the field after the harvest. Because of their high organic matter and nutritional content, these residues are essential for preserving soil health, increasing agricultural output, and reducing negative environmental effects. The management of crop residues varies significantly across the globe, influenced by regional agricultural practices, climatic conditions, and economic factors. Annually, the world produces approximately 4 billion tons of crop residues, with rice, wheat, corn, and sugarcane being the major contributors. Asia, particularly due to its intensive rice and wheat farming, accounts for a substantial share of these residues (Kannoj et al. 2023). In India, the production of crop residue in dry matter increased from 80 million tonnes in 1950–1951 to 520 million tonnes in 2017–2018, as reported in the Economic Survey of India (2020). In many parts of the world, especially in developing countries, crop residues are often burned in the fields. India, for instance, around 92 million tons of crop residues are burned each year, predominantly in the states of Punjab, Haryana, and Uttar Pradesh. This practice leads to significant air pollution, releasing particulate matter and GHGs like CO_2, CH_4, and N_2O. Specifically, crop residue burning in India contributes to the emission of 149 million tons of CO_2, 9 million tons of CO, 0.25 million tons of particulate matter, 0.07 million tons of black carbon, and 1 million tons of N_2O annually (Babu et al. 2023a; Kannoj et al. 2023).These emissions not only exacerbate global warming but also pose serious public health risks, contributing to respiratory and cardiovascular diseases. However, crop residues also possess significant economic value. When properly utilized, they can enhance soil fertility and reduce the reliance on chemical fertilizers (Singh et al. 2016a, 2017). For example, incorporating crop residues into the soil can contribute up to 4.5 tons of organic carbon per hectare annually. Biomass recycling practices involve the sustainable reuse and conversion of organic materials, such as agricultural residues, forestry by-products, and food waste, into valuable resources like energy, fertilizers, and soil amendments (Fig. 8.1).

These practices are critical in promoting environmental sustainability, enhancing soil health, and reducing reliance on non-renewable resources.

8.3 Biomass Recycling Technologies

There are several recycling practices that farmers can adopt to maximize the benefits of crop biomass:

8.3.1 Mulching

Mulching, a method of managing crop residue, involves spreading a layer of either organic or inorganic material over the naked soil surface. This technique provides a range of benefits (Fig. 8.2) such as enhancing soil health, conserving moisture in the lean-season, reducing weed growth, and boosting crop yields (Kumar et al. 2023b). It is commonly employed in both traditional and modern farming systems, playing a key role in sustainable agriculture. The recommended thickness of the mulch layer varies depending on the material used, with a 2–4 inch layer being ideal for most organic mulches. Coarser materials, like wood-chips, may require a thicker layer. Mulching is usually performed after planting and before dry or hot seasons begin to ensure optimal moisture retention and weed control (Kumar et al. 2018; Joseph et al. 2023). Mulch layers need to be replenished periodically as they decompose or get compacted. Regular inspection and maintenance ensure continued effectiveness.

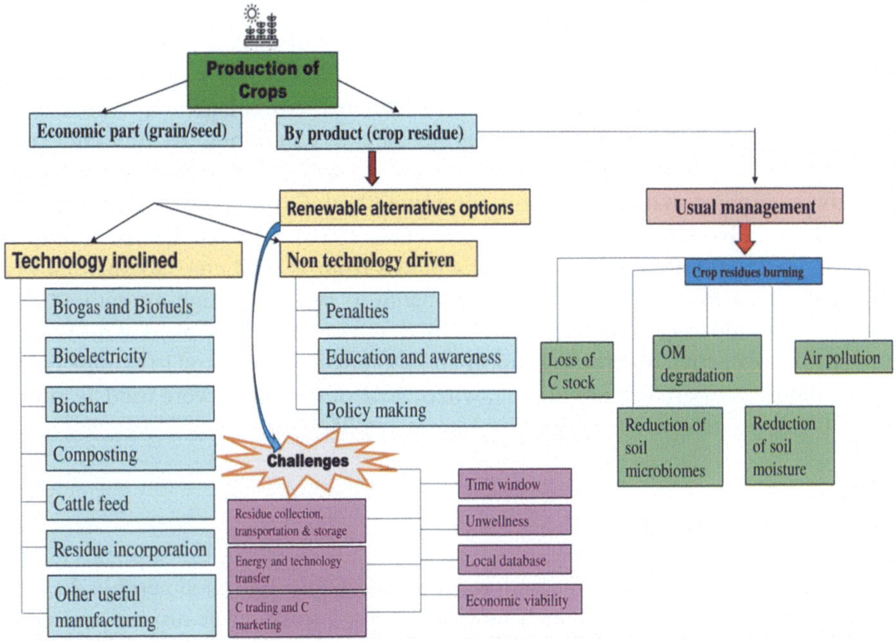

Fig. 8.1 Flow-diagram of crop residues recycling

Fig. 8.2 Effect of application of mulching on soil health

8.3.1.1 Benefits of Mulching

Mulching helps prevent soil erosion by shielding the soil surface from the effects of wind and water. It conserves moisture by reducing soil evaporation, making more water available for crops (Verma et al. 2024). Mulch also regulates soil temperature, keeping it cooler in hot weather and warmer in cold conditions. By blocking sunlight, mulch suppresses weed germination and growth. As it breaks down, mulch contributes organic matter to the soil, improving both fertility and structure (Singh et al. 2021a, b; Kumar et al. 2023b). Khybri et al. (1980) observed that use of mulch @ 4 Mg ha^{-1} in maize rows reduced the soil loss from 17.8 Mg ha^{-1} to only 5.4 Mg ha^{-1} and runoff from 37% to 15% as compared to normal ploughing without any mulch at 8% land slope (Table 8.1). In Dehradun, grass mulch applied at 4 Mg ha^{-1} proved more effective in reducing runoff and soil erosion compared to canopy cover from a high-density maize population, where 50% of the plants were used as mulch after 30 days of growth as reported by Narain et al. (2000).

8.3.2 Composting

Compost is a nutrient-rich soil supplement made by the controlled aerobic decomposition of organic materials such as manure, yard trash, crop residue and food wastes. In addition to recycling organic waste, compost application improves soil health, encourages sustainable farming, and reduces its negative effects on the environment (Puri et al. 2020; Singh et al. 2021a). Composting is a versatile practice that can be implemented in diverse settings, ranging from backyard gardens to commercial farms and industrial operations.

8.3.2.1 Types of Composting Systems

Various composting systems cater to different scales and needs:

Backyard composting: Small-scale composting systems suitable for home gardeners. Common methods include compost bins, tumblers, and piles.

Vermicomposting: Decomposes organic waste using earth worms. In addition to producing nutrient-rich worm castings, vermicomposting is appropriate for small-scale indoor or outdoor composting.

Aerated static pile (ASP) composting: Involves creating large compost piles with forced aeration. ASP systems are efficient for managing large volumes of organic waste in commercial or municipal settings.

Table 8.1 Effect of mulching on runoff, soil loss and grain yield of maize

Treatment	Runoff (%)	Soil loss (Mg ha^{-1})	Maize yield (kg ha^{-1})
Control (no mulch)	36.9	17.8	2433
Mulch 2 Mg ha^{-1}	17.9	7.3	2317
Mulch @ 4 Mg ha^{-1}	14.7	5.4	2395

Source: Adapted from Khybri et al. (1980)

Windrow composting: To maintain aeration, organic materials have been placed in long rows, or windrows, and turned occasionally. This method is widely used in large-scale agricultural and municipal composting operations.

In-Vessel composting: Encloses organic materials in a container or vessel, allowing for controlled temperature, moisture, and aeration. In-vessel composting is efficient and reduces odors, making it suitable for urban areas and institutions.

8.3.2.2 Benefits of Composting

By converting organic waste into nutrient-rich organic matter, composting improves soil fertility and lowers the demand for synthetic fertilisers by releasing vital elements gradually. By improving soil structure, compost increases porosity and aggregation, which supports better root growth, water infiltration, and aeration, resulting in healthier plants. The heat produced during composting can eliminate pathogens and weed seeds, decreasing pest and disease risks. Additionally, compost boosts the soil's moisture retention, reducing the frequency of irrigation and helping plants endure drought. It also stabilizes soil, lowering erosion risks by improving structure and increasing organic matter, which is vital in erosion-prone areas. Compost acts as a carbon sink, storing carbon in the soil and reducing greenhouse gas emissions to help slow down climate change. Additionally, composting lowers waste management expenses and methane emissions by keeping organic waste out of landfills. Finally, compost fosters a habitat for beneficial soil microorganisms essential for nutrient cycling, disease suppression, and organic matter decomposition.

In a maize-wheat rotation system, the integrated application of farmyard manure (FYM) with 50% of the recommended NPK fertilizers (50% NPK + 50% FYM) achieved the highest values for energy ratio (7.3), human energy profitability (142.4), energy productivity (0.22 kg MJ^{-1}), and energy profitability (6.3 MJ ha^{-1}) compared to other treatment methods, according to a study by Singh et al. (2016a) in Dehradun. Nevertheless, the highest energy intensiveness (24.61 MJ US $\$^{-1}$) was maintained by *in situ* green manure of sunnhemp and inorganic fertilizers on an equal NPK basis (50 + 50 GM). Additionally, this treatment showed lower environmental loading ratio (3.74), higher emergy yield ratio (2.66), and lower emergy investment ratio (0.60). This approach yielded a higher environmental sustainability index (0.71) than other treatments, enhancing both the ecological and financial sustainability of the production system. However, aside from specific energy aspects, FYM alone could not match other fertilization options in terms of energy budgeting and synthesis. The study results indicate that innovative integrated nutrient management, combining chemical fertilizers with organic manures—specifically FYM for energy budgeting and green manure (GM) for energy synthesis—can serve as viable, eco-friendly soil conservation strategies. This approach could help non-OPEC countries, which rely heavily on imported phosphorus and potassium fertilizers, by cutting chemical fertilizer costs by 50%. Additionally, the results showed that in different years, 50 + 50 (FYM) could retain significantly higher ($p < 0.05$) maize (18–74%) and wheat (10–77%) production than 100-0. However, in some years, 50 + 50 (GM) outperformed 100-0 in terms of soil loss (13–50%) and runoff (16–40%) decrease, which increased the amount of natural resources conserved.

According to the study's findings, using chemical fertilizers in conjunction with organic manures especially FYM or GM may be a practical and environmentally responsible way to conserve soil (Singh et al. 2017).

8.3.3 Direct Incorporation

Incorporating crop residues directly into the soil is a sustainable farming practice that entails plowing or tilling leftover plant materials, such as stalks, leaves, and roots, back into the soil after harvest. This approach boosts soil organic matter, improves soil fertility, and supports sustainable agricultural practices (Singh et al. 2019; Verma et al. 2024; Kumar et al. 2023a, b). Known for its positive impact on soil health, water retention, and nutrient cycling, direct incorporation is a key element of integrated soil fertility management.

8.3.3.1 Methods of Direct Incorporation

Plowing: Plowing is the most common method of incorporating crop residues. A plow is used to turn the soil, burying the residues and mixing them with the soil. This method is effective but can be labor-intensive and may disturb soil structure if not done properly.

Disking: Disking involves using a disk harrow to chop and mix crop residues into the soil. This method is less invasive than plowing and helps retain more soil structure while effectively incorporating residues.

Chisel plowing: Chisel plowing uses a chisel plow with narrow shanks to penetrate the soil and mix residues without completely turning the soil over. This method minimizes soil disturbance and is beneficial for conserving soil moisture and structure.

No-Till farming: No-till farming leaves crop residues on the soil surface, where they decompose naturally and are gradually incorporated into the soil by biological activity. This method reduces soil disturbance and erosion while enhancing soil organic matter (Sharma et al. 2014).

8.3.3.2 Benefits of Direct Incorporation

One of the primary benefits of residue incorporation is the improvement of soil organic matter content (Verma et al. 2024). A study conducted in Uttarakhand reported that incorporating wheat and rice residues into the soil increased organic carbon content by 0.2–0.5% over two growing seasons, contributing to improved soil structure and water-holding capacity (Sharma et al. 2022). Moreover, residue incorporation has been shown to reduce the reliance on chemical fertilizers. Research in Himachal Pradesh demonstrated that direct incorporation of paddy straw reduced the need for nitrogen fertilizers by 15–20% without compromising crop yields (Pathak et al. 2023). This reduction in chemical fertilizer use not only lowers production costs for farmers but also mitigates the environmental risks associated with excessive fertilizer application, such as nutrient runoff and groundwater contamination. Direct incorporation of crop residues plays a crucial role in combating soil erosion by enhancing soil structure and increasing surface cover. Studies have shown that fields with residue incorporation experience 30–40% less soil erosion compared to fields where residues are removed or burned (Singh et al. 2023a, b). This reduction in soil erosion is vital for maintaining the productivity of agricultural lands in the Himalayas and for preventing the siltation of rivers and reservoirs, which is a major environmental concern in the region. Residue incorporation also contributes to environmental sustainability by reducing greenhouse gas emissions. Residue burning, a common practice in some Himalayan areas, releases significant amounts of carbon dioxide (CO_2), methane (CH_4), and nitrous oxide (N_2O), all of which are potent greenhouse gases. By incorporating residues into the soil, these emissions can be significantly reduced. For example, research in the northeastern Himalayas showed that incorporating rice residues reduced CO_2 emissions by approximately 25% compared to burning (NESAC 2023). This reduction in emissions not only helps in mitigating climate change but also improves air quality, which is a growing concern in many parts of the Indian Himalayas (Kumar et al. 2022, 2023a, b). Incorporation of crop residues also stabilizes the soil surface, reducing erosion from wind and water, which is crucial for preserving soil health and preventing topsoil loss (Singh et al. 2021a). Furthermore, directly incorporating crop residues eliminates the need for burning a practice that contributes to air pollution and greenhouse gas emissions (Singh et al. 2021b; Kumar et al. 2023b) and helps mitigate climate change by sequestering carbon in the soil. The decomposition of organic matter also provides habitat and food for soil microorganisms, earthworms, and other beneficial organisms, enhancing soil biodiversity and promoting a healthy soil ecosystem (Akram et al. 2019; Gogoi et al. 2021).

8.3.4 Biochar Production

Biochar is a carbon-rich, stable material created through the pyrolysis of organic biomass in a low-oxygen environment. It is a type of charcoal primarily used as a soil amendment to improve soil health, sequester carbon, and address climate change (Gupta et al. 2019; Gudade et al. 2022a). The production and application of biochar have garnered global attention due to its potential environmental, agricultural, and economic benefits (Harder et al. 2020). Biochar is produced by heating organic materials, such as agricultural residues, wood chips, manure, and other biomass, at high temperatures (typically between 300 and 700 °C) in a controlled, oxygen-limited environment. This pyrolysis process breaks down the organic matter into a stable carbon form, releasing gases and producing bio-oil as by-products. The resulting biochar is a porous, black substance that retains much of the original carbon content and structure of the biomass (Hou et al. 2017; Gudade et al. 2022b).

8.3.4.1 Benefits of Biochar

Biochar has gained attention for its ability to improve soil properties, particularly in degraded and nutrient-poor soils common in the Himalayan region. When applied to the soil, biochar enhances soil fertility by increasing the soil's organic carbon content, improving nutrient retention, and promoting beneficial microbial activity. In the Indian Himalayas, studies have shown that biochar can significantly increase soil organic carbon levels. For instance, a study conducted in Uttarakhand found that the application of biochar at a rate of 10 tons per hectare resulted in a 25–30% increase in soil organic carbon after one cropping season (Sharma et al. 2023). This increase in organic carbon is crucial for improving soil structure, water-holding capacity, and overall soil fertility, particularly in acidic soils where nutrient availability is limited (Das et al. 2019). Biochar application increased the cation exchange capacity of soils by 20–35%, leading to improved nutrient retention and reduced leaching losses (Negi et al. 2022). This is particularly important in the Himalayan region, where high rainfall can lead to significant nutrient runoff and soil erosion. It also increases microbial biomass and activity by 15–25%, contributing to enhanced soil fertility and resilience against soil-borne diseases (Bhatt et al. 2022). These benefits make biochar a valuable tool for restoring degraded soils and promoting sustainable agriculture in the Himalayan region (Gudade et al. 2022a). The production and application of biochar in the Indian Himalayas can play a significant role in reducing the region's carbon footprint. A study estimated that the application of biochar at a rate of 5–10 tons per hectare across the Himalayan states could sequester up to 1.5 million tons of CO_2 annually (ICAR 2023). This level of carbon sequestration could make a substantial contribution to India's climate change mitigation goals (Gupta et al. 2019).Due to its high stability, biochar can remain in the soil for hundreds to thousands of years, effectively sequestering carbon and helping to mitigate climate change by reducing greenhouse gas levels in the atmosphere. Biochar helps reduce emissions of methane (CH_4) and nitrous oxide (N_2O), two potent greenhouse gases associated with traditional agricultural practices. Research conducted in Sikkim found that biochar application reduced methane emissions from rice paddies by 15–20% and nitrous oxide emissions by 25–30%, compared to conventional practices without biochar (NESAC 2023). These reductions are particularly important in the context of the Indian Himalayas, where agriculture is a significant source of greenhouse gas emissions (Gudade et al. 2022b).

8.3.5 Residue Removal and Utilization

Residue removal and utilization encompass the management practices involved in harvesting, processing, and applying crop residues in agriculture. These residues, including stalks, leaves, husks, and roots left after harvest, are crucial for soil health, nutrient cycling, erosion control, and overall agricultural sustainability (Singh et al. 2016b, 2018a; Berardi et al. 2020). Implementing effective residue management strategies is vital for maximizing agricultural productivity while preserving soil fertility and ecosystem resilience (Singh et al. 2018b; Cheng et al. 2020).

8.3.5.1 Methods of Residue Removal and Utilization

Harvesting and collection: Crop residues are collected during harvest using combine harvesters or specialized equipment. Depending on the crop and farming system, residues can be left on the field or removed for further processing.

Direct incorporation: Residues are incorporated into the soil through tilling or plowing right after harvest. This approach boosts soil organic matter, improves nutrient availability and soil structure, and helps reduce soil erosion (Roy et al. 2023a, b).

Surface mulching: Residues are left on the soil surface as mulch to conserve moisture, suppress weeds, and protect the soil from erosion. This method is particularly suitable for no-till or conservation agriculture practices (Singh et al. 2014).

Animal feed: Residues such as straw and hay are used as feed for livestock, providing additional income for farmers and reducing waste. Livestock digestion can further process residues into valuable manure for soil fertility.

Bioenergy production: Crop residues, especially lignocellulosic materials like corn stover and wheat straw, can be converted into biofuels or bioenergy using technologies such as biomass gasification or ethanol production. According to a report by the Centre for Science and Environment, approximately 10 kg of agricultural residue can produce 1 kg of compressed biogas.

8.3.5.2 Benefits of Residue Removal and Utilization

Residue removal, when managed appropriately, can play a significant role in restoring soil health in the Indian Himalayas. However, it is essential to balance residue removal with practices that return organic matter to the soil to prevent nutrient depletion. Das et al. (2019) stated that the removing a portion of the residues for use in composting or bioenergy production, while leaving the rest to decompose in the field, can help maintain soil organic matter levels and prevent soil degradation. When residues are removed and composted, the resulting organic compost can be applied back to the soil, enriching it with essential nutrients and organic matter. Studies in the Himachal Pradesh region have shown that the application of compost derived from crop residues increased soil organic carbon by 15–20% and improved soil fertility, particularly in soils that were previously nutrient-deficient (Sharma et al. 2023). The use of compost also improves soil structure, water retention, and microbial activity, which are crucial for sustaining agricultural productivity in the mountainous terrains of the Himalayas. Mulching, another practice involving the use of removed residues, has been shown to significantly reduce soil erosion and improve moisture conservation. In the northeastern Himalayas, research has demonstrated that mulching with paddy straw reduced soil erosion by 25–30% and improved soil moisture content by 10–15% during the dry season (NESAC 2023). This practice not only conserves soil but also creates a more favorable environment for crop growth, particularly in rain-fed agricultural systems. For instance, the conversion of crop residues into biogas has been shown to significantly reduce methane emissions from traditional biomass burning practices. A study in Uttarakhand estimated that converting the region's annual crop residue surplus into biogas could prevent the release of approximately 2 million tons of CO_2 equivalent per year

(Pathak et al. 2023). This reduction in emissions is crucial for the region, which is particularly vulnerable to the impacts of climate change, such as glacier melting and altered rainfall patterns. Moreover, the production of biochar from crop residues, as mentioned earlier, provides a stable form of carbon that can be sequestered in the soil, thereby reducing atmospheric CO_2 levels. The integration of biochar into soil management practices not only enhances soil health but also contributes to long-term carbon sequestration. Research in Sikkim showed that applying biochar made from maize residues increased soil organic carbon by 30% and sequestered approximately 1.2 tons of CO_2 per hectare annually (Bhatt et al. 2023). The adoption of sustainable residue management practices are essential for the long-term sustainability of agriculture and environmental conservation in the Himalayas (Kumar et al. 2024).

8.3.6 Agro-Geotextiles

Biomass recycling through agro-geotextiles is an innovative approach that combines sustainable agricultural practices with the benefits of geotextiles, which are permeable fabrics used in civil engineering and environmental projects (Bhattacharya et al. 2012). Agro-geotextiles are made from natural fibers derived from agricultural residues or plants, such as jute, coir (coconut fiber), sisal, or other biodegradable materials. These geotextiles are used for soil conservation, erosion control, and enhancing soil fertility while recycling biomass in an eco-friendly way. Jute and coir-based agro-geotextiles are extensively used in India for soil conservation and erosion control in riverbanks, roadsides, and hilly terrains. The Indian government has promoted the use of these geotextiles through various initiatives, recognizing their environmental benefits (Bhattacharya et al. 2009). In hilly or mountainous areas, agro-geotextiles help stabilize slopes by holding the soil in place, especially during heavy rainfall. This prevents landslides and soil degradation. These geotextiles are used in the construction of road and railway embankments to provide reinforcement, reduce soil erosion, and improve drainage. Agro-geotextiles are also employed in agriculture and landscaping to improve soil health, control weeds, and conserve moisture. They act as mulch, protecting the soil from temperature fluctuations and reducing water evaporation (Dev et al. 2019). Agro-geotextiles are biodegradable, reducing the environmental impact associated with the use of synthetic geotextiles. After their useful life, they decompose naturally, returning organic matter to the soil. s agro-geotextiles break down, they release nutrients back into the soil, enhancing its fertility and promoting healthy plant growth. This can reduce the need for chemical fertilizers (Gaballah et al. 2020). The use of natural fibers in agro-geotextiles contributes to carbon sequestration, as the plants used for fiber production absorb carbon dioxide from the atmosphere during their growth (Capasso et al. 2019).

The prefix of agro-geo-textile (AGT), ‘agro’ means field crops, ‘geo’ mean earth and the ‘textile’ means fabric. Therefore, according to the definition, the agro-geotextile is defined as follows: “A permeable geo-synthetic comprised solely of agriculturally important textiles. Geotextiles are used with foundation, soil, rock, earth, or any other geotechnical engineering-related material as an integral part of human-made project, structure, or system. AGT or erosion controlling mats can be prepared from the combination of strong stem and stiff leaves of a perennial natural grass *Arundo donax* (AD)in the foothills of the Indian Himalayas. When its culms reach a height of 15–20 ft and develop green compact leaf biomass, AGT can be prepared at an economically viable price by tying with locally available sources, such as rope or plant fibers. With rough texture, high water-holding capacity, and more weight when wet, AD as AGT has important properties that guarantee physical contact between the soil and geotextiles, so helping to limit soil erosion. (Singh et al. 2019). AD-AGT emplacement is an inexpensive method since it is naturally growing in stream beds next to the main farming land (Photo 8.1).After biodegradation, it recycles nutrients into the soil, and its rapid growth and increased density make it useful for increasing surface roughness and decreasing soil loss. This improves the soil's aggregate stability by encouraging the test crops' vegetative and root development. (Roy et al. 2023a). Reduced soil erosion in the emplaced plots under AD-AGT also contributes to increased system productivity and SOC sequestration (Roy et al. 2023b). AD-AGT biomass contributed 0.24–0.90 MG C ha^{-1} $year^{-1}$ to the soil over different treatments of the study. Thus, the use of AD-AGT in soil has a significant effect on soil erosion control and SOC sequestration and, hence, could be adopted in the foothills of the IHR and similar agro-ecologies (Photo 8.2).

Photo 8.1 *Arundo donax* (Giant reed) grass

Photo 8.2 AD-AGT emplaced in between two maize rows to control soil erosion

8.3.7 Industrial Byproducts

Biomass recycling through industrial byproducts involves the repurposing of organic waste and residues generated by various industries, such as agriculture (straw, husks, shells, and other residues from crops like rice, wheat, corn, and sugarcane), forestry, food processing, and bioenergy production, into valuable resources (Iglesias et al. 2018). This approach not only helps in managing waste but also contributes to the circular economy by transforming byproducts into useful materials, energy, and soil amendments. Agricultural residues can be converted into bioenergy, biofuels, compost, and bio-based materials like paper, board, and bioplastics (Kumari et al. 2019). Forestry byproducts are used for producing bioenergy (e.g., pellets, briquettes), biochar, and in the manufacture of wood-based products like fiberboard and particleboard. Pulp and paper mills generate large quantities of lignin, black liquor, and sludge. These byproducts can be converted into bioenergy, bioplastics, and soil conditioners. Lignin, for example, can be used to produce bio-based adhesives and carbon fibers (Mandal et al. 2020). Industrial byproducts are used as feedstock for bioenergy production, either through direct combustion or conversion into biofuels like ethanol and biodiesel. This reduces dependence on fossil fuels and promotes the use of renewable energy sources. Recycling industrial byproducts reduces the volume of waste sent to landfills or incinerators, mitigating environmental pollution and associated disposal costs. The recycling of industrial byproducts creates new markets and revenue streams for industries, particularly in the production of bioenergy, biofuels, and bioproducts (Qureshi et al. 2020).

The industrial solid wastes can be divided based on the nature of the industry as follows:

- Energy industry
- Agro-industry
- Mineral industry

All these three different types of industries produce by-products with the potential of getting recycled as an alternate fertilizer, as a soil amendment or as an organic input depending on the composition and source of the by-product. The agro industrial by-products like bagasse, press mud, spent wash, food processing wastes, tea industry wastes etc. are prospective fertilizer sources along with high organic C content. Additionally, these materials can be used as soil amendments (Dotaniya et al. 2016). The industrial by-products give an opportunity to utilize the waste material generating valuable resources and bridge the gap between the demand and supply of the plant nutrients in the agricultural sector (Kainthola et al. 2019). The energy-based industries like coal; steel etc. produces fly ash, basic slag all of which are potent soil ameliorants (Skousen et al. 2013). Similarly, phosphor-gypsum, a by-product of the fertilizer industry is a source of essential plant nutrient, particularly phosphorus. The North Western Himalayan economy is an agrarian economy where agriculture forms the major livelihood of the population. Though extensive literature on use of different industrial by-products in these areas is limited, but there is scope of utilizing the same to enhance crop productivity and boost soil health. Besides improving soil health, these alternate fertilizer materials can significantly reduce the emission of greenhouse gases like carbon dioxide and nitrous oxide (Rittl et al. 2018).

Sugar industry is second major agricultural industry in the India ranking next to the textile industry. In the North Western Himalayan region, except for Uttarakhand no other state has a major share in area and production in terms of sugarcane. In India, Uttarakhand ranks 10th in terms of area, production and productivity during 2022–2023 (https://sugarcane.dac.gov.in/pdf/StatisticsAPY.pdf) and the sugar industry by-product can be a potential input in the agriculture sector, particularly the foothills and lower Himalayas where the sugar industries are dominantly located. The two important by-products of this industry are the sugarcane bagasse and press mud. Press mud or filter cake is generated as a by-product of the purification of sugar by carbonation or sulphitation process (Ghulam et al. 2012). Crushing of 100 tonnes of sugarcane produces around 3 tonnes of press mud (Gupta et al. 2019) and 45–50 Mt of bagasse. These are mainly rejected waste material of sugarcane industries that cause problem of storage and pollutes the surrounding sugar mills on accumulation (Bhosale et al. 2012). Press mud application besides having beneficial impact on soil in terms of organic matter and nutrient addition also improves the P supplying capacity of soil. The organic acids produced during decomposition of press mud helps to mobilize the P and improves its availability (Dotaniya et al.

2013). Generally, the carbonation press mud contains around 60% $CaCO_3$, which can be used as amelioration for acid soils (Tiwari et al. 1998). In the North Western Himalayan Region where the soil tends to be acidic in nature, carbonation press mud can be an effective ameliorant. Roy et al. (2022) reported that application of press mud and bagasse improved the soil pH in the moderately acidic soils of NW Himalayas over control by 9.4% and 10% respectively over control (pH 5.27) in a maize-wheat cropping system. The high organic C content and high nutrient content particularly Ca, Mg (Partha and Sivasubramanian 2006) makes them a very effective alternate fertilizer source.

Besides the agro-industrial wastes, by-products like fly ash, basic slag etc. bears a positive influence on the soil health. The fly ash is derived from the coal combustion for generation of electricity and almost 750 million tonnes of fly ash is generated globally per annum (Izquierdo and Querol 2012). The fly ash is characterized by low bulk density, high surface area, high water holding capacity and higher concentration of plant nutrients. Blending fly ash with different locally available materials like gypsum, lime, FYM, press mud, biochar etc. for application as an alternate fertilizer source and soil conditioner has proven beneficial as reported by several researchers (Chaudhary et al. 2011). The fly ash has an alkaline pH and forms an effective amendment for the acid soils which are dominant in the North Western Himalayas. The yield of several root crops increased significantly (up to 62%) under the influence of fly ash amendment in acid soils with pH 5.0 (Tripathi et al. 2010). Another emerging industrial by-product with very less application in agriculture is the brick kiln dust. The brick kiln dust has similar characteristics of fly ash and is generated in huge quantity as India is the leading clay brick producer after China (Maithel et al. 2012). The high Ca and Mg content, low bulk density makes it a potent amendment for acid soils. The nematocidal properties of brick kiln dust were reported by Rizvi and Khan (2011) while Raghav and Khan (2002) observed enhanced tomato growth and yield with its application. Roy et al. (2022) utilized brick kiln dust along with arbuscular mychorrhizal fungi as soil amendment in acid soils of Himalayan foot hills and obtained promising results in a maize wheat cropping system. The different industrial by-products offer a great alternative to be used as a soil conditioner and plant nutrient source in the Himalayan agriculture and help in its sustenance (Table 8.2).

8.3.8 Municipal Solid-Wastes

Biomass recycling through municipal solid wastes (MSW) involves the recovery and transformation of organic materials found in household and urban waste streams into valuable products like compost, biogas, and bioenergy (Ragaert et al. 2017). Municipal solid waste includes a variety of materials, such as food scraps, yard trimmings, paper, and other organic residues, which can be effectively recycled to reduce landfill use, mitigate environmental impact, and contribute to renewable

Table 8.2 Use of industrial by-products and municipal solid wastes for enhancing crop productivity and soil health

Location	Soil Type	Crop	Amendment	Impact	References
Palampur, Himachal Pradesh	Alfisols	Wheat	By-product from medicinal plant industry	10–11.5% increase in wheat yield	Mahajan et al. (2002)
Palampur, Himachal Pradesh	Alfisols	Wheat	Biological sludge	30% increase in yield	Mahajan et al. (2002)
Dehradun, Uttarakhand	Alfisols	Maize-wheat	Press mud, bagasse and brick kiln dust @ 5–10 Mg ha^{-1}	Improvement in soil pH by 9–21% over control	Roy et al. (2022)
Kharagpur, India	Acidic Laterites	Rice	Fly ash @ 10 Mg ha^{-1} + Farm Yard Manure	Yield increase by 6.2%	Rautaray et al. (2003)
Srinagar, Jammu and Kashmir			Composting of Municipal solid waste	Total organic C increased to 16.5%	Hameed et al. (2021)

energy production (Rajca et al. 2020). The organic components of MSW can be incinerated to produce energy or subjected to advanced thermal processes like gasification or pyrolysis to generate biofuels and synthetic gases. Waste-to-energy processes are particularly effective in reducing waste volume while recovering energy (Roy et al. 2023a). Waste-to-energy facilities convert biomass in MSW into electricity, heat, or fuel, providing a renewable energy source and reducing landfill use. Certain organic materials, particularly food waste and plant matter, can be converted into biofuels like ethanol or biodiesel through fermentation or transesterification. This process enables the transformation of organic waste into a renewable energy source. Biofuels produced from MSW can be used to power vehicles or in other energy applications, contributing to the reduction of fossil fuel dependency and greenhouse gas emissions (Adhikari and Pandey 2020; Roy et al. 2023b). Anaerobic digestion and composting are regulated methods of handling organic waste that reduce methane emissions, a major greenhouse gas. Furthermore, reducing reliance on fossil fuels via the use of biogas generated by anaerobic digestion as a renewable energy source contributes to the reduction of greenhouse gas emissions (Behera et al. 2015; Sharma et al. 2020). Compost derived from organic municipal solid waste (MSW) enhances soil structure, fertility, and water retention, which supports healthier plant growth and reduces soil erosion, fostering more sustainable land management practices (Singh et al. 2019). Biomass recycling from MSW recovers valuable nutrients and organic matter, which can be reused in agricultural and landscaping applications. This approach closes the nutrient cycle, decreasing reliance on synthetic inputs and supporting a circular economy (Sharma et al. 2020).

8.3.9 Integrated Farming Systems: A Close-Loop Waste Recycling Approach

Integrated Farming Systems (IFS) are agricultural systems that combine multiple agricultural enterprises such as crop production, livestock rearing, fish farming, and agroforestry in a synergistic manner (Kumar et al. 2022). The key principle of IFS is the recycling of waste products from one enterprise as inputs for another, creating a closed-loop system that minimizes waste and enhances resource use efficiency (Behera and France 2014). This approach is particularly beneficial in the Indian Himalayan region, where resources are often limited, and sustainable land use is essential (Behera and France 2016; Behera et al. 2018; Babu et al. 2023a, b).In an IFS, crop residues, animal manure, and other organic wastes are recycled within the farm to maintain soil fertility and reduce the need for external inputs. Crop residues that would otherwise be burned or discarded are used as feed for livestock, whose manure is then composted and returned to the fields as organic fertilizer. This recycling of nutrients within the farm system helps to maintain soil health, reduce dependency on chemical fertilizers, and lower production costs (Behera and France 2014; Babu et al. 2021). A study conducted in Himachal Pradesh demonstrated the effectiveness of IFS in enhancing soil fertility and agricultural productivity. The study found that farms practicing IFS had 20–25% higher soil organic carbon levels compared to conventional farms, primarily due to the regular incorporation of organic matter from composted manure and crop residues (Thakur et al. 2022). This increase in soil organic carbon contributed to improved soil structure, water retention, and nutrient availability, resulting in higher crop yields and better resilience to climate variability.

The close-loop waste recycling system inherent in IFS offers significant environmental benefits, particularly in terms of reducing greenhouse gas emissions and enhancing biodiversity (Sahoo et al. 2019). By recycling organic wastes within the farm, IFS reduces the need for external inputs such as synthetic fertilizers, which are often associated with high energy use and greenhouse gas emissions (Babu et al. 2021). Moreover, the diversified nature of IFS promotes biodiversity, which enhances ecosystem services such as pest control, pollination, and soil health (Kumar et al. 2022; Fatima et al. 2023). In the northeastern Himalayas, research has shown that IFS can reduce methane and nitrous oxide emissions by 15–20% compared to conventional farming practices, primarily due to the reduced use of chemical fertilizers and the incorporation of organic matter into the soil (NESAC 2023). This reduction in emissions is crucial for mitigating climate change in the Himalayan region, which is particularly vulnerable to its impacts. Economically, IFS provides farmers with multiple sources of income, thereby reducing risk and enhancing livelihood security. A farmer practicing IFS may derive income from crop sales, livestock products, fish farming, and agroforestry, ensuring a steady income stream even if one enterprise fails due to adverse weather conditions or market fluctuations. Studies in Uttarakhand have shown that farmers practicing IFS had 30–40% higher income stability compared to those relying solely on monocropping (Singh et al. 2023b).

8.3.10 Challenges in Sustainable Crop Residue Management

8.3.10.1 Technical Constraints

Effective management of crop residues is crucial for preserving soil health, boosting agricultural productivity, and reducing environmental impact. However, the implementation of these sustainable practices often faces significant technical challenges and machinery requirements. These barriers can impede the adoption of optimal methods, particularly in regions with limited access to advanced agricultural technologies. Many farmers, particularly in developing areas, do not have access to the advanced machinery necessary for effective residue management. This includes equipment for collecting, processing, and incorporating residues into the soil. The high initial cost of purchasing such machinery, like mulchers, balers, and specialized tillage tools, can be prohibitive for small-scale farmers. Additionally, integrating new machinery into existing farming practices can be difficult, requiring modifications to current systems and workflows, which demands time, effort, and additional resources. The effective use of advanced machinery also requires specialized knowledge and training, which many farmers may lack. This can limit their ability to operate and maintain the new equipment efficiently. Furthermore, efficient residue management depends on proper infrastructure for transporting and storing crop residues. Without adequate facilities, the timely and effective use of residues can be hindered. Advanced machinery often depends on fuel and energy, which can be scarce or costly in some regions. A reliable energy supply is crucial for the consistent operation of residue management equipment. Additionally, factors like soil type, moisture levels, and topography can influence the performance of machinery, necessitating equipment that can adapt to diverse environmental conditions. Weather conditions, such as excessive rainfall or drought, can also affect the feasibility of certain residue management practices, requiring machinery that is resilient and adaptable to varying climates.

8.3.10.2 Machinery Requirements

Balers are crucial for collecting and compacting crop residues into bales, making them easier to handle, store, and transport. These machines are vital for efficiently managing large quantities of residues. Choppers and shredders cut crop residues into smaller pieces, aiding faster decomposition and easier soil incorporation. This chopping process also reduces pest habitats and enhances nutrient cycling. Mulchers distribute the chopped residues evenly across the field, creating a protective cover that helps retain soil moisture, reduce erosion, and improve soil health. Tillage tools like rotary tillers and disc harrows incorporate residues into the soil, promoting decomposition and nutrient release while preparing the seedbed for the next planting season and preserving soil structure. No-till drills enable direct seeding into residue-covered soil without prior tillage, preserving soil structure, reducing erosion, and maintaining soil moisture. Strip tillers create narrow tilled strips for seeding while leaving the rest of the field undisturbed, balancing residue management with minimal soil disturbance. Precision agriculture tools, including sensors and monitoring devices like soil moisture sensors and remote sensing equipment, monitor field conditions and optimize residue management. GPS-guided machinery and automated systems improve the precision and efficiency of residue management, reducing labor and enhancing outcomes. Compost turners aerate and mix compost piles, speeding up the decomposition of crop residues into nutrient-rich compost. Systems that convert crop residues into bioenergy, such as biomass boilers for heat and anaerobic digesters for biogas, offer renewable energy solutions while reducing waste. Addressing technical challenges and fulfilling machinery requirements are key to sustainable crop residue management. Overcoming these challenges requires investing in advanced equipment and providing farmers with training and support. Ensuring access to suitable machinery, adapting it to local conditions, and integrating it into existing farming practices are essential steps to maximize the benefits of sustainable residue management. By tackling these technical and logistical barriers, we can improve soil health, boost agricultural productivity, and reduce the environmental impact of farming.

8.3.10.3 Socio-economic Barriers and Farmer Adoption

In the Himalayan region of India, socio-economic barriers and farmer adoption issues have significantly impacted biomass recycling efforts, which are crucial for soil and environmental sustainability. Here are some key factors:

Economic constraints: Many farmers in the Himalayan region operate on a small scale and face financial constraints. The cost of implementing biomass recycling systems, such as composting facilities or advanced recycling technologies, can be prohibitive. This limits their ability to invest in sustainable practices.

Lack of awareness and education: There is often a lack of awareness about the benefits of biomass recycling and how to effectively implement it. Farmers may not be familiar with the techniques or the positive impact on soil health and crop yield, leading to lower adoption rates.

Traditional practices: Farmers in the region may adhere to traditional agricultural practices that do not incorporate modern recycling methods. Shifting from established practices to new methods requires education, demonstration, and sometimes financial incentives.

Infrastructure challenges: The region's challenging terrain and poor infrastructure can make it difficult to transport and process biomass. Limited access to technology and resources further hinders the effective implementation of recycling practices.

Policy and support: There may be insufficient government policies or support programs that encourage biomass recycling. Without strong policy frameworks, incentives, or subsidies, farmers may lack the motivation or resources to adopt new practices.

Cultural factors: In some cases, cultural attitudes towards waste and recycling can influence adoption rates. If recycling is not a culturally embedded practice, there might be resistance to change.

8.3.10.4 Cultural and Social Factors

Cultural and social factors can significantly affect the adoption of biomass recycling practices in the Himalayan region of India. Some of these factors are given below.

Traditional beliefs and practices: In many Himalayan communities, traditional farming methods are deeply rooted and resistant to change. Farmers may be reluctant to adopt biomass recycling because it deviates from established practices, which they believe are more effective or culturally appropriate.

Perception of waste: The perception of what constitutes waste and how it should be handled can vary culturally. In some areas, there may be a preference for burning biomass or using it for other purposes, rather than recycling it. This can be due to beliefs about the effectiveness of such practices or a lack of understanding of recycling benefits.

Community norms and social influence: Social norms and peer influence play a significant role in the adoption of new practices. If biomass recycling is not widely accepted or practiced within the community, individual farmers may be less likely to adopt it due to fear of social ostracism or skepticism from peers.

Lack of local champions: Effective adoption of new practices often requires local leaders or champions who advocate for change and demonstrate its benefits. In the absence of influential figures promoting biomass recycling, there may be less motivation for widespread adoption.

Gender roles and responsibilities: In some areas, gender roles can impact the adoption of new practices. If women, who may be primarily responsible for managing household waste or small-scale farming activities, are not involved in decision-making processes or lack access to information and resources, this can hinder the adoption of biomass recycling.

Education and awareness: Cultural and social factors also influence education and awareness levels. If communities lack access to information about the benefits and methods of biomass recycling, they may not recognize its importance or potential impact on soil and environmental sustainability.

8.3.10.5 Policy and Institutional Barriers

Policy and institutional barriers can significantly hinder the adoption of biomass recycling practices in the Himalayan region of India. These barriers often stem from gaps in policy frameworks, lack of institutional support, and systemic challenges. Some of these factors are summarized hereunder:

Insufficient policy frameworks: The lack of comprehensive and supportive policies for biomass recycling can be a major barrier. Without clear guidelines, incentives, or regulations that promote recycling, farmers and communities may lack the motivation or resources to implement these practices.

Limited financial incentives: There may be a lack of financial incentives such as subsidies, grants, or low-interest loans to support the adoption of biomass recycling technologies. Without financial support, the initial investment required for recycling infrastructure can be prohibitive for many farmers.

Weak institutional support: Institutions responsible for environmental protection, agricultural development, and waste management may not have strong mandates or resources dedicated to biomass recycling. This can result in inadequate support and guidance for implementing recycling practices at the local level.

Fragmented implementation: Policies and programs related to biomass recycling may be fragmented or poorly coordinated across different government departments and agencies. This lack of integration can lead to inefficiencies and gaps in implementation.

Inadequate research and development: There may be insufficient investment in research and development to create and disseminate effective biomass recycling technologies tailored to the specific needs of the Himalayan region. Without innovative solutions, recycling efforts may be less effective or practical.

Regulatory challenges: Regulations governing waste management and recycling may be outdated or not well-suited to the local context. For instance, regulations may not address the unique challenges of the Himalayan terrain or may not incentivize sustainable practices effectively.

Lack of capacity building: There may be a shortage of programs aimed at building local capacity and expertise in biomass recycling. Without training and resources for local authorities, farmers, and community members, the adoption and management of recycling practices can be hindered.

Inconsistent enforcement: Even if policies are in place, inconsistent enforcement and monitoring can undermine their effectiveness. Without proper oversight, compliance with recycling practices may be low, and the intended benefits may not be realized.

8.3.10.6 Market Access and Infrastructure

Market access and infrastructure issues can significantly impact the effectiveness of biomass recycling efforts in the Himalayan region of India. Some of these factors are given below:

Limited infrastructure for collection and processing: In the Himalayan region, the rugged terrain and challenging weather conditions can make it difficult to develop and maintain infrastructure for collecting, transporting, and processing biomass.

Poor transportation networks: Inadequate transportation infrastructure, such as poorly maintained roads or lack of access to remote areas, can hinder the movement of biomass materials. This can lead to delays and increased costs, making it less feasible for farmers and businesses to engage in recycling activities.

Lack of recycling facilities: There may be a shortage of local recycling facilities or processing units capable of handling biomass materials. Without accessible facilities, farmers may be forced to either burn or discard biomass rather than recycle it.

Market access challenges: Farmers may face difficulties accessing markets for recycled biomass products. If there are no established markets or demand for compost, biochar, or other recycled biomass products, there is little incentive for farmers to invest in recycling practices.

High costs of technology and equipment: The cost of technology and equipment necessary for efficient biomass recycling can be prohibitive, particularly in remote or economically disadvantaged areas. High initial investment and maintenance costs can deter adoption of recycling practices.

Economic viability: The economic viability of recycling operations may be undermined by high costs and low market prices for recycled products. If the financial returns from recycling do not outweigh the costs, farmers and businesses may be less likely to participate.

Fragmented supply chains: Inefficient or fragmented supply chains can disrupt the collection, processing, and distribution of recycled biomass products.

8.3.11 Policy and Institutional Support

To advance biomass recycling and enhance soil and environmental sustainability in the Himalayan region of India, comprehensive policy and institutional support is essential. Development of integrated biomass management policies (Fig. 8.3) that address all aspects of biomass handling, from collection to utilization, tailored to the region's unique challenges is imperative. Set clear sustainability goals to reduce waste, improve soil health, and enhance environmental quality, aligning with national and regional policies. Provide financial subsidies and grants to support biomass recycling facilities and adoption of new technologies by small and medium enterprises and farmers. Facilitate access to low-interest loans for infrastructure and equipment, and create market-based incentives, such as carbon credits, to boost the

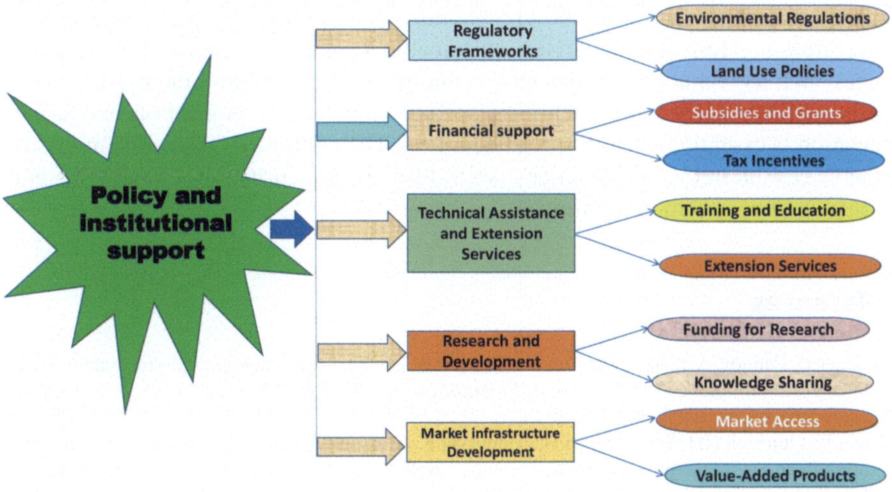

Fig. 8.3 Policy and institutional support for the implementation of effective biomass recycling

value and demand for recycled biomass products. Implement training programs for farmers, local businesses, and officials on best practices for biomass recycling. Establish knowledge-sharing platforms to foster collaboration among research institutions, government agencies, and communities to drive innovation and practical solutions. Invest in infrastructure crucial for biomass recycling, including collection centers, transportation networks, and processing facilities. Prioritize projects that address the Himalayan terrain's specific challenges and encourage public-private partnerships to leverage resources and expertise. Set clear regulations and standards for biomass recycling, including quality and environmental safeguards. Ensure regulations are practical and enforceable, and develop mechanisms for monitoring and compliance to uphold these standards. Fund research into biomass recycling technologies suited to the region and support pilot projects to test and refine new methods. Use these initiatives to gather data and build evidence for broader adoption. Involve local communities in developing and implementing recycling initiatives, ensuring their perspectives are considered. Conduct awareness campaigns to educate on the benefits of recycling and showcase successful examples to encourage wider adoption. These strategies will help to overcome barriers and promote effective biomass recycling in the Himalayan region.

8.4 Conclusion

Sustainable biomass recycling is crucial for improving the soil health and resilience in the face of climate change. It involves practices like direct incorporation, mulching, composting, and biochar production, which improve soil organic matter content, nutrient availability, structure improvement, and water retention. This supports soil fertility and microbial activity, vital for agricultural productivity. In the context of climate change, residues act as a natural buffer against soil erosion, improve soil moisture retention, and mitigate greenhouse gas emissions by promoting carbon sequestration. However, achieving sustainable residue management faces challenges such as technological barriers, economic constraints, and the need for supportive policies. Integrating scientific research, technological innovation, and farmer education is essential for widespread adoption. The continued advancement of sustainable residue recycling practices is vital for food security, agricultural sustainability, and ecosystem protection.

References

Adhikari P, Pandey A (2020) Bioprospecting plant growth promoting endophytic bacteria isolated from Himalayan yew (*Taxus wallichiana*Zucc.). Microbiol Res 239:126536. https://doi.org/10.1016/j.micres.2020.126536

Akram U, Quttineh NH, Wennergren U, Tonderski K, Metson GS (2019) Enhancing nutrient recycling from excreta to meet crop nutrient needs in Sweden—a spatial analysis. Sci Rep 9:10264. https://doi.org/10.1038/s41598-019-46706-7

Babu S, Mohapatra KP, Das A, Yadav GS, Singh R, Chandra P, Avasthe RK, Kumar A, Devi MT, Singh VK, Panwar AS (2021) Integrated farming systems: climate resilient sustainable food production system in the Indian Himalayan region. In: Venkatramanan V, Shah S, Prasad R (eds) Exploring synergies and trade-offs between climate change and the sustainable development goals. Springer, Singapore

Babu S, Das A, Singh R, Mohapatra KP, Kumar S, Rathore SS et al (2023a) Designing an energy-efficient, economically feasible, and environmentally robust integrated farming system model for sustainable food production in the Indian Himalayas. Sustain Food Technol 1:126–142. https://doi.org/10.1039/D2FB00016D

Babu S, Singh R, Avasthe RK, Rathore SS, Kumar S, Das A, Layek J, Sharma V, Wani OA, Singh VK (2023b) Conservation tillage and diversified cropping enhance system productivity and eco-efficiency and reduce greenhouse gas intensity in organic farming. Front Sustain Food Syst 7:1114617. https://doi.org/10.3389/fsufs.2023.1114617

Behera UK, France J (2014) Developmental farm model for revolutionizing agricultural growth. ICAR News 20(3):5–6

Behera UK, France J (2016) Integrated farming systems for the livelihood security of small and marginal farmers in India and other developing countries. Adv Agron 138:235–282

Behera UK, Babu A, Kaechele H, France J (2015) Energy self-sufficient sustainable integrated farming systems for livelihood security under a changing climate scenario in an Indian context: a case-study approach. CAB Rev 10(19):11

Behera UK, Bhargavi B, Meena SL, Bhattacharyya R, Singh R, Singh RK, Singh VK, Gupta N, Rana KS (2018) Research Bulletin on Integrated farming system model for livelihood security of small and marginal farmers. Division of Agronomy, Indian Agricultural Research Institute, New Delhi

Berardi P, Almeida MF, Lopes M, Maia DL, Dias J (2020) Analysis of Portugal's refuse derived fuel strategy, with particular focus on the northern region. J Clean Prod 277:123262. https://doi.org/10.1016/j.jclepro.2020.123262

Bhatt J, Chauhan S, Rana R (2022) Biochar and microbial activity: implications for soil health in the Indian Himalayas. Mountain Res Dev 42(4):89–98

Bhatt J, Chauhan S, Rana V (2023) Economic resilience through integrated farming systems in the Northeastern Himalayas. J Rural Dev Econ 9(4):145–155

Bhattacharyya R, Fullen MA, Davies K, Booth CA (2009) Utilizing palmleaf geotextile mats to conserve loamy sand soil in the United Kingdom. Agric Ecosyst Environ 130:50–58. https://doi.org/10.1016/j.agee.2008.11.015

Bhattacharyya R, Yi Z, Yongmei L, Li T, Panomtaranichagul M, Peukrai S, Booth CA (2012) Effects of biological geotextiles on above ground biomass production in selected agro-ecosystems. Field Crop Res 126:23–36

Bhattacharyya R, Pal S, Lal R, Mandal B, Chandran P (2015) Soil erosion in India: a review. J Indian Soc Soil Sci 63(2):251–264

Bhattacharyya R, Bhatia A, Ghosh BN, Santra P, Mandal D, Kumar G, Singh R, Madhu M, Ghosh A, Mandal AK, Paul R, Datta A, Sharma PC, Mondal UK, Jha P, Anil KS, Lalitha M, Kumar M, Panwar NR et al (2023) Soil degradation and mitigation in agricultural lands in the Indian Anthropocene. Eur J Soil Sci:e13388. https://doi.org/10.1111/ejss.13388

Bhosale PR, Chonde SG, Raut PD (2012) Studies on extraction of sugarcane wax from press mud of sugar factories from Kolhapur district, Maharashtra. J Environ Res Dev 6(3A):615–620

Capasso I, Lirer S, Flora A, Ferone C, Cioffi R, Caputo D (2019) Reuse of mining waste as aggregates in fly ash-based geopolymers. J Clean Prod 220:65–73. https://doi.org/10.1016/j.jclepro.2019.02.164

Chaudhary SK, Rai UN, Mishra K, Huang HG, Yang XE, Inouhe M, Gupta DK (2011) Growth and metal accumulation potential of *Vigna radiata* L. grown under fly-ash amendments. Ecol Eng 37:1583–1588

Cheng K, Hao W, Wang Y, Yi P, Zhang J, Ji W (2020) Understanding the emission pattern and source contribution of hazardous air pollutants from open burning of municipal solid waste in China. Environ Pollut 263:114417. https://doi.org/10.1016/j.envpol.2020.114417

CPCB (2020) Air quality monitoring report. Central Pollution Control Board. https://cpcb.nic.in/

Das S, Lee SH, Kumar P, Kim KH, Lee SS, Bhattacharya SS (2019) Solid waste management: scope and the challenge of sustainability. J Clean Prod 228:658–678. https://doi.org/10.1016/j.jclepro.2019.04.323

Dev S, Saha S, Kurade MB, Salama ES, El-Dalatony MM, Ha GS (2019) Perspective on anaerobic digestion for biomethanation in cold environments. Renew Sustain Energy Rev 103:85–95. https://doi.org/10.1016/j.rser.2018.12.034

Dotaniya ML, Datta SC, Biswas DR, Meena BP (2013) Effect of solution phosphorus concentration on the exudation of oxalate ions by wheat (*Triticum aestivum* L.). Proc Int Acad Sci India Sec B Biol Sci 83(3):305–309

Dotaniya ML, Datta SC, Biswas DR, Dotaniya CK, Meena BL, Rajendiran S, Lata M (2016) Use of sugarcane industrial by-products for improving sugarcane productivity and soil health. Int J Recycl Org Waste Agric 5:185–194

Fatima A, Singh VK, Babu S, Singh RK, Upadhyay PK, Rathore SS, Kumar B, Hasanain M, Parween H (2023) Food production potential and environmental sustainability of different integrated farming system models in northwest India. Front Sustain Food Syst 7:95946

Food and Agriculture Organization (2019) Rural demographics and sustainable development in the Indian Himalayan Region (IHR). FAO, Rome. https://www.fao.org/sustainability/en/

Gaballah ES, Abdelkader TK, Luo S, Yuan Q, El-Fatah Abomohra A (2020) Enhancement of biogas production by integrated solar heating system: a pilot study using tubular digester. Energy 193:116758. https://doi.org/10.1016/j.energy.2019.116758

Ghulam S, Khan MJ, Usman K, Shakeebullah. (2012) Effect of different rates of press mud on plant growth and yield of lentil in calcareous soil. Sarhad J Agric 28(2):249–252

Gogoi B, Borah N, Baishya A, Dutta S, Nath DJ, Das R et al (2021) Yield trends, soil carbon fractions and sequestration in a rice-rice system of North-East India: effect of 32 years of INM practices. Field Crop Res 272:108289. https://doi.org/10.1016/j.fcr.2021.108289

Gudade BA, Malik GC, Das A, Babu S, Banerjee M, Kumar A, Singh R, Bora SS (2022a) Performance of French bean crop as influenced by biochar levels and INM practices in acid Inceptisol of Meghalaya. Ann Agric Res 43(2):84–89

Gudade BA, Malik GC, Das A, Babu S, Kumar A, Singh R, Bhupenchandra I (2022b) Effect of biochar levels and integrated nutrient-management practices on agro-physiological performance and productivity of maize (*Zea mays*). I. J Agron 67(4):380–385

Gupta AK, Negi M, Nandy S, Alatalo JM, Singh V, Pandey R (2019) Assessing the vulnerability of socio-environmental systems to climate change along an altitude gradient in the Indian Himalayas. Ecol Indic 106:105512. https://doi.org/10.1016/j.ecolind.2019.105512

Hameed M, Bhat RA, Pandit BA, Ramzan S, Dijoo ZK, Wani MA (2021) Qualitative assessment of compost engendered from municipal solid waste and green waste by indexing method. J Air Waste Manage Assoc 72(2):210–219

Harder R, Wielemaker R, Molander S, Öberg G (2020) Reframing human excreta management as part of food and farming systems. Water Res 175:115601. https://doi.org/10.1016/j.watres.2020.115601

Hou N, Wen L, Cao H, Liu K, An X, Li D (2017) Role of psychotropic bacteria in organic domestic waste composting in cold regions of China. Bioresour Technol 236:20–28. https://doi.org/10.1016/j.biortech.2017.03.166

ICAR (2015) Soil health card scheme: nutrient status of Indian soils. Indian Council of Agricultural Research. https://www.manage.gov.in/publications/reports/shc.pdf

ICAR (2018) Impact of crop residue burning on soil health. Indian Council of Agricultural Research. https://iiss.icar.gov.in/eMagazine/v2i2/6.pdf

ICAR (2023) Annual report on agricultural production and residue management in Indian Himalayan states. Indian Council of Agricultural Research. https://icar.org.in/sites/default/files/2023-07/ICAR-Annual-Report-2022-23-English.pdf

Iglesias M, Marguí E, Camps F, Hidalgo M (2018) Extractability and crop transfer of potentially toxic elements from Mediterranean agricultural soils following long-term sewage sludge applications as a fertilizer replacement to barley and maize crops. Waste Manag 75:312–318. https://doi.org/10.1016/j.wasman.2018.01.024

Izquierdo M, Querol X (2012) Review article leaching behavior of elements from coal combustion fly ash: an overview. Int J Coal Geol 94:54–66

Joseph A, Bhattacharyya R, Biswas DR, Das TK, Bandyopadhyay KK, Dey A et al (2023) Long-term adoption of bed planted conservation agriculture based maize/ cotton-wheat system enhances soil organic carbon stabilization within aggregates in the Indo-Gangetic Plains. Front Environ Sci 11:1216242. https://doi.org/10.3389/fenvs.2023.1216242

Kainthola J, Kalamdhad AS, Goud VV (2019) A review on enhanced biogas production from anaerobic digestion of lignocellulosic biomass by different enhancement techniques. Process Biochem 84:81–90

Kannoj, Tomar M, Choudhary R, Bhardwaj S (2023) Crop residue management. Agric Sci Res Rev XIV

Khybri ML, Bhardwaj SP, Prasad SN, Sewa R (1980) Effect of minimum tillage and mulch on soil and water loss in maize on 8% slope. Annual report. CSWCRTI, Derhadun, p 128

Kumar A, Behera UK, Shiva-Dhar SL, Bhatiya A, Meena MC, Gupta G, Singh RK (2018) Effect of tillage, crop residue and phosphorus management practices on the productivity and profitability of maize cultivation in Inceptisols. Indian J Agric Sci 88:1558–1167

Kumar A, Babu S, Singh S, Saha S, Bhatt M, Devi EL, Laha R, Mishra VK (2022) Integrated Organic Farming System (IOFS): an holistic approach for food, nutrition and livelihood security of marginal and small farmers in terraces land of Sikkim Himalayas. Agricolation 2(2):11–17

Kumar A, Behera UK, Dhar S, Babu S, Singh R, Upadhyay PK, Saha S, Devadas R, Kumar A, Gupta G, Singh RK, Gudade BA, Karan S, Verma G (2023a) Deciphering the role of phosphorus management under conservation agriculture-based wheat production system. Front Sustain Food Syst 7:1235141. https://doi.org/10.3389/fsufs.2023.1235141

Kumar A, Behera U, Dhar S, Shukla L, Singh R, Babu S, Sharma VK, Upadhyay PK, Bairwa RK, Gupta G, Kumar A, Singh S, Bhandari N, Qureshi AA, Gudade BA, Verma G, Kumar R, Kumar A (2023b) Conservation tillage and microbially mediated integrated phosphorus management enhance productivity, profitability, and energy use efficiency of maize. J Agric Sci:1–44. https://doi.org/10.1017/S0021859623000266

Kumar A, Singh R, Babu S, Avasthe RK, Das A, Pandey A, Gudade BA, Devadas R, Saha S, Rathore SS (2024) Integrated organic nutrient management: a sustainable approach for cleaner maize (*Zea mays* L.) production in the Indian Himalayas. Org Agric. https://doi.org/10.1007/s13165-024-00471-8

Kumari K, Kumar S, Rajagopal V, Khare A, Kumar R (2019) Emission from open burning of municipal solid waste in India. Environ Technol 40:2201–2214. https://doi.org/10.1080/09593330.2017.1351489

Mahajan KK, Kumar S, Dev SP, Bhardwaj KKR, Gupta SP (2002) Evaluation of industrial wastes in wheat (*Triticum aestivum*)-maize (*Zea mays*) cropping system in mid-hills sub-humid zone of Himachal Pradesh. Indian J Agric Sci 72(5)

Maithel S, Lalchandani D, Malhotra G, Bhanware P, Uma R, Ragavan S (2012) Brick kilns performance assessment; a roadmap for cleaner brick production in India. Shakti Sustainable Energy Foundation and Climate Works Foundation, New Delhi

Mandal D, Giri N, Srivastava P (2020) The magnitude of erosion-induced carbon (C) flux and C-sequestration potential of eroded lands in India. Eur J Soil Sci 71:151–168. https://doi.org/10.1111/ejss.12886

MoEFCC (2014) State of environment report: soil erosion and land degradation. Ministry of Environment, Forest and Climate Change, Government of India. https://moef.gov.in/uploads/2019/09/Final-SoER-2015-India-.pdf

Narain P, Samra JS, Singh RK, Sindhwal NS (2000) Erosion losses and USLE parameters in relation to land cover management. J Indian Soc Soil Sci 48(4):1–7

Negi P, Thakur R, Bhatt R (2022) Enhancing soil fertility with biochar: a study from Himachal Pradesh. Agric Res 11(3):234–242

NESAC (2018) Land use and land cover assessment in North East India. North Eastern Space Applications Centre. https://nesac.gov.in/assets/resources/2020/12/EVALUATING-LAND-

USE-LAND-COVER-CHANGE_IMPACTS-ON-RAINFALL-FORECAST-OVER-THE-NORTH-_EASTERN-REGION-OF-INDIA-USING-THEWEATHER_RESEARCH-AND-FORECASTING-MODEL.pdf

NESAC (2023) Agricultural residue management in North Eastern India: a regional perspective. North Eastern Space Applications Centre. https://nesac.gov.in/assets/resources/2024/08/Annual-Report-2022-23-English-High-Res.pdf

Partha N, Sivasubramanian V (2006) Recovery of chemicals from pressmud—a sugar industry waste. Indian Chem Eng 48(3):160–163

Pathak H, Singh R, Joshi R (2023) Bioenergy from crop residues: a climate change mitigation strategy for the Indian Himalayas. J Renew Energy Environ Sustain 8(1):87–95

Puri K, Joshi R, Singh V (2020) Open garbage dumps near protected areas in Uttarakhand: an emerging threat to Asian Elephants in the Shivalik Elephant Reserve. J Threat Taxa 12:16571–16575. https://doi.org/10.11609/jott.4392.12.11.16571-16575

Qureshi MS, Oasmaa A, Pihkola H, Deviatkin I, Tenhunen A, Mannila J et al (2020) Pyrolysis of plastic waste: opportunities and challenges. J Anal Appl Pyrolysis 152:104804. https://doi.org/10.1016/j.jaap.2020.104804

Ragaert K, Delva L, Van Geem K (2017) Mechanical and chemical recycling of solid plastic waste. Waste Manag 69:24–58. https://doi.org/10.1016/j.wasman.2017.07.044

Raghav D, Khan AA (2002) Impact of industrial particulate pollutants applied to soil on growth and yield of tomato. Thai J Agric Sci 35:187–194

Rajca P, Poskart A, Chrubasik M, Sajdak M, Zajemska M, Skibinski A et al (2020) Technological and economic aspect of refuse derived fuel pyrolysis. Renew Energy 161:482–494. https://doi.org/10.1016/j.renene.2020.07.104

Rautaray SK, Ghosh BC, Mittra BN (2003) Effect of fly ash, organic wastes and chemical fertilizers on yield, nutrient uptake, heavy metal content and residual fertility in a rice–mustard cropping sequence under acid lateritic soils. Bioresour Technol 90:275–283

Rittl TF, Butterbach-Bahl K, Basile CM, Pereira LA, Alms V, Dannenmann M, Cerri CE (2018) Greenhouse gas emissions from soil amended with agricultural residue biochars: effects of feedstock type, production temperature and soil moisture. Biomass Bioenergy 117:1–9

Rizvi R, Khan AA (2011) Influence of brick-kiln dust on penetration and development of Meloidogyne javanica in eggplant roots and its impact on plant growth and yield. Nematolo Med 9(1):35–39

Roy T, Mandal U, Yadav D, Mandal D (2022) Potential of different agro-industrial wastes for sustainable soil management and reclamation of soil acidity in the North-west Himalayan region. In: Abstract book of LMPFRS. Indian Association of Soil and Water Conservationists, Dehradun

Roy P, Bhattacharyya R, Singh RJ, Sharma NK, Kumar G, Madhu M, Biswas DR, Ghosh A, Das S, Joseph AM, Das TK, Kumar SN, Jat SL, Saharawat YS, Jha P (2023a) Impact of agro-geotextiles on soil aggregation and organic carbon sequestration under a conservation-tilled maize-based cropping system in the Indian Himalayas. Front Environ Sci 11:1309106. https://doi.org/10.3389/fenvs.2023.1309106

Roy R, Bhattacharyya R, Biswas DR, Singh RJ, Das TK, Sharma DK, Yadav S, Ann Maria Joseph AM, Jha P (2023b) Effect of using Agro-geotextiles on soil carbon sequestration in the Indian Himalayas. Indian. J Agric Sci 93(7):768–773

Sahoo AK, Pattanaik P, Haldar D, Mohanty UC (2019) Integrated farming system: a climate smart agriculture practice for food security and environment resilience. Int J Trop Agric 37:193–201

Sharma A, Bhatt R (2017) Micronutrient status in soils of central Himalayas: a review. J Soil Water Conserv 16(3):207–214

Sharma NK, Ghosh BN, Mandal D, Singh RJ, Mishra PK (2014) Conservation agriculture for resource conservation in North-Western Himalayan region. In: Somasundaram J, Choudhary RS, Rao AS, Hati KM, Sinha NK, Coumar MV (eds) Conservation agriculture for carbon sequestration and sustaining soil health. New India Publishing Agency, New Delhi, pp 105–126. ISBN: 978-93-83305-32-2

Sharma V, Giri A, Thakur S, Pant D (2020) Resource recovery from inert municipal waste. In: Varjani S, Pandey A, Gnansounou E, Khanal SK, Raveendran S (eds) Current developments in biotechnology and bioengineering. Elsevier, Cambridge, pp 251–262

Sharma R, Thakur R, Singh S (2022) Impact of crop residue incorporation on soil organic carbon and fertility in the Himalayan region. J Soil Water Conserv 17(2):120–128

Sharma R, Singh V, Chauhan S (2023) Estimation of crop residue availability and management practices in Indian Himalayas. J Mountain Agric 12(1):45–60

Singh R, Sharma E, Rai SC (2010) Soil characteristics and fertility status in relation to land-use and slope position in a Himalayan watershed. Catena 81(1):68–75

Singh RJ, Ghosh BN, Sharma NK, Dadhwal KS, Patra, Sridhar Deshwal JS (2014) In-situ green manuring in maize-wheat cropping system to check soil degradation and carbon improvement in sloppy soils of Himalaya. Popular Kheti 2(4):50–53. (Available online at www.popularkheti.info)

Singh RJ, Ghosh BN, Sharma NK, Patra S, Dadhwal KS, Mishra PK (2016a) Energy budgeting and energy synthesis of rainfed maize-wheat rotation system with different soil amendment applications. Ecol Indic 61(2):753–765. https://doi.org/10.1016/j.ecolind.2015.10.026

Singh RJ, Meena RL, Sharma NK, Kumar S, Kumar K, Kumar D (2016b) Economics, energy, and environmental assessment of diversified crop rotations in sub-Himalayas of India. Environ Monit Assess 188:79. https://doi.org/10.1007/s10661-015-5085-2

Singh RJ, Ghosh BN, Sharma NK, Patra S, Dadhwal KS, Meena VS, Deshwal JS, Mishra PK (2017) Effect of seven years of nutrient supplementation through organic and inorganic sources on productivity, soil and water conservation, and soil fertility changes of maize-wheat rotation in north-western Indian Himalayas. Agric Ecosyst Environ 249:177–186. https://doi.org/10.1016/j.agee.2017.08.024

Singh R, Babu S, Avasthe RK, Yadav GS, Chettri TK, Singh A (2018a) Effect of organic mulches and vermicompost on productivity, profitability and energetic of mustard (*Brassica campestris*) in popcorn (*Zea mays everta*)-mustard cropping system in rainfed Sikkim Himalaya. Indian J Agric Sci 88(11):87–91. https://doi.org/10.56093/ijas.v88i11.84916

Singh R, Babu S, Avasthe RK, Singh M, Yadav GS, Chettri TK (2018b) Residual effect of organic amendments on growth, productivity, economics and agri-energetics of local popcorn (*Zea mays everta*) in toria (*Brassica campestris*)—popcorn cropping system under mid hills of Sikkim Himalayas. Indian J Agric Sci 88(12):1887–1892

Singh RJ, Deshwal JS, Sharma NK, Ghosh BN, Bhattacharyya R (2019) Effects of conservation tillage based agro-geo-textiles on resource conservation in sloping croplands of Indian Himalayan Region. Soil Tillage Res 191:37–47. https://doi.org/10.1016/j.still.2019.03.012

Singh R, Babu S, Avasthe RK, Yadav GS, Das A, Mohapatra KP, Kumar A, Singh VK, Chandra P (2021a) Crop productivity, soil health, and energy dynamics of Indian Himalayan intensified organic maize-based systems. Int Soil Water Conserv Res 9(2):260–270

Singh R, Babu S, Avasthe RK, Meena RS, Yadav GS, Das A, Mohapatra KP, Rathore SS, Kumar A, Singh C (2021b) Conservation tillage and organic nutrients management improve soil properties, productivity, and economics of a maize-vegetable pea system in the Eastern Himalayas. Land Degrad Dev:1–18. https://doi.org/10.1002/ldr.4066

Singh R, Kumar A, Babu S, Avasthe R, Das A, Rathore SS, Kumar S, Singh C, Sharma V, Bhupenchandra I (2023a) Development of organic nutrients management system for profitable and soil-supportive French bean (*Phaseolus vulgaris* L.) farming in north-eastern Himalayas, India. Front Sustain Food Syst 7:1115521. https://doi.org/10.3389/fsufs.2023.1115521

Singh V, Bhat J, Chauhan S (2023b) Soil erosion control through crop residue management in the Indian Himalayas. Mountain Res Dev 43(1):75–82

Singh RJ, Kumar G, Sharma NK, Deshwal JS, Madhu M (2024) Extreme rainfall storm-induced surface runoff and sediment dynamics of conservation tillage-based agro-geotextiles emplaced on sloping croplands of the Indian Himalayan region. Phys Chem Earth Parts A/B/C 135:103644. https://doi.org/10.1016/j.pce.2024.103644

Skousen J, Yang JE, Lee JS, Ziemkiewicz P (2013) Review of fly ash as a soil amendment. Geosyst Eng 16(3):249–256

Thakur A, Kumari S, Sinai, Borker S, Prashant SP, Kumar A, Kumar R (2021) Solid waste management in Indian Himalayan region: current scenario, resource recovery, and way forward for sustainable development. Front Energy Res 9:609229. https://doi.org/10.3389/fenrg.2021.609229

Thakur R, Sharma A, Singh V (2022) Enhancing soil fertility and productivity through integrated farming systems in Himachal Pradesh. J Mountain Agric 14(3):165–175

Tiwari RJ, Bangar KS, Nema GK, Sharma BK (1998) Long term effect of press mud and nitrogenous fertilizers on sugarcane and sugar yield on a typic chromustert. J Indian Soc Soil Sci 46:243–245

Tripathi RC, Jha SK, Ram LC (2010) Impact of fly ash application on tracemetal content in some root crops, energy sources. Part A: Recov Util Environ Effects 32:576–589

Verma G, Dhaka AK, Singh B, Kumar A, Choudhary AK, Kumar A, Kamboj NK, Hasanain M, Singh S, Bhupenchandra I, Shabnam, Sanwal P, Kumar S (2024) Productivity, soil health, and carbon management index of soybean-wheat cropping system under double zero-tillage and natural-farming based organic nutrient management in north-Indian plains. Sci Total Environ 917:170418. https://doi.org/10.1016/j.scitotenv.2024.170418

WHO (2018) Ambient (outdoor) air pollution. World Health Organization. https://www.who.int/news-room/fact-sheets/detail/ambient-(outdoor)-air-quality-and-health

Zhang F, Zhao Y, Wang D, Yan M, Zhang J, Zhang P et al (2020) Current technologies for plastic waste treatment: a review. J Clean Prod 282:124523. https://doi.org/10.1016/j.jclepro.2020.124523

Soil Carbon Sequestration Potential in Achieving the Land Degradation Neutrality (LDN)

9

Priyanka Saha, Sougata Roy, Suchismita Satapathy, Sabyasachi Koley, Alekhya Gunturi, and Ayan Sarkar

Abstract

Industrialization, urbanization, and unsustainable farming practices pose a serious threat to the global land, causing 25% of it to degrade and 24 billion tonnes of soil to be lost annually. This degradation impacts 3.2 billion people and incurs annual costs ranging from US$6.3–10.6 trillion. In order to combat climate change, agroecosystems must reduce their rapid greenhouse gas (GHG) emissions. Soil carbon sequestration emerges as a critical solution to address these challenges. Preserving and enhancing soil organic carbon (SOC) is essential for global food security and combating land degradation, particularly in India, where mismanagement has caused SOC declines of 0.2–0.4% in major food grain producing states like Punjab and Haryana. Achieving targets of Sustainable Development Goal, especially land degradation neutrality (LDN), is pivotal, with India to restore 26 Mha degraded land by 2030. Restoring degraded agricultural land through carbon sequestration involves employing various agronomic techniques such as zero tillage, mulching, cover cropping, residue retention, agroforestry, and green and brown manuring etc. Conservation tillage practices, capable of sequestering 0.4–0.8 Pg C $year^{-1}$ globally, is crucial for reducing agricultural emissions. Ecological intensification in land management enhances food security and SOC sequestration through essential ecosystem processes, empha-

P. Saha (✉)
ICAR-Krishi Vigyan Kendra, West Garo Hills, Meghalaya, India

S. Roy
International Potato Centre (CIP-CGIAR), Patna, Bihar, India

S. Satapathy · S. Koley
Banaras Hindu University, Varanasi, Uttar Pradesh, India

A. Gunturi · A. Sarkar
ICAR-Indian Agricultural Research Institute, New Delhi, India

© The Author(s), under exclusive license to Springer Nature Singapore Pte Ltd. 2025

S. Babu et al. (eds.), *Ecological Solutions to Agricultural Land Degradation*,
https://doi.org/10.1007/978-981-96-3392-0_9

sizing the need for integrated sustainable practices in global land management strategies.

Keywords

Agroforestry · Biochar · Conservation agriculture · GHG · Resource use efficiency

Abbreviations

BECCS	Bioenergy with Carbon Capture and Storage
BM	Brown Manuring
CA	Conservation agriculture
CCAC	Climate and Clean Air Coalition
CCS	Carbon capture and storage
CT	Conventional tillage
DAS	Days after sowing
FAO	Food and Agriculture Organization
FYM	Farmyard manure
GHG	Greenhouse Gases
Gt	Gigaton (billion metric tons or 10^{15} grams or a Tonne)
IEA	International Energy Agency
IFS	Integrated farming systems
IPCC	Intergovernmental Panel on Climate Change
IUCN	International Union for Conservation of Nature
IUCN	International Union for Conservation of Nature and Natural Resources
LDN	Land degradation neutrality
Mg	Mega grams (one million grams or 1000 kilograms or a Tonne)
MT	Minimum tillage
NBMMP	National Biogas and Manure Management Programme
NEA	Nuclear Energy Agency
OECD	Organization for Economic Co-operation and Development
OF	Organic farming
PAU	Panjab Agriculture University
Pg	Peta gram (billion metric tons or 10^{15} grams or a Tonne)
PTR	Puddled transplanted rice
RUE	Resource use efficiency
SDG	Sustainable development goal
SHC	Soil Health Card
SOC	Soil organic carbon
SOM	Soil organic matter
TOC	Total organic carbon
UN	United Nations
UNCCD	United Nations Convention to Combat Desertification
ZT	Zero Tillage

9.1 Introduction

Land degradation, is refers to the deterioration in land quality due to anthropological activities, resulting in poor ecological health, biological productivity, and human well-being. Over 3.2 billion people's livelihood security is directly impacted by the ~6 billion hectares of degraded land that exist worldwide. Globally, land degradation is estimated to have a moderate to severe impact on 52% of agricultural land. Furthermore, the majority of degraded lands worldwide (40%) are found in rural, high-poverty areas. By 2030, there will be a 50%, 45%, and 30% growth in the world's food, energy, and water demands, respectively. Meeting these needs sustainably hinges on conserving and restoring land productivity (UNCCD 2021). The importance of Land Degradation Neutrality (LDN) is underscored by projections that by 2050, global crop yields could reduce by 10% due to land degradation and changing climate, with some regions facing up to a 50% reduction (UNCCD 2021). Additionally, land degradation is forecasted to drive a 30% increase in world food prices over the next 25 years. Given the projected increase in the world's population and food consumption by 2050, sustainable management, conservation, and restoration of land resources are critical for transitioning to sustainable food production, necessitating a minimum 75% reduction in current yield gaps.

Land degradation is primarily driven by unsustainable agricultural practices, deforestation, desertification, soil erosion, waterlogging, salinization, and mining activities. These activities contribute to GHG emissions and reduce carbon uptake, exacerbating climate change impacts. One-third of all greenhouse gas emissions worldwide come from forestry, agriculture, and other land uses (IPCC 2007). Since the Industrial Revolution, there has been a notable increase in the concentration of the three main greenhouse gases: carbon dioxide (CO_2), methane (CH_2), and nitrous oxide (N_2O), driving climate change concerns.

Economically, global land degradation costs an estimated US$85.8 trillion annually, with ecosystem services losses ~ US$6 trillion per year, equivalent to over 10% of the global gross product (Chasek 2021). Hence, sustainable agricultural practices like diversifying crop varieties, using compatible ground cover, managing crop residues, minimizing pesticide use, and adopting minimum tillage or no-till practices must be promoted at larger scale (Yadav et al. 2020a, 2021a; Das et al. 2022a). These practices enhance soil health and climate change resilience, essential for ensuring food, water, and energy balance (Babu et al. 2023a, b; Yadav et al. 2024). Preserving soil as a vital carbon sink is pivotal in lessening the effects of climate change (Wani et al. 2023a).

Implementing effective soil-conservation and land-use practices is imperative to enhance soil's role in carbon sequestration (Rathore et al. 2022; Yadav et al. 2021b). Practices like minimum tillage, crop rotation, and the use of organic amendments (e.g., farmyard manure, biochar, compost, green manuring) enhance soil quality, increase soil organic carbon (SOC) stocks, and boost agricultural productivity (Singh et al. 2021a; Babu et al. 2020b; Raj et al. 2022). The SOC improves soil structure, water retention, nutrient availability, and microbial activity, among other soil and ecological activities (Bengtsson et al. 2019; Yadav et al. 2019; Das et al. 2020). Thus, maintaining global food security, reducing the effects of climate change, strengthening soil health, and increasing agricultural output all depend on the management of SOC (Avasthe et al. 2023; Das et al. 2022b).

9.2 Concept of Land Degradation Neutrality (LDN)

The United Nations Convention to Combat Desertification (UNCCD) defined the idea of Land Degradation Neutrality (LDN) during the 2012 Rio+20 meeting. Land degradation neutrality, or LDN, is described as "a state whereby the quantity and quality of land resources necessary to support ecosystem functions and services and enhance food security remain stable or increase within specified temporal and spatial scales and ecosystems" by the UNCCD. The goal of LDN is to preserve or improve the quantity and quality of land resources in order to sustain ecosystem functioning, increase food security, and strengthen resistance to the effects of climate change (Kust et al. 2018). It calls for striking a balance so that restoration initiatives counteract the rate of land degradation and ensuring no net loss of productive land. To achieve the LDN targets prevent the further degradation of healthy land, reduce the existing degraded land through sustainable management practices and restoration of the degraded land.

The UNCCD outlines five objectives for LDN: (a) sustaining ecosystem services, (b) enhancing land productivity for food security, (c) increasing land resilience, (d) integrating with broader social and economic goals, and (e) promoting inclusive land governance. In India, LDN targets aim to reclaim 26 Mha of degraded land by 2030, focusing on both forested and non-forest areas impacted by deforestation, intensive agriculture, soil erosion, and wetland depletion. This initiative not only aims to restore land productivity but also plays a pivotal role in carbon sequestration efforts. By restoring degraded lands, LDN contributes to the absorption of carbon dioxide from the environment, thereby mitigating impact of changing climate. The UNCCD underscores the importance of 19 principles for effective implementation of LDN. These principles advocate for the preservation of natural capital, uphold human rights and sovereignty, and promote integrated land-use planning that balances social, economic, and ecological sustainability. Local knowledge and participatory approaches are crucial components in ensuring the success of LDN initiatives. They foster inclusive governance and support sustainable land management practices, which are essential for achieving the dual goals of land restoration and climate change mitigation (UNCCD 2023). Thus, LDN not only addresses the urgent need to restore degraded lands but also aligns with global efforts to sequester carbon, combat land degradation, and build resilience against climate change impacts.

9.3 Role of Soil Carbon in Land Degradation Neutrality and Climate Change Mitigation

Climate change significantly impacts managed systems like forestry, pastures, and crops, influencing the atmospheric CO_2 levels and the worldwide terrestrial carbon sink (Betts et al. 2007). The vulnerability of organic carbon reservoirs to these changes highlights the critical role of sustainable land management and climate mitigation (Wani et al. 2022). In accordance to the IPCC's Third Assessment Report (TAR), land-use decisions and practices such as afforestation-reforestation and agricultural techniques like nitrogen fertilization and tillage affect carbon stocks and fluxes. While elevated CO_2 levels initially promote soil organic carbon storage, nutrient limitations may constrain this benefit (Allard et al. 2008; van Groenigen et al. 2006). However, uncertainties persist regarding the impact of extreme events like European heatwave in 2003, which caused substantial soil carbon losses (Ciais et al. 2008). Additionally, tropospheric ozone pollution can impede carbon sequestration by altering plant productivity and litter composition. Modelling studies indicate diverse future trends in carbon storage across managed lands, underscoring the need for integrated approaches that align carbon sequestration with emissions reduction and land-use policies for sustainable outcomes (Felzer et al. 2005; Rosenzweig and Tubiello 2007).

9.3.1 Status and Threats to Soil Organic Carbon Reservoir

Soil organic carbon (SOC) is pivotal for meeting global needs like LDN, climate change mitigation, water resource enhancement, biodiversity conservation, and poverty reduction, while also supporting new industries and economic growth (Lal et al. 2007). SOC plays a significant role in storing carbon, thereby improving soil health and ecosystem functions essential for food production and climate resilience. Soil organic matter (SOM) contains ~50–58% SOC, which enhances soil fertility and nutrient availability crucial for agricultural productivity (Pribyl 2010). Globally, the soil carbon pool (Fig. 9.1), predominantly composed of SOC, is estimated at 1550 Gt, surpassing both the biotic and atmospheric carbon pools (Batjes 1996). The distribution of SOC varies significantly across climatic zones, with northern latitudes holding the highest densities. Tropical regions, such as the Amazon and Congo basins, also store substantial amounts of carbon due to their humid forests (Deng and Shangguan 2017). However, SOC stocks are vulnerable to land use changes and climate impacts, with significant losses observed globally, particularly in tropical and temperate regions (FAO and ITPS 2015). Land use practices like deforestation, agricultural intensification, and peatland drainage contribute to annual SOC losses ranging from 0.7 to 2.1 Gt C, exacerbating GHG emissions and global warming (Lal 2010).

Climate change further threatens SOC stability, with warmer and drier conditions projected to accelerate SOC loss through increased soil respiration (Bond-Lamberty and Thomson 2010). Studies suggest that a 1 °C temperature rise could

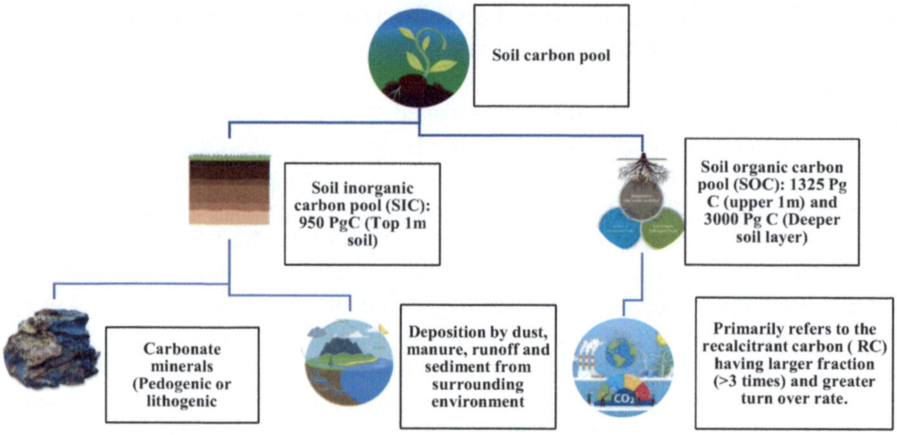

Fig. 9.1 Showing soil carbon pool status. (Modified from Ladha et al. 2016)

release over 30 Gt of soil carbon, highlighting the urgency of understanding and mitigating these impacts globally (Crowther et al. 2016). Therefore, safeguarding SOC through alternate land management practices is crucial for maintaining ecosystem-based services, mitigating climate change, and ensuring long-term agricultural sustainability (Lal 2004).

9.3.2 The Impact of Organic Carbon on Soil Health

Soil health encompasses the overall vitality and functionality of soil as a dynamic ecosystem, critical for agricultural productivity and environmental balance. Various perspectives define soil health, emphasizing its role in sustaining plants, animals, and humans. It reflects soil's multifaceted ability to cycle nutrients, maintain biodiversity, regulate water dynamics, and moderate climate (Karlen et al. 1997). This paradigm shift broadens the understanding of soil beyond fertility to include its intricate relationships with agriculture and the environment, vital for sustainable land management and environmental stewardship. Soil health involves the physical, chemical, and biological properties essential to its functionality. SOC, derived from organic matter decomposition, serves as a prime indicator of soil health (Babu et al. 2020a; Gayan et al. 2023). Continuous cultivation practices deplete SOC, contributing to soil degradation, reduced microbial diversity, nutrient deficiencies, and low productivity (Kumawat et al. 2023). Understanding factors influencing SOC dynamics such as vegetation type, altitude, and climate provides insights into soil health (Wani et al. 2022, 2023a). Altitudinal variations significantly affect SOC content by influencing temperature, moisture, and soil processes (Wani et al. 2023a; Babu et al. 2020b). SOC plays a crucial role in maintaining soil structure, nutrient retention, and water-holding capacity (Vogel et al. 2018; Yadav et al. 2019, 2020b). It acts as an insulator, stabilizing soil temperature and supporting microbial activity, thereby enhancing soil resilience to climate variability (Zhu et al. 2019). Increased SOC content improves soil porosity, infiltration rates, and reduces surface runoff and erosion (Zuazo and Pleguezuelo 2009; Raj et al. 2022). SOC also binds with heavy metals and pollutants, reducing their mobility and toxicity, thereby safeguarding ecosystem health (Sudarshan et al. 2022).

9.4 Concept of Carbon Sequestration

Carbon fixation through photosynthesis is a natural biological process that removes CO_2 from the atmosphere (Dahlman et al. 2001), converting carbon into inert substances in what is known as carbon sequestration, crucial for mitigating global warming. Given the urgent projections of potential CO_2 emissions doubling in the coming centuries, photosynthetic carbon fixation has been extensively studied for its role in long-term CO_2 sequestration (Dahlman et al. 2001). Strategies to reduce global atmospheric carbon levels include lowering GHG emissions, capturing and storing carbon for extended periods, and advancing alternative energy sources. The subsequent approach is long-term storage and carbon capture, involves collecting and preserving CO_2 using various methods (Zeng 2008). However, the relatively low concentrations of CO_2 in the atmosphere compared to nitrogen and oxygen present thermodynamic challenges for effective capture. Once captured, stored carbon must be securely deposited in enduring reservoirs, including water bodies, geological formations, chemical processes, and terrestrial structures. The adoption of carbon sequestration methods to mitigate CO_2 emissions has gained wide acceptance due to environmental concerns (Table 9.1). Due to direct and indirect contributions from human activity, the cycle of terrestrial carbon is greatly impacted,

Table 9.1 Definition of carbon sequestration by different sources

Source	Definition
IPCC (2001)	Carbon sequestration by terrestrial ecosystems is the cumulative release of carbon dioxide (CO_2) from the atmosphere or the inhibition of CO_2 releases from terrestrial environments into the atmosphere.
Izaurralde et al. (2007)	Carbon sequestration is the process of managing land in a way that increases ecosystem C storage while decreasing the level of heterotrophic respiration, increasing net primary productivity, or both.
Sundermeier et al. (2005)	Soil carbon sequestration is the process of absorbing carbon dioxide from the atmosphere into the soil by means of crop residues and other organic matter in a way that is not immediately released again. This transfer of carbon, often known as "sequestering," not only enhances soil health and long-term agronomic production but also helps balance emissions from burning fossil fuels as well as other carbon-emitting activities.
Johnson et al. (2007)	The net removal of CO_2 from the atmosphere into long-lasting carbon storage tanks is known as terrestrial carbon sequestration. These pools may consist of existing biomass (e.g., trees), long-term biomass products (e.g., lumber), persistent biomass in soils (e.g., perennial roots and microbes), or recalcitrant organic and inorganic C in soils and beneath ground conditions (West and Post 2002).

accounting for around 50% of CO_2 sequestration (Luo et al. 2015; Schmitz et al. 2014). Fossil fuel consumption is a significant contributor to anthropogenic CO_2 emissions (Holtsmark 2015), which are expected to rise with increasing global energy demands, particularly in developing nations. Elevated atmospheric CO_2 levels exacerbate the greenhouse effect, leading to global warming and environmental degradation (Manisalidis et al. 2020). Carbon capture and storage (CCS) involve the absorption and long-term CO_2 storage from exhaust emissions, crucial for mitigating these harmful effects (Raza et al. 2019; Kumar et al. 2022). Biological systems such as peatlands and seawater naturally sequester carbon (Brouns et al. 2014; Raven 2011), while autotrophic organisms convert CO_2 through photosynthesis into organic carbon (Hu et al. 2019). Intentional separation and disposal of CO_2 into non-atmospheric reservoirs are integral to carbon sequestration efforts.

9.5 Types of Carbon Sequestration

The process of removing carbon dioxide (CO_2) from the atmosphere or rerouting it from emission sources and storing it in aquatic environments, terrestrial habitats, and geological formations is known as carbon sequestration. The global carbon cycle retained a balance between CO_2 absorption and emission prior to human-caused emissions. However, current CO_2 sinks are insufficient to offset the rising emissions from human activities. There are mainly four types of carbon sequestration processes. Researchers often categorize these strategies into biological and non-biological processes, considering various criteria like sequestration techniques, involved organisms, and potential carbon sinks to account for the diversity of approaches (Table 9.2).

Non-biological sequestration Non-biological sequestration methods involve capturing atmospheric carbon through chemical and physical reactions (Lal 2008). This strategy involves chemical, oceanic, and geological sequestration techniques, each of which can be further subdivided as detailed in Table 9.2.

Biological sequestration Biological sequestration utilizes living organisms such as microbes and plants to accumulate and store carbon through natural processes. Human-designed methods often mimic these natural processes. Sustainability efforts emphasize land-based sequestration and the restoration of carbon in oceanic reservoirs. Soil carbon sequestration and phyto-sequestration are key components, vital for capturing and storing carbon to meet environmental and climate objectives.

Plants absorb CO_2, transporting 1% to soil for storage via roots, in organic and inorganic forms. Soil carbon sequestration and phyto-sequestration are linked, affected by climate, temperature, precipitation, clay, minerals, moisture, and

Table 9.2 Types of carbon sequestration and their description

Type of carbon sequestration	Description	References
Non-biological sequestration		
Oceanic sequestration	Oceanic sequestration leverages the oceans as storage reservoirs for atmospheric CO_2. Due to their vast pool of inorganic carbon, about 50 times greater than the carbon content in the atmosphere, oceans can significantly store atmospheric CO_2. Higher salt levels and elevated temperatures notably reduce CO_2 solubility in oceanic waters	Smith et al. (2016), Raven and Falkowski (1999)
Geological sequestration	Capturing CO_2 from industrial sources and injecting it underground into geological structures for long-term storage is known as geological sequestration. This method, used in petroleum-related industries, reduces GHG emissions and aids oil recovery. Geological formations like coal seams, deep saline aquifers, and oil and gas reservoirs may store CO_2. Petroleum companies have utilized this technique since the 1970s to enhance oil extraction. However, safety, stability, and public acceptance remain challenges. Saline aquifers, in particular, are effective for CO_2 storage due to their porosity and brine saturation.	Kovscek and Cakici (2005), Bergman and Winter (1995)
Chemical sequestration	Chemical sequestration techniques are non-biological methods that rely on specific chemical processes. One approach involves mineral carbonation, where rocks undergo chemical weathering to convert atmospheric CO_2 into stable carbonate minerals. This reaction occurs naturally with mineral oxides of metals present in rocks, forming carbonates that securely trap CO_2 for extended periods (Maroto-Valer et al. 2005; Lal 2008). Despite its current industrial efficiency limitations, natural mineral carbonation reactions are feasible for commercial applications.	Huijgen and Comans (2003), Maroto-Valer et al. (2005), Lal (2008)
Biological sequestration		
Terrestrial sequestration	Terrestrial sequestration involves the absorption of CO_2 by forests, grasslands, and soils through photosynthesis, storing carbon in both above-ground biomass and soil organic matter. This process, termed phyto-sequestration and soil-C sequestration, is essential for long-term carbon storage, widely adopted due to its appeal for mitigating carbon emissions. Ecosystems act as significant carbon sinks on earth, accumulating carbon through photosynthesis and storing it in living and decaying organisms. Reduces levels of CO_2 in the atmosphere, lowers emissions of greenhouse gases, and contributes to mitigating climate change.	Post et al. (2009), Lal (2008), Smith et al. (2016)

(continued)

Table 9.2 (continued)

Type of carbon sequestration	Description	References
Ocean fertilization	It is a biological method, boosts carbon sequestration by stimulating phytoplanktonic photosynthesis. However, this approach, akin to direct injection, raises concerns about potential ecological impacts, including the release of methane and nitrogen monoxide, both potent greenhouse gases. Therefore, while promising, the efficacy of ocean fertilization remains variable and contingent upon mitigating unintended ecological consequences.	Yang et al. (2008)
Bioenergy with carbon capture and storage	It integrates biomass energy generation with the captured and stored of CO_2 emitted during biomass combustion, with the goal of achieving negative emissions. While it holds promise for substantial CO_2 removal, it faces hurdles such as ensuring sustainable biomass production and sufficient storage capacity.	Keith et al. (2009)

texture. Managing ecosystems to restore soil and increase biomass helps reduce net CO_2 emissions, crucial for enhancing carbon sequestration.

9.6 Agronomic Approaches for Soil Organic Carbon Sequestration

SOC sequestration, which lowers greenhouse gas (GHG) concentrations in the atmosphere by absorbing atmospheric carbon dioxide and storing it in soil organic matter, is essential for preventing climate change. According to the Global Methane Assessment, rice production contributes significantly to anthropogenic methane emissions, accounting for about 8% worldwide and nearly 40% of total emissions (Gupta 2021). Additionally, paddy rice fields contribute 9–11% of agricultural GHG emissions, with 11% of global agricultural N_2O emissions originating from these fields (IPCC 2014). Effective agronomic practices are crucial for enhancing SOC sequestration, described here under (Fig. 9.2).

9.6.1 Zero-Till Farming

Globally, soils hold 3.5% of total carbon, over 1.7% in the atmosphere and 1% in biota. Unsustainable practices like frequent tilling and residue removal deplete SOC in croplands, threatening food security and environmental quality (Yadav et al. 2021a; Ngangom et al. 2020). The global SOC loss is projected at 21.08 ± 0.91 g C/m^2 per year (Lal 2023). SOC below 1.5–2.0% in the root zone drives soil degradation and reduces ecosystem capacity (Babu et al. 2023b). Hence, effective SOC management is essential for restoring soil health and productivity. Zero tillage (ZT) has emerged as a crucial strategy in modern agriculture to combat soil degradation,

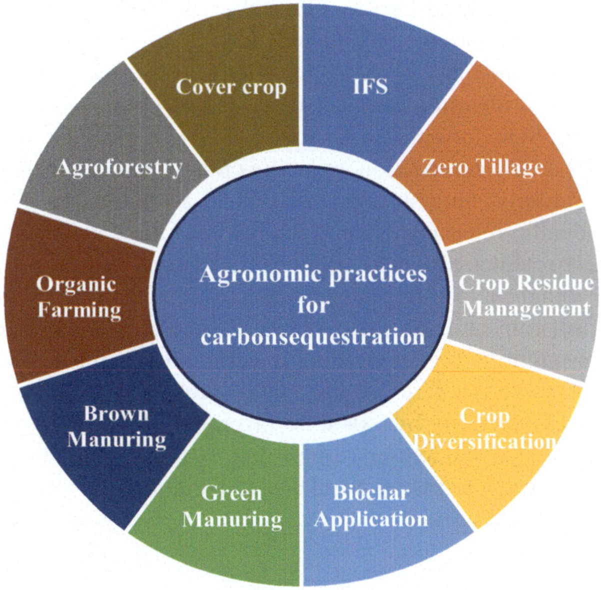

Fig. 9.2 Major Agronomic approaches for sequestering SOC

carbon depletion, and inefficiencies in resource use such as water, nutrients, and energy (Das et al. 2013; Das et al. 2018; Yadav et al. 2021a, b). Defined by the FAO as a method to conserve natural resources through minimal soil disturbance, permanent organic soil cover, and crop diversification, under an umbrella term conservation agriculture (CA) has seen significant global adoption. The adoption of CA practices, particularly conservation tillage and no-till systems, enhances soil physical properties like aggregation and reduces bulk density, while also promoting SOC sequestration. Globally, conservation tillage practices alone could sequester carbon at rates ranging from 0.4 to 0.8 petagrams (Pg) of carbon per year (Babu et al. 2023c). Transitioning from conventional to no-till systems could further increase SOC by 0.1–1.0 megagrams (Mg) of carbon per hectare annually (Lal 2023). In rice-wheat-mungbean cropping systems with residue management, CA practices can save irrigation water by 25–30%. Beyond soil benefits, CA reduces reliance on energy-intensive tillage, thereby lowering GHG emissions associated with fossil fuel use which ultimately helps in carbon sequestration indirectly (Ansari et al. 2023; Yadav et al. 2020).

9.6.2 Retention/ Incorporation of Crop Residue

Crop residue retention is a pivotal strategy in CA aimed at preserving leftover crop components rather than tilling them into the soil (Rusinamhodzi 2015). This practice significantly enhances soil carbon levels, crucial for improving soil structure, nutrient retention, and overall fertility (Turmel et al. 2015; Yadav et al. 2020b). It also encourages the adoption of complementary conservation practices such as cover cropping and mulching. Cover crops contribute organic matter through biomass, with their roots releasing exudates that soil microbes use as a carbon source, thereby promoting SOC formation (Austin et al. 2017). The root systems of cover crops aid in soil particle compaction, reducing erosion and enhancing organic matter retention. Similarly, mulching supports SOC accumulation as the materials decompose gradually, releasing organic carbon into the soil (Zhang et al. 2022; Singh et al. 2021b). Mulch also helps retain soil moisture, fostering microbial activity that further enhances organic matter breakdown and SOC levels (Shen et al. 2016). In contrast, conventional tillage practices accelerate organic matter decomposition by exposing it to oxygen, leading to carbon losses through CO_2 emissions (Reddy et al. 2021). In a ZT rice-maize system, retaining crop residue as mulch increased system productivity by 3.0–9.3% (average 6.4%) over 5 years compared to systems without residue retention (Singh et al. 2016). Minimizing or eliminating tillage slows down organic matter breakdown, facilitating greater SOC retention besides reducing the soil erosion.

9.6.3 Diversification of Cropping System Through Bioenergy Crop

Diversification of cropping systems, particularly through crop rotation and inclusion of bioenergy crops, is a viable strategy to mitigate the uncertainties posed by climate change in agriculture (Rathore et al. 2022). This approach involves expanding crop varieties within rotations or planting annual and perennial crops in spaces between woody crops, enhancing soil carbon sequestration (Yang et al. 2021; Almagro et al. 2023). Diverse crop rotations, incorporating species with varied root structures and nutrient requirements, contribute to SOC levels and overall soil health. Integrating green manures into these rotations further boosts SOC by continuously supplying organic matter (Sharma et al. 2017). Bioenergy crops, such as sugarcane, oil crops, and maize, are critical in this diversification strategy due to their renewable energy potential and significant biomass accumulation. In developing economies, bioenergy crops rank as the fourth-largest electricity source, illustrating their substantial role in energy production (Jansson et al. 2010). These crops not only help reduce atmospheric CO_2 levels by potentially replacing fossil fuels but also enhance C sequestration through root and soil absorption of carbon (Bransby et al. 1998). However, their effectiveness and profitability depend on factors like production rates, fertilization, and harvest management (Walia et al. 2022).

9.6.4 Biochar for Carbon Sequestration

Biochar application has garnered attention due to its potential to improve soil health, improve fertility, immobilize nutrients, and store carbon (Babu et al. 2023c).

Biochar, a carbon-rich, porous material derived from plant biomass through pyrolysis, is stable and resistant to decomposition, thus sequestering carbon in soil for several thousands of years (Terrestrial Carbon Sequestration Report, 2011). This process not only boosts SOC levels but also supports microbial activity by providing a stable substrate. Biochar positively impacts soil physicochemical characteristics by lowering bulk density, decreasing nutrient leaching, and improving water storage and cation exchange capacities (Wani et al. 2023b; Babu et al. 2023c). In Punjab, India, biochar has proven effective for rice farmers, enhancing crop yields by up to 10% when applied at 5 $Mgha^{-1}$ and contributing to nitrogen savings (Gupta et al. 2020). This demonstrates biochar's dual role in carbon sequestration and sustainable agriculture. By converting biomass into biochar, significant carbon can be captured in soils, mitigating global warming and improving soil health (Post et al. 2009). Despite biochar's promising benefits, its impact on SOC accumulation varies with soil type, climate, and crop management. Biochar production methods, such as pyrolysis, gasification, and hydrothermal carbonization, yield by products like bio-oil and syngas while capturing carbon that would otherwise be released into the atmosphere.

9.6.5 Green Manuring

Green-manure crops, grown either in-situ or ex-situ and incorporated into the soil, significantly enhance soil physio-chemical, and biological properties.

They protect against soil from erosion by mitigating raindrop impact and reducing surface runoff velocity, thereby preserving soil organic matter. Decomposition of green manure releases organic products like polysaccharides, which act as binding agents, forming large, stable aggregates that improve soil structure (Zhang et al. 2023). This aggregation process reduces soil bulk density, increases porosity, and enhances water-retention capacity. In a meta-analysis, Ma et al. (2019) concluded that green manure reduces bulk density by ~5.6%, increases microbial biomass carbon (MBC) by around 28%, and boosts soil enzyme activity by 14–39%. The vegetative cover from green manure also enhances soil infiltration rates and moisture retention. Additionally, these crops, through their biological nitrogen fixation (BNF) adds significant nitrogen to the soil, along with other essential nutrients (Table 9.3). Green manure aids in SOC sequestration through four main mechanisms (Fig. 9.3).

9.6.6 Brown Manuring

The "no-till" version of green manuring known as "brown manuring" (BM) uses a selective herbicide (2,4-D) to kill legume plants before they blossom instead of cultivating them. Current BM technique used for rice entails growing *Sesbania bispinosa* alongside the crop as a co-culture for the first 25–30 days after sowing (DAS) and then knocking it down with a selective herbicide 2,4-D, which supplied up to 35 $kgha^{-1}$ nitrogen (Maitra and Zaman 2017). When chlorophyll is broken

Table 9.3 Potential of green manuring in soil organic carbon sequestration in various cropping system

Green manure on SOC sequestration	References
Green manure-rice rotation increased soil organic carbon stocks by 1.62 mg CO_2eq per hectare annually, surpassing the increase observed in fallow-rice rotation.	Gao et al. (2013)
Long-term use of ryegrass/milk vetch green manure in rice-based systems increased the soil carbon pool index while decreasing the labile carbon pool and carbon management index.	Hou et al. (2022)
Green manuring, particularly with *kharif* cowpea and Amaranthus panicum, enhances soil organic carbon sequestration in vertisols	Tamboli et al. (2011)
Legume green manuring enhances soil organic carbon (SOC) stocks by 15–23% over fallow	Yao et al. (2019)
PTR (puddled transplanted rice) system with 25% wheat residue and green manuring before rice, followed by ZT wheat with 100% rice residue as mulch maximize the SOC sequestration over conventional practice.	Sharma et al. (2021)
Application of wheat residue (6 mg ha^{-1}) and green manure (20 mg ha^{-1}) during rice cultivation, along with nitrogen fertilizer (120 kg N ha^{-1}) significantly increased SOC concentration, quantity, and sequestration rate over wheat cultivation without rice residue	Aulakh (2001)
SOC sequestration potential of green manuring is higher in 0–20 cm than sub surface soil.	Ma et al. (2019)
In eastern India's rice-wheat systems, FYM sequestered 26%, paddy straw 18%, and green manure 8% of carbon, highlighting varying effectiveness in soil organic carbon sequestration.	Ghosh et al. (2018)

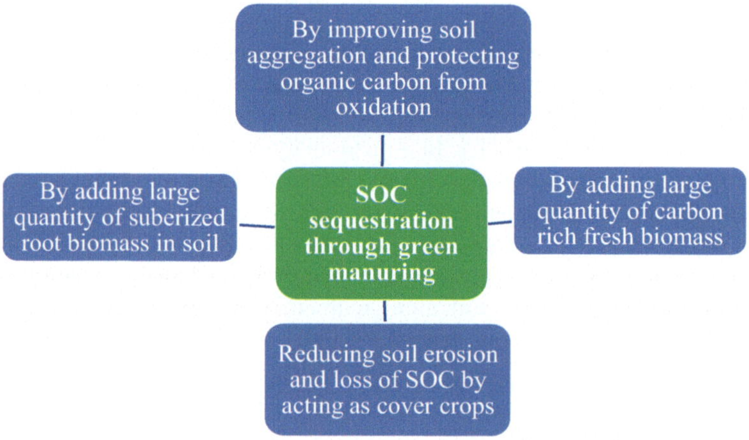

Fig. 9.3 Principal mechanism of SOC sequestration by green manuring

down, *Sesbania* leaves and debris turn dark brown or yellow, which is why they are called BM. In order to contribute organic manure, the dead *Sesbania* leftovers are left lying in the field alongside the crop of choice until they break down naturally. Essentially, *Sesbania* provides interference as a living cover during the pre-killing time and stimulation as a dead material for mulch during the post-killing era. BM lowers pesticide dosage and soil residue while also reducing weed interference through allelopathy or smothering effects (Oyeogbe et al. 2017). In agricultural production, combining organic and green manures can provide nutrients and organic carbon to increase crop productivity and preserve or enhance soil health (Barrios 2007). Many areas have used green manuring with *Sesbania* spp., with differing degrees of success (Sharma and Ghosh 2000). Crop residues, such as rice straw, offer benefits like temperature regulation and nutrient delivery but face competing uses like cattle feed (Ladha et al. 2016). Alternative biomass sources, such as perennial leguminous tree leaves and twigs, are viable mulching options, though costly (Mulvaney et al. 2017; Yadav et al. 2018). Legume co-culture, or BM, produces nutrient-rich in-situ biomass, enhancing soil health and SOC (Lal 2016; Deng and Shangguan 2017). BM improves crop biomass production, with mulched plots showing higher SOC pools (29.7–29.8 $Mgha^{-1}$ vs. 29.0 $Mgha^{-1}$), sequestration rates (391–428 kg ha per year vs. 221 kg ha per year), and carbon retention efficiency (6.64–6.94% vs. 4.66%) compared to non-mulched plots (Yadav et al. 2019). Using cowpea as a cover crop doubled the carbon sequestration rate to 478 kg C ha per year. In India, line planting *Sesbania* with rice increased SOC above the baseline by 0.28% (Pradhan et al. 2020). BM stabilizes soil carbon and supports beneficial microbial communities, enhancing SOC through the decomposition of organic materials like sesbania leaves and roots (Keerthi et al. 2022).

9.6.7 Cover Crop and Organic Mulching

In conventional farming, raindrops on bare soil disrupt aggregates, block pores, and reduce water infiltration, leading to erosion and runoff (Yadav et al. 2023, 2024). Cover crops like cowpea, and sweet potato help mitigate these issues by managing nutrients and reducing soil erosion (Yadav et al. 2019, 2024). Leguminous cover crops provide nitrogen for subsequent crops, minimizing fertilizer needs (Singh et al. 2021b), while grasses reduce nitrogen leaching and loss. A combination of legumes and grasses offers both benefits (Ranells and Wagger 1997). These crops improve soil infiltration, intercept rain, and prevent erosion, while also retaining soil moisture and adding organic matter when decomposed, enhancing soil quality. Cover crops form stable soil aggregates, enhance water absorption, resist erosion, reduce nutrient runoff, lower water pollution, and reduce the farm's carbon footprint (Yadav et al. 2021b, 2023; Singh et al. 2021a). Mulching, covering topsoil with plant materials, further enhances soil structure by promoting organisms like earthworms, which create pores for better water infiltration and reduced surface runoff. As mulch decomposes, it increases, stabilizes soil structure, and prevents erosion. Organic mulches decompose over time, enhancing soil water retention and nutrient content (Ranjan et al. 2017), and improving water use efficiency by suppressing weed growth (Ossom et al. 2001). Organic mulches significantly increase SOC by providing steady organic inputs while reducing mineralization rates (Saroa and Lal 2003; Zahed et al. 2022). Enhancing SOC in soils is essential for reducing atmospheric CO_2 levels, improving soil quality and productivity, and mitigating CO_2 emissions.

9.6.8 Agroforestry Measures and High Biomass Farming

High biomass farming involves cultivating fast-growing, high-yield plant species alongside trees to maximize carbon capture and storage, enhancing soil carbon sequestration and overall ecosystem productivity. Agroforestry, integrates agricultural and forestry practices to create sustainable and carbon-efficient land-use systems. Through afforestation and reforestation efforts, agroforestry effectively sequesters carbon, mitigates emissions, and enhances resilience to climate impacts. Unlike natural forests, agroforestry systems offer diverse benefits such as erosion control, improved soil fertility, water management, and biodiversity enhancement, thereby reducing pressure on natural forests for resources. Studies suggest that agroforestry systems have greater potential to sequester carbon compared to traditional field crops or pastures, owing to increased aboveground and belowground carbon storage facilitated by trees (Haile et al. 2008; Yadav et al. 2021c).The effectiveness of agroforestry in carbon sequestration varies with the type of system deployed. Agro-silvicultural systems, combining crops and trees, act as net sinks of GHGs emission, whereas agro-silvopastoral systems may be GHG sources due to management practices like tillage and fertilization (Kandji et al. 2006). IPCC (2007) identifies agroforestry as crucial for synergizing adaptation and mitigation efforts, with potential carbon sequestration ranging from 1.1 to 2.2 Pg C in terrestrial ecosystems over the next 50 years. The carbon storage capacity of agroforestry systems is influenced by socio-economic factors and environmental conditions (Table 9.4). In humid tropics, these systems can sequester over 70 $Mgha^{-1}$ of C in the top 20 cm of soil, varying across species and geographical regions (Mutuo et al. 2005; Newaj and Dhyani 2008). Tree species characteristics, such as deciduous or evergreen nature and growth rates, also impact carbon sequestration rates. Specific agroforestry systems, such as silvo-pastoral in Chile and block plantations in India, demonstrate varying sequestration potentials, emphasizing the role of system design and management in optimizing carbon storage (Dube et al. 2011; Chauhan et al. 2015; Romero et al. 1991).

9.6.9 Organic Farming

Organic farming plays a vital role in carbon sequestration, mitigating climate change and enhance soil health. The practice of organic farming can significantly impact SOC stocks, particularly in the context of climate change. Increased soil carbon turnover and soil organic matter mineralization is typically correlated with rising temperatures and consistent precipitation patterns. To maintain or increase SOC levels, organic farming practices emphasize the supplying of organic matter, through crop residues, cover crops, and pasture fodder, while minimizing soil cultivation to reduce mineralization (Chenu et al. 2019; Spiegel et al. 2018). Organic farming stands out due to its low-input, ecologically sustainable methods that ensure food availability while promoting soil carbon sequestration. Long-term studies reveal that applying farmyard manure (FYM) significantly enhances soil fertility

Table 9.4 Carbon sequestration potential of different agroforestry practices in different regions

Agroforestry	Location	Tree sp.	Age (Years)	C sequestration rate (Mg Cha^{-1} $year^{-1}$)	Source
Silvi-pasture	India	*Albizia Amara* + *Dichrostachys cinerea* + *Leucaena leucocephala*	8	6.72	NRCAF (2007)
	Chile	*Pinus ponderosa*	18	12.44	Dube et al. (2011)
	Costa Rica	*Acacia mangium* + *Arachispintoi*	10–16	10.81	Dube et al. (2011)
Agri-silviculture	India	*Alnus nepaliensis* + wheat, mustard, lentil in winter season; *Alnus nepaliensis* + pigeon pea, cow pea in rainy season	5	0.256	Maikhuri et al. (2000)
	Brazil	*Theobroma cacao* L.	30	3.12	
Block plantation	India	*Eucalyptus tereticornis*	3.5	4.40	Chauhan et al. (2015)
		Populus deltoides	6	9.24	Chauhan et al. (2015)
Farm forestry	India	*Pinus* spp.	30	7.1	Hooda et al. (2007)
Alley cropping	North America	*Populus deltoides*	15	6.63	Peichl et al. (2006)
Home garden	India	Mixed tree species	71	1.60	Saha et al. (2009)
	India	Bamboo		1.46	Nath and Das (2011)
Boundary plantation	India	*Populus deltoides*	6	5.45	Chauhan et al. (2015)
Multistrata agroforestry	Ethiopia	Mixed tree species	32–54	2.17–3.66	Negash et al. (2022)

Table 9.5 Carbon sequestration potential (kg C ha^{-1} $year^{-1}$) of different organic sources

Organic farming practices	Carbon sequestration potential (kg C $ha^{-1}year^{-1}$)	References
Organic farming practices (average)	287 ± 102	Autret et al. (2019), Gattinger et al. (2012)
Cattle manure	160	Blanco-Canqui et al. (2020)
Green manure (alfalfa)	180	Blanco-Canqui et al. (2020)
Farmyard manure	292 ± 132	Bolinder et al. (2020)
Solid manure (temperate climate) 5–10 t ha^{-1} y^{-1}	160	Don et al. (2024)
Compost (average) (a) 8 t ha^{-1} y^{-1} (b) 14 t ha^{-1} y^{-1} (c) 20 t ha^{-1} y^{-1}	714 ± 404 (a) 115 (b) 558 (c) 1021	Minasny et al. (2017), Poulton et al. (2018)

and carbon sequestration across various soils and climates (Li et al. 1994). Application of manure has been demonstrated to boost soluble carbon concentrations and reduce carbon turnover rates compared to control or fertilized plots (Gregorich et al. 1998; Yadav et al. 2013). Additionally, FYM promotes the development and maintenance of soil macroaggregates (Whalen and Chang 2002). Research by Kukal and Benbi (2009) demonstrated that FYM plots in rice-wheat and maize-wheat systems had higher SOC sequestration rates than those with NPK fertilizers. The rice-wheat system, in particular, showed a threefold higher sequestration rate due to prolonged soil submersion and greater biomass production. Parker et al. (2002), also observed higher SOC content with poultry litter in a cotton-rye system. Organic farming enhances SOC stocks through the use of organic fertilizers, crop diversification with legumes, and the establishment of cover crops (Babu et al. 2020b; Babu et al. 2023a). These practices not only sequester carbon but also boosts long-term sustainability of soil health, making organic farming a vital strategy in the fight against climate change (Table 9.5).

9.6.10 Integrated Farming System (IFS) for Carbon Sequestration

Integrated farming systems (IFS) combine practices like crop production, livestock rearing, and agroforestry to enhance productivity and environmental health, significantly contributing to carbon sequestration. These systems maximize resource use efficiency (RUE) and minimize waste; for instance, crop residues feed livestock, and livestock manure enriches soil, reducing synthetic fertilizer need (Pretty et al. 2018). IFS sequester more carbon than monoculture systems, effectively reducing greenhouse gas emissions. The diverse species in IFS increase biodiversity, soil health, and climate resilience (Dury et al. 2013). Diverse farm enterprises like crops and livestock synergistically enhance food production, livelihoods, and ecosystem services (Fatima et al. 2023). This holistic approach balances productivity and sustainability by linking soil, water, food, and energy. IFS with efficient soil and water practices minimize land degradation and agriculture waste, sustaining higher productivity and profitability (Saha et al., 2022). Mixed production systems retain resources economically, ensuring livelihood and nutritional security while reducing environmental pressure (Garnett et al. 2013). IFS significantly increase productivity and improves farm-waste management. Thus, IFS reduce carbon emissions and increases production, thereby preventing climate change.

9.7 Soil Carbon Dynamics and GHG Fluxes: Implications for Land Degradation Neutrality

Enhancing SOC through agricultural practices plays a pivotal role in LDN goals by promoting sustainable land management. Initially, SOC levels can rise with agricultural modifications, but this accumulation stabilizes as SOC reaches a balance between generation and respiration (Chenu et al. 2019). This equilibrium marks a critical phase in assessing the effectiveness of these practices in mitigating climate change under the LDN framework. For instance, reducing tillage can significantly increase SOC reserves, especially in arid regions. Compared to traditional plowing, these methods can sequester about 3 mgha-1C, or 11 mg CO_2 ha-1, in the topsoil during a 30-year period (Lugato et al. 2018). This results in an annual storage of 0.1 Mg C ha^{-1}, contributing positively to LDN targets by enhancing soil fertility and resilience. However, it's essential to consider the trade-offs involved, particularly regarding GHG emissions such as N_2O. Reduced tillage techniques could result in 0.2 mg CO_2eqha-1 more annual N2O emissions (Guenet et al. 2021). With time, this creates a balance between increased CO_2 absorption and additional N_2O emissions, influencing the overall climate mitigation impact within the LDN framework. While SOC sequestration stabilizes once equilibrium is reached, ongoing N_2O emissions underscore the complexity of achieving long-term climate benefits. Consequently, in order to ensure efficient land use strategies which, promote both soil health as well as global sustainability goals, evaluating agricultural techniques within LDN necessitates a careful evaluation of temporal dynamics and potential environmental implications.

9.8 Policy Interventions

Government policies play a pivotal role in advancing carbon sequestration efforts towards achieving LDN. India has committed to significant targets under global initiatives like the Bonn Challenge, aiming to regenerate 26 million hectares of land that have been degraded by 2030. This commitment, raised from an initial target of 21 million hectares, underscores India's dedication to environmental neutrality and ecosystem restoration (IUCN 2021). Supporting these goals, India's Ministry of Agriculture& Farmers Welfare launched the Soil Health Card (SHC) scheme to provide farmers with tailored soil health information and recommendations, promoting sustainable soil management practices. Additionally, initiatives like the National Biogas and Manure Management Programme (NBMMP) promote the use of biogas plants to convert organic waste into bio-manure, enhancing soil fertility and crop yields (MNRE 2018). Policies on customized fertilizers and nutrient management further emphasize sustainable agricultural practices, aiming to optimize fertilizer use based on soil health assessments (Chaudhari 2018). Through these integrated efforts and commitments under global agreements like the Paris Climate Change Agreement, India aims to enhance carbon sequestration, mitigate GHG emissions, and achieve sustainable land management practices conducive to LDN. In South Asia, India holds a prominent position in agriculture, yet SOC sequestration efforts lag behind international concerns. Farmers lack awareness of soil health's crucial role in food security, nutrition, sustainability, and economic stability. Challenges include financial constraints, inadequate knowledge dissemination, and subsidies favouring carbon-depleting practices. Government policies predominantly support large-scale agribusiness, neglecting rural and small-scale farmers who lack demonstrations of successful soil-C practices. To address these issues, India must promote sustainable soil-C sequestration through awareness campaigns, resource mobilization, and policy reforms. Policy reforms should redirect subsidies towards carbon-rich strategies, agroforestry, and no-till practices, known to enhance SOC. Financial incentives and educational support for farmers are crucial for fostering sustainable agricultural practices and securing India's agricultural future (OECD/IEA/NEA 2015).

9.9 Conclusion

In order to address the global issues like food security, desertification, and climate change, soil carbon sequestration is essential, intricately linked to preserving SOC. Agricultural lands face degradation from factors such as erosion, intensive cultivation, and poor management, necessitating sustainable restoration strategies. Agronomic techniques like mulching, conservation agriculture, and cover cropping offer effective pathways to rejuvenate degraded soils economically and sustainably. Regenerative farming practices tailored to local agroecosystems, such as agroforestry and reduced tillage, play crucial roles in enhancing SOC levels and mitigating land degradation. Policy reforms are essential to promote widespread adoption of these strategies and reduce carbon footprints across diverse agricultural landscapes. These reforms serve as a roadmap for achieving land degradation neutrality and underline India's commitment to sustainable land management and climate resilience. By integrating scientific understanding with practical solutions and aligning policies. This comprehensive approach underscores the urgency of collaborative efforts and policy alignment to ensure the resilience and sustainability of agricultural systems worldwide.

References

Allard V, Ourcival JM, Rambal S, Joffre R, Rocheteau A (2008) Seasonal and annual variation of carbon exchange in an evergreen Mediterranean forest in southern France. Glob Chang Biol 14(4):714–725

Almagro M, Re P, Díaz-Pereira E, Boix-Fayos C, Sánchez-Navarro V, Zornoza R, Martínez-Mena M (2023) Crop diversification effects on soil organic carbon and nitrogen storage and stabilization is mediated by soil management practices in semiarid woody crops. Soil Tillage Res 233:105815

Ansari MA, Ravisankar N, Ansari MH, Babu S, Layek J, Panwar AS (2023) Integrating conservation agriculture with intensive crop diversification in the maize-based organic system: impact on sustaining food and nutritional security. Front Nutr 10:1137247. https://doi.org/10.3389/fnut.2023.1137247

Aulakh MS (2001) Impacts of integrated management of crop residues, fertilizer N and green manure on productivity, nitrate leaching, carbon sequestration, denitrification and N2O emissions in rice-wheat system. In: Proceedings of 12th world fertilizer congress, Beijing, China

Austin EE, Wickings K, McDaniel MD, Robertson GP, Grandy AS (2017) Cover crop root contributions to soil carbon in a no-till corn bioenergy cropping system. GCB Bioenergy 9(7):1252–1263

Autret B, Beaudoin N, Rakotovololona L, Bertrand M, Grandeau G, Gréhan E et al (2019) Can alternative cropping systems mitigate nitrogen losses and improve GHG balance? Results from a 19-yr experiment in northern France. Geoderma 342:20–33

Avasthe RK, Devi SH, Bhupenchandra I, Amit K, Chongtham SK, Babu S, Raghavendra S, Anup D, Gudade BA, Bora SS (2023) Agro-tactics for reducing carbon footprint in agricultural production systems: a review. Indian J Agron 68(2):115–125. https://doi.org/10.59797/ija.v68i2.332

Babu S, Mohapatra KP, Yadav GS, Lal R, Singh R, Avasthe RK, Das A, Chandra P, Gudade BA, Kumar A (2020a) Soil carbon dynamics in diverse organic land use systems in North Eastern Himalayan ecosystem of India. Catena. https://doi.org/10.1016/j.catena.2020.104785

Babu S, Singh R, Avasthe RK, Yadav GS, Mohapatra KP, Selvan T, Das A, Singh VK, Valente D, Petrosillo I (2020b) Soil carbon dynamics in Indian Himalayan intensified organic rice-based cropping sequences. Ecol Indic. https://doi.org/10.1016/j.ecolind.2020.106292

Babu S, Singh R, Avasthe RK, Kumar S, Rathore SS, Singh VK, Ansari MA, Valente D, Petrosillo I (2023a) Soil carbon dynamics under organic farming: impact of tillage and cropping diversity. Ecol Indic. https://doi.org/10.1016/j.ecolind.2023.109940

Babu S, Singh R, Avasthe R, Rathore SS, Kumar S, Das A, Layek J, Sharma V, Wani OA, Singh VK (2023b) Conservation tillage and diversified cropping enhance system productivity and eco-efficiency and reduce greenhouse gas intensity in organic farming. Front Sustain Food Syst 7:1114617. https://doi.org/10.3389/fsufs.2023.1114617

Babu S, Singh R, Kumar S, Rathore SS, Yadav D, Yadav SK, Yadav V, Ansari MA, Das A, Rajanna GA, Wani OA, Raj R, Yadav DK, Singh VK (2023c) Biochar implications in cleaner agricultural production and environmental sustainability. Environ Sci Adv. https://doi.org/10.1039/D2VA00324D

Barrios E (2007) Soil biota, ecosystem services and land productivity. Ecol Econ 64(2):269–285

Batjes NH (1996) Total carbon and nitrogen in the soils of the world. Eur J Soil Sci 47(2):151–163

Bengtsson A, Bengtsson J, Sedin M, Sjöholm E (2019) Carbon fibers from lignin-cellulose precursors: effect of stabilization conditions. ACS Sustain Chem Eng 7(9):8440–8448

Bergman PD, Winter EM (1995) Disposal of carbon dioxide in aquifers in the US. Energy Convers Manag 36(6–9):523–526

Betts RA, Boucher O, Collins M, Cox PM, Falloon PD, Gedney N et al (2007) Projected increase in continental runoff due to plant responses to increasing carbon dioxide. Nature 448(7157):1037–1041

Blanco-Canqui H, Kaiser M, Hergert GW, Creech CF, Nielsen R, Maharjan B et al (2020) Can char carbon enhance soil properties and crop yields in low-carbon soils? J Environ Qual 49(5):1251–1263

Bolinder MA, Crotty F, Elsen A, Frac M, Kismányoky T, Lipiec J et al (2020) The effect of crop residues, cover crops, manures and nitrogen fertilization on soil organic carbon changes in agroecosystems: a synthesis of reviews. Mitig Adapt Strateg Glob Chang 25:929–952

Bond-Lamberty B, Thomson A (2010) Temperature-associated increases in the global soil respiration record. Nature 464(7288):579–582

Bransby DI, McLaughlin SB, Parrish DJ (1998) A review of carbon and nitrogen balances in switchgrass grown for energy. Biomass Bioenergy 14(4):379–384

Brouns K, Verhoeven JT, Hefting MM (2014) Short period of oxygenation releases latch on peat decomposition. Sci Total Environ 481:61–68

Chasek P (2021) Is it the end of the COP as we know it? An analysis of the first year of virtual meetings in the UN environment and sustainable development arena. Int Negot 28(1):37–68

Chaudhari PP (2018) Integrated nutrient management in greengram [Vignaradiata (L.) Wilczek] cultivars. IJCS, 6(2):943–945

Chauhan SK, Sharma R, Singh B, Sharma SC (2015) Biomass production, carbon sequestration and economics of on-farm poplar plantations in Punjab, India. J Appl Nat Sci 7(1):452–458

Chenu C, Angers DA, Barré P, Derrien D, Arrouays D, Balesdent J (2019) Increasing organic stocks in agricultural soils: knowledge gaps and potential innovations. Soil Tillage Res 188:41–52

Ciais P, Borges AV, Abril G, Michel Meybeck G, Folberth DH, Janssens IA (2008) The impact of lateral carbon fluxes on the European carbon balance. Biogeosciences 5(5):1259–1271

Crowther TW, Todd-Brown KE, Rowe CW, Wieder WR, Carey JC, Machmuller MB et al (2016) Quantifying global soil carbon losses in response to warming. Nature 540(7631):104–108

Dahlman RC, Jacobs GK, Metting FB Jr (2001) What is the potential for carbon sequestration by the terrestrial biosphere. In: First National Conference on Carbon Sequestration, pp 14–17

Das TK, Bhattacharyya R, Sharma AR, Das S, Saad AA, Pathak H (2013) Impacts of conservation agriculture on total soil organic carbon retention potential under an irrigated agro-ecosystem of the western Indo-Gangetic Plains. Eur J Agron 51:34–42

Das TK, Saharawat YS, Bhattacharyya R, Sudhishri S, Bandyopadhyay KK, Sharma AR, Jat ML (2018) Conservation agriculture effects on crop and water productivity, profitability and soil organic carbon accumulation under a maize-wheat cropping system in the North-western Indo-Gangetic Plains. Field Crop Res 215:222–231

Das A, Layek J, Ramkrushna GI, Basavaraj S, Lal R, Krishnappa R, Yadav GS, BabuSubhash and Ngachan SV. (2020) Conservation tillage and residue management improves soil properties under upland rice–rapeseed system in subtropical Eastern Himalayas. Land Degrad Dev. https://doi.org/10.1002/ldr.3568

Das A, BabuSubhash DM, Sanjeev K, Singh R, Avasthe RK, Rathore SS, Yadav SK, Singh VK (2022a) Restoring soil carbon in marginal land of Indian Himalayas: impact of crop intensification and conservation tillage. J Environ Manag. https://doi.org/10.1016/j.jenvman.2022.115603

Das A, BabuSubhash SR, Kumar S, Rathore SS, Singh VK, Datta M, Yadav SK, Wani OA, Yadav D (2022b) Impact of live mulch- based conservation tillage on soil properties and productivity of summer maize in Indian Himalayas. Sustain For 14:12078. https://doi.org/10.3390/su141912078

Deng L, Shangguan ZP (2017) Afforestation drives soil carbon and nitrogen changes in China. Land Degrad Dev 28(1):151–165

Don A, Seidel F, Leifeld J, Kätterer T, Martin M, Pellerin S, Chenu C (2024) Carbon sequestration in soils and climate change mitigation—Definitions and pitfalls. Global Change Biology 30(1):e16983

Dube F, Thevathasan NV, Zagal E, Gordon AM, Stolpe NB, Espinosa M (2011) Carbon sequestration potential of silvopastoral and other land use systems in the Chilean Patagonia. In: Carbon sequestration potential of agroforestry systems: opportunities and challenges. Springer, Dordrecht, pp 101–127

Dury J, Garcia F, Reynaud A, Bergez JE (2013) Cropping-plan decision-making on irrigated crop farms: A spatio-temporal analysis. Eur J Agron 50:1–10

FAO & ITPS (2015) Status of the world's soil resources (SWSR)—main report. Food and agriculture Organization of the United Nations and Intergovernmental Technical Panel on Soils, Rome. Available online: http://www.fao.org/3/a-i5199e.pdf

Fatima A, Singh VK, Babu Subhash, Upadhyay PK, Rathore SS, Kumar B, Hasanain M, Parween H (2023) Food production potential and environmental sustainability of different farming system models in northwest India. Front Sustain Food Syst. https://doi.org/10.3389/fsufs.2023.959464

Felzer B, Reilly J, Melillo J, Kicklighter D, Sarofim M, Wang C et al (2005) Future effects of ozone on carbon sequestration and climate change policy using a global biogeochemical model. Clim Chang 73:345–373

Gao JS, Xu MG, Hua DC, Jing H, Dong CW, Bai ZX et al (2013) Effects of long-term rice-rice-green manure cropping rotation on rice yield and soil fertility. Acta Agron Sin 39(2):343

Garnett T, Appleby MC, Balmford A, Bateman IJ, Benton TG, Bloomer P et al (2013) Sustainable intensification in agriculture: premises and policies. Science 341(6141):33–34

Gattinger A, Muller A, Haeni M, Skinner C, Fliessbach A, Buchmann N et al (2012) Enhanced top soil carbon stocks under organic farming. Proc Natl Acad Sci 109(44):18226–18231

Gayan A, Borah P, Nath D, Kataki R (2023) Soil microbial diversity, soil health and agricultural sustainability. In: Sustainable agriculture and the environment. Academic, pp 107–126

Ghosh A, Bhattacharyya R, Meena MC, Dwivedi BS, Singh G, Agnihotri R, Sharma C (2018) Long-term fertilization effects on soil organic carbon sequestration in an Inceptisol. Soil Tillage Res 177:134–144

Gregorich EG, Rochette P, McGuire S, Liang BC, Lessard R (1998) Soluble organic carbon and carbon dioxide fluxes in maize fields receiving spring-applied manure. J Environ Qual 27(1):209–214. American Society of Agronomy, Crop Science Society of America, and Soil Science Society of America

Guenet B, Gabrielle B, Chenu C, Arrouays D, Balesdent J, Bernoux M et al (2021) Can N2O emissions offset the benefits from soil organic carbon storage? Glob Chang Biol 27(2):237–256

Gupta S (2021) Carbon sequestration in cementitious matrix containing pyrogenic carbon from waste biomass: a comparison of external and internal carbonation approach. J Build Eng 43:102910

Gupta RK, Hussain A, Sooch SS, Kang JS, Sharma S, Dheri GS (2020) Rice straw biochar improves soil fertility, growth, and yield of rice–wheat system on a sandy loam soil. Exp Agric 56(1):118–131

Haile SG, Nair PR, Nair VD (2008) Carbon storage of different soil-size fractions in Florida silvopastoral systems. J Environ Qual 37(5):1789–1797

Holtsmark B (2015) Quantifying the global warming potential of CO2 emissions from wood fuels. GCB Bioenergy 7(2):195–206

Hooda N, Gera M, Andrasko K, Sathaye J, Gupta M, Vasistha H et al (2007) Community and farm forestry climate mitigation projects: case studies from Uttaranchal, India. Mitig Adapt Strateg Glob Chang 12:1099–1130

Hou P, Xue L, Wang J, Petropoulos E, Xue L, Yang L (2022) Green manure amendment in paddies improves soil carbon sequestration but cannot substitute the critical role of N fertilizer in rice production. Agronomy 12(7):1548

Hu C, Li M, Qiu J, Sun YP (2019) Design and fabrication of carbon dots for energy conversion and storage. Chem Soc Rev 48(8):2315–2337

Huijgen WJJ, Comans RNJ (2003) Carbon dioxide sequestration by mineral carbonation. Literature. Review

Intergovernmental Panel on Climate Change (IPCC) (2001) Climate Change 2001: the scientific basis. In: Houghton JT et al (eds)., 881 pp Contribution of working group I to the third assessment report of the Intergovernmental Panel on Climate Change. Cambridge University Press, New York

Intergovernmental Panel on Climate Change (IPCC) (2007) Climate change 2007: the scientific basis. In: Solomon S et al (eds) Contribution of working group I to the fourth assessment report of the Intergovernmental Panel on Climate Change. Cambridge University Press, New York

IPCC (2014) Summary for policymakers. In: Field CB et al (eds) Climate Change 2014: impacts, adaptation, and vulnerability. Part A: global and sectoral aspects. Contribution of working group II to the fifth assessment report of the Intergovernmental Panel on Climate Change. Cambridge University Press, Cambridge/New York, pp 1–32

IUCN (2021) The Bonn Challenge/ECCA30 initiative welcomes three new pledges from the region. https://iucn.org/news/eastern-europe-and-central-asia/202110/bonn-challengeecca30-initiative-welcomes-three-new-pledgesregion (Accessed on July, 2024)

Izaurralde RC, Williams JR, Post WM, Thomson AM, McGill WB, Owens LB, Lal R (2007) Long-term modeling of soil C erosion and sequestration at the small watershed scale. Clim Chang 80:73–90

Jansson J, Marell A, Nordlund A (2010) Green consumer behavior: determinants of curtailment and eco-innovation adoption. J Consum Mark 27(4):358–370

Johnson JMF, Franzluebbers AJ, Weyers SL, Reicosky DC (2007) Agricultural opportunities to mitigate greenhouse gas emissions. Environ Pollut 150(1):107–124

Kandji ST, Verchot LV, Mackensen J, Boye A, Van Noordwijk M, Tomich TP et al (2006) Opportunities for linking climate change adaptation and mitigation through agroforestry systems. In: World agroforestry into the future. World Agroforesty Centre, Nairobi, pp 113–122

Karlen DL, Mausbach MJ, Doran JW, Cline RG, Harris RF, Schuman GE (1997) Soil quality: a concept, definition, and framework for evaluation (a guest editorial). Soil Science Society of America Journal, 61(1):4–10

Keerthi MG, Prend CJ, Aumont O, Lévy M (2022) Annual variations in phytoplankton biomass driven by small-scale physical processes. Nat Geosci 15(12):1027–1033

Keith H, Mackey BG, Lindenmayer DB (2009) Re-evaluation of forest biomass carbon stocks and lessons from the world's most carbon-dense forests. Proc Natl Acad Sci 106(28):11635–11640

Kovscek AR, Cakici MD (2005) Geologic storage of carbon dioxide and enhanced oil recovery. II. Cooptimization of storage and recovery. Energy convers Manag 46(11–12):1941–1956

Kukal SS, Benbi DK (2009) Soil organic carbon sequestration in relation to organic and inorganic fertilization in rice–wheat and maize–wheat systems. Soil Tillage Res 102(1):87–92

Kumar SS, Mir SA, Wani OA, Babu Subhash YM, Bhat MA, Hussain N, Wani AA, Kumar R, Devideen Y, Dar SR (2022) Land-use systems regulate carbon geochemistry in the temperate Himalayas, India. J Environ Manag. https://doi.org/10.1016/j.jenvman.2022.115811

Kumawat A, Yadav D, Srivastava P, Babu Subhash, Singh D, Vishwakarma DK, Sharma VK, Madhu M (2023) Restoration of agroecosystems with conservation agriculture for food security to achieve UN-sustainable development goals. Land Degrad Dev. https://doi.org/10.1002/ldr.4677

Kust G, Andreeva O, Lobkovskiy V, Telnova N (2018) Uncertainties and policy challenges in implementing land degradation neutrality in Russia. Environ Sci Pol 89:348–356

Ladha JK, Tirol-Padre A, Reddy CK, Cassman KG, Verma S, Powlson DS et al (2016) Global nitrogen budgets in cereals: A 50-year assessment for maize, rice and wheat production systems. Sci Rep 6(1):1–9

Lal R (2004) Soil carbon sequestration in India. Clim Chang 65(3):277–296

Lal R (2008) Sequestration of atmospheric CO2 in global carbon pools. Energy Environ Sci 1(1):86–100

Lal R (2010) Enhancing eco-efficiency in agro-ecosystems through soil carbon sequestration. Crop Sci 50:S-120

Lal R (2016) Soil health and carbon management. Food Energy Secur 5(4):212–222

Lal R (2023) Farming systems to return land for nature: It's all about soil health and re-carbonization of the terrestrial biosphere. Farm Syst 1(1):100002

Lal R, Follett RF, Stewart BA, Kimble JM (2007) Soil carbon sequestration to mitigate climate change and advance food security. Soil Sci 172(12):943–956

Li C, Frolking S, Harriss R (1994) Modeling carbon biogeochemistry in agricultural soils. Glob Biogeochem Cycles 8(3):237–254

Lugato E, Leip A, Jones A (2018) Mitigation potential of soil carbon management overestimated by neglecting N2O emissions. Nat Clim Chang 8(3):219–223

Luo Y, Keenan TF, Smith M (2015) Predictability of the terrestrial carbon cycle. Glob Chang Biol 21(5):1737–1751

Ma X, Wang C, Dong B, Gu G, Chen R, Li Y et al (2019) Carbon emissions from energy consumption in China: its measurement and driving factors. Sci Total Environ 648:1411–1420

Maikhuri RK, Semwal RL, Rao KS, Singh K, Saxena KG (2000) Growth and ecological impacts of traditional agroforestry tree species in central Himalaya, India. Agrofor Syst 48:257–271

Maitra S, Zaman A (2017) Bio-drainage in waterlogged areas for agricultural sustainability. Int J Bioresour Sci 4(2):79–84

Manisalidis I, Stavropoulou E, Stavropoulos A, Bezirtzoglou E (2020) Environmental and health impacts of air pollution: a review. Front Public Health 8:14

Maroto-Valer MM, Tang Z, Zhang Y (2005) CO2 capture by activated and impregnated anthracites. Fuel Process Technol 86(14–15):1487–1502

Minasny B, Malone BP, McBratney AB, Angers DA, Arrouays D, Chambers A et al (2017) Soil carbon 4 per mille. Geoderma 292:59–86

MNRE (2018) Annual report 2018-29, Government of Indian, https://mnre.gov.in/en/annual-reports-2018-19/ (Accessed on July, 2024)

Mulvaney MJ, Wood CW, Balkcom KS, Kemble J, Shannon DA (2017) No-till with high biomass cover crops and invasive legume mulches increased total soil carbon after three years of collard production. Agroecol Sustain Food Syst 41(1):30–45

Mutuo PK, Cadisch G, Albrecht A, Palm CA, Verchot L (2005) Potential of agroforestry for carbon sequestration and mitigation of greenhouse gas emissions from soils in the tropics. Nutr Cycl Agroecosyst 71:43–54

Nath AJ, Das AK (2011) Carbon storage and sequestration in bamboo-based smallholder homegardens of Barak Valley, Assam. Curr Sci:229–233

Negash M, Kaseva J, Kahiluoto H (2022) Perennial monocropping of khat decreased soil carbon and nitrogen relative to multistrata agroforestry and natural forest in southeastern Ethiopia. Reg Environ Chang 22(2):38

Newaj R, Dhyani SK (2008) Agroforestry systems for carbon sequestration: present status and scope. Indian J Agrofor 10(1)

Ngangom B, Das A, Lal R, Ramkrushna GI, Layek J, Basavaraj S, Babu Subhash YGS, Ghosh PK (2020) Double mulching improves soil properties and productivity of maize-based cropping system in eastern Indian Himalayas. Int Soil Water Conserv Res. https://doi.org/10.1016/j.iswcr.2020.07.001

NRCAF (2007) Perspective plan: vision 2025. NRCAF, Jhansi

OECD/IEA/NEA (2015) Aligning Policies for a Low-carbon Economy. https://www.oecd-nea.org/jcms/pl_14756/projected-costsof-generating-electricity-2015-edition?details=true (Accessed on July, 2024)

Ossom EM, Pace PF, Rhykerd RL, Rhykerd CL (2001) Effect of mulch on weed infestation, soil temperature, nutrient concentration, and tuber yield in Ipomoea batatas (L.) Lam. in Papua New +Guinea. Trop Agric 78(3):144

Oyeogbe AI, Das TK, Bhatia A, Singh SB (2017) Adaptive nitrogen and integrated weed management in conservation agriculture: impacts on agronomic productivity, greenhouse gas emissions, and herbicide residues. Environ Monit Assess 189:1–11

Parker MA, Nyakatawa EZ, Reddy KC, Reeves DW (2002) Soil carbon and nitrogen as influenced by tillage and poultry litter in North Alabama. In: Making conservation tillage conventional: Building a future on 25 years of research, Auburn, Alabama, pp 283–287

Peichl M, Thevathasan NV, Gordon AM, Huss J, Abohassan RA (2006) Carbon sequestration potentials in temperate tree-based intercropping systems, southern Ontario, Canada. Agrofor Syst 66:243–257

Post WM, Amonette JE, Birdsey R, Garten CT Jr, Izaurralde RC, Jardine PM et al (2009) Terrestrial biological carbon sequestration: science for enhancement and implementation. In: Carbon sequestration and its role in the global carbon cycle, vol 183. American Geophysical Union, Washington, DC, pp 73–88

Poulton P, Johnston J, Macdonald A, White R, Powlson D (2018) Major limitations to achieving "4 per 1000" increases in soil organic carbon stock in temperate regions: evidence from long-term experiments at Rothamsted research, United Kingdom. Glob Change Biol 24(6):2563–2584

Pradhan A, Chandrakar T, Dixit A, Kerketta AK (2020) Brown manuring effects on soil and yield of rice (Oryzasativa) under rainfed midland farming system. Indian J Agric Sci 90(9):1810–1812

Pretty J, Toulmin C, Williams S (2018) Sustainable intensification in African agriculture. Int J Agric Sustain 9(1):5–24

Pribyl DW (2010) A critical review of the conventional SOC to SOM conversion factor. Geoderma 156(3–4):75–83

Raj R, Das TK, Chakraborty D, Bhattacharyya R, BabuSubhash GP, Kumar V, Ekka U, Sen S, Ghosh S, Roy A, Sharma T (2022) Soil physical environment and active carbon pool in rice–wheat system of South Asia: impact of long-term conservation agriculture practices. Environ Technol Innov. https://doi.org/10.1016/j.eti.2022.102966

Ranells NN, Wagger MG (1997) Grass–legume bicultures as winter annual cover crops. Agronomy Journal, 89(4):659–665

Ranjan P, Patle GT, Prem M, Solanke KR (2017) Organic Mulching-A Water Saving Technique to Increase the Production of Fruits and Vegetables. Current Agriculture research journal 5(3)

Rathore SS, Babu Subhash SK, Singh R, Yadav SK, Singh VK, Singh C (2022) Designing energy cum carbon-efficient environmentally clean production system for achieving green economy in agriculture. Sustain Energy Technol Assess. https://doi.org/10.1016/j.seta.2022.102190

Raven JA (2011) Praeger review: effects on marine algae of changed seawater chemistry with increasing atmospheric CO2. In: Biology and environment: proceedings of the Royal Irish Academy. Royal Irish Academy, pp 1–17

Raven JA, Falkowski PG (1999) Oceanic sinks for atmospheric CO2. Plant Cell Environ 22(6):741–755

Raza A, Gholami R, Rezaee R, Rasouli V, Rabiei M (2019) Significant aspects of carbon capture and storage–a review. Petroleum 5(4):335–340

Reddy V, Rathore SS, Babu S, Shekhawat K, Singh VK (2021) In-situ green-manuring for enhanced rice (Oryzasativa) productivity under north eastern hill region. Indian J Agron 66(3):370–372

Romero F, Chana C, Montenegro J, Sanchez L A, Guevara G (1991) Productividad de Gliricidiasepium y Erythrinaberteroanaencercasvivasmanejadasbajotresfrecuencias de podaen la zona atlantica de Costa Rica

Rosenzweig C, Tubiello FN (2007) Adaptation and mitigation strategies in agriculture: an analysis of potential synergies. Mitig Adapt Strateg Glob Chang 12:855–873

Rusinamhodzi L (2015) Crop rotations and residue management in conservation agriculture. Conserv Agric:21–37

Saha SK, Nair PK, Nair VD, Kumar BM (2009) Soil carbon stock in relation to plant diversity of homegardens in Kerala, India. In: Adv Agrofor. Springer, Dordrecht, pp 53–65

Saha P, Bohra JS, Nayak H, Singh T, Barman A (2022) Diversification of rice (Oryzasativa)-based cropping system of Varanasi for enhanced productivity and employment generation. Indian J Agric Sci 92(8):1026–1028

Saroa GS, Lal R (2003) Soil restorative effects of mulching on aggregation and carbon sequestration in a Miamian soil in Central Ohio. Land Degrad Dev 14(5):481–493

Schmitz OJ, Raymond PA, Estes JA, Kurz WA, Holtgrieve GW, Ritchie ME et al (2014) Animating the carbon cycle. Ecosystems 17:344–359

Sharma AR, Ghosh A (2000) Effect of green manuring with Sesbaniaaculeata and nitrogen fertilization on the performance of direct-seeded flood-prone lowland rice. Nutr Cycl Agroecosyst 57:141–153

Sharma P, Laor Y, Raviv M, Medina S, Saadi I, Krasnovsky A et al (2017) Green manure as part of organic management cycle: effects on changes in organic matter characteristics across the soil profile. Geoderma 305:197–207

Sharma M, Kaushal R, Kaushik P, Ramakrishna S (2021) Carbon farming: prospects and challenges. Sustainability 13(19):11122

Shen Y, Chen Y, Li S (2016) Microbial functional diversity, biomass and activity as affected by soil surface mulching in a semiarid farmland. PLoS One 11(7):e0159144

Singh AK, Rai A, Singh N (2016) Effect of long-term land use systems on fractions of glomalin and soil organic carbon in the Indo-Gangetic plain. Geoderma 277:41–50

Singh R, BabuSubhash ARK, Meena RS, Yadav GS, Das A, Mohapatra KP, Rathore SS, Kumar A, Singh C (2021a) Conservation tillage and organic nutrients management improve soil properties, productivity, and economics of the maize-vegetable pea system in the Eastern Himalayas. Land Degrad Dev. https://doi.org/10.1002/ldr.4066

Singh R, Subhash B, Avasthe RK, Yadav GS, Das A, Mohapatra KP, Kumar A, Singh VK, Chandra P (2021b) Crop productivity, soil health, and energy dynamics of Indian Himalayan intensified organic maize-based systems. Int Soil Water Conserv Res 9(2):260–270. https://doi.org/10.1016/j.iswcr.2020.11.003

Singh D, Mishra AK, Patra S, Dwivedi AK, Ojha CSP, Singh VP, Mariappan S, BabuSubhash SN, Yadav D, Ojasvi PR, Gopal K, Madhu M, Sena DR, Chand L, Kumar S (2023) Effect of long-term tillage practices on runoff and soil erosion in sloping croplands of Himalaya, India. Sustain For 15:8285. https://doi.org/10.3390/su15108285

Smith P, Davis SJ, Creutzig F, Fuss S, Minx J, Gabrielle B et al (2016) Biophysical and economic limits to negative CO2 emissions. Nat Clim Chang 6(1):42–50

Spiegel H, Mosleitner T, Sandén T, Zaller JG (2018) Effects of two decades of organic and mineral fertilization of arable crops on earthworms and standardized litter decomposition. Die Bodenkultur: J Land Manag Food Environ 69(1):17–28

Sudarshan S, Shekhawat K, Rathore SS, Singh RK, Kumar V (2022) Nitrogen-management options for different crop-establishment methods in wheat under rice (Oryzasativa)-wheat (Triticumaestivum) cropping system. Indian J Agron 67(2):197–200

Sundermeier A, Reeder R, Lal R (2005) Soil carbon sequestration fundamentals, Rep. No. OSU factsheet AEX-510–05. OSUE, Columbus

Tamboli BD, Shelke SR, Bagwan IR, Indi DV, Kadam JR, Deshpande AN (2011) Effect of green manuring on dryland rabi sorghum productivity and soil quality in vertisol. Indian J Dryland Agric Res Dev 26(2):33–37

Turmel MS, Speratti A, Baudron F, Verhulst N, Govaerts B (2015) Crop residue management and soil health: a systems analysis. Agric Syst 134:6–16

UNCCD (2021) UNCCD, CBD and UNFCCC Joint Liaison Group. Retrieved from https://www.unccd.int/convention/about-convention/unccd-cbd-and-unfccc-joint-liaison-group

UNCCD (2023). Retrieved from https://www.unccd.int/events/desertification-drought-day/2023

van Groenigen KJ, Six J, Hungate BA, de Graaff MA, Van Breemen N, Van Kessel C (2006) Element interactions limit soil carbon storage. Proc Natl Acad Sci 103(17):6571–6574

Vogel LE, Pot V, Makowski D, Garnier P, Baveye PC (2018) To what extent do uncertainty and sensitivity analyses help unravel the influence of microscale physical and biological drivers in soil carbon dynamics models? Ecol Model 383:10–22

Walia SS, BabuSubhash GRS, Kaur T, Kohima N, Panwar AS, Yadav DK, Ansari MA, Ravishankar N, Kumar S, Kaur K, Ansari MH (2022) Designing resource-efficient and environmentally safe cropping systems for sustainable energy use and economic returns in indo-Gangetic Plains, India. Sustain For 14:14636.10.3390/su142114636

Wani OA, Kumar S, Hussain N, Wani AA, BabuSubhash AP, Rashid M, Popescu SM, Mansoor S (2022) Multi-scale processes influencing global carbon storage and land-carbon-climate nexus: A critical review. Pedosphere. https://doi.org/10.1016/j.pedsph.2022.07.002

Wani OA, Sharma V, Kumar SS, BabuSubhash SKR, Rathore SS, Marwaha S, Ganai NA, Dar SR, Yeasin M, Raghavendra S, Tomar J (2023a) Climate plays a dominant role over land management in governing soil carbon dynamics in North Western Himalayas. J Environ Manag. https://doi.org/10.1016/j.jenvman.2023.117740

Wani OA, Akhter F, Kumar SS, BabuSubhash KRH, Mir SA, Mahdi SS, Malik AR, Bangroo S, Gaafar A-RZ, Popescu SM, Rathore SS (2023b) Mitigating soil erosion through biomass-derived biochar: exploring the influence of feedstock types and pyrolysis temperature. Land 12(12):2111. https://doi.org/10.3390/land12122111

Whalen JK, Chang C (2002) Macroaggregate characteristics in cultivated soils after 25 annual manure applications. Soil Sci Soc Am J 66(5):1637–1647

Yadav SK, Singh Y, Yadav MK, Subhash B, Singh K (2013) Effect of organic nitrogen sources on yield, nutrient uptake and soil health under rice (*Oryzasativa*) based cropping sequence. Indian J Agric Sci 83(2):170–175

Yadav GS, Das A, Lal R, Babu S, Meena RS, Patil SB et al (2018) Conservation tillage and mulching effects on the adaptive capacity of direct-seeded upland rice (Oryzasativa L.) to alleviate weed and moisture stresses in the North Eastern Himalayan region of India. Arch Agron Soil Sci 64(9):1254–1267

Yadav GS, Das A, Lal R, Babu S, Datta M, Meena RS et al (2019) Impact of no-till and mulching on soil carbon sequestration under rice (Oryzasativa L.)-rapeseed (Brassica campestris L. var. rapeseed) cropping system in hilly agro-ecosystem of the eastern Himalayas, India. Agric Ecosyst Environ 275:81–92

Yadav GS, Das A, BabuSubhash MKP, Lal R, Rajkhowa DJ (2020a) Potential of conservation tillage and altered land configuration to improve soil properties, carbon sequestration and productivity of maize based cropping system in eastern Himalayas, India. Int Soil Water Conserv Res. https://doi.org/10.1016/j.iswcr.2020.12.003

Yadav GS, BabuSubhash DA, Mohapatra KP, Singh R, Avasthe RK, Roy S (2020b) No till and mulching enhance energy use efficiency and reduce carbon footprint of a direct-seeded upland rice production system. J Clean Prod. https://doi.org/10.1016/j.jclepro.122700

Yadav GS, BabuSubhash DA, Datta M, Mohapatra KP, Singh R, Singh VK, Rathore SS, Chakraborty M (2021a) Productivity, soil health, and carbon management index of Indian Himalayan intensified maize-based cropping systems under live mulch-based conservation tillage practices. Field Crop Res. https://doi.org/10.1016/j.fcr.2021.108080

Yadav GS, Das A, Kandpal BK, BabuSubhash LR, Datta M, Das B, Singh R, Singh VK, Mohapatra KP, Chakraborty M (2021b) The food-energy-water-carbon nexus in a maize-maize-mustard cropping sequence of the Indian Himalayas: an impact of tillage-cum-live mulching. Renew Sust Energ Rev. https://doi.org/10.1016/j.rser.2021.111602

Yadav GS, Kandpal BK, Das A, BabuSubhash MKP, Devi AG, Devi HL, Chandra P, Singh R, Barman KK (2021c) Impact of 28-year-old agroforestry systems on soil carbon dynamics in Eastern Himalayas. J Environ Manag. https://doi.org/10.1016/j.jenvman.2021.111978

Yadav D, Singh D, BabuSubhash MM, Singh DV, Mandal D, Rathore AC, Sharma VK, Singhal V, Kumawat A, Yadav DK, Yadav RK, Kumar S (2023) Intensified cropping reduces soil erosion and improves rainfall partitioning and soil properties in the marginal land of the Indian Himalayas. Int Soil Water Conserv Res. https://doi.org/10.1016/j.iswcr.2023.10.002

Yadav D, Singh D, Madhu M, Singh DV, Islam S, Kumar G, Gupta AK, Mandal D, BabuSubhash BA, Kumawat A, Dubey R, Pati K, Bhargavi B, Yadav DK (2024) Interplanted crops and rainfall variability regulate soil erosion in rainfed maize of the Indian Himalayas. Land Degrad Dev. https://doi.org/10.1002/ldr.5221

Yang H, Xu Z, Fan M, Gupta R, Slimane RB, Bland AE, Wright I (2008) Progress in carbon dioxide separation and capture: A review. J Environ Sci 20(1):14–27

Yang X, Steenhuis TS, Davis KF, van der Werf W, Ritsema CJ, Pacenka S et al (2021) Diversified crop rotations enhance groundwater and economic sustainability of food production. Food Energy Secur 10(4):e311

Yao Z, Zhang D, Liu N, Yao P, Zhao N, Li Y et al (2019) Dynamics and sequestration potential of soil organic carbon and total nitrogen stocks of leguminous green manure-based cropping systems on the loess plateau of China. Soil Tillage Res 191:108–116

Zahed Z, Mufti S, Kumar SS, Wani OA, Mushtaq F, Rasool R, BabuSubhash AI, Gaber A, Hossain A (2022) Organic and inorganic mulches combination improve the productivity, quality and profitability of rainfed potato in the temperate Himalayan region. GesundePflanzen 74:1109–1122. https://doi.org/10.1007/s10343-022-00650-1

Zeng N (2008) Carbon sequestration via wood burial. Carbon Balance Manag 3:1–12

Zhang F, Li S, Yue S, Song Q (2022) The effect of long-term soil surface mulching on SOC fractions and the carbon management index in a semiarid agroecosystem. Soil Tillage Res 216:105233

Zhang ZH, Jun NIE, Liang H, Wei CL, Yun WANG, Liao YL et al (2023) The effects of co-utilizing green manure and rice straw on soil aggregates and soil carbon stability in a paddy soil in southern China. J Integr Agric 22(5):1529–1545

Zhu D, Ciais P, Krinner G, Maignan F, JornetPuig A, Hugelius G (2019) Controls of soil organic matter on soil thermal dynamics in the northern high latitudes. Nat Commun 10(1):3172

Zuazo VHD, Pleguezuelo CRR (2009) Soil-erosion and runoff prevention by plant covers: a review. Sustain Agric:785–811

Nature-Based Solutions for Food Security and Soil Sustainability

10

Seema Sangwan, Garima Saxena, and Radha Prasanna

Abstract

Food security refers to the consistent access of individuals to adequate, secure, and healthy food that completes their nutritional demand for an agile and strong life. Soil sustainability refers to a living ecosystem which provides a healthy habitat to support the growth of humans, animals, and plants. Both concepts are interconnected, with soil health being crucial for agricultural productivity and, consequently, food security. However, challenges such as soil degradation, including erosion, compaction, nutrient depletion, and pollution, pose significant threats to soil sustainability. Human actions like desertification, overgrazing, improper farming applications etc. accelerate soil degradation, reducing it's potential to support crops and preserve water. Climate change exacerbates these issues through increased temperatures and extreme weather events. Additionally, modern agricultural practices, including monoculture and excessive use of chemical inputs, further degrade soil health. Urbanization also intensifies these challenges by limiting arable land and increasing food demand in cities. Economic inequality worsens food insecurity, especially as climate change disproportionately impacts vulnerable communities dependent on agriculture. Nature-based solutions (NBS) offer sustainable strategies for addressing these challenges by leveraging natural processes. Agroforestry, conservation agriculture, and wetland restoration are examples of NBS that ameliorate soil potency, prevent erosion, and increase crop yields. Additionally, biofertilizers contribute to soil health by reducing dependency on chemical fertilizers and increasing crop resilience to climate change. Together, these solutions provide holistic approaches to improving both food security and soil sustainability, fostering more resilient and sustainable agricultural systems.

S. Sangwan (✉) · G. Saxena · R. Prasanna
Division of Microbiology, ICAR-Indian Agricultural Research Institute, New Delhi, India

© The Author(s), under exclusive license to Springer Nature Singapore Pte Ltd. 2025

S. Babu et al. (eds.), *Ecological Solutions to Agricultural Land Degradation*,
https://doi.org/10.1007/978-981-96-3392-0_10

Keywords

Nature based solution · Soil sustainability · Food security · Agriculture · Circular food systems

Abbreviations

NBS	Nature-based solutions
FAO	The Food and Agriculture Organization
GHG	Green house gas
CA	Conservation agriculture
BFs	Biofertilizers
SOM	Soil organic matter
CEC	Cation exchange capacity

10.1 Introduction

Food security is the state in which all individuals have consistent access to sufficient, safe, and nutritious food that meets their dietary needs for an active and healthy life. Soil sustainability, on the other hand, refers to the capacity of soil to continue functioning as a vital living ecosystem that supports plants, animals, and humans (FAO 2017a, b). Both food security and soil sustainability are deeply intertwined. Soil well-being is a crucial component of agricultural productivity, which is essential for ensuring food security. However, numerous challenges threaten both soil sustainability and food security, leading to a complex global issue that requires integrated solutions. For example, soil degradation includes erosion, compaction, nutrient depletion, and pollution, is one of the most significant threats to soil sustainability. Human activities such as clearing of plants, overfeeding to animals, and inappropriate agricultural applications accelerate soil degradation, reducing the soil's capacity to support crops and conserve water (Lal 2015). Climate change exacerbates soil degradation through alteration in precipitation patterns, greater heat, and substantial weather situations such as floods and droughts (Schmidt et al. 2011). Modern cultivation approaches like monocropping and excessive use of chemical fertilizers and pesticides deplete essential nutrients, reduce soil organic matter, and disrupt the soil microbiome which leads to the reduction in soil productivity and structure over the time (Tilman et al. 2002). Urbanization further exacerbates this challenge by reducing the availability of fertile soil and increasing food requirements in urban settings (Godfray et al. 2010). Economic disparities contribute significantly to food insecurity. This inequality is often exacerbated by the influence of climate change and environmental degradation, which disproportionately affect poorer communities that rely heavily on agriculture (FAO 2021).

Nature-based solutions (NBS) offer sustainable approaches to addressing food security and soil sustainability challenges by leveraging natural processes and ecosystem services. NBS harness natural actions and ecological assistance to mark environmental and societal challenges. For instance, agroforestry integrates trees with crops, enhancing soil fertility and preventing erosion, which directly contributes to both soil health and increased agricultural productivity (Jose 2009). Conservation agriculture, which includes practices like minimal soil disturbance, soil cover and crop rotation, improves soil structure and nutrient cycling while also boosting crop yields and resilience to climate change (Kassam et al. 2018). Wetland restoration further supports these goals by improving water quality and maintaining soil moisture, crucial for sustaining agricultural activities (Mitsch and Gosselink 2015). Biofertilizers offer a trustworthy and environment beneficial solution to the challenges of food security and soil sustainability. Through increasing soil fertility, enhancing soil structure, and decreasing the dependency on chemical fertilizers, biofertilizers promote long-term strength of agronomic systems. Their capacity in increasing agricultural yields, lessen costs, and enhancing resilience to climate change makes them a crucial component of nature-based solutions for global food security (Bhattacharyya and Jha 2012; Rashid et al. 2016). Together, these NBS can potentially provide holistic, sustainable strategies that improves both food security and soil sustainability.

10.2 Classification of Nature-Based Solutions

NBS can be categorized (Fig. 10.1) into four main approaches:

10.2.1 Agroecological Practices

10.2.1.1 Agroforestry

Agroforestry combines trees and shrubs into agricultural environments, enhancing ecosystem services and agricultural productivity (Jose 2009). Agroforestry provides additional food sources (e.g., fruits, nuts) and income, diversifying production systems and reducing risks associated with crop failures (FAO 2017a, b). It improves soil fertility, prevents erosion, enhances water retention, and increases biodiversity (Nair 2014). For example: Alley cropping-crops are grown between rows of trees, exemplifies agroforestry. Agroforestry enhances the below and above biomass, root and rhizoids production, litter fall, organic carbon sequestration in soil which play a pivotal role to reduce GHG emission from environment (Getnet et al. 2023). The total biomass carbon storage varied from 7.12 to 57.03 Mg C ha^{-1} in diverse agroforestry systems under different water regimes in India (Newaj et al. 2016).

10.2.1.2 Conservation Agriculture

Conservation agriculture (CA) rely on minimal soil disruption, preserving soil cover, and practicing crop rotation to sustain agricultural productivity (Kassam

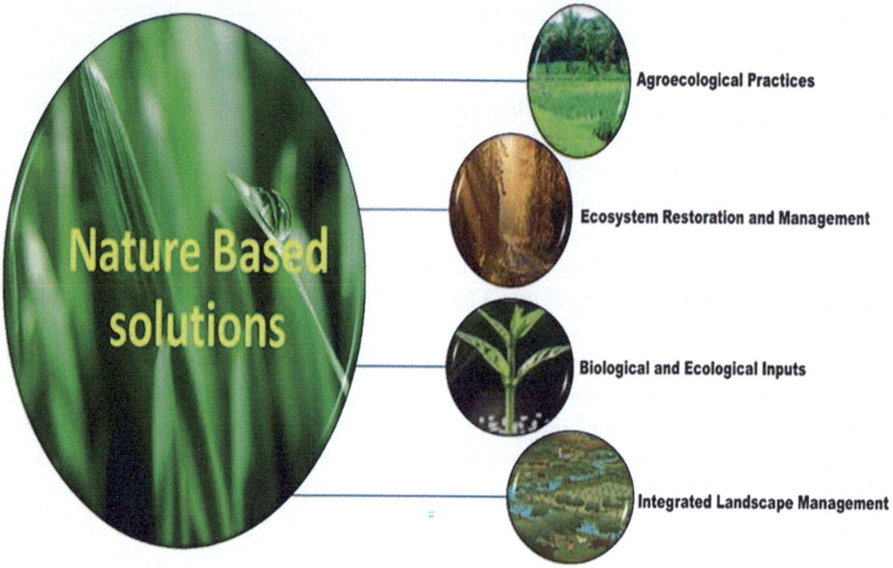

Fig. 10.1 Categorization of Nature-based solutions for food security and soil sustainability

et al. 2009). CA increases crop yields over time by improving soil health and water-use efficiency (Pretty 2008). It improves soil structure, reduces erosion, and enhances organic matter content (Lal 2015). No-till farming, combined with cover cropping and crop rotation, is a common CA practice. Carbon and water footprints are the environmental indicators which can trace the biological loads (Galli et al. 2012). CA can lessen the greenhouse gas emission (GHGE) from agricultural fields due to decreased application of chemical pesticides and fertilizers. Furthermore, permanent soil cover, diverse plantation, and mulch layers can shrink soil erosion which ultimately minimize GHGE and maximize soil organic carbon accumulation under CA (Freitaga et al. 2024). In CA fields, rainwater easily gets penetrate due to superior soil structure which decreases the irrigation rates by up to 50% and enhances crop yield (Palm et al. 2014). NBS can significantly improve water availability and quality in elevated rainfall areas with nearly no surface runoff. Thus, CA is a better NBS which can contribute to enhanced functioning of water cycles.

10.2.1.3 Organic Farming

Organic farming avoided chemicals, and depends on natural resources like compost, green manure, and biopesticides (Lotter 2003). Organic farming produces high-quality, nutritious food and supports rural livelihoods (Reganold and Wachter 2016). It enhances soil health through organic inputs, promotes biodiversity, and reduces pollution (Mäder et al. 2002).

10.2.2 Ecosystem Restoration and Management

10.2.2.1 Wetland Restoration

Wetland restoration involves rehabilitating degraded wetlands to restore their ecological functions (Mitsch and Gosselink 2015). Wetlands support aquaculture and agriculture by improving water quality and availability (Zedler and Kercher 2005). Restoration enhance groundwater recharge, maintain soil moisture, and reduce soil degradation (Verhoeven and Setter 2010).

10.2.2.2 Reforestation and Afforestation

Reforestation and afforestation involve cultivation of trees in deforested or degenerated areas to reimpose forest ecosystems (Lamb 2014). These practices provide timber, non-timber forest products, and improve microclimates (Chazdon 2008). Both reforestation and afforestation prevent erosion, improve soil structure, and increase carbon sequestration (Zomer et al. 2008). Community-managed forests on degraded lands to restore ecosystem services are perfect example to highlight the importance of reforestation and afforestation as nature-based solution.

10.2.2.3 Riparian Buffer Zones

Riparian buffer zones are vegetated regions containing water bodies that help protect water quality and stabilize riverbanks (Gregory et al. 1991). These zones protect water quality, which is vital for irrigation and livestock (Naiman and Décamps 1997). Riparian buffer zones prevent erosion, filter pollutants, and enhance habitat connectivity (Sweeney and Newbold 2014). For example, grass strips along streams reduce sediment runoff from agricultural fields.

10.2.3 Biological Inputs

10.2.3.1 Biofertilizers

Biofertilizers (BFs) are natural inputs containing living microorganisms that enhance soil nutrient availability (Vessey 2003). Biofertilizers increase crop yields and reduce reliance on chemical fertilizers, making farming more sustainable (Bhattacharyya and Jha 2012). BFs ameliorate soil fertility, increase microbial activity, and encourage sustainable minerals cycling (Adesemoye and Kloepper 2009). Application of microbes fixing nitrogen (*Rhizobium*), solubilizing phosphorous and potash is recommended as form of nature-based solution for sustainable agriculture (Fig. 10.2) as BFs have been estimated to increase the crop production and yield by10–40% (Stewart and Roberts 2012). From 2015 to 2020, biofertilizer market has been reported to enlarge at a composite growth rate of 14% which will exceed to the 1.88 billion by the end of 2025 (Daniel et al. 2022).

10.2.3.2 Biopesticides

Biopesticides are natural pest control agents originated from microorganisms, plants, or minerals (Copping and Menn 2000). They reduce crop losses to pests

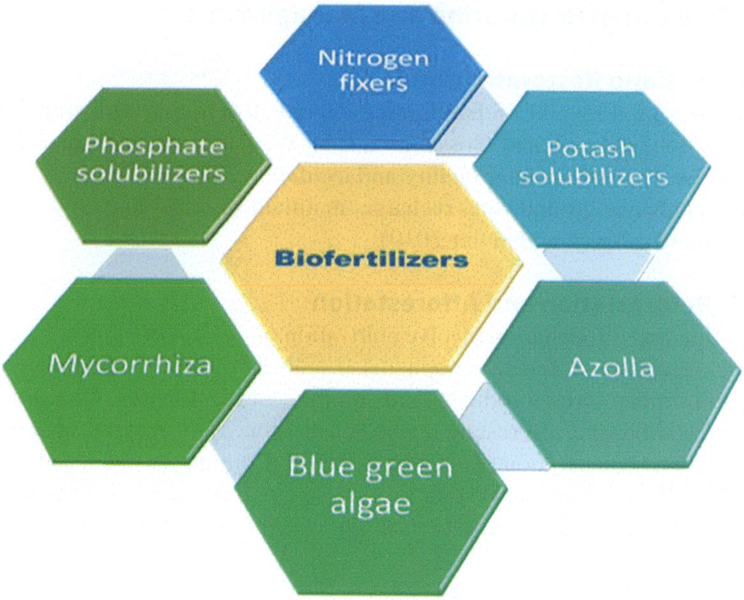

Fig. 10.2 Different types of biofertilizers as nature-based solutions in food production

while minimizing health risks associated with chemical pesticides (Glare et al. 2012). Biopesticides reduce soil contamination and preserve beneficial soil organisms (Damalas and Koutroubas 2018). *Bacillus thuringiensis* (Bt) was successfully employed as a biopesticide to restrict insect pests while application of essential oils such as rosemary, lemongrass, peppermint, thyme etc. (Jyotsna et al. 2024) and commercialized botanical pesticides- neem, *Pyrethrum*, *Sabadilla* (Leng et al. 2014) are also some common nature-based solutions. A minimum of 15 microorganisms and 970 bioformulations have been indexed in Central Insecticides Board and Registration Committee which compose around 5% of the Indian pesticide market (Kumar et al. 2019).

10.2.3.3 Green Manure and Cover Crops

Green manure and cover crops are cultivated and incorporated into the soil to upgrade soil fertility and productivity (Sarrantonio and Gallandt 2003). These practices enhance soil fertility, leading to better crop yields (Snapp et al. 2005). Green manure and covers crop enhance organic matter, reduce weeds, and improves soil structure (Creamer and Baldwin 2000). Planting leguminous cover crops like clover or vetch fixes atmospheric nitrogen and improves organic matter contents in soil. Furthermore, cover crops protect soil upper layer from water and air erosion, maintain soil moisture, increase soil porosity and organic carbon matter, lower soil mass, promote soil aggregation and structure (Haruna et al. 2020). *Medicago sativa*, *Crotolaria juncea*, *Trifolium* spp., are some common cover crops which generally improves the soil properties (Teasdale et al. 2007).

10.2.4 Integrated Landscape Management

Watershed management involves coordinated management of land, water, and resources within a watershed to sustain ecosystem services (Calder 2000). It ensures a reliable water supply for agriculture and reduces flood risks (Falkenmark and Rockström 2006). Watershed management reduces soil erosion, improves water retention, and enhances landscape resilience (Tiwari et al. 2008). Terracing and contour farming are implemented in hilly areas to reduce runoff and soil erosion.

10.3 Nature Based Solution (NBS) and Agricultural Production

Nature-based solutions positively affect agricultural production by promoting sustainable practices that enhance ecosystem services, improve soil health, water management, pest control, biodiversity, and climate resilience. Implementing NBS can lead to more adaptable and prolific agricultural networks. Some of the important advantages of NBS are mentioned below.

10.3.1 Improved Soil Health

NBS such as agroforestry, cover cropping, and crop rotation revamp soil structure, boost minerals cycling, and better organic matter content. This leads to enhanced soil fertility and, higher agricultural yields (Lal 2011).

10.3.2 Improved Water Management

Techniques such as afforestation, rainwater harvesting, use of wetlands etc. help regulate water availability, reducing the impact of droughts and floods on agricultural productivity (Abhilash et al. 2016).

10.3.3 Pest and Disease Control

NBS such as intercropping and the use of natural predators reduce reliance on chemical pesticides by naturally controlling pest populations, leading to healthier crops and reduced production costs (Pretty and Bharucha 2015).

10.3.4 Biodiversity and Ecosystem Services

Biodiversity enhancement through NBS (e.g., creating habitat corridors and conserving pollinator species) boosts pollination and natural pest control, directly benefiting agricultural yields (Kremen and Miles 2012).

10.3.5 Climate Resilience

Nature-based strategies like agroecology and organic farming increase the flexibility of cultivation systems to climate variability by improving microclimate regulation and decreasing greenhouse gas emissions (Altieri et al. 2015).

10.4 Nature Based Solutions for Circular Food Systems

NBS are crucial for enhancing circular food systems as they mimic natural processes to improve sustainability, resource efficiency, and resilience. By promoting practices like composting, agroforestry, and crop rotation, NBS enable nutrient cycling within agricultural systems, lower the demand for chemical fertilizers and reducing waste. Water conservation is another key benefit, with practices such as rainwater harvesting and constructed wetlands capturing, filtering, and recycling water for irrigation, ensuring efficient water use. Organic waste management through biogas production and composting transforms agricultural waste into valuable resources like renewable energy and nutrient-rich fertilizers, reducing landfill waste and closing nutrient loops (Pretty and Bharucha 2015). NBS also contribute significantly to soil regeneration and fertility; techniques such as cover cropping and decreased tillage enhance soil organic matter and structure, leading to healthier, more productive soils that require fewer external inputs. Moreover, biodiversity is enhanced through NBS like polyculture, agroecology, and habitat restoration, which support ecosystem aids like pest control and pollination. This biodiversity increases the durability of food systems to ecological changes, pests, and diseases, ensuring sustainable food production. NBS also play a crucial role in climate change reduction and transformation by sequestering carbon through practices like agroforestry and reforestation, while improving the strength of agricultural techniques to utmost weather conditions (Altieri et al. 2015). These practices lower greenhouse gas emissions and enhance water retention and protect against soil erosion. By reducing dependency on external inputs, NBS also contribute to local economic benefits, providing sustainable livelihoods and increasing food security for smallholder farmers. Additionally, integrating waste as a resource in circular food systems—such as using livestock manure to fertilize crops and converting crop residues into animal feed—creates closed-loop systems that enhance efficiency and reduce environmental impacts. These combined benefits of NBS make them vital for transitioning to more sustainable, circular food systems.

10.5 Nature Based Solution and Ecosystem Services

NBS are instrumental in enhancing ecosystem services, which are essential for maintaining the health and conservation of our environment. These services include nutrient cycling, water purification, carbon sequestration, and biodiversity support. For instance, agroforestry and crop rotation are key NBS practices that significantly improve nutrient cycling by integrating trees and diverse crops into farming systems. Trees in agroforestry systems capture deep soil nutrients and return them to the surface through leaf litter, enriching soil productivity and reducing the requirement for chemical fertilizers (Yu et al. 2022). Furthermore, constructed wetlands and riparian buffer zones are effective NBS for water refinement, as they sieve out pollutants, such as nitrogen and phosphorus, from agricultural runoff, thus preventing eutrophication in aquatic ecosystems (Ferreira et al. 2023). Moreover, NBS participate a critical role in carbon sequestration, a critical ecosystem service for mitigating climate change. Practices like afforestation, agroforestry, and cover cropping increase carbon storage in biomass and soils, thereby reducing greenhouse gas emissions and improving soil health (Bossio et al. 2020). Biodiversity support is another significant benefit of NBS, where practices like polyculture and habitat restoration create diverse habitats that attract beneficial species, improve pollination, and increase ecosystem resilience (Dainese et al. 2019). Overall, the incorporation of NBS into field management not only bolsters the provision of these ecosystem services but also upgrades the flexibility of ecosystems to environmental changes, securing long-term sustainability and productivity.

10.6 Nature Based Solution and Environmental Co-benefits

NBS have significant co-benefits for climate and atmospheric conditions by leveraging natural methods to mark climate change and improve air quality. For instance, reforestation and afforestation are prominent NBS that intensify carbon capturing, removal and storing atmospheric CO_2 in forest biomass and soils. This method not only alleviates climate change by reducing greenhouse gas concentrations but also helps regulate local temperatures and improve air quality (Grimm et al. 2023). Urban green spaces, including parks and green roofs, further contribute by cooling urban heat islands through evapotranspiration and shading, which reduces the need for air conditioning and lowers energy consumption (Barton et al. 2023). Additionally, these green spaces absorb pollutants like particulate matter and nitrogen dioxide, improving air quality and reducing respiratory problems among urban populations (Hartig et al. 2023). Agroforestry practices provide climate resilience by stabilizing soil, enhancing water retention, thereby supporting more sustainable land management and mitigating climate impacts (Zomer et al. 2022). Furthermore, wetland restoration, an important NBS, enhances atmospheric regulation by sequestering carbon in peat and wetland soils while also filtering pollutants from water, contributing to overall climate and environmental health. Collectively, these NBS address immediate climate challenges and offer long-term benefits for atmospheric stability and urban liveability.

10.7 Nature Based Solution for Soil Health Build-Up

NBS remarkably enhance soil health by addressing its chemical, physical, and biological characteristics. Cover cropping involves growing plants between main crop cycles to cover the soil, which reduces soil erosion, ameliorates its structure, and improves water infiltration. These crops also help in building soil organic matter (SOM), which provides better soil aggregation and reduced compaction (Lal 2016). Decreased tillage lowers soil disturbance, maintaining soil structure and preventing erosion, while agroforestry integrates trees with crops to stabilize soil and improve its physical properties through root systems that enhance soil structure and reduce erosion (Yu et al. 2022). Composting and incorporation of organic amendments improve soil properties by adding essential nutrients. This lessens the dependency on synthetic fertilizers and enhances nutrient availability. Green manures enhance soil's cation exchange capacity (CEC) and nutrient-holding capacity, thus making nutrients more approachable to plants (Altieri et al. 2015). Cover cropping also contributes to soil chemical health by adding organic matter that increases soil fertility and nutrient cycling (Bossio et al. 2020). Organic farming practices promote soil biological health by avoiding synthetic chemicals and enhancing biodiversity. These practices support a diverse soil microbial community, which is crucial for mineral cycling and organic matter decay. Mulching with organic materials such as straw or leaves provides a habitat for beneficial microorganisms and earthworms, enhancing soil biological activity and promoting a healthy soil ecosystem (Dainese et al. 2019).

10.8 Biofertilizers: An Emerging NBS

BFs offer a range of benefits for sustainable agriculture by enhancing soil fertility and plant overall health through natural processes and that's why play a potent role as NBS. BFs contain living microorganisms and when adding to the soil or plants, they can encourage plant growth by maximising the availability of minerals.

10.8.1 Key Benefits of Biofertilizers

10.8.1.1 Soil Fertility Improvement

Biofertilizers ameliorate soil by fixing atmospheric nitrogen, decomposing organic matter, and mobilizing nutrients that are otherwise inaccessible to plants. Nitrogen-fixing bacteria (e.g., *Rhizobium* spp.) and *arbuscular mycorrhizae* fungi are common types of biofertilizers that enhance soil nutrient accessibility and improve plant health and growth (Meshram et al. 2020). These microorganisms transform nutrients into a bioavailable form, hence minimizing the need for chemical fertilizers. Employment of algal biofertilizers magnify the beneficial microbe's community in soil which improve soil richness (Mishra et al. 2013) also they reduce soil salinity by the transition of minerals from soil to plants (Saadatnia and Riahi 2009). Biofertilizers like *arbuscular mycorrhizae* fungi establishes symbiotic associations with plant roots, which improves soil structure by enhancing root growth and soil aggregation. This symbiosis increases soil porosity and reduces compaction, leading to better water infiltration and root development (Fall et al. 2022). Due to filamentous structure of algal biofertilizers, they enlarge the soil pore size and play a pivotal role in increment of moisture holding capacity through the production of adhesive and viscous materials (Roger and Reynaud 1982). BFs contribute to mineral- nutrient cycle by degrading organic matter and liberating vital nutrients back into the soil. For example, phosphorus-solubilizing bacteria transform insoluble phosphorus into soluble forms that plants can readily absorb, thus enhancing phosphorus availability (Bashan and de-Bashan 2023). PSB like *Pseudomonas*, *Bacillus*, *Arthrobacter*, *Escherichia* have been documented to solubilize 30–35 kg of inorganic P ha^{-1} (Gaur et al. 2004). Similarly, nitrogen fixing bacteria convert atmospheric nitrogen into ammonia and play a crucial role in nitrogen- mineral cycle. *Azotobacter* can fix 20–40 kg N ha^{-1} $year^{-1}$ (Thomas and Singh 2019), *Azolla-Anabaena* can fix 30–60 kg N ha^{-1} $year^{-1}$ (Kollah et al. 2016) while *Rhizobium* can contribute 50–300 kg N ha^{-1} $year^{-1}$ (Brahmaprakash and Sahu 2012) in nitrogen cycle.

10.8.1.2 Reduction of Environmental Pollution

By reducing the requirement for chemical fertilizers, biofertilizers help in lower the probability of nutrient turn out and leaching, which can turn water pollution and eutrophication. Their application promotes more efficient nutrient use, thus minimizing environmental impacts (Van Der Heijden et al. 2015). Biofertilizers are capable of degrading contaminants from polluted soil and water by the process of bioremediation. *Pseudomonas*, *Arthrobacter*, *Citrobacter*, *Enterobacter* can reduce Pb, Cr, Ni, Cu concentrations by 7–37%. *Mesorhizobium*, *Rhizobium*, *Bradyrhizobium* have been documented as metal purifier diazotroph biofertilizers (Rizvia et al. 2022).

10.8.1.3 Promotion of Plant Health

BFs enhance the plant resistance to diseases and stress by reviving plant defence mechanisms and uplifting whole plant vigour. For instance, certain biofertilizers produce growth hormones and antibacterial/ antifungal molecules that defend plants from pathogens and pests (Bashan et al. 2023).

10.8.2 Application of BFs

To maximize the use of biofertilizers as NBS, effective application and management strategies are crucial. These strategies ensure that biofertilizers provide optimal benefits for soil health, crop productivity, and environmental sustainability. Here's a comprehensive approach to their application and management that includes. BFs must be selected on the basis of soil type, crop requirements, and specific nutrient needs. Different microorganisms (e.g., P solubilizing bacteria, nitrogen-fixing bacteria, and arbuscular mycorrhizae fungi) serve distinct functions, so matching the right type is essential (Meshram et al. 2020). Ensure high-quality formulations that contain viable and effective microorganisms. They can be in the form of granules, powders, or liquid suspensions. QC measures should be in place to maintain the potency and efficacy of biofertilizers (Bashan and de-Bashan 2023). Application of BFs at key growth stages is crucial to maximize their effectiveness. For example, application of BFs during planting or early growth stages when nutrient demands are high (Bashan et al. 2023). Optimum dose is also important for harnessing the potential of BFs Over-application can lead to imbalances in soil microbial communities or competition for nutrients (Lal 2016). Combined application of BFs with other organic applications, like composting and decreased tillage, to enhance overall soil health and sustainability (Altieri et al. 2015). Moreover, inclusion of BFs into comprehensive nutrient management plans is crucial for achieving the balanced and sustainable nutrient management (Van Der Heijden et al. 2015). There are three common methods of biofertilizers applications namely ***seed treatment:*** This method enhances early growth and improves seedling establishment by directly providing microorganisms to the root zone (Van Der Heijden et al. 2015), ***soil application:*** Incorporate biofertilizers into the soil through broadcasting, drilling, or mixing with compost. This method ensures that microorganisms are distributed throughout the root zone and can interact with plant roots (Bossio et al. 2020) and ***fertigation:*** This technique allows for precise application and better distribution of biofertilizers in the root zone, especially for large-scale farming operations (Fall et al. 2022).

10.8.3 Challenges and Suggestive Strategies for the Promotion of Biofertilizers

Adopting biofertilizers as an NBS in agriculture comes with several challenges. Addressing these challenges requires a multifaceted approach involving education, infrastructure improvements, and supportive policies. Farmers and agricultural practitioners often lack understanding of the benefits and application methods of biofertilizers. This knowledge gap can lead to underutilization or improper use of biofertilizers (Bashan et al. 2023). Hence, providing practical demonstrations and success stories can help illustrate their benefits and application methods (Bashan et al. 2023). The initial cost of biofertilizers can be higher compared to synthetic fertilizers as they may be needed in large amount to meet the desired results, and their availability may be limited in certain regions, especially in developing countries (Meshram et al. 2020). Hence, subsidies or financial incentives are crucial to reduce the initial cost of biofertilizers and make them more accessible, particularly for smallholder and resource-limited farmers. Developing cost-effective formulations can also help lower costs (Meshram et al. 2020). The efficacy of biofertilizers can change according to soil conditions, crop types, and climatic factors. This variability can lead to inconsistent results and reluctance to adopt them. Acquire research to improve the effectiveness of biofertilizers across various soil types and crops. Developing new formulations and application methods tailored to specific conditions can enhance their performance and reliability (Fall et al. 2022). The processing, quality, and formulation of biofertilizers can vary widely, and there is often a lack of standardization in production and application practices. This inconsistency can affect the reliability and effectiveness of biofertilizers (Barton et al. 2023). Establish and enforce standards for biofertilizer production and quality control. Certification systems can help in securing that biofertilizers manage good quality and efficacy requirements, increasing trust and adoption (Barton et al. 2023). Integrating biofertilizers with existing agricultural practices and synthetic fertilizers can be challenging, especially if there is a lack of knowledge about how to combine them effectively (Bossio et al. 2020). Hence, proper guidelines on how to use biofertilizers alongside synthetic fertilizers can help various stakeholders' transition smoothly (Bossio et al. 2020). Inadequate policies and regulations regarding the registration, use, and promotion of biofertilizers can hinder their widespread adoption (Van Der Heijden et al. 2015). Governments can play a key role by creating favourable conditions for research, development, and adoption of BFs (Van Der Heijden et al. 2015). Advanced collaborations between research institutions, industry, and government agencies is highly required to promote biofertilizer development and dissemination (Bashan and de-Bashan 2023).

10.9 Future Prospects and Challenges of NBS

The future prospects of NBS are promising as they offer sustainable and multifaceted approaches to describe world environmental challenges like ecological and climate change, ecosystem loss, and water scarcity. NBS, including reforestation, wetland restoration, and sustainable agricultural practices, are increasingly recognized for their potential to furnish ecological services, enhance resilience, and support biodiversity (BenDor et al. 2015). Advances in remote sensing and modeling techniques can further improve the ability to monitor and quantify the effectiveness of NBS, leading to more precise and impactful interventions (Pereira et al. 2023). However, challenges remain, such as the need for long-term funding, the integration of NBS into existing policy frameworks, and addressing socio-economic barriers that affect implementation (Miller et al. 2023). Furthermore, ensuring that NBS are equitably applied and do not exacerbate social inequalities is crucial for their success. As research and policy continue to evolve, addressing these challenges while leveraging technological and scientific advancements will be key to maximizing the potential of NBS for a sustainable future.

10.10 Conclusion

NBS offer a compelling pathway to enhancing food security and soil sustainability by harnessing natural processes to address key agricultural challenges. By integrating applications like cover cropping, agroforestry, and organic farming, NBS improve soil health through enhanced nutrient cycling, increased soil organic matter (SOM), and better soil structure, which in turn supports more resilient and productive agricultural systems. These solutions also bolster food security by enhancing crop productivity, decreasing dependence on synthetic inputs, and encouraging biodiversity, which promote ecosystem's stability and suppleness against pests and diseases. The holistic approach of NBS not only dwells instant agricultural demands also provides long-term benefits by mitigating climate change impacts, improving water management, and restoring degraded lands. Although, successful implementation of NBS faces array of challenges like limited awareness, high initial costs, and variability in effectiveness. However, by addressing these challenges through targeted education, financial incentives, and supportive policies NBS can be potentially utilize for achieving sustainable food systems and soil health. Overall, NBS represent a transformative approach to ensuring food security and soil sustainability.

References

Abhilash PC, Dubey RK, Tripathi V, Gupta VK (2016) Enhanced agricultural sustainability through nature-based solutions. COSUST 23:35–42

Adesemoye AO, Kloepper JW (2009) Plant-microbes interactions in enhanced fertilizer-use efficiency. Appl Soil Ecol 42(1):105–111

Altieri MA, Nicholls CI, Henao A, Lana MA (2015) Agroecology and the design of climate change-resilient farming systems. ASD 35(3):869–890

Barton H, Grant M, Guise R (2023) The role of urban green infrastructure in improving air quality. Environ Res Lett 18(4):043501

Bashan Y, De-Bashan LE (2023) Plant growth-promoting bacteria: a review of their mechanisms and potential applications. Appl Soil Ecol 177:104954

Bashan Y, De-Bashan LE, Moreno MT (2023) The role of biofertilizers in plant health: an overview. J Plant Growth Regul 42:1–14

BenDor TK, Livengood A, Lester TW, Davis A, Yonavjak L (2015) Defining and evaluating the ecological restoration economy. Restoration Ecology 23(3):209–19

Bhattacharyya PN, Jha DK (2012) Plant growth-promoting rhizobacteria (PGPR): emergence in agriculture. World J Microbiol Biotechnol 28(4):1327–1350

Bossio DA, Cook-Patton SC, Ellis PW, Fargione J, Sanderman J, Smith P, Wood S, Zomer RJ, Unger M, Emmer IM, Griscom IW (2020) The role of soil carbon in natural climate solutions. Nat Sustain 3:391–398

Brahmaprakash G, Sahu PK (2012) Biofertilizers for sustainability. J Indian Inst Sci 92:37–62

Calder IR (2000) The blue revolution: land use and integrated water resources management. Routledge

Chazdon RL (2008) Beyond deforestation: restoring forests and ecosystem services on degraded lands. Science 320(5882):1458–1460

Copping LG, Menn JJ (2000) Biopesticides: a review of their action, applications, and efficacy. Pestic Sci 54(2):182–196

Creamer NG, Baldwin KR (2000) An evaluation of summer cover crops for use in vegetable production systems in North Carolina. HortScience 35(4):600–603

Dainese M, Martin EA, Aizen MA, Albrecht M, Bartomeus I, Bommarco R, Garibaldi LA (2019) A global synthesis reveals biodiversity-mediated benefits for crop production. Sci Adv 5(10):eaax0121

Damalas CA, Koutroubas SD (2018) Current status and recent developments in biopesticide use. Agriculture 8(1):13

Daniel AI, Fadaka AO, Gokul A, Bakare OO, Aina O, Fisher S, Burt AF, Mavumengwana V, Keyster M, Klein A (2022) Biofertilizer: the future of food security and food safety. Microorganisms 10(6):1220

Falkenmark M, Rockström J (2006) The new blue and green water paradigm: breaking new ground for water resources planning and management. J Water Resour Plan Manag 132(3):129–132

Fall AF, Nakabonge G, Ssekandi J, Founoune-Mboup H, Apori SO, Ndiaye A, Badji A, Ngom K (2022) Roles of Arbuscular Mycorrhizal fungi on soil fertility: contribution in the improvement of physical, chemical, and biological properties of the soil. Front Fungal Biol 3:723892

FAO (2017a) The future of food and agriculture: trends and challenges. Food and Agriculture Organization, Rome

FAO (2017b) Agroforestry for food security and nutrition. Food and Agriculture Organization, Rome

FAO (2021) The state of food security and nutrition in the world

Ferreira CSS, Kasanin-Grubin M, Kapovi M, Solomun S, Sushkova M, Minkina T, Zhao W, Kalantari Z (2023) Wetlands as nature-based solutions for water management in different environments. Curr Opin Environ Sci 33:100476

Freitaga M, Friedrichb T, Kassam A (2024) The carbon footprint of conservation agriculture. IJAS 2331949:1–20

Galli A, Wiedmann T, Ercin E, Knoblauch D, Ewing B, Giljum S (2012) Integrating ecological, carbon and water footprint into a "footprint family" of indicators: definition and role in tracking human pressure on the planet. Ecol Indic 16:100–112

Gaur R, Noam S, Johri B, Rossi P, Aragno M (2004) Diacetylphloroglucinol-producing pseudomonads do not influence AM fungi in wheat rhizosphere. Curr Sci 88:453–457

Getnet D, Mekonnen Z, Anjulo A (2023) The potential of traditional agroforestry practices as nature-based carbon sinks in Ethiopia. NBS 4:100079

Glare TR, Caradus JR, Gelernter WD, Jackson TA, Keyhani NO, Köhl J, Stewart A (2012) Have biopesticides come of age? Trends Biotechnol 30(5):250–258

Godfray HCJ, Beddington JR, Crute IR, Haddad L, Lawrence D, Muir JF, Pretty J, Robinson S, Thomas SM, Toulmin C (2010) Food security: the challenge of feeding 9 billion people. Science 327:812–818

Gregory SV, Swanson FJ, McKee WA, Cummins KW (1991) An ecosystem perspective of riparian zones. BioSci 41(8):540–551

Grimm NB, Faeth SH, Golubiewski NE, Redman CL, Wu J, Bai X (2023) The ecology of cities and towns: urban ecosystems and their services. Annu Rev Ecol Evol Syst 54:295–321

Hartig T, Mitchell R, De Vries S, Frumkin H (2023) Nature and health. Annu Rev Public Health 44:1–23

Haruna SI, Anderson SH, Udawatta RP, Gantzer CJ, Phillips NC, Cui S, Gao Y (2020) Improving soil physical properties through the use of cover crops: a review. Age 3:e20105

Jose S (2009) Agroforestry for ecosystem services and environmental benefits: an overview. Agrofor Syst 76:1–10

Jyotsna B, Patil S, Prakash YS, Rathnagiri P, Kishor PBK, Jalaja N (2024) Essential oils from plant resources as potent insecticides and repellents: current status and future prospectives. Biocatal Agric Biotechnol 61:103395

Kassam A, Friedrich T, Shaxson F, Pretty J (2009) The spread of conservation agriculture: justification, sustainability and uptake. Int J Agric Sustain 7(4):292–320

Kassam A, Friedrich T, Derpsch R (2018) Global spread of conservation agriculture. Int J Environ Stud 76(1):29–51

Kollah B, Patra AK, Mohanty SR (2016) Aquatic microphylla *Azolla*: a perspective paradigm for sustainable agriculture, environment, and global climate change. Environ Sci Pollut Res 23:4358–4369
Kremen C, Miles A (2012) Ecosystem services in biologically diversified versus conventional farming systems: benefits, externalities, and trade-offs. Ecol Soc 17(4):40
Kumar KK, Sridhar J, Murali- Baskaran RK, Senthil- Nathan S, Kaushal P, Dara SK, Arthurs S (2019) Microbial biopesticides for insect pest management in India: current status and future prospects. J Invertebr Pathol 165:74–81
Lal R (2011) Soil health and climate change: an overview. In: Singh B, Cowie A, Chan K (eds) Soil health and climate change, Soil biology. Springer, Berlin/Heidelberg, pp 1–24
Lal R (2015) Restoring soil quality to mitigate soil degradation. Environ Sci Pol 7(5):5875–5895
Lal R (2016) Biochar and soil carbon sequestration. In: Guo M, He Z, Uchimiya SM (eds) Agricultural and environmental applications of biochar: advances and barriers. Soil Science Society of America, Inc, Madison, pp 175–197
Lamb D (2014) Large-scale forest restoration, 1st edn. Routledge
Leng P, Zhang Z, Pan G, Zhao M (2014) Applications and development trends in biopesticides. Afr J Biotechnol 10:19864–19873
Lotter DW (2003) Organic agriculture. J Sustain Agric 21(4):59–128
Mäder P, Fließbach A, Dubois D, Gunst L, Fried P, Niggli U (2002) Soil fertility and biodiversity in organic farming. Science 296(5573):1694–1697
Meshram NK, Sagar W, Manoj (2020) Biofertilizer and their role in sustainable agriculture-a review. Outlook Agric 9:127–130
Miller J, Vesuviano G, Wallbank J, Fletcher D, Jones L (2023) Hydrological assessment of urban Nature-Based Solutions for urban planning using Ecosystem Service toolkit applications. Landsc Urban Plan 234:104737
Mishra DJ, Singh R, Mishra UK, Kumar SS (2013) Role of bio-fertilizer in organic agriculture: a review. Res J Recent Sci 2:39–41
Mitsch WJ, Gosselink JG (2015) Wetlands, 5th edn. Wiley
Naiman RJ, Décamps H (1997) The ecology of interfaces: riparian zones. Annu Rev Ecol Syst 28:621–658
Nair PKR (2014) Grand challenges in agroecology and land use systems. Front Environ Sci 2:1–5
Newaj R, Chavan S, Alam B (2016) Biomass and carbon storage in trees grown under different agroforestry systems in semi-arid region of central India. Indian For 142(7):642–648
Palm C, Blanco-Canqui H, DeClerck F (2014) Conservation agriculture and ecosystem services: an overview. Agric Ecosyst Environ 187:87–105
Pereira P, Yin C, Hua T (2023) Nature-based solutions, ecosystem services, disservices, and impacts on well-being in urban environments. Curr Opin Environ Sci Health 33: 100465
Pretty J (2008) Agricultural sustainability: concepts, principles and evidence. Philos Trans R Soc B Biol Sci 363(1491):447–465
Pretty J, Bharucha ZP (2015) Integrated pest management for sustainable intensification of agriculture in Asia and Africa. Insects 6(1):152–182
Rashid MI, Mujawar LH, Shahzad T, Almeelbi T, Ismail IM, Oves M (2016) Bacteria and fungi can contribute to nutrients bioavailability and aggregate formation in degraded soils. Microbiol Res 183:26–41
Reganold JP, Wachter JM (2016) Organic agriculture in the twenty-first century. Nat Plants 2:15221
Rizvia A, Ahmedb B, Umara S, Khan MS (2022) Bacterial biofertilizers for bioremediation: a priority for future research. In: Soni R, Suyal DC, Yadav AN, Goel R (eds) Trends of applied microbiology for sustainable economy. Academic, pp 565–612
Roger PA, Reynaud PA (1982) Free living blue green algae in tropical soils. In: Dommergues YR, Diem HG (eds) Microbiology of tropical soils and plant productivity. Martinus Nijhoff Publisher, La Hague, pp 147–168
Saadatnia H, Riahi H (2009) Cyanobacteria from paddy fields in Iran as a biofertilizer in rice plants. Plant Soil Environ 55:207–212

Sarrantonio M, Gallandt ER (2003) The role of cover crops in North American cropping systems. J Crop Prod 8:53–74
Schmidt M, Torn M, Abiven S, Dittmar T, Guggenberger G et al (2011) Persistence of soil organic matter as an ecosystem property. Nature 478:49–56
Snapp SS, Swinton SM, Labarta R, Mutch D, Black JR, Leep R, O'Neil K (2005) Evaluating cover crops for benefits, costs, and performance within cropping system niches. J Agron 97(1):322–332
Stewart W, Roberts T (2012) Food security and the role of fertilizer in supporting it. Proc Eng 46:76–82
Sweeney BW, Newbold JD (2014) Streamside buffer width needed to protect stream water quality, habitat, and organisms: a literature review. JAWRA 50(3):560–584
Teasdale JR, Brandsaeter LO, Calegari A, Neto FS, Upadhyaya MK, Blackshaw RE (2007) Cover crops and weed management. In: Upadhyaya MK, Blackshaw RE (eds) Non-chemical weed management: principles, concepts and technology. CABI, Wallingford, pp 49–64
Thomas L, Singh I (2019) Microbial biofertilizers: types and applications. In: Giri B, Prasad R, Wu QS, Varma A (eds) Biofertilizers for sustainable agriculture and environment. Springer, Cham, pp 1–19
Tilman D, Cassman K, Matson P, Naylor R, Polasky S (2002) Agricultural sustainability and intensive production practices. Nature 418:671–677
Tiwari KR, Sitaula BK, Nyborg ILP (2008) Determinants of farmers' adoption of improved soil conservation technology in a middle mountain Watershed of Central Nepal. Environ Manag 42:210–222
Van der Heijden MGA, Martin FM, Selosse MA, Sanders IR (2015) Mycorrhizal ecology and evolution: the past, the present, and the future. New Phytol 205(4):1406–1423
Verhoeven JTA, Setter TL (2010) Agricultural use of wetlands: opportunities and limitations. Ann Bot 105(1):155–163
Vessey JK (2003) Plant growth promoting rhizobacteria as biofertilizers. Plant Soil 255(2):571–586
Yu N, Liu J, Ren B, Zhao B, Liu P, Gao Z, Zhang J (2022) Long-term integrated soil-crop management improves soil microbial community structure to reduce GHG emission and increase yield. Front Microbiol 13:1024686
Zedler JB, Kercher S (2005) Wetland resources: status, trends, ecosystem services, and restorability. Ann Rev Environ Res 30:39–74
Zomer RJ, Trabucco A, Coe R, Place F (2008) Trees on farm: analysis of global extent and geographic patterns of agroforestry. ICRAF Working Paper 89
Zomer RJ, Trabucco A, Bossio DA, Verchot LV (2022) Climate change mitigation: a spatial analysis of global land suitability for clean development mechanism afforestation and reforestation. Agric Ecosyst Environ 126:67–80

Ecosystem Restoration for Achieving the Land Degradation Neutrality

11

Anjali Patel, Sanjeev Kumar, Subhash Babu, Rajiv K. Singh, S. S. Rathore, and Kapila Shekhawat

Abstract

The degradation of our planet's forests, rangelands, and agricultural lands poses an enormous risk to sustainable development, biodiversity conservation and our capacity to mitigate and adapt over climate change. This degradation undermines the actual foundations of human well-being, including water security, food and livelihoods, and exacerbates issues such as conflict, poverty and displacement. The basic causes for degrading land are multifaceted, including overharvesting of natural resources, overexploitation for agriculture and the broader forces of economic development and populace increase. In retort to this global challenge, the idea of LDN (Land Degradation Neutrality) become familiarized at the Rio + 20 Earth Summit and later enshrined within the UNSDGs (Target 15.3). LDN objects to break or reverse land degradation, ensuring that land, now and in the future, is managed sustainably. It is not an additional burden but an opportunity to integrate conservation and development in a way that enhances the value of conservation efforts. Further, ecosystem restoration is the method of returning degraded environments to a more natural state, is emerging as a key strategy for achieving LDN. It offers a powerful tool for mitigating and adapting the climate change impacts and ecological disasters. By focusing upon restoration, conservation and sustainable management, we can create more resilient ecosystems that support biodiversity and human needs. To realize the full potential of LDN, we must adopt integrated ecosystem management approaches that ensure sustainability at all levels and avoid unintentional consequences. The conservation

A. Patel (✉) · S. Babu · R. K. Singh · S. S. Rathore · K. Shekhawat
ICAR-Indian Agricultural Research Institute, New Delhi, India

S. Kumar
Department of Soil Science & Agricultural Chemistry, Sardar Vallabhbhai Patel University of Agriculture and Technology, Meerut, Uttar Pradesh, India

© The Author(s), under exclusive license to Springer Nature Singapore Pte Ltd. 2025

S. Babu et al. (eds.), *Ecological Solutions to Agricultural Land Degradation*,
https://doi.org/10.1007/978-981-96-3392-0_11

sector is already equipped with a range of tools and practices to deliver LDN, and these must be applied in a gender-responsive manner to ensure inclusivity and effectiveness. As the world embarks on an era of ecological restoration, with Governments committing to ambitious restoration projects, it is authoritative that we understand the significance of LDN within a broader sustainable development and conservation agendas.

Keywords

Ecosystem · Land use system · LDN · Restoration

Abbreviation

$	Dollar
%	Percent
°C	Degree Celsius
e.g.	For example
ABM	Agent-based models
AFR	Africa Forest Landscape Restoration Initiative
ANR	Assisted Natural Regeneration
CA	Cellular Automata
CBD	Convention on Biological Diversity
CEM	Commission on Ecosystem Management
CO_2	Carbon dioxide
CoE-SLM	Centre of Excellence on Sustainable Land Management
COP12	Conference of the Parties (COP 12), November 2006 in Nairobi, Kenya
EBA	Ecosystem-Based Adaptation
ELD	Economics of Land Degradation Initiative
ESG	Environmental, Social, and Governance
EU	European Union
FAO	Food and Agriculture Organization
Fig.	Figure
FLR	Forest Landscape Restoration
GDP	Gross Domestic Product
GEF	Global Environment Facility
GHGs	Green House gases
GIS	Geographical Information System
GtC	Gigatonnes of carbon
$GtCO_2$	Gigatonnes of carbon dioxide
i.e.	That is
ICAR	Indian Council of Agricultural Research
IPCC	Intergovernmental Panel on Climate Change
IPLCs	Indigenous Peoples and Local Communities
IRS	Indian Remote Sensing Satellites

ISIMIP	The Inter-Sectoral Impact Model Intercomparison Project
IUCN	International Union for Conservation of Nature
IWRM	Integrated Water Resource Management
JFMCs	Joint Forest Management Committees
LDN	Land Degradation Neutrality
MEA	Millennium Ecosystem Assessment
Mha	Million hectares
MRV	Monitoring, Reporting and Verification
NAAS	National Academy of Agricultural Sciences
NAMAs	Nationally Appropriate Mitigation Actions
NBSAP	National Biodiversity Strategies and Action Plans
NDCs	Nationally Determined Contributions
NGOs	Non-governmental organizations
Rio+20	United Nations Conference taken on Sustainable Development, held in Rio de Janeiro, Brazil in June 2012
SAC	Space Applications Centre
SDGs	Sustainable Development Goals
SER	Society for Ecological Restoration
SOC	Soil organic carbon
SSM	Sustainable soil management
UN	United Nations
UNCCD	United Nations Convention to Combat Desertification
UNFCCC	United Nations Framework Convention on Climate Change
UNU-EHS	United Nations University Institute for Environment and Human Security
USA	United States of America
VCS	Verified Carbon Standard
WWF	World Wide Fund for Nature

11.1 Introduction

Land covers about 29% of the Earth's surface, totalling 149 million square kilometres, however, only 71% land is considered inhabitable (UNCCD 2019). Global land helps over 7.8 billion humans, along with countless other life bureaucracy, which includes an anticipated 0.28 million plant species, 5.6 million animal species, and 0.61 million fungi species (Mora et al. 2011). Degraded land has emerged as a sizable international issue (IPCC 2019). The key roots of the deprivation are human doings and climate variability, which have severe impacts across dry sub-humid, semi-arid, and arid regions (UNCCD 2022). About 40% and 20% of land area globally is considered affecting diverse landscapes such as croplands, drylands, wetlands, forests, and grasslands and degraded land, respectively. Land degradation states to the ongoing decline in the capability of ecologies to offer crucial services over a prolonged period (MEA 2005). The UNCCD specifically describes desertification as degradation of land in arid regions inclusive sub-humid, semi-arid, and arid regions, resultant in loss of economic and biological efficiency (UNCCD 2022). This degradation not only diminishes natural resource's productivity but also poses a thoughtful danger to biodiversity (Shao et al. 2016).

Various countries have made commitments underneath global agreements like the Rio Conventions and the Bonn mission to repair a thousand Mha of degraded area via 2030 to tackle alarming conditions. Land restoration involves halting degradation or rehabilitating degraded land, usually via strategies which includes safeguarding natural processes, and reforestation (UNU-EHS 2023). Restoration is a cost-effective and recognized approach to combat biodiversity loss and climate change, safeguard essential natural services, maintain land productivity, reduce the risks of disasters and pandemics, and strengthen the flexibility of both communities and ecosystems contrary to climate tremors and environmental stresses. An ecosystem comprises all living organisms in a particular area interacting with each other and their non-living surroundings. Each organism within an ecosystem has a specific role or niche. Ecosystem restoration involves aiding the restoration of ecosystems that have been destroyed or damaged, in addition to preserving those which remain intact. Restoration can apply to various types of ecosystems, such as forests, farmlands, urban areas, wetlands, freshwater environments, grasslands, peatlands, oceans, and coastal regions. Nearly anyone, including governments, development agencies, businesses, communities, and individuals, can initiate restoration efforts. This practice is complex and dynamic, requiring a well-considered mix of best practices to achieve effective and lasting results. Successful restoration relies on understanding ecosystems, planning, implementing, and managing the process continuously.

11.1.1 Global Statistics on Land Degradation

Land degradation characterize by the deterioration of land's economic or biological functions of land under rainfed, irrigated, range, pasture, forest, and woodlands because of a broad spectrum of natural and man-made causes such as water and wind-accelerated soil erosion, deterioration of soil's chemical, biological, and physical characteristics, and the consequent long-term native vegetation loss (UNCCD 2012). The transformation of once-fertile land into dry desert is known as desertification, and it has far-reaching consequences for human health due to a cascade of interconnected processes. Globally, 33% of the land has been under a severe form of degradation affecting about 3 billion world's population which is likely to worsen in the future due to increased food demand (Mekouar 2020) and about 100 Mha of land has already been lost in world as a result of different land deprivation forms (UNCCD 2019). The ELD (Economics of Land Degradation Initiative) estimates that land degradation results in annual lack of ecosystem services valued between \$6.3 trillion and \$10.6 trillion globally, that is equivalent to world's 10–17% GDP (ELD 2019). Further, about 40% and 9% of agricultural lands experience moderate degradation and severe degradation, respectively which can impact adversely global food productivity by 13% (Mohamed et al. 2019). Erosion by water and wind, deterioration by vegetation, salinity, water logging, and other variables all contribute to agricultural land degradation. Human activities and the extensive exploitation of land have degraded around 25% of ice-free land area of Earth. However, estimates of global land degradation vary, the Global Environment Facility (GEF) estimates 25% of land is degraded, while the "World Atlas of Desertification" suggests a staggering 75%, and the Global Land Outlook places the figure at 40% (UNCCD 2022). Regional impacts also vary, with Sub-Saharan Africa experiencing 5% land degradation due to human activities, Southern America 17% and South Asia suffering the most (~41% of its land), 70% of that is severely affected. Southeast Asia is followed by Western Asia with 20% of its land degraded, 75% of that is severely impacted (FAO 2021b).

In response to this growing crisis, major international bodies like the UNCCD (United Nations Convention to Combat Desertification), CBD (Convention on Biological Diversity), and UNFCCC (United Nations Framework Convention on Climate Change) have identified restoration as a key approach to reach their goals. The Bonn Challenge, as an instance, aims to brings back 350 M ha of deforested land globally by 2030, with over 60 countries and jurisdictions pledging their support (IUCN 2020). Initiative 20 × 20, associated with the Bonn Challenge, targeted the restoration of 20 M ha in the Caribbean and Latin America by 2020, aiming for 50 M ha by 2030. The AFR100 (Africa Forest Landscape Restoration Initiative), which was launched at Paris's Global Landscapes Forum in 2015, has established a goal to restore 100 M ha throughout Africa through 2030. These efforts are part of a broader movement, including the Global Partnership on Landscape and Forest Restoration, that brings together individuals, communities, organizations, and governments and with the shared objective of restoring degraded and deforested lands worldwide. Over past two decades, changes in land usage have significantly reduced the value of this vital resource, with current losses estimated at $4.3 to $20 trillion per year. The ELD report suggests that effectively tackling land degradation could boost global income by an additional $75.6 trillion annually (ELD 2019). Furthermore, the UN Decade on Ecosystem Restoration, running from 2021 to 2030, seeks to break ecosystem degradation, and restore them to support universal sustainability goals, fostering a robust global movement to accelerate restoration efforts (Global Environment Facility 2024).

11.1.2 Land Degradation Scenario in India

In India, promoting restoration in respect to address land degradation is crucial due to its significant influence on the well-being and the economy of millions. With 2.4% of the Earth's land area supports 18% of the global population, India experiences immense pressure on its natural resources (SAC 2021). Out of country's total 328.7 million hectares land, about 228.3 million hectares are characterized as semi-arid, arid, and dry sub-humid, regions, with 83.7 million hectares suffering from desertification. Agriculture, which employs 54.6% of India's workforce, is a cornerstone of the economy. According to Ministry of Agriculture and Farmers Welfare

(2021), 139.4 million hectares of land are used for cultivation, with 68.65 million hectares (49.2%) being irrigated, while the remaining 70.8 million hectares (50.8%) rely on rainfall. Rainfed areas, often resource-deficient and prone to land degradation, experience low productivity owing to climate's unpredictable nature.

The India's Land Degradation Graphics highlights that nine states namely, Rajasthan, Gujarat, Maharashtra, Karnataka, Odisha, Ladakh, Madhya Pradesh, Telangana, and Jharkhand account for 80% of the country's degraded lands. Since 2003–2005, fourteen states have experienced over a 10% rise in land degradation, including Mizoram (188%), Punjab (80%), Arunachal Pradesh (47%), and Assam (46%), among others. Un-irrigated forest areas and agricultural land contribute to 60% of the country's over-all degraded land (SAC 2021). This escalating degradation poses significant challenges to the agricultural sustainability and the livelihoods of millions who depend on it.

Out of country's entire land area (328.73 M ha), over 120.72 million hectares are degraded annually (ICAR/NAAS 2010). Further, about 251.71 million Indians comprising 18.39% of the population were exposed to land degradation (UNCCD 2019). More than 30% of land is degraded in 13 Indian states leading to a decrease in health, biodiversity, and production (Statista Research Department 2024). Water and wind erosion together account for 70% of land degradation, whereas salinity accounts for 15% of the overall degraded area (Mythili and Goedecke 2016). Approximately 50% of the central, northeastern, and southern regions are badly altered by land degradation. The challenge of providing for the anticipated 60% increase in demand for food, feed, and fiber (FAO 2015) urges to reverse land degradation for a sustainable land system. SDG 15 of the UNs also emphases on achieving land degradation neutrality.

11.1.3 Environmental, Social and Economic Consequences

Global land resources are increasingly under pressure due to agriculture and land use alteration. The primary drivers of land degradation worldwide include unsustainable resource management, overgrazing, over-exploitation, inappropriate irrigation practices, mining, pollution, nutrient depletion, excessive use of agrochemicals, deforestation, quarrying, and urban expansion. These issues are particularly acute in world's driest regions, where the impacts are most severe. Drylands cover ~40% of Earth's land zone is home to around two billion human population. The majority of those who rely on these dryland landscapes live in developing countries, wherever effects of drought and land degradation are most felt, especially among women and children, who are particularly vulnerable (Global Environment Facility 2024).

11.1.4 Concept of LDN

LDN (Land Degradation Neutrality) is defined by the UNCCD as a state where the quality and quantity of land resources, essential for supporting ecosystem services and functions while enhancing food security, persist stable or improve. Land degradation is a significant universal environmental issue (Qadir et al. 2014), not only contributes to climate change but also accelerates the nitrogen and SOC loss (Das et al. 2022). The concept of LDN was initially introduced as "zero net land degradation" at the 2012 Rio+20 UN Conference on Sustainable Development. LDN objects to re-establish already degraded land and to prevent further deprivation, caused by environmental or human factors. The Convention Parties describe LDN as a specific condition where land resources remain stable or improve, ensuring they can continue to backing food security and ecosystem functions within specific spatial and temporal scales (Mukherjee et al. 2018).

Land degradation management has multiple assistances, including climate change adaptation, mitigation, enhanced food security, biodiversity preservation, and sustainable livelihoods. Attaining LDN involves three key activities: maintaining healthy land to avoid new land degradation, minimizing prevailing degradation through sustainable land management practices, and restoring degraded lands to more productive or natural state. LDN gives a significant possibility aimed at the IUCN preservation and improvement sectors to address one of the maximum urgent demanding situations to sustainable development and conservation. It's far a striving aim which allows international locations to trail development in the direction of environmental sustainability throughout their whole terrestrial region and is an ambition that all conservation participants must assist (Fig. 11.1).

The concept of LDN was introduced at the 2012 UN Conference on Sustainable Development (Rio+20) to tackle the pressing challenge of increasing the fuel, fibre, and food production to encounter future demands without more degrading limited land resources. LDN envisions a world where human activities have a neutral or even positive impact on the land. The GEF has supported this vision through its LDN portfolio, which includes 67 projects approved between 2016 and 2022 during the GEF-6, and GEF-7 funding cycles. These projects, implemented across 56 countries and involving nine different GEF agencies, are in various stages of progress, from initial planning to mid-term evaluations. The total investment in these initiatives amounts to \$320 million in GEF grants, with over \$2.5 billion in co-financing (Global Environment Facility 2024).

11.1.5 Global and Regional Targets for Achieving LDN

The 12th session of the Parties' Conference to the UNCCD COP12 adopted 35 decisions addressing land degradation, drought, and desertification during 2015. These decisions included strategies for pursuing LDN within the framework of the SDGs and aligning UNCCD goals with the SDGs. As a follow-up, the Global Mechanism and the UNCCD Secretariat launched the LDN Target Setting Program, which is now part of the SDGs as target 15.3, aiming to acquire a world's neutral land degradation by 2030.

WWF (World Wide Fund for Nature) indorses Parties to the UNCCD to:

- Expedite and expand the coordinated execution of restoration pledges under the Rio Conventions, such as the UNCCD's LDN targets, upcoming national action

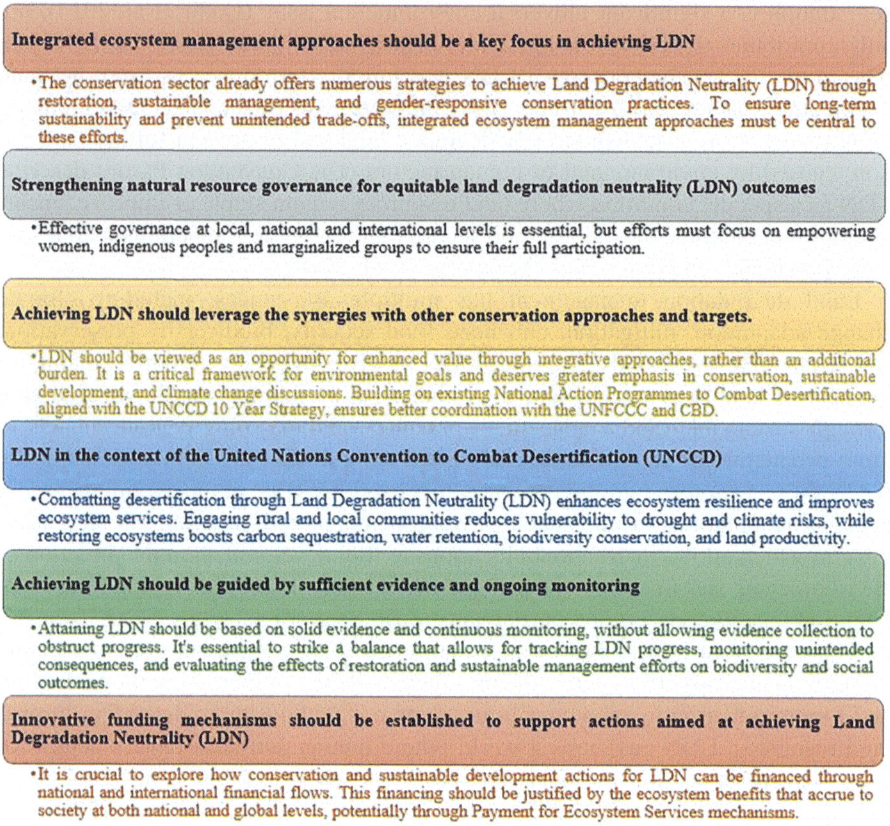

Fig. 11.1 Striving goal for environmental sustainability: tracking progress across terrestrial areas

plans within the Post-2020 Global Biodiversity Framework of the CBD, and the NDCs (Nationally Determined Contributions) and longstanding policies of the UNFCCC, as well as national initiatives focused on achieving the UN SDGs. This strategy aims to optimize effect and efficiency while encouraging collaborative efforts among the private sector, governments, and other stakeholders.

- Establish a strong connection between Land Degradation Neutrality (LDN), Nature Positive initiatives, and Nature-based Solutions. LDN serves as a crucial foundation for achieving an Eco-Restorative World by 2030, as highlighted in the modern Global Land Viewpoint by the UNCCD, and plays a vital role in addressing broader societal goals through Nature-based Solutions, particularly in restoration efforts.
- Create inclusive and holistic policy frameworks to facilitate the restoration of dishonoured natural or semi-natural bionetworks, including rangeland and agricultural areas. These policies must focus on strengthening restoring ecosystem services, supporting rural livelihoods, food security, and avoiding or minimizing the transformation of more natural ecosystems for agricultural production.

- Indorse incentives and sector-specific circumstances to finance and implement restoration efforts throughout the economy. This approach must incorporate helpful land tenure policies and guarantee meaningful contribution from IPLCs (Indigenous Peoples and Local Communities), as well as from youth and women.

WWF urges the UNCCD Secretariat to strengthen associations amongst investors on data restoration to monitor advancement on commitments under pertinent Contexts, Partnerships, and Conventions. This effort will enhance coordination and alignment, improve monitoring, and assess impact as the UN Decade on Ecosystem Restoration advances.

11.1.6 India's Pledge for Attaining LDN

India has shown significant leadership in tackling land degradation, first committing at the 21st Conference of the Parties (CoP) of the UNFCCC to restore 13 M ha of deforested or degraded land by 2020, with a further 8 M ha targeted for restoration by 2030. In 2019, as chair of CoP 15 of the UNCCD, India further raised its ambition, revising its goal to reinstate 26 M ha by 2030 (UNCCD 2019).

11.1.7 The Role of Ecosystem Restoration in Achieving LDN

Restoration holds significant potential for both climate mitigation and adaptation, making it a key nature-based solution and an integral part of NDCs (Nationally Determined Contributions). Landscape and forest restoration, is particularly recognized as a vital eco-based climate solution, with strong scientific backing for its role in mitigating climate change. To live with a 1.5 °C pathway, an additional 100–1000 $GtCO_2$ have to be removed from atmosphere alongside achieving net-zero carbon emissions. Landscape and forest restoration can play a crucial role in meeting these removal targets. Some studies estimate that reforestation and afforestation could have an annual mitigation potential of up to 10.1 $GtCO_2$, while others indicate that tree restoration could sequester between 133.2 and 276.2 GtC. Additionally, restoring native wetlands and forests is considered 1/5th of the most cost-efficient climate solutions. Despite varying definitions of restoration, its importance as a climate action is undeniable (WWF 2022).

Integrating and harmonizing national motion plans underneath the Rio Conventions into a unified, single strategy offering a valuable prospect to align restoration commitments, targets, and goals. This approach would enable the realization of multiple benefits simultaneously and maximize returns on investment. Greater coherence would enhance departmental and sectoral harmonization, generate a pipeline of investment-prepared programs and schemes, and involve both the civil society and private sector to achieve cost-efficient, mutual gains through land restoration.

While numerous promising pledges under various initiatives and conventions are there, the shift from commitment to execution is critical for achieving the LDN targets, under the UNCCD and advancing the UN Decade for Restoring Ecosystems. However, tracking global restoration progress is challenging, since countries apply varied metrics. Enhancing reporting consistency and alignment in data, baselines, and indicators is essential for increasing the effect of hands-on restoration efforts. As the lead agency responsible for overseeing SDG target 15.3, the UNCCD has a distinctive function in promoting improved synchronization between reported data and restoration commitments across various Bonn Challenge and conventions (WWF 2022).

11.1.8 Examples of Successful Ecosystem Restoration Contributing to LDN

Though forest and landscape restoration are widely recognized for its importance, the role of other ecosystems often goes undervalued. Savannahs and grasslands, for instance, have been discovered to retain nearly as much carbon as forests, accounting for approximately one-third of global terrestrial carbon reserves. Peatlands are even more remarkable, as they are the most carbon-rich terrestrial ecosystems, that can store up to two-fold carbon per hectare as timberlands. Despite covering <3% of the globe's surface, peatlands accumulate $1/3^{rd}$ of the planet's soil organic carbon, averaging about carbon sequestration rate of 1375 tonnes per hectare. Mangroves also have significant carbon storage potential, with their ability to sequester carbon being approximately 50% greater than that of tropical rainforests (470 tonnes of carbon per hectare in comparison to 320 tonnes per hectare). Currently, mangrove ecosystems hold >21 gigatons CO_2, but the yearly destruction of mangroves releases 24 Mt of CO_2, roughly equivalent to the emissions of 5 million cars in a year. Additionally, rivers serve as a key component of the carbon cycle, transporting 200 Mt of carbon to the ocean annually (Singh et al. 2023).

11.2 Drivers and Causes of Land Degradation

The alteration of woodlands, forests, and rangelands into croplands is a leading cause of land degradation (MEA 2005; Bai et al. 2008), a problem that is expected to worsen with climate change, as it intensifies freshwater scarcity. If global population growth continues, coupled with increasing food demand and ongoing waste, food production might need to rise by 70–100%, potentially driving further deforestation and rangeland clearance if sustainable farming practices are neglected (FAO 2009). The most apparent causes of land degradation, such as agricultural expansion and excessive resource extraction, are often rooted in deeper issues like population growth, economic development, insecure land tenure, and inadequate policies (Geist and Lambin 2004).

Tackling land degradation necessitates an understanding of underlying causes and strong political commitment to implement solutions. However, this challenge is compounded by disparities in knowledge and perspectives on land's function, which is seen as an ecosystem, a landscape, a planning unit, or a cultural concept, depending on the context. These varying interpretations, along with the complexity of land systems, contribute to the implementation of policies and investments that can inadvertently drive land degradation.

In drylands, crop production is increasingly taking place in regions with lower rainfall and inferior soils, where the sustainability is uncertain, leading to land desertion. Government often prioritizes expanding agricultural areas over improving productivity or reducing post-harvest losses. Traditional sustainable land use practices, like crop rotations, agroforestry, and mobile pastoralism, are being replaced by less sustainable methods like mechanized, fertilizer-dependent farming, continuous cropping, and sedentary livestock ranching. The shift towards less sustainable farming systems is determined by mix of technological, economic, cultural, policy factors, and demographic, with many countries lacking adequate policy assistance to counteract this trend. Declining or weak governance and natural resource rights further exacerbate the problem, leading to discouragements for justifiable land stewardship and the degradation of previously well-managed lands. For instance, many pastoralist communities have lost traditional communal resource management practices due to the privatization of land, unclear legal rights, and exclusion from key resources. The harm of land and reduced admittance to natural capitals have negative impacts for sustainable land use, as critical areas for dry season pasturing are converted for agriculture, urban development, or conservation purposes (McGahey et al. 2008; Behnke et al. 1993; United Nations 2011; Davies et al. 2010).

11.2.1 Natural Drivers

Various environmental conditions and natural hazards can trigger this process, as illustrated in the accompanying Fig. 11.2.

Sometimes, natural hazards are so intense that they create unproductive land even without human activity. Examples include naturally saline soils found in dry interior basins or regions with natural erosion features, like bad-lands. These areas are often referred to as problem soils.

Land degradation becomes a significant issue when land shortages drive the expansion of agriculture into areas with these natural hazards. These hazards are the underlying, or predisposing, conditions for land degradation.

11.2.2 Anthropogenic Causes

Several factors drive land degradation worldwide, including:

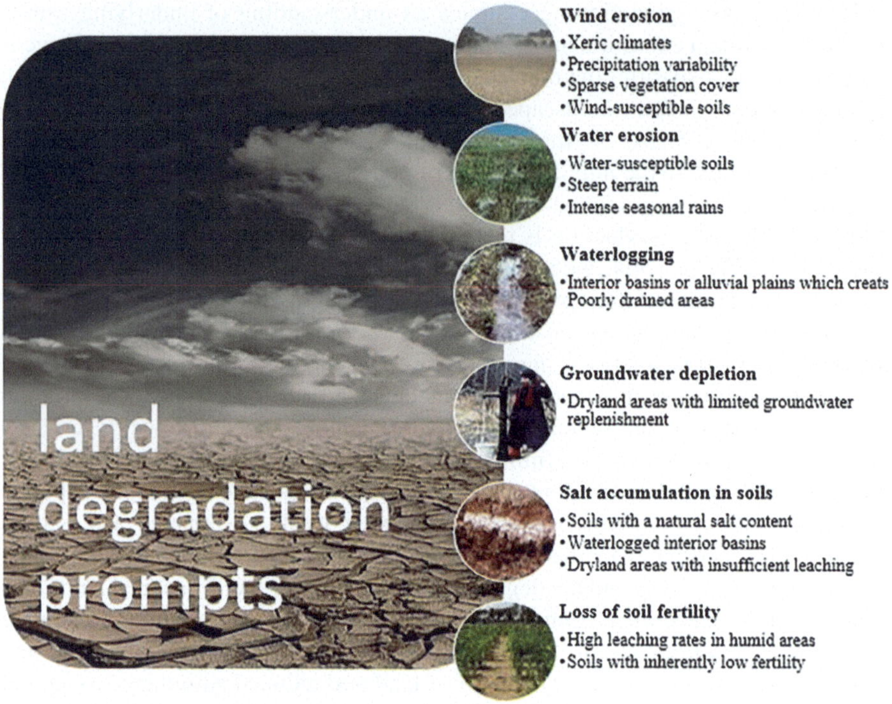

Fig. 11.2 Triggers of land degradation

Unsustainable Resource Management Over-exploitation of natural resources, clearing vegetation, depleting nutrients, overgrazing, poor irrigation practices, excessive agrochemical use, mining, pollution, and urban sprawl contribute significantly to land degradation.

Deforestation and Land Use Change Deforestation not only degrades land directly but also exacerbates erosion, particularly when the cleared land has steep slopes or easily erodible soils. Overcutting of forests for timber, fuel, and other products can lead to degradation, especially when it's not managed properly.

Overgrazing and Unsustainable Agriculture Overgrazing occurs when livestock graze pastures beyond their capacity, leading to reduced vegetation cover and increased soil erosion. This loss of vegetation and erosion reduces soil quality and its resistance to further erosion. Additionally, modern agricultural practices, including mechanization, can harm soil and water resources, ultimately decreasing land productivity.

Urbanization and Infrastructure Development Rapid urban growth, road construction, mining, and industrial activities contribute to land degradation by converting valuable agricultural land into built environments.

Unbalanced Fertilizer Use To compensate for declining soil fertility, farmers often use fertilizers. However, an overreliance on nitrogenous fertilizers can lead to imbalances with other essential nutrients like phosphorus, sulphur, and zinc, further degrading soil quality.

Shifting Cultivation In north-eastern India shifting cultivation has led to water erosion and declining soil fertility.

11.2.3 Socio-economic Causes

Poverty and Land Tenure Insecurity Despite significant economic progress in many regions, improvements in the welfare of rural poor remain questionable. Many farmers still live close to or below the poverty line, which often leads to land degradation. Richer farmers typically manage their land better than poorer ones, though more research is needed to confirm this. Additionally, agricultural growers are hesitant to devote in land preservation if their property rights are insecure. Issues like tenancy and open-access resources contribute to this insecurity.

Land Shortage Historically, land could be expanded into cultivation to address food shortages or poverty. However, in much of South Asia, this option is no longer viable. For example, between 1980 and 1990, agricultural land changes were minimal in India, Pakistan, Sri Lanka, and Afghanistan, with some regions like Bangladesh even seeing a decrease. In Nepal, land expansion came through deforestation and farming on less sustainable slopping lands.

Market Pressures and Global Trade Impacts Small land holdings create intense economic pressures on farmers to meet immediate food and income needs. This urgency often means that resources are not available for land care practices like green manuring or soil conservation. Economic pressures are exacerbated by modern communications and rising aspirations, which drive greater demands for income and can lead to less sustainable farming practices.

11.3 Principles and Approaches of Ecosystem Restoration

Ecosystem restoration involves helping damaged or destroyed ecosystems recover and protecting those that are still healthy. By restoring ecosystem, we can enhance its biodiversity, which in turn provides numerous benefits, such as richer soils, increased timber and fish yields, and better greenhouse gas storage. Restoration can be achieved in various ways, like actively planting new vegetation or removing pressures to allow nature to heal itself. However, it's not at all time practical or feasible to restore an ecological system to its original condition.

11.3.1 Principles of Ecosystem Restoration

11.3.1.1 Ecological Integrity and Sustainability

The principle (Fig. 11.3) highlights that every restoration effort, regardless of its size, contributes to the global objectives of conserving life on planet. These objectives are component of the UN SDGs, aiming to enhance lives universally by ensuring clean water, protecting biodiversity, eliminating poverty, and more. Ecosystem restoration is crucial in achieving several of these global goals (FAO et al. 2021).

Almost any activity that benefits nature could be considered ecological restoration, so this principle aims to provide clear guidelines for participants. Land restoration is essential for reversing degradation and achieving LDN. It's important to carefully choose where and when to carry out restoration efforts. A common strategy is to incorporate restoration into land-use planning. As outlined in the proposal, "the activity should lead to an overall improvement in environmental well-being and integrity, biodiversity, and human health, encompassing the sustainable provision of services and goods." Restoration efforts can target a extensive range of degraded ecosystems, including semi-natural, natural, cultural, and urban seascapes and landscapes. The UN has also outlined specific types of activities for these efforts, such as returning ecosystems to their original, minimizing harmful socio-environmental impacts, pre-degraded conditions, and addressing threats like pollution and contamination.

11.3.1.2 Adaptive Management and Resilience

Restoring ecosystems isn't just about covering up past damage and pretending it didn't happen. According to this principle, all restoration efforts must tackle the underlying causes of degradation and biodiversity loss. Without addressing these root causes, any seemingly successful short-term fixes might fail over time. One effective approach is to integrate sustainable practices that both support biodiversity and reduce the negative impacts of urban development. For instance, agroforestry provides a practical solution by recognizing the necessity of agriculture while shifting away from harmful practices. It promotes regenerative methods that enhance environmental health without compromising livelihoods (Fransen 2022).

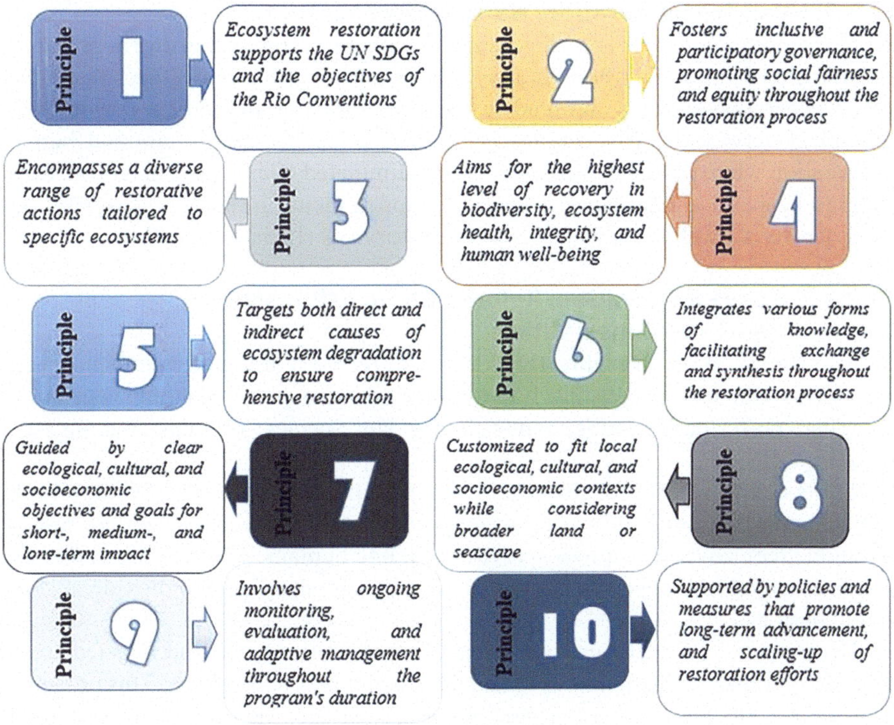

Fig. 11.3 Guiding Principles for Ecosystem Restoration

11.3.2 Approaches to Ecosystem Restoration

Ecosystem-based approaches are widely used for restoring and managing land and ecosystems in a way that aligns with natural processes. These methods leverage nature to address both development and environmental challenges. For instance, EBA (Ecosystem-Based Adaptation) includes a variety of activities designed to minimize susceptibility and strengthen adaptability to climate change. EBA can involve strategies such as IWRM (Integrated Water Resource Management) and FLR (Forest Landscape Restoration), which utilize ecosystem-based approaches to enhance community resilience, including both men and women. Similarly, nature-based solutions to Disaster Risk Reduction emphasis on harnessing ecosystem services to help communities prepare for, manage, and get well from disasters. This involves making decisions that consider both present and forthcoming human demands and the ecological needs of bionetworks. In areas suffering from land degradation, practices like landscape restoration and sustainable land management can mitigate the risks and impacts of floods and droughts.

11.3.2.1 Passive Restoration: Natural Regeneration

Passive restoration focuses on eliminating the human activities causing degradation and allowing land to recover naturally. This low-cost approach leverages the land's own regenerative abilities and tends to be more effective in landscapes with high ecological networking (Crouzeilles et al. 2017a, b). It requires minimal human intervention when persistent turbulences are eliminated (Benayas et al. 2009). Sites undergoing passive restoration often see improvements in species diversity and plant density, as well as enhanced ecosystem services (Piaia et al. 2020).

11.3.2.2 Active Restoration: Reforestation, Afforestation, and Soil Conservation

Active restoration involves directly aiding land recovery through various interventions. This can include placing hollow logs or wood piles, planting vegetation, dredging, prescribed burning, reintroducing key species, or managing invasive species.

Reforestation is vital for restoring ecological balance. It helps rebuild wildlife habitats, conserve soil, and ensure water security. By reforestation, we can restore essential green cover for various species, including humans. Additionally, reforestation is essential in combating climate change by capturing carbon dioxide, a chief greenhouse gas.

Afforestation, on the other hand, creates opportunities to expand forested areas, potentially transforming landscapes and developing new ecosystems. This proactive strategy not only addresses environmental issues but also supports future economic growth. In India, afforestation can drive the development of forestry, eco-tourism, and green energy industries, improve air quality, and create beneficial microclimates for agriculture.

11.3.2.3 ANR (Assisted Natural Regeneration)

ANR, also known as managed regrowth, involves the human effort to protect and nurture natural tree seedlings in forested areas. ANR combines elements of active planting with passive restoration by helping trees and native plants recover naturally. This is achieved by removing obstacles and threats to their growth while utilizing local knowledge and traditional practices. For instance, seedlings are shielded from dense undergrowth and highly flammable plants like *Imperata* grass. Although there is no standardized definition or methodology for ANR, its primary aim is to enhance forest productivity. Studies show that ANR is an affordable, eco-based approach to restoring large land areas and delivering vital ecosystem amenities such as clean water and fertile soil, both essential to human well-being (Chazdon et al. 2022).

11.3.2.4 Integrated Approaches

The program aims to deliver variety of environmental and social benefits by using joined approach to re-establish disturbed ecosystems. It will concentrate on restoring kinds of ecosystem that offer significant benefits, including:

- **Converted or degraded environments** such as headwaters, peatlands, watersheds, grasslands, wetlands, shrublands, riverine forests, estuaries, ecological networks, coastal areas (like coastal marine ecosystems), mangroves, native woodlands, and steppingstone habitats.
- **Degraded natural landscapes** like dry-lands, grass-lands, pastures, and forests, putting a variety of cost-effective, best methods like assisted natural regeneration and natural regeneration to re-establish their services and functions.
- **Deteriorated agricultural systems** in diverse ecological landscapes with highly efficient ecological assistances, achieved through financing for sustainable land management practices, including agro-ecological diversification, rangeland restoration, and agro-silvo-pastoral systems.

11.3.3 Case Studies of Successful Restoration Projects

Restoration efforts span diverse ecosystems, including reforestation in forests to enhance biodiversity, wetland rehabilitation to improve water quality and wildlife habitat, grassland recovery to support native species and prevent erosion, and dryland restoration to combat desertification and sustain local livelihoods (Indian ESG Network 2022).

11.3.3.1 The Great Green Wall, Africa

- **Project overview**: The Great Green Wall project seeks to address desertification in Africa's Sahel region by establishing a diverse mix of trees, plants, and sustainable land management techniques along an 8000-kilometer corridor.
- **Achievements**:
- ***Reforestation and land restoration:*** The project have revitalized over 18 M ha of deteriorated landscapes in countries such as Senegal, Ethiopia, and Nigeria.
- ***Economic benefits:*** It has generated employment and income through sustainable agriculture and forestry, positively impacting millions of individuals.
- ***Biodiversity enhancement:*** The rise in vegetation has boosted soil fertility, diminished erosion, and created new habitats for wildlife.

11.3.3.2 Chesapeake Bay Restoration, USA

- **Project overview**: The Chesapeake Bay, the largest estuary in the United States, has struggled with severe pollution and habitat loss. Since the 1980s, the Chesapeake Bay Program, a collaborative effort among federal and state agencies, local governments, and NGOs has been dedicated to restoring the estuary's health.
- **Achievements**:
- ***Pollution reduction:*** Substantial decreases in nitrogen, phosphorus, and sediment pollution have been accomplished by enhancing agricultural methods, upgrading wastewater treatment systems, and improving stormwater management.

- ***Habitat restoration:*** Substantial area under wetlands and submerged land have been rehabilitated, leading to better water quality and creating habitats for fish and wildlife.
- ***Community engagement:*** The program has effectively involved local communities and stakeholders in restoration initiatives, promoting a sense of responsibility and care for the environment.

11.3.3.3 Loess Plateau Watershed Rehabilitation, China

- **Project overview**: The Loess Plateau, once among the most degraded regions in China, has seen a dramatic turnaround thanks to a major watershed rehabilitation project that began in the 1990s.
- **Achievements**
- ***Soil and water conservation:*** Terracing, reforestation, and the building of check dams have greatly diminished soil erosion and enhanced water retention.
- ***Agricultural improvement:*** Adopting sustainable farming practices has boosted crop yields and enhanced food security for local communities.
- ***Economic development:*** The project has elevated millions out of poverty by generating sustainable livelihoods and enhancing living conditions.

11.3.3.4 Aral Sea Restoration, Kazakhstan

- **Project overview**: The Aral Sea, which was formerly the 4th largest lake in the world, experienced dramatic shrinkage due to poor water management practices. In response, Kazakhstan launched the Kok-Aral Dam project to help restore a portion of the Northern Aral Sea.
- **Achievements**
- ***Increased water level:*** Building the Kok-Aral Dam has elevated water levels and revived parts of the Northern Aral Sea, which has, in turn, rejuvenated local fisheries.
- ***Biodiversity recovery:*** The rise in water levels has brought back fish species and revitalized local biodiversity.
- ***Economic revival:*** The revival of the fishery industry has generated employment opportunities and enhanced food security for local communities.

11.3.3.5 Atlantic Forest Restoration Pact, Brazil

- **Project overview**: The Brazil's Atlantic Forest, renowned for its incredible biodiversity, has experienced extensive deforestation. The Atlantic Forest Restoration Alliance, a consortium of NGOs, government agencies, and private sector partners, is committed to rehabilitating 15 M ha of this forest by 2050.
- **Achievements**
- ***Reforestation efforts:*** More than 700,000 hectares of forest have been rejuvenated through efforts in reforestation and natural regeneration.

- ***Carbon sequestration:*** The rehabilitated forest regions are capturing substantial amounts of carbon, aiding in climate change reduction.
- ***Biodiversity conservation:*** The restoration work has improved biodiversity hotspots, including many that are native to the Atlantic Forest.

11.4 Strategies and Practices for Achieving LDN through Restoration

LDN represents the balance between areas of land that are degraded and those that are restored, with restored land compensating for the sustained degradation within the sustained land use. Achieving LDN can be hindered by land degradation, while restoration efforts promote it (Fig. 11.4). There are critical thresholds for both ecosystem restoration and land degradation (Jiang et al. 2020). Factors like unstainable human actions or severe weather events can reduce land function and cause land degradation. Sustainable land use practices can help prevent reaching these thresholds. For example, agronomic techniques such as mulching and intercropping are used to maintain land health and prevent degradation. Studies have reviewed the practical impacts of sustainable land management (Liniger et al. 2019; Giger et al. 2018; Stavi and Lal 2015).

11.4.1 Assessing and Mapping Degraded Lands

To evaluate land degradation, various methods are employed, including field monitoring, observations, expert and land user opinions, productivity estimates, measurements, modelling, and remote sensing. These approaches are applied through utilizing either quantitative or qualitative, or a combined measure. Table 11.1 affords an overview of the systems evaluated, the technique of assessment used, the parameters, processes, factors, and the values or units of measurements for various levels of land degradation assessment, ranging from global to national, regional, local, and field/farm levels. The summary outlines the diverse assessment methods utilized for analysing land degradation procedures across the various levels and systems.

11.4.2 Restoration Planning, and Prioritization

Efficient planning is crucial for achieving the project goals, with the creation of a restoration strategy serving as the foundation for the complete ecological restoration progression. For the plan to succeed, it should stem from well-versed problem-solving at all stages and incorporate a participatory progression that engages all rights holders and stakeholders influenced by the renovation, as well as specialists in ecosystem restoration and degradation. A common vision should be established, along with restoration targets (such as desired cultural, ecological, and

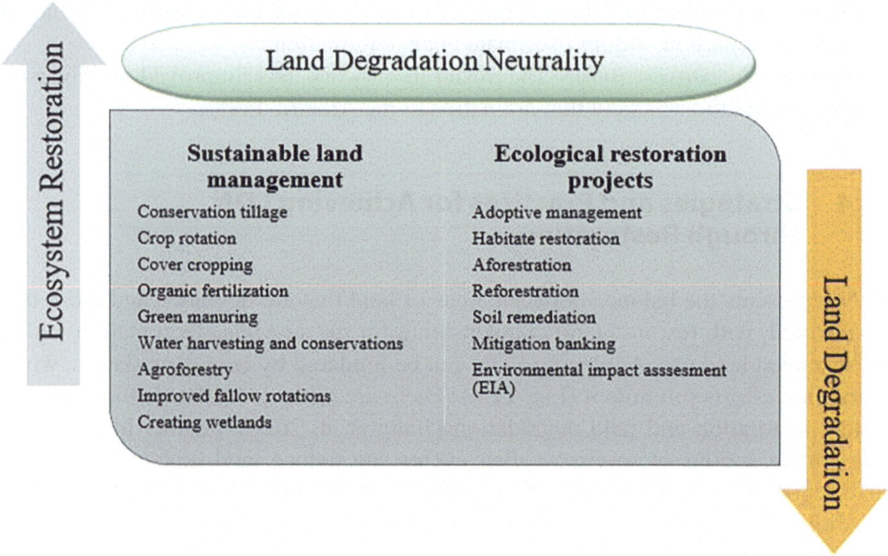

Fig. 11.4 The interconnection between land degradation, ecosystem restoration and land degradation neutrality

socioeconomic conditions) and specific project goals, all grounded in fair and transparent negotiations.

11.4.3 Implementation of Restoration Interventions

Ecosystem restoration requires experimentation and innovation, which can often result in unforeseen outcomes and challenges. As a result, initial trials are usually necessary to guide decision-making, such as species selection and their arrangement or to test various approaches throughout the project, allowing for adaptive management. Additionally, a range of restoration events may be desirable to halt or prevent degradation and to modify both abiotic and biotic surroundings that facilitate natural rejuvenation and the introduction of translocated species. For example, it is often vital to eliminate threats, like grazing, that may harm newly grown flora. Whenever viable, practitioners should take advantage of opportunities for natural recovery through procedures like ecological succession. Nevertheless, if natural regeneration is not an option, further interventions may be required, such as altering abiotic conditions to aid in the establishment or relocation of species to the site. In many instances, employing restoration methods that reduce social impacts like implementing justifiable management options for soil, water, vegetation, and land, along with promoting alternative livelihood options like ecotourism can help prevent further degradation by addressing socio-economic needs. It is also crucial to ensure that any changes to abiotic or biotic conditions are implemented carefully to minimize or avoid unintended negative effects and to mitigate them when they

Table 11.1 Research approach for LDN assessment

Approach	Level	Indicator	Reference
Signalling approach	Worldwide	Forest volume, forest cover, and biological diversity	Ptichnikov and Martynyuk (2020)
		Aeolian erosion	Chappell et al. (2019)
		Aquatic erosion	Wuepper et al. (2021)
	Area of zone	Erosion rate and cumulative soil loss	Tsymbarovich et al. (2020)
		Relative humidity	Kuderina et al. (2020)
		Soil pollution and ravine formation	Speranza et al. (2019)
Survey approach		Spatial survey	Von Maltitz et al. (2019)
		Agricultural surveys, on-site assessments and community-based mapping	Crossland et al. (2018)
		Cross-verification or triangulation	Nzuza et al. (2021)
		Professional anecdotal evidence	Garcia et al. (2019)

occur. Additionally, effectively communicating and reporting about project implementation is vital for raising awareness, maintaining provision from the donors, government, and public, and fostering association and shared learning among restoration practitioners.

11.4.4 Financing and Incentivizing Ecosystem Restoration

Developing affordable, sustainable, and realistic plans for enduring management, along with securing sufficient funding (whether direct or through generated revenue), is crucial, especially given the typically limited personnel and budget resources available during this project phase. Planning and budgeting must account for long-term management needs that can be presumed, such as ongoing elimination of offensive species, site defence, maintenance of natural disturbance patterns, and management of hydrological regimes. However, it is important to recognize that some management needs may be unpredictable due to the dynamic nature of ecosystems and the possibility of unexpected events. Although ongoing management occurs after implementation, these actions should be budgeted and planned for during the restoration's planning segment and should include provisions for both anticipated and adaptive management strategies.

11.4.5 Evaluation and Monitoring

Ecosystem restoration is a long-standing endeavour, full of uncertainties regarding the best ways to reach project goals. Climate change increases more uncertainties, making it harder to predict how ecosystems will respond to restoration efforts. Because of this, it's essential to evaluate and monitor the project eventually. This process helps to understand: (1) whether the strategic restoration events were carried out as intended (implementation monitoring); (2) how effectively the targets for restoration were achieved and if the objectives and goals of project were fulfilled (effectiveness monitoring); and (3) the overall impact on cultural, ecological, or socio-economic factors, both negative and positive (effects monitoring). When done properly, evaluation and monitoring allow for adaptive management, enabling successful actions to be scaled up, ineffective strategies to be abandoned, and new, promising methods to be incorporated (FAO et al. 2023).

11.5 Monitoring, Reporting and Verification (MRV)

MRV is essential for our universal efforts to address climate change. It allows nations to fulfil transnational reporting obligations, such as Biennial Update Reports, National GHG Inventories, and National Communications. It similarly plays a crucial role in demonstrating progress towards initiatives like Intended Nationally Determined Contributions, NAMAs (Nationally Appropriate Mitigation Actions), and Low Emission Development Strategies. Additionally, MRV provides the critical information, nations need to support their sustainable economic development goals and wider climate change.

11.5.1 Importance of MRV in LDN

MRV process is relevant for field-level projects that implement designated SSM (Sustainable Soil Management) operations in specific cultivated areas underneath defined circumstances. This phase aims to confirm that the project's activities comply with the essential necessities of the protocol.

- **Monitoring:** It is aimed to regularly show that the implemented practices on SSM are successfully sequestrating CO_2 from atmosphere over the short run, sequestering carbon in soils in the medium run, and decreasing greenhouse gas (GHG) releases relative to a baseline scenario. This phase comprises three main monitoring activities: soil sampling, SOC modelling, and estimates of GHG emissions.
- **Reporting:** The reporting phase focuses on making information available to diverse users and promoting public announcement and regular verification. Project stakeholders are required to document their emissions reductions in inventories and standardized set-ups, which are subsequently verified by an self-governing third-party assessor. Reporting must conform to specific formats, many of which can be found online, including those provided by the Certified Carbon Standard (VCS-VERRA 2019) for the VCS Program (https://verra.org/project/vcs-program/).
- **Verification:** Verification entails independently assessing the consistency and accurateness of the information reported or the methodologies employed to produce it. Additionally, it ensures quality assurance and control by providing

response on monitoring techniques, measurement, and enhancements in reporting practices.

The SAC of ISRO, in collaboration with 19 institutions, tracks and compiles desertification inventories throughout India using IRS data within a GIS. The "Desertification and Land Degradation Atlas of India" provides maps showing the status of desertification and land degradation in all states based on processes, severity, and land use. These reports help identify areas most impacted by desertification and degradation of land. A four-tier monitoring system is suggested to oversee advancement in attaining LDN within the country.

Local bodies like *Gram Sabhas* and Joint Forest Management Committees (JFMCs) will conduct social audits of local operations, with information made publicly available. RS-GIS and third-party monitoring will regularly assess UNCCD indicators for desirable changes in land productivity, carbon stocks, and land cover. A GIS driven web-based Integrated Management Information System Platform, should be developed at the CoE-SLM (Centre of Excellence for Sustainable Land Management) to integrate various LDN programs. This platform will ensure timely funding, robust financial planning, and efficient use of resources, and will create reports on the progress of both public and private projects across the country.

11.6 Challenges and Barriers to Ecosystem Restoration for LDN

A variety of external factors can impact conditions within a restoration area and influence the success of restoration efforts (Fig. 11.5). These aspects comprise (but are not limited to) the degree of historical habitat loss and degradation, which affects both the feasibility of achieving the restoration goals and the availability of species in the surrounding landscape that can repopulate restored areas. Other significant influences include pollution, the impacts of offensive species and overexploitation, and, increasingly, the impacts of climate change. Smaller sites, in particular, may struggle with small species populations that are subject to genetic issues and an increased risk of local extinction (Park et al. 2022).

11.6.1 Technical and Ecological Challenges

Public perception of landscapes, ideas of what is considered "natural," and the value people place on the land can pose challenges to restoration efforts at various scales. This ranges from overgrazed yet cherished national parks to the excessive mowing of roadside verges for a tidier appearance. This issue is especially pronounced when passive restoration methods are used, as these areas may be viewed as neglected or abandoned. These values and attitudes also shape the broader societal conversation around restoration, which in turn impacts political decisions and policy-making (Park et al. 2022).

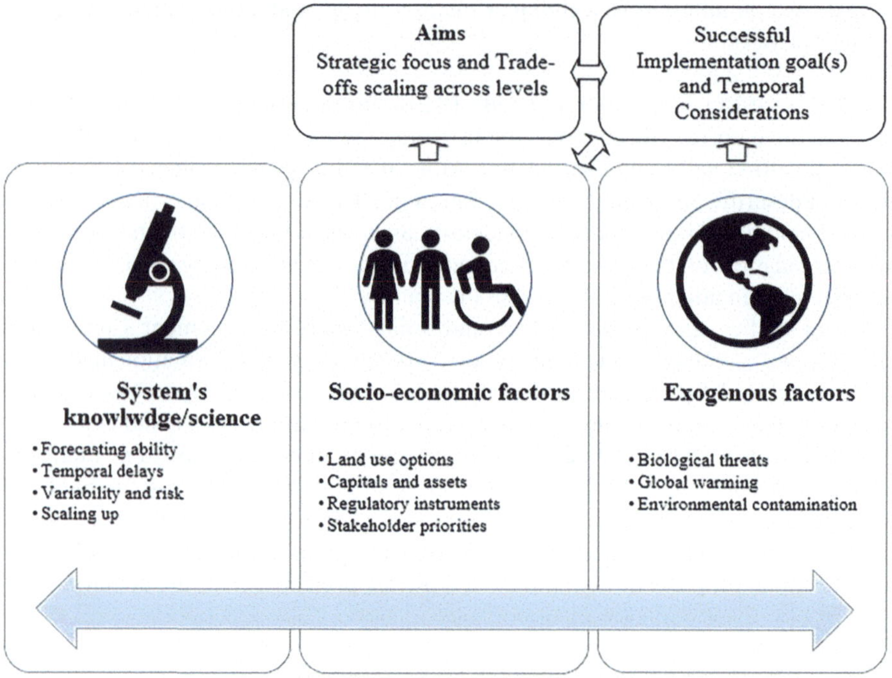

Fig. 11.5 Agenda for ecological renovation achievement: participating multi-scale drivers

11.6.2 Socioeconomic and Institutional Barriers

A major challenge is how to allocate larger areas for ecosystem restoration while still maintaining or even boosting agricultural productivity to address the needs of an expanding population. Dedicating land for restoration can reduce the space available for other uses, such as food production and forestry, although some practices, like agroecology, are exceptions. Evidence from marine environments shows that properly implemented Marine Protected Areas can lead to significant increases in fish populations and overall biodiversity. While many country-specific policies promote restoration for biodiversity conservation and ecosystem service provision, numerous agricultural policies continue to encourage and reward practices that harm the environment, like the EU's Common Agricultural Policy, which supports farmers who cultivate drained peatlands (Park et al. 2022).

11.6.3 Overcoming Barriers

Numerous restoration initiatives fail because they neglect the needs and interests of local investors who are most directly involved with the land (Brancalion and Holl 2020; Fleischman et al. 2020). Restoration practitioners should not only navigate the intricate ecological progressions that affect ecosystem recovery but also consider the varied economic, political, and social frameworks, where restoration occurs. It is crucial to enhance our understanding of the situations, promoting restoration successfully and to recognize how these may differ in different locations.

Ecological systems can naturally recover in farming landscapes where there is little burden to convert land, often due to rural-to-urban migration or/and changes in agrarian policies (Crouzeilles et al. 2020; Chazdon et al. 2020). However, encouraging restoration is more challenging in regions where the opportunity cost of giving up land conversion or deserting prevailing agricultural land is high. To ensure that restored habitats remain in place, landholders need to be remunerated for such costs. Designing effective schemes i.e. 'payment for ecosystem services' can be complex, especially when land tenure or access rights are uncertain, as these uncertainties can undermine restoration incentives or shift land degradation to alternative fields (Ford et al. 2020). For instance, in some regions, halting agricultural activities can lead to the damage of land rights for landholders, and naturally regenerating areas may be vulnerable to land grabs (Holl et al. 2022). Moreover, even though indigenous peoples often manage intact ecosystems best, their land rights are frequently contested by national or regional governments (FAO 2021a; Kruid et al. 2021; Haenssgen et al. 2022). Understanding the complex interconnections among (inter-)national, regional, and local governance structures that influence land use is crucial for effective ecosystem restoration with the complete involvement of the people who rely on these lands.

For restoration projects to be effective, it is essential to co-design them with individuals who are most affected by their implementation (Waring et al. 2023). This can be accomplished through 'top-down' strategies, in which a central authority, such as a national government, spearheads the restoration initiatives while consulting various shareholders, or 'bottom-up' strategies, where regional communities lead the necessary changes (Reed et al. 2018). Apiece approach comes with its own set of advantages and disadvantages. For example, China's 'Grain-for-Green' program, recognized as the chief reforestation effort globally, was a top-down initiative that compensated landholders for planting trees on degraded land to mitigate soil erosion (Hua et al. 2016). While this program significantly increased forest cover, much of it consists of monocultures with low biodiversity and has led to higher atmospheric water loss compared to inherent grasslands, paying to local water deficiencies (Schwärzel et al. 2020).

Conversely, a bottom-up tactic is typified by the 'regreening' initiatives in the Sahel region that began in the 1980s. By utilizing traditional knowhow, small-scale farmers encouraged the natural regrowth of vegetations on their properties, which contributed to mitigating desertification and enhancing food security for about three million population (Magrath 2020). High-resolution satellite imagery now reveals that 1.8 billion trees are flourishing in regions once considered prime examples of dryland desertification (Brandt et al. 2020). However, it is crucial to recognize that these re-established lands continue to be actively managed for agricultural purposes. This movement sought to improve carbon sequestration and biodiversity within cultivated sceneries, but it did not aim for or attain the complete restoration of unmanaged innate habitats. While carbon capture and biodiversity improvement in agrarian contexts is essential, certain parts of the landscape need to be entirely reinstated to prevent the ongoing reduction in biodiversity.

In past two decades, the FLR (Forest Landscape Restoration) method has developed to address how ecosystem restoration and protection can be well-adjusted with forestry and agriculture on a landscape-scale (César et al. 2021). A key element of this measure is participatory planning, which involves association among landowners, local communities, NGOs, and governments to optimize numerous ecosystem services concurrently (Aguiar et al. 2021). This collaborative approach could pave the way for significant ecosystem regeneration: with appropriate institutional support and favourable market conditions, millions of hectares in smallholder-dominated landscapes could be restored (Shyamsundar et al. 2022; Busch et al. 2019). However, since FLR is a comparatively new concept, there is an urgent need for additional information on where and when these intrusions are effective and how restored landscapes can be managed adaptively with time amid changing socioeconomic conditions.

11.7 Future Directions and Opportunities

The future trajectory of LDN remains uncertain owing to the incomplete understanding of human impacts and climate change. Dryland ecosystems are at danger of continuing and expanding to degrade (Bryan et al. 2018), making the future achievement of LDN in these areas seem unlikely. Several critical questions need to be addressed: (1) What will the upcoming tendency of LDN be? (2) Will ecological restoration remain essential for tackling land degradation in the future? (3) How widespread will the need for ecological engineering be, and in which locations will it be necessary? Given the worsening intense human activities and climate change, an urgent prerequisite is needed to predict the future trends of LDN. Although there are limited studies on forthcoming LDN prognoses, with just two articles providing scenario assessments (Xu and Zhang 2021) and projections (Schulze et al. 2021), more research is required.

Land-use models are frequently used to forecast upcoming land cover and land use. These models encompass a variety of approaches, including statistical methods and machine/data-driven learning, grid-based models, sectoral economic models, region-specific economic models, multi-agent systems (MAS), and combined-method models (Gomes et al. 2021). They have wide-ranging applications and are backed by an array of regional (Liao et al. 2020) and global (Chen et al. 2020) long-term data sets that provide essential information for forecasting LDN. Leveraging prevailing ISIMIP data alongside land-use model outputs can aid in projecting LDN, establish a scientific foundation for ecological engineering, and facilitate the evaluation and monitoring of Sustainable Development Goals (SDGs).

11.8 Conclusion

Land or soil degradation is significantly inclined by both human activities and natural factors. Immediate action is essential to break these degradation progressions and re-establish affected lands. Despite the progress in scientific understanding and technological innovation, much of this information remains fragmented and lacks a unified approach. Consequently, conducting regular assessments of soil and land degradation—considering its type, severity, extent, and causes is crucial. Human activities can result in both positive and adverse impacts on land, necessitating further research to understand these effects and their trade-offs. The effectiveness of mitigation efforts is often limited by differing perspectives on land degradation and related challenges. The "Guidelines for Convergent Approach for Greening India" advocate for a collaborative and joined strategy for restoring and developing degraded landscapes. This tactic supports eco-restoration project's effective planning, implementation, and monitoring, thereby promoting sustainable forest and natural resource management in vulnerable regions. Coordinated international efforts are crucial for prioritizing investments in effective LDN programs. Addressing food security, environmental balance and land degradation in a holistic manner is key to achieving LDN targets.

References

Aguiar S, Mastrángelo ME, Brancalion PHS, Meli P (2021) Transformative governance for linking forest and landscape restoration to human well-being in Latin America. Ecosyst People 17:523–538. https://doi.org/10.1080/26395916.2021.1976838

Bai ZG, Dent DL, Olsson L, Schaepman ME (2008) Proxy global assessment of land degradation. Soil Use Manag 24:223–234. https://doi.org/10.1111/j.1475-2743.2008.00169.x

Behnke RH, Scoones I, Kerven C (1993) Range ecology at disequilibrium: new models of natural variability and pastoral adaptation in African savannas. Overseas Development Institute, London. https://doi.org/10.1002/ldr.3400050108

Benayas JMR, Newton AC, Daiz A, Bullock JM (2009) Enhancement of biodiversity and ecosystem services by ecological restoration: a meta-analysis. Science 325(5944):1121–1124. https://doi.org/10.1126/science.1172460

Brancalion PHS, Holl KD (2020) Guidance for successful tree planting initiatives. J Appl Ecol 57:2349–2361. https://doi.org/10.1111/1365-2664.13725

Brandt M, Tucker CJ, Kariryaa A, Rasmussen K, Abel C, Small J, Chave J, Rasmussen K, Hiernaux P, Diouf AA, Kergoat L, Mertz O, Lgel C, Gieseke F, Schoning J, Li S, Melocik K, Meyer J, Sinno S, Romero E, Glennie E, Montagu A, Dendoncker M, Fensholt R (2020) An unexpectedly large count of trees in the West African Sahara and Sahel. Nature 587:78–82. https://doi.org/10.1038/s41586-020-2824-5

Bryan BA, Gao L, Ye YQ, Sun XF, Connor JD, Crossman ND, Stafford-Smith M, Wu JG, He CY, Yu DY, Liu ZF, Li A, Huang QX, Ren H, Deng XZ, Zheng H, Niu JM, Han GD, Hou XY (2018) China's response to a national land-system sustainability emergency. Nature 559(7713):193–204. https://doi.org/10.1038/s41586-018-0280-2

Busch J, Engelmann J, Cook-Patton SC, Griscom BW, Kroeger T, Possingham H, Shyamsundar P (2019) Potential for low-cost carbon dioxide removal through tropical reforestation. Nat Clim Chang 9:463–466. https://doi.org/10.1038/s41558-019-0485-x

César RG, Belei L, Badari CG, Viani RAG, Gutierrez V, Chazdon RL, Brancalion PHS, Morsello C (2021) Forest and landscape restoration: a review emphasizing principles, concepts, and practices. Land 10:28–22. https://doi.org/10.3390/land10010028

Chappell A, Webb NP, Leys JF, Waters CM, Orgill S, Eyres MJ (2019) Minimising soil organic carbon erosion by wind is critical for land degradation neutrality. Environ Sci Pol 93:43–52. https://doi.org/10.1016/j.envsci.2018.12.020

Chazdon RL, Lindenmayer D, Guariguata MR, Crouzeilles R, Rey Benayas JM, Lazos Chavero E (2020) Fostering natural forest regeneration on former agricultural land through economic and policy interventions. Environ Res Lett 15:043002. https://doi.org/10.1088/1748-9326/ab79e6

Chazdon R, Calixto B, Oliveira M, Messinger J, Alves J, Calmon M, Anderson W (2022) The benefits and power of assisted natural regeneration. World Resources Institute

Chen GZ, Li X, Liu XP, Chen YM, Liang X, Leng JY, Xu XC, Liao WL, Qiu YA, Wu QL, Huang KN (2020) Global projections of future urban land expansion under shared socioeconomic pathways. Nat Commun 11(1):537. https://doi.org/10.1038/s41467-020-14386-x

Crossland A, Winowiecki LA, Pagella T, Hadgu K, Sinclair F (2018) Implications of variation in local perception of degradation and restoration processes for implementing land degradation neutrality. Environ Dev 28:42–54. https://doi.org/10.1016/j.envdev.2018.09.005

Crouzeilles R, Ferreira MS, Chazdon RL, Lindenmayer DB, Sansevero JBB, Monteiro L, Iribarrem A, Latawiec A, Strassburg BBN (2017a) Ecological restoration success is higher for natural regeneration than for active restoration in tropical forests. Sci Adv 3(11):e1701345. https://doi.org/10.1126/sciadv.1701345

Crouzeilles R, Ferreira MS, Chazdon RL, Lindenmayer DB, Sansevero JBB, Monteiro L, Iribarrem A, Latawiec AE, Strassburg BBN (2017b) Ecological restoration success is higher for natural regeneration than for active restoration in tropical forests. Sci Adv 3:1–8. https://doi.org/10.1126/sciadv.1701345

Crouzeilles R, Maurenza D, Prieto PV, Barros FSM, Jakovac C, Ferreira MS, Chazdon RL, Lindenmayer DB, Brancalion PHS, Ceccon E, Adams C, Lazos-Chavero E, Monteiro L, Junqueira AB, Strassburg BBN, Guariguata MR (2020) Associations between socio-environmental factors and landscape-scale biodiversity recovery in naturally regenerating tropical and subtropical forests. Conservation Letters 14(2): https://doi.org/10.1111/conl.12768

Das S, Chatterjee S, Rajbanshi J (2022) Response of soil organic carbon to conservation practices including climate-smart agriculture in tropical and subtropical regions: a metaanalysis. Sci Total Environ 805:150428. https://doi.org/10.1016/j.scitotenv.2021.150428

Davies J, Niamir-Fuller M, Kerven C, Bauer K (2010) Extensive livestock production in transition: the future of sustainable pastoralism. In: Steinfeld et al (eds) Livestock in a changing landscape, volume 1: drivers, consequences and responses. Island Press

ELD (2019) New report reveals the true cost of land degradation. https://blog.cabi.org/2015/09/25/new-report-reveals-the-true-cost-of-land-degradation/

FAO (2009) Global agriculture towards 2050. Paper presented to high-level expert forum on "how to feed the world in 2050", Rome, 12–13 October 2009. FAO, Rome

FAO (2015) State of food insecurity in the world. Retrieved from: http://www.fao.org/3/a--i4671e.pdf

FAO (2021a) Forest governance by indigenous and tribal peoples. FAO, Santiago, pp 1–21. https://doi.org/10.4060/cb2953en

FAO (2021b) The state of the worlds land and water resources for food and agriculture-systems at breaking point. Synthesis Report, Rome. https://doi.org/10.4060/cb7654en

FAO, IUCN CEM, SER (2021) Principles of ecosystem restoration to guide the United Nations decade 2021–2023. FAO, Rome

FAO, SER, IUCN CEM (2023) Standards of practice to guide ecosystem restoration. A contribution to the United Nations Decade on Ecosystem Restoration. Summary report. FAO, Rome. https://doi.org/10.4060/cc5223en

Fleischman F, Basant S, Chhatre A, Coleman EA, Fischer HW, Gupta D, Gunelarp B, Kashwan P, Khatri D, Muscarella R, Powers JS, Ramprasad V, Rana P, Solorzano CR, Veldman JW

(2020) Pitfalls of tree planting show why we need people-centered natural climate solutions. Bioscience 70:947–950. https://doi.org/10.1093/biosci/biaa094
Ford SA, Jepsen MR, Kingston N, Lewis E, Brooks TM, MacSharry B, Mertz O (2020) Deforestation leakage undermines conservation value of tropical and subtropical forest protected areas. Glob Ecol Biogeogr 29:2014–2024. https://doi.org/10.1111/geb.13172
Fransen B (2022) 10 principles of ecosystem restoration. Ecomatcher
Garcia CL, Teich I, Gonzalez-Roglich M, Kindgard AF, Ravelo AC, Liniger H (2019) Land degradation assessment in the Argentinean Puna: comparing expert knowledge with satellite-derived information. Environ Sci Pol 91:70–80. https://doi.org/10.1016/j.envsci.2018.10.018
Geist HJ, Lambin EF (2004) Dynamic causal patterns of desertification. Bioscience 54(9):817–829. https://doi.org/10.1641/0006-3568(2004)054[0817:DCPOD]2.0.CO;2
Giger M, Liniger H, Sauter C, Schwilch G (2018) Economic benefits and costs of sustainable land management technologies: an analysis of Wocat's global data. Land Degrad Dev 29(4):962–974. https://doi.org/10.1002/ldr.2429
Global Environment Facility (2024) Land degradation
Gomes E, Inácio M, Bogdzevič K, Kalinauskas M, Karnauskaitė D, Pereira P (2021) Future land-use changes and its impacts on terrestrial ecosystem services: a review. Sci Total Environ 781:Article 146716. https://doi.org/10.1016/j.scitotenv.2021.146716
Haenssgen MJ, Lechner AM, Rakotonarivo S, Leepreecha P, Sakboon M, Chu TW, Auclair E, Vlaev I (2022) Implementation of the COP26 declaration to halt forest loss must safeguard and include indigenous people. Nat Ecol Evol 6:235–236. https://doi.org/10.1038/s41559-021-01650-6
Holl KD, Ashton MS, Bukoski JJ, Culbertson KA, Curran SR, Harris TB, Potts MD, Valverde YL, Vincent JR (2022) Redefining "abandoned" agricultural land in the context of reforestation. Front For Glob Change 5:1–6. https://doi.org/10.3389/ffgc.2022.933887
Hua F, Wang X, Zheng X, Fisher B, Wang L, Zhu J, Tang Y, Yu DW, Wilcove DS (2016) Opportunities for biodiversity gains under the world's largest reforestation programme. Nat Commun 7:12717. https://doi.org/10.1038/ncomms12717
ICAR/NAAS (2010) Degraded and wastelands of India: status and spatial distribution. Indian Council of Agricultural Research, and National Academy of Agricultural Sciences, New Delhi
Indian ESG Network (2022). https://in.linkedin.com/company/indian-esg-network?trk=article-ssr-frontend-pulse_publisher-author-card
IPCC (2019) Climate change and land: an IPCC special report on climate change, desertification, land degradation, sustainable land management, food security and greenhouse gas fluxes in terrestrial ecosystems
IUCN (2020) International Union for conservation of nature. Restore our future: Bonn Challenge. https://www.bonnchallenge.org/
Jiang C, Zhang HY, Zhao LL, Yang ZY, Wang XC, Yang L, Wen ML, Geng SB, Zeng Q, Wang J (2020) Unfolding the effectiveness of ecological restoration programs in combating land degradation: achievements, causes, and implications. Sci Total Environ 748:141552. https://doi.org/10.1016/j.scitotenv.2020.141552
Kruid S, Macedo MN, Gorelik SR, Walker W, Moutinho P, Brando PM, Casanho A, Alencar A, Baccini A, Coe MT (2021) Beyond deforestation: carbon emissions from land grabbing and forest degradation in the Brazilian amazon. Front For Glob Change 4. https://doi.org/10.3389/ffgc.2021.645282
Kuderina TM, Suslova SB, Lunin VN, Kudikov AV (2020) Atmospheric moisture as a factor of land degradation neutrality in forest-steppe landscapes. Arid Ecosyst 10(2):156–160. https://doi.org/10.1134/S2079096120020079
Liao WL, Liu XP, Xu XY, Chen GZ, Liang X, Zhang HH, Li X (2020) Projections of land use changes under the plant functional type classification in different SSP-RCP scenarios in China. Sci Bull 65(22):1935–1947. https://doi.org/10.1016/j.scib.2020.07.014
Liniger H, Harari N, Lynden GV, Fleiner R, Leeuw JD, Bai ZG, Critchley W (2019) Achieving land degradation neutrality: the role of SLM knowledge in evidence-based decision-making. Environ Sci Pol 94:123–134. https://doi.org/10.1016/j.envsci.2019.01.001

Magrath J (2020) Regreening the Sahel: a quiet agroecological revolution. https://doi.org/10.21201/2020.5662

McGahey D, Davies J, Barrow E (2008) Pastoralism as conservation in the horn of Africa: effective policies for conservation outcomes in the drylands of Eastern Africa. Ann Arid Zone 46:353–377

MEA (2005) Ecosystems and human well-being: desertification synthesis. Millennium ecosystem assessment. World Resources Institute, Washington, DC

Mekouar MA (2020) 15. Food and Agriculture Organization of the United Nations (FAO). Yearbook of International Environmental Law 31(1):326–340. https://doi.org/10.1093/yiel/yvab061

Ministry of Agriculture and Farmers Welfare (2021) Annual report 2020–21. Department of Agriculture, Cooperation & Farmers' Welfare, Government of India, New Delhi. https://agricoop.nic.in/Documents/annual-report-2020-21.pdf

Mora C, Tittensor DP, Adl S, Simpson AG, Worm B (2011) How many species are there on earth and in the ocean? PLoS Biol 9(8):e1001127. https://doi.org/10.1371/journal.pbio.1001127

Mohamed E, Belal AA, Ali RR, Saleh A, Hendawy EA (2019) Land degradation. The soils of Egypt:159–174

Mukherjee I, Das SK, Kumar A (2018) Atmospheric CO 2 level and temperature affect degradation of Pretilachlor and Butachlor in Indian soil. Bull Environ Contam Toxicol 100(6):856–861. https://doi.org/10.1007/s00128-018-2340-6

Mythili G, Goedecke J (2016) Economics of land degradation in India. In: Nkonya E, Mirzabae A, von Braun J (eds) Economics of land degradation and improvement—a global assessment for sustainable development. Springer, Cham, pp 431–469

Nzuza P, Ramoelo A, Odindi J, Kahinda JMM, Lindeque L (2021) A triangulation approach for assessing and mapping land degradation in the Lepellane catchment of the greater Sekhukhune District, South Africa. S Afr Geogr J:514–538. https://doi.org/10.1080/03736245.2021.2000481

Park K, McCarthy B, Harris J (2022) Ecosystem restoration: Securing biodiversity, complexity and resilience. Environ Sci 33(3):2–86. https://www.the-ies.org/resources/ecosystem-restoration-securing

Piaia BB, Rovedder APM, Procknow D, Camargo B, Gazzola MD, Croda JP, Stefanella MM (2020) Natural regeneration as an indicator of ecological restoration by applied nucleation and passive restoration. Ecol Eng 157:105991. https://doi.org/10.1016/j.ecoleng.2020.105991

Ptichnikov AV, Martynyuk AA (2020) Adaptation of international indicators of land degradation neutrality for the assessment of forest ecosystems in arid conditions in Russia. Arid Ecosyst 10(2):127–134. https://doi.org/10.1134/S2079096120020109

Qadir M, Quillérou E, Nangia V, Murtaza G, Singh M, Thomas RJ, Drechsel P, Noble AD (2014) Economics of salt-induced land degradation and restoration. Nat Resour Forum A United Nations Sustain Dev J 38(4). https://doi.org/10.1111/1477-8947.12054

Reed MS, Vella S, Challies E, de Vente J, Frewer L, Hohenwallner-Ries D, Huber T, Neumann RK, Oughton EA, Sidolidel Ceno J, van Delden H (2018) A theory of participation: what makes stakeholder and public engagement in environmental management work? Restor Ecol 26:S7–S17. https://doi.org/10.1111/rec.12541

SAC (2021) Desertification and land degradation atlas of India (Assessment and analysis of changes over 15 years based on remote sensing). Desertification and Land Degradation: 282. ISBN: 978-93-82760-39-9

Schulze K, Malek Z, Verburg PH (2021) How will land degradation neutrality change future land system patterns? A scenario simulation study. Environ Sci Pol 124:254–266. https://doi.org/10.1016/j.envsci.2021.06.024

Schwärzel K, Zhang L, Montanarella L, Wang Y, Sun G (2020) How afforestation affects the water cycle in drylands: a process-based comparative analysis. Glob Chang Biol 26:944–959. https://doi.org/10.1111/gcb.14875

Shao Y, Lunetta R, Wheeler B, Iiames JS, Campbell JB (2016) An evaluation of time-series smoothing algorithms for land-cover classifications using MODIS-NDVI multi-temporal data. Remote Sens Environ 174:258–265. https://doi.org/10.1016/j.rse.2015.12.023

Shyamsundar P, Cohen F, Boucher TM, Kroeger T, Erbaugh JT, Waterfield G, Clarke C, Cook-Patton SC, Garcia E, Juma K, Kaur S, Leisher C, Miller DC, Oester K, Saigal S, Siikamaki J, Sills EO, Thaung T, Trihadmojo B, Veigna F, Zhang XX (2022) Scaling smallholder tree cover restoration across the tropics. Glob Environ Change 76:102591. https://doi.org/10.1016/j.gloenvcha.2022.102591

Singh S, Giri K, Mishra G, Kumar M, Singh RK, Pandey S, Mullick M, Sharma R (2023) Pathways to achieve land degradation neutrality in India. Indian Council of Forestry Research and Education, Dehradun

Speranza CI, Adenle A, Boillat S (2019) Land degradation neutrality-potentials for its operationalisation at multi-levels in Nigeria. Environ Sci Pol 94:63–71. https://doi.org/10.1016/j.envsci.2018.12.018

Statista Research Department (2024) Share of land under degradation in India as of 2019, by state. https://www.statista.com/statistics/1426607/india-land-degradation-by-state/

Stavi I, Lal R (2015) Achieving zero net land degradation: challenges and opportunities. J Arid Environ 112:44–51. https://doi.org/10.1016/j.jaridenv.2014.01.016

Tsymbarovich P, Kust G, Kumani M, Golosov V, Andreeva O (2020) Soil erosion: an important indicator for the assessment of land degradation neutrality in Russia. Int Soil Water Conserv Res 8(4):418–429. https://doi.org/10.1016/j.iswcr.2020.06.002

UNCCD (2012) United Nations Convention to Combat Desertification (UNCCD) in Bonn, Germany

UNCCD (2019) Land climate: key message from the IPCC special report on climate change and land by United Nations convention to combat desertification

UNCCD (2022) The global land outlook, 2nd edn, Bonn

United Nations (2011) The state of food and agriculture 2010–2011. Women in Agriculture: Closing the gender gap for development Food and Agriculture Organization of the FAO Rome. http://www.fao.org/docrep/013/i2050e/i2050e00.htm

UNU-EHS (2023) Land restoration: 5 key elements of reviving our Earth. United Nations University-Institute for Environment and Human Security, Bonn

VCS-VERRA (2019) Verified carbon standard. Protocols by Verra. Available at: https://verra.org/project/vcs-program/registry-system/

Von Maltitz GP, Gambiza J, Kellner K, Rambau T, Lindeque L, Kgope B (2019) Experiences from the South African land degradation neutrality target setting process. Environ Sci Pol 101:54–62. https://doi.org/10.1016/j.envsci.2019.07.003

Waring BG, Gurgel A, Köberle AC, Paltsev S, Rogelj J (2023) Natural climate solutions must embrace multiple perspectives to ensure synergy with sustainable development. Front Clim 5. https://doi.org/10.3389/fclim.2023.1216175

Wuepper D, Borrelli P, Panagos P, Lauber T, Crowther T, Thomas A, Robinson DA (2021) A 'debt' based approach to land degradation as an indicator of global change. Glob Chang Biol 27(21):5411–5413. https://doi.org/10.1111/gcb.15830

WWF (2022) Towards nature positive—the role of restoration for land degradation neutrality. WWF, Gland

Xu DY, Zhang XY (2021) Multi-scenario simulation of desertification in North China for 2030. Land Degrad Dev 32(11):1060–1074. https://doi.org/10.1002/ldr.3746

Crop Diversification for Land Restoration and Food Security

12

Ruchika Choudhary, Arvind Verma, Kannoj, Manish Tomar, Rupa Kumpatla, and Tony Manoj Kumar Nandipamu

Abstract

Monocultural practices exacerbate the issue of land degradation resulting in the reduction of soil nutrients, causing erosion, decline in organic matter and heightened vulnerability to pests and diseases. Therefore, it is crucial to adopt a well-rounded approach that combines sustainable practices and effective resource management along with strategic policy measures that ensure long-term food security and land restoration. The balanced strategy will address not just the current difficulties in agriculture but also other bigger aims that are environmental and socio-economic. Crop diversification is recognized as a green agricultural production tactic due to its ecological, economic and environmental benefits. Crop diversification enhances crop output through two approaches: horizontal, by adding more crops to existing systems and vertical, by processing crops into value-added products. While horizontal diversification boosts production and

R. Choudhary (✉) · A. Verma
Department of Agronomy, Maharana Pratap University of Agriculture and Technology, Udaipur, Rajasthan, India

Kannoj
Forage Section, Department of Genetics and Plant Breeding, CCS Haryana Agricultural University,
Hisar, Haryana, India

M. Tomar
Department of Agronomy, CCS Haryana Agricultural University, Hisar, Haryana, India

R. Kumpatla
Department of Soil Science, G. B. Pant University of Agriculture & Technology, Pantnagar, Uttarakhand, India

T. M. K. Nandipamu
Department of Agronomy, G. B. Pant University of Agriculture & Technology, Pantnagar, Uttarakhand, India

© The Author(s), under exclusive license to Springer Nature Singapore Pte Ltd. 2025

S. Babu et al. (eds.), *Ecological Solutions to Agricultural Land Degradation*,
https://doi.org/10.1007/978-981-96-3392-0_12

cropping intensity, vertical diversification focuses on economic returns. Crop diversification can potentially reduce greenhouse gas (GHG) emissions and increase plant biomass, soil health, protein production and microbial biodiversity. By lowering fertilizers and other production inputs, it can minimize environmental costs and boost farmers' incomes. Therefore, diversified cropping can serve as a viable approach in achieving sustainable food production and land restoration under a diverse ecosystem.

Keywords

Climate resilience · Food security · Soil health · Sustainable agriculture

Abbreviations

FAO	Food and Agriculture Organization
UN	United Nations
GHG	Greenhouse gas emissions
N	Nitrogen
Al	Aluminium
C	Carbon
CH_4	Methane
N_2O	Nitrous oxide
Ca	Calcium
P	Phosphorus
K	Potassium
CO_2	Carbon dioxide

12.1 Introduction

Addressing the food needs of the rising population remains daunting since the global population is expected to exceed 11 billion by 2050 (Sharma et al. 2024). The galloping population is a pressing challenge for resource management, food security, sustainable development and policy planning. Worldwide land degradation is projected to affect 20–40% of the land area and nearly half of the human population (World Bank 2023). During 2023, ~9.1% of individuals were undernourished and 28.9% of the population had food insecurity. Severe food insecurity grew from 9.1% in 2019 to 10.6% in 2020. Both these issues can be tackled with higher efficiency, which can be done only by managing the land and water resources wisely (FAO 2023). Population, food production, land availability and land degradation across various continents are given in Table 12.1. Agriculture in developing nations

Table 12.1 Continent-wise data of different parameters

Continent	Population (2023, Billion)	Land availability (Million km²)	Food production	Land degradation (% of total land)
Asia	4.74	44.58	2.3 billion	39
Africa	1.45	30.37	520 million	65
North America	0.60	24.71	980 million	25%
Europe	0.74	22.14	1.1 billion	23
South America	0.44	17.84	910 million	35%
Antarctica	0.00	14.00	N/A	0%
Oceania	0.04	8.53	85 million	13%

Source: FAO (2023)

like India, Pakistan, Sri Lanka etc. underwent major changes following the emergence of modern agricultural technology in the late 1960s. Cereal-dominated agriculture was formerly necessary in the developing world due to increased food grain demand. Since the Green Revolution, the shortage of pulse production or crop diversification led to prioritizing rice and wheat ignoring the country's food security. With the driving shift in food security and the importance of multi-front dietary patterns avoiding malnutrition, efforts should channel the diversification of cereals with cropping systems that enhance nutritional supply, climate resilience and resource use efficiency while lowering greenhouse gas (GHG) emissions without sacrificing calorie output are necessary (Davis et al. 2019). Long-term predictions indicate that by 2030 ~ 40% of dietary energy needs will be fulfilled by commodities other than food grains. This energy must be derived from horticulture, livestock, fisheries and other non-foodgrain commodities, demonstrating the importance of diverse farming, especially incorporating non cereal-dominant production system enterprises. Crop diversification can be intensified by applying current technologies such as agricultural mechanization, optimal crop nutrition irrigation, use of high-quality seeds and other planting materials and the adoption of recent technologies like protected cultivation and conservation agriculture.

Growers choose mono-culture farming because of its efficiency: employing specialised technologies enhances planting and harvesting, consequently optimizing the yields and decreasing labor expenses. Furthermore, the knowledge and abilities necessary are not complex with a single crop or animal rearing. In return, however, intense monoculture farming has increased dependency on agrochemicals, elevated energy consumption for agrichemical manufacture, and increased greenhouse gas emissions from heavy fertilizer use. With new technologies and sophisticated production techniques emerging, growers began adopting high-yield plant varieties, mechanization and agrochemicals. However, most of the social, financial and environmental difficulties presented by intensive monocropping systems would have been averted if crop diversity were introduced at farm and value chain levels (Pretty and Bharucha 2014). Crop diversification can be done through crop rotation, cover crops, agroforestry systems and intercropping (Wezel et al. 2014). Crop diversification can enable soil's optimal health, reduce chemical inputs reliance and boost resilience to pests and diseases to produce a more secure and sustainable setup of food production.

The continued reliance on single cropping, combined with increased input usage has a variety of consequences, including decreased soil fertility, degraded soil health, increased GHG emissions and depletion of both above and below-ground biodiversity (Das et al. 2022a). These changes pose significant global challenges to food and nutritional security, emphasizing the critical need to reevaluate our land management practices and improve ecosystem services such as carbon (C) sequestration, soil fertility and microbial population promotion. The present need is to maintain soil fertility using a diversified cropping system and utilize the best crop management practices to ensure soil fertility, profitability and resource use efficiency (Singh et al. 2021a; Yadav et al. 2021). Diversified crop rotations boost system productivity by up to 38%, improve GHG balance by 88% and reduce N_2O emissions by 39%. Including legumes rotations raises soil organic carbon by 6–8% and enhances soil microbial activity (Yang et al. 2024; Singh et al. 2021b). Diversified cropping systems aim to enhance ecosystem functions and services, reducing reliance on single crops and agronomic inputs while supporting high yields and climate change mitigation. They protect soil from erosion and boost CO_2 capture. Increasing agroecosystem biodiversity through practices like intercropping and habitat diversification, can improve ecosystem services without lowering yields. However, some diversified systems had negative impacts on crop yields and climate regulation (Tamburini et al. 2020). In semiarid regions, managing crop diversification is challenging due to potential competition for water and nutrients (Morugán-Coronado et al. 2020; Rathore et al. 2022). Successful inter-cropping requires choosing species suited to local conditions, compatible with available machinery and economically viable (Singh et al. 2021c). Realizing the benefits of crop diversification needs serious efforts to maintain a balance between resource management and post-harvest strategies to conserve the produce to alleviate malnutrition and maintain food security.

12.2 Crop Diversification

The word 'diverge', which means to move or stretch in a distinct direction from a common point, is the root of the term 'diversification'. Crop diversification can thus be defined as moving away from a single crop's dominance in a given location and toward the production of a wide variety of crops to satisfy growing demand. It is also the strategy of cultivating more than one type of agroecologically suitable crops from the same or dissimilar species in a single place throughout time and/or space (Paroda 2022).

Diversification at the macro level frequently entails an economy's secondary and tertiary sectors becoming more prominent than the primary sector in the long run, with respect to their contributions to national revenue and workforce distribution. Diversification can be grouped into three basic types: employment diversification, resource diversification, and agricultural diversification. It comprises four components of growth: area, yield, pattern and pricing. In the agriculture industry, growth can be influenced by shifts in land use (altering production costs), variation in input prices (altering input acquisition cost), and advancements in technology and input intensification (which modify out prices.

Additionally, diversity or the cropping pattern mix can influence growth by altering and utilization of resources. To enhance agricultural yields and optimize revenue generation at local, regional and national scales, two principal forms of diversification are essential *viz.*, vertical diversification, which encompasses the expansion into agri-business activities and horizontal diversification, which involves the adoption of multi- cropping practices (Fig. 12.1).

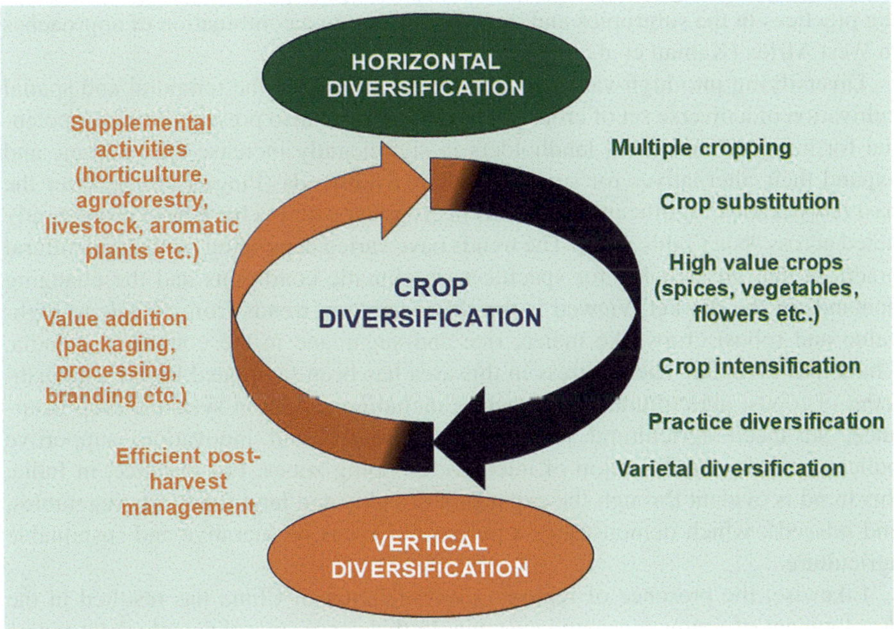

Fig. 12.1 Crop diversification and its components

12.2.1 Crop Diversification Patterns

Current agricultural production systems struggle to meet food demands while preserving biodiversity simultaneously. Conventional agricultural production practices often exacerbate land degradation, greenhouse gas emissions, biodiversity loss, poor diets and sustainability issues (Babu et al. 2023). With a projected 35–56% increase in demand for agricultural commodities from 2010 to 2050, there is growing pressure to intensify agricultural practices or expand cropland (Van Dijk et al. 2021). Sustainable intensification means boosting production per unit while minimizing environmental harm; has the potential to transform agriculture. Diversified farming can potentially enhance food production by ~10% (Zhao et al. 2022), economic output by ~20% (Himmelstein et al. 2016; Sánchez et al. 2022), reduce groundwater depletion by 19% (Yang et al. 2021), mitigate carbon emissions by 11% and enhance biodiversity conservation by ~20% (Beillouin et al. 2021; Jones et al. 2021; Babu et al. 2020a). Despite its benefits, diversified farming is occasionally criticized for yielding lower outputs and profits compared to simplified systems. Nevertheless, the yields from diversified farming often exhibit superior quality and can command higher prices in niche markets. Approximately 47% of the world's land is suitable for economically viable diversified systems, especially in the global North (Kamau et al. 2023). Diversification has the potential to support sustainable intensification through several strategies: expanding cropland in Europe, intensifying practices in the subtropics and tropics or adopting a combination of approaches in West Africa (Kamau et al. 2023).

Diversifying into high-value crops not only allows for the temporal and spatial cultivation of a diverse set of crops simultaneously but also presents a unique potential for marginal and small landholders to significantly increase their income and expand their alternatives for sustaining their livelihoods (Pingali 2015). Over the past few decades, significant alterations in cropping patterns have been prominently noted across Asia (Table 12.2). The trends have varied depending on the agricultural practices that are suitable for specific agro-climatic conditions and the changing demands of the market. Viewed is the shift visible in trends from cereals to high-value and robust crops like maize, rice and sugarcane in the countries of India, China and Vietnam. The progress in this area has been facilitated by the establishment of robust agricultural infrastructure including irrigation systems, crop insurance, advanced agricultural technologies, research and innovation, supportive policies and the identification of intensive cropping zones. For instance, in India, this trend is evident through the expansion of cultivated land for fruits, vegetables, and oilseeds, which demonstrates a purposeful focus on lucrative and sustainable agriculture.

Likewise, the presence of regional diversification in China has resulted in the development of unique cropping systems. In this sense, rice-rice and wheat-maize systems have flourished in the Yangtze River basin and the North China Plain, respectively (Fei et al. 2020). There is a rising interest in combining aquaculture with rice cultivation, which has demonstrated promising potential for enhancing production and promoting ecological sustainability. Rice-based cropping systems in

Table 12.2 Diversification opportunities of cropping systems in Asian countries (some examples)

Country	Existing cropping systems	Opportunities for diversification
India	Rice-wheat, Rice-maize, sugarcane-wheat	Rice-pulse, maize-legume, sugarcane-pulse
Pakistan	Wheat-Rice, cotton-wheat, sugarcane-wheat	Wheat-maize, cotton-legume, sugarcane-vegetable
Bangladesh	Rice-Rice, jute-rice	Rice-fish, Rice-pulses, Rice-vegetable
Nepal	Rice-wheat, maize-millet	Rice-mustard, maize-pulse, wheat-millet
Sri Lanka	Rice-rice, tea-rubber	Rice-pulses, tea-intercropping, rubber-cocoa
Bhutan	Rice-wheat, maize-millet	Rice-vegetable, maize-pulse, wheat-buckwheat
China	Rice-rice, wheat-maize	Rice-aquaculture, wheat-legume, Rice-vegetable
Indonesia	Rice-rice, maize-soybean	Rice-fish, maize-cassava, Rice-pulses
Vietnam	Rice-Rice, rice-maize	Rice-shrimp, Rice-aquaculture, Rice-vegetable
Thailand	Rice-rice, cassava-maize	Rice-fish, cassava-Pigeonpea, Rice-pulses
Philippines	Rice-Rice, corn-rice	Rice-aquaculture, corn-legume, Rice-vegetable

Source: Timsina et al. (2010)

China, Bangladesh, Indonesia and Vietnam have undergone diversification, primarily through the implementation of rice-fish systems, intercropping of cereals with a wide range of legumes and other similar strategies (Zhu et al. 2024). In South Asia, short-duration types of pulse, in rice fallows are successful (Mishra et al. 2022). Mixed cropping systems with pulses and oilseeds could be successful in enhancing soil fertility and augmenting farm income in nations of both Central and Southeast Asia, where these crops could be inserted into current cereal-based systems with a variety of potentials (Yadav et al. 2024; Mirdoraghi et al. 2024).

12.3 Factors Determining Crop Diversification

12.3.1 Infrastructure

Markets and road networks are major determinants of agricultural diversity. The most significant factor influencing the pattern of diversification is the market's notable changes in relative prices. Indeed, market pressures frequently promote diversification toward high-value commodities such as oilseeds, as shown in both India and Indonesia (Schreinemachers et al. 2021). However, global price fluctuations have the potential to destabilize domestic producers, of which majority are smallholders. However, for growers to react effectively, a strong market infrastructure supported by robust institutional arrangements is required to ensure that accurate market information reaches to growers (Fan and Brzeska 2016).

12.3.2 Resource Endowments

Agro-climatic factors within a region significantly influence diversification practices. Specific soils and water regimes may be optimal for particular crops, while physical and geographical attributes *viz.*, drainage characteristics, elevation, and slopes, alongside other climatic elements including rainfall, temperature, humidity, and sunlight intensity, must be considered in production decisions. Efforts to alter these natural and physical conditions to accommodate different cropping patterns can often be exceedingly challenging and economically unfeasible (Banerjee and Banerjee 2014).

12.3.3 Technology

Crop diversification is constrained by climatic, land, and socio-economic factors and is heavily dependent on technological advancements. The technology embedded in fertilizer, seed, draught power can enhance the advantages of one crop relative to another.

12.3.4 Irrigation

Effective water management and policy support encourage diversification and sustainability in farming practices. The availability of a consistent supply of water has greatly improved irrigation, resulting in increased production and the ability to produce a variety of crops, reduced reliance on rainfall, lowering drought concerns and enabling the use of modern irrigation technologies.

12.3.5 Policy Issues

Policy initiatives profoundly affect cropping patterns and the degree of diversification. In India and other emerging countries, policymakers have prioritized food security by focussing on the cultivation of foodgrain crops. Government policies impacting crop diversification include tax and tariff regulations, pricing strategies, trade policies and policies related to public expenditure and agrarian reforms.

12.3.6 The Interactive Nature of Enterprises

Crop production is closely related with activities like dairying, which influences cropping patterns through the demand for feed and fodder, leading to crop diversification. Consequently, a comprehensive approach to crop diversification must consider a broader context by analyzing the interrelated aspects of crop production and other agricultural enterprises at the grassroots level.

12.3.7 Shift in Consumption

Changes in demand patterns provide a strong stimulus to diversification. Rising urbanization and evolving life styles have shifted food consumption preferences from staple grains to non-staple foods. As populations migrate from rural area to urban centres, substantial structural transformations in consumption patterns occur for various reasons. First, there are more varieties of food in the urban markets compared to what is given to rural populations. In addition, city people are exposed to foreign cultures' eating patterns distinct from the local population. Most foods that require less time to prepare fit easily into the quick pace of urban living. Furthermore, metropolitan employment is frequently less physically demanding therefore there is a reduced caloric consumption. Finally, urban inhabitants do not generally generate their food and hence their consumption decisions are not restricted by production restrictions.

12.4 Crop Diversification for Land Restoration

12.4.1 Crop Diversification's Role in Restoring Degraded Lands

Monocropping and cultivation of the same crops over the years create perilous soil conditions including loss of soil fertility, altercations in soil chemistry and make soil susceptible to various forms of degradation. Soil degradation is the obliteration of the intrinsic properties of soil caused by natural or human processes resulting in the loss of land's production capacity, soil fertility and abatement or annihilation of soil microhabitat (Maximillian et al. 2019). The primary drivers of soil degradation are unsustainable land use practices, erosion, reduction in organic matter content, reduced mineral-associated and particulate organic carbon, over-fertilization, salinization, nutrient depletion and biodiversity loss (Onet et al. 2019; Musa et al. 2024; Zhang et al. 2024; Yadav et al. 2024). Adoption of energy and input-intensive crop rotations *viz*., rice-rice, rice-wheat, sugarcane and tea cultivation cause problems like sublayer compaction, salt accumulation and soil degradation (Yang et al. 2022). Studies showed that approximately 50% of the rice fields had degradation problems and wheat yields in rice-wheat systems have reduced by 18% than sole cultivation (Jin et al. 2022). Degraded soils, lack organic matter and essential nutrients and are unable to support crop growth. Without appropriate management or conservation efforts, these soils become irreversibly unproductive, leading to significant ecological and socioeconomic repercussions. Multiple approaches are considered as a part of remediation techniques for degraded soils. One of the techniques that garnered attention is use of crop diversification as it is more effective, sustainable, technically applicable and can restore biodiversity. Crop diversification encompassing both vertical and horizontal diversification, can alleviate the ill effects on soil by enhancing the intrinsic characteristics of the soil and bolstering an array of ecosystem functions.

12.4.1.1 Restoring the Salt-Affected Soils

Salt-affected soils majorly occur in arid and semi-arid areas where the rainfall is low and inadequate irrigation facilities prevent the leaching down of the accumulated salts (Fig. 12.2). Salinity impairs crop development in all stages and results in the loss of around 16.84 million tonnes of agricultural output including pulses, oilseeds, cereals and cash crops (Kumar and Sharma 2020). Therefore, effective remediation and management of these lands is necessary for meeting global food demand. Many technologies and agronomic practices have proven beneficial in the reclamation of salt-affected soils, out of which diversification like going for winter cropping in coastal saline soils with monocropping going for salt-tolerant varieties incorporation of crop residues and agroforestry systems is proven successful.

Halophytes due to their salt-tolerant mechanisms contribute majorly to the restoration of the salt-affected soils (Liang and Shi 2021). Furthermore, incorporating salt or salinity-tolerant varieties in an intercropping system can improve the properties of the salt-affected soils through mechanisms such as phytoremediation where the plants accumulate salts in their tissues (Shankar and Evelin 2019; Tallou et al. 2024). Diversified cropping systems such as crop rotations, mixed cropping, residue incorporation and growing cover crops can reduce soil degradation and promote the sequestration of carbon (Semmartin et al. 2023). Integrated alfalfa with wheat stabilized the soil structure in sodic soils by maintaining the ions balance through its nutrient absorption (Su et al. 2024). Similarly, certain crops like *Suaeda salsa*, alfalfa, *Salicornia, Sarcocornia and Solanum pennelli,* when grown as an intercrop or cover crop exhibit the ability to ameliorate the saline soils by absorbing salts, in this process the leaf spongy cell's vacuole accumulate Na^{+} and sequester them in their above-ground biomass (Egea et al. 2023).

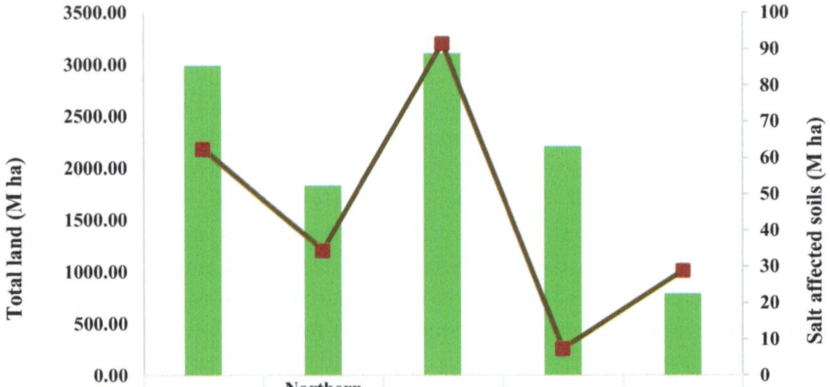

	Africa	Northern America	Asia	Europe	Australia
Total land	2992.17	1834.69	3111.64	2213.03	795.54
Salt affected soils	62.55	34.55	91.6	7.5	29

Fig. 12.2 Extent of salt-affected soils in the world (FAO 2022)

12.4.2 Restoring the Compacted Soils

Soil compaction is a form of degradation where bulk density, soil strength, porosity and hydraulic properties are generally retarded. Overuse of machinery, short crop rotations and intensive cultivation in agriculture contributed to enhanced food production but also increased soil compaction which led to reduced fertility. Compaction can reduce the crop yield, it may be from crop to crop, in wheat yield reduction may be nearly 38% whereas in corn it can be as high as 50% (Shaheb et al. 2021). Thick and deep roots enhance soil structure by creating channels as they grow and upon decomposing, these roots leave behind open pores known as bio-spores which further improve soil permeability. An effective cover crop or bio-tillage should possess deeper roots, rapid growth and quick decomposition thus increasing macro porosity and enhancing pore structure (Cercioglu et al. 2018). Acting as bio-tillage, these roots reduce compaction and promote the growth of the subsequent crop roots improving overall soil structure (Zhang and Peng 2021). These bio-pores facilitate the preferential transport of water and air thereby augmenting soil aeration and attenuating surface runoff and soil erosion (Zhang et al. 2019). Crops with multi-branch taproots (eg. chicory and sugarbeet) develop more voluminous bio pores than herringbone taproot crops like alfalfa (Uteau et al. 2013) thereby helping it to penetrate the compacted layer and improve soil structure which plays a crucial role in maintaining soil fertility Incorporating crops with tap roots like lucerne, chicory and lupins in crop rotation and cover crops like sunhemp, millets, sorghum has shown to reduce soil compaction (Kumar et al. 2018). As these crops reduce compaction, it helps the succeeding crops to deal with the penetration resistance of the soil.

12.4.3 Crop Diversification for Acid Soil Management

Acid soils, characterized by a pH value of <5.5 are predominantly found in tropical and subtropical areas and constitute 30% of the total area globally (Rahman et al. 2018). Intensive cultivation of crops like rice, wheat, sugarcane and tea often leads to excessive application of synthetic inputs, disrupting the soil pH and exacerbating soil acidification (Ni et al. 2018; Zhu and Shen 2023; El Chami et al. 2020). These soils are associated with Al, Mn toxicities and primary nutrients & Ca deficiencies affecting crop growth (Rahman et al. 2024; Jing et al. 2024). Crop diversification like intercropping, green manure application and the cultivation of cover crops have emerged as an effective strategy for ameliorating aluminium (Al) toxicity and enhancing phosphorus (P) availability in soils (Zhang et al. 2024). Diversified cropping improves both the quantity and quality of crop residues by improving shoot and root biomass. This augmented organic matter addition into the soil affects the pH and fertility of the soil (Ghimire and Bista 2016). Varietal diversification is also a promising approach to overcoming the yield reduction challenges in acid soil (Kumar et al. 2023). Studies also revealed that the addition of green manure crops like cowpea, *Leucaena* and *Cassia* are effective in detoxifying Al toxicity in acidic soils which is facilitated by the exudates and polymeric substances released during the decomposition of these crops (Jaiswal et al. 2018). The practice of *Gliricidia* and *Leucaena* alley or using N fixing trees as hedge trees add a huge amount of organic matter which helps in making the P available by reducing the Al saturation and toxicity by the release of low-weight organic substances (Hombegowda et al. 2022). These findings suggest that crop diversification ameliorates soil acidity by increasing organic matter and residue inputs which neutralizes the acidic conditions and helps to maintain the pH in addition to this, these diverse cropping systems support diverse microbial communities (Fu et al. 2021)

12.4.4 Crop Diversification Improves Soil Quality

Continuous practice of cereal- cereal cropping systems, mainly rice-rice, rice-wheat and sorghum-wheat systems leads to soil nutrient exhaustion and imbalance compromising its fertility and health. Nutrient imbalances can be alleviated by using multiple approaches among which crop diversification has garnered significant attention owing to its capacity to optimize nutrient cycling and rectify imbalances by integrating crops with diverse nutrient demands (Maitra et al. 2021; Das et al. 2022b). Amongst all nutrients, nitrogen (N) is dispensable for the advancement and growth of the crops. However, the majority of crops rely on the assimilation of soil nitrogen and certain clades, particularly legumes, possess the potential to fix atmospheric nitrogen (Kumawat et al. 2022). For example, the N transfer from legume to cereal in wheat-kidney bean intercropping is 6% higher whereas in wheat-faba bean intercropping N sharing was 13% higher than in monoculture (Chapagain and Riseman 2015). Legumes and non-legume crop intercropping have complemented ecological niches enabling nitrogen transfer and less competitive interactions (Thilakarathna et al. 2016). Concerning phosphorus, crop diversity augments soil fertility through rhizosphere phosphorus (P) mobilization in phosphorus-deficient soils (Yu et al. 2022). Legumes possess a greater ability to release protons than cereals which makes the rhizosphere more acidic and facilitates the release and absorption of insoluble phosphorus. Crops like *Urochloa ruziziensis* in crop rotation serve a crucial part in the phosphorus cycle by mobilizing unavailable phosphorus for uptake by other plants (Baptistella et al. 2020). Potassium (K) level is typically high in saline soils hence growing forage crops may aid in reducing K concentrations in these soils (Biswas and Biswas 2014). Intercropping groundnut with maize followed by peas with in situ residue retention significantly ($p < 0.05$) increased the availability of soil micro and macronutrients, soil organic carbon (SOC), carbon stratification ratio and soil enzyme activities compared to monocropping (Ansari et al. 2022). Likewise, the induction of additional crops, particularly legumes, into the rice-maize system significantly enhanced carbon recycling, SOC content, carbon pool, sequestration rate, carbon pool index (CPI), carbon management index (CMI) and carbon retention efficiency (CRE) in marginal lands of the Indian Himalayas (Babu et al. 2020b). Few example of effect of diversification in existing cropping systems on land restoration given in Table 12.3.

12.4.5 Crop Diversification Role in Erosion Control and Water Management

Soil erosion serves as one of the key causes of land degradation and worldwide, soil erosion left roughly ten million hectares of agricultural land barren (Gachene et al. 2020). Soil erosion leads to the loss of the uppermost fertile layer thereby leading to reduced soil productivity impairing the soil infiltration capacity, drainage leading to water logging, reduced aeration, and decreased soil fertility. A study showed a nutrient loss of 16.6–39.5 kg C, 5.5–29.8 kg N, 3.9–16.4 kg P, and 5.2–14.6 kg K per hectare per year, respectively in sole potato cropping due to erosion (Nyawade et al. 2019). Soil erosion is a big concern in places with high slopes, river banks, and coastal banks. Crop diversification can attenuate soil erosion by alleviating the effect of wind and water. A worldwide meta-analysis indicated that ground cover management showed a 25.5% reduction in runoff and 91.5% soil loss, apart from this it also reduced the loss of total nitrogen, ammonical nitrogen, nitrate nitrogen and total phosphorus by 25.5%, 90.7%, 74.6% and 81%, respectively compared to the soil (Zhang et al. 2020). Diversification with a combination of shallow fibrous roots in the surface soil with deeply penetrating roots was preferred to reduce soil erosion (Yadav et al. 2024). This may be due to the formation of bio-pores which move water and air preferentially, promote soil aeration and reduce surface runoff

Table 12.3 Effect of diversification in existing cropping systems on land restoration

Existing cropping system	Effect on soil quality	Diversification in existing systems	Remediation effect	References
Rice-wheat, Cotton-based cropping system	Exhaustive in nature, lead to depletion of SOC and soil structure, multi-nutrient deficiencies	Legumes-wheat, Pearlmillet-wheat, Cotton-wheat, Maize- wheat	Increased carbon sequestration and the level of nutrients through root turnover and leaf litter	Moharana et al. (2017), Brar et al. (2023)
Sugarcane-based cropping system	Decrease in the carbon and nitrogen pool	Sugarcane+legume; Sugarcane+oilseeds (mainly mustard); Sugarcane+ vegetables	Rhizodesposits of other crops help in the solubilization of essential nutrients and augment their availability	Singh et al. (2021c), Shukla et al. (2022)
Maize and sorghum based	Exhaustive in nature, soil erosion, reduced soil fertility, decline in total factor productivity	Maize+cowpea-garden pea; Mazie+cowpea-barley Mazie+ okra; Sorghum+greengram-wheat; Mazie-chickpea	Minimizes soil compaction, protects the soil from soil erosion, increases the organic matter and N content	Jain et al. (2015), Sharma et al. (2017), Zhang et al. (2021)

and soil erosion (Zhang et al. 2019). A study in maize-wheat crop rotation in the Indian Himalayan region showed that conservation agriculture practices such as growing vegetation strips and using weed mulching lowered the runoff coefficients and soil loss approximately by 45 and 54% respectively (Ghosh et al. 2015). The reduced soil loss is attributed to the increased biomass provided by vegetation strips, which enhances infiltration and reduces runoff. The increased canopy cover, resilient canopy, and root growth in agroforestry systems reduce the direct impact of raindrops and improve infiltration capacities compared to sole agriculture crops (Zhang et al. 2024). Vegetative cover is a crucial factor in minimizing the initial stage of erosion (splash, sheet and interill) but for the advanced stage of erosion (rill and gully type) root structure of the plant plays a key role (Qian et al. 2024; Moustakidis et al. 2024).

12.4.6 Crop Diversification's Role in Soil Biodiversity Restoration

Biodiversity, referred to as species richness of species within an area, is adversely affected by land degradation. This process led to a shift in plant diversity, which in turn directly effects the microbial diversity. For instance, the Shannon diversity indices of bacteria and fungi in degraded grasslands differ significantly specifically, fungi exhibit greater sensitivity to degradation compared to bacteria (Wu et al. 2021). Similarly, in another study on the response of various soil microbial taxa to salt stress, *Archaea* were affected by the sodium content and soil EC while the bacteria remained unaffected (Zhang et al. 2024). Many studies showed growing similar crops annually increases soil-borne pathogens and reduces the beneficial microbes. For example, a study on tea monoculture showed a reduction in beneficial fungal species and an enhancment in the pathogenic fungal species (Arafat et al. 2019). Agricultural practices like intensification, monoculture and deforestation to increase the agricultural lands are reasons for the declining biodiversity hence alteration in agricultural practices is necessary for conserving biodiversity (Udawatta et al. 2019). Agroforestry is a better practice than monoculture, as it improves the microbial biomass, basal respiration and enzymatic activity leading to better biodiversity in the soil (Babu et al. 2020c). Cultivation of alder + large cardamom system registered higher soil enzymatic reactions and SOC in the Sikkim Himalayan region of India over annual crop cultivations (Babu et al. 2020c). In Indo-Gangetic plains, 5–11% higher microbial counts were reported in the maize-wheat system than in rice-wheat (Choudhary et al. 2018). This inferred that the soil microbial diversity under similar crops or cropping system cultivation was reduced however diversification in the cropping system helps in the establishment of different niches in the rhizosphere leading to the secretion of a wide range of exudates, extra polymeric substances which favours greater biodiversity.

12.5 Crop Diversification's Role in Food Security

Food security refers to the state in which individuals consistently have the means to obtain enough food that is both safe and nutritious to satisfy their nutritional requirements and personal food choices enabling them to lead healthy and active lives. Despite numerous existing efforts, food security remains the primary concern confronting the globe today. The Food and Agricultural Organization estimates that the global population suffering from undernourishment in 2021 is around 768 million, representing an increase of 117.7 million compared to 2019. Furthermore, around 928 million individuals were facing acute food insecurity, out of which 471 million and 347 million were inhabitants of Asia and Africa, respectively (FAO 2022). Consequently, it is necessary to implement focused interventions with sustainable solutions to address food insecurity, particularly in regions that are disproportionately affected.

12.5.1 Yield Stability and Resilience

Production stability and ensured yields are the key components are desired traits for smallholder farmers. Smallholder farmers depend heavily on seasonal yields for both their economic well-being and food supply. As a result, fluctuations in crop yields can be particularly challenging, directly affecting the availability of food and income needed for essential household needs. This makes crop diversification crucial as it can help farmers better manage food insecurity by potentially increasing yields (Cowger and Weisz 2008). Diversification also provides stability and acts as a form of insurance, allowing farmers to rely on other crops if one fails. For instance, intercropping with crops like tobacco sugarcane, maize, wheat, potato and broad bean has been shown to boost yields in some cases by 33.2–84.7% (Li et al. 2009).

Developing resilient and economical agricultural systems is vital in the regions, where most people rely heavily on agricultural products like food, fuel and fodder for their livelihoods. Most farmers are smallholders typically owning less than 2 hectares of land, a figure anticipated to shrink considerably due to ongoing land fragmentation and unchecked urbanization. These farmers practice low-resource agriculture (Altieri et al. 2012), making them susceptible to climate change impacts, as they possess little resources to invest in costly coping strategies (Lin 2011). Crop diversification is widely recognized as one of the most ecologically sustainable, economical and feasible strategies for reducing uncertainty in agriculture, particularly for small-scale farmers (Avasthe et al. 2023). This approach involves growing more than one type of crop, either from the same species or distinct species within a specific area. By increasing both temporal and spatial biodiversity on the farm, crop diversification enhances resilience, allowing an agroecosystem to recover to its original productive state after disturbances (Njeru 2013). Additionally, it strengthens crop resilience and boosts farmers' incomes, ultimately improving the livelihood security of small and marginal growers.

12.5.2 Pest and Disease Management

Growing one crop or monoculture increases the incidence of pest and disease outbreaks because it lacks the diversity of animal and plant species that naturally control pests and diseases through predation and antagonistic behaviour. This often leads to the need for greater amounts of pesticides and herbicides, which contribute to environmental pollution compared to more diverse farming practices. The heavy use of agricultural chemicals also reduces the availability of insects and worms which are essential food sources for birds and beneficial organisms. Monoculture, therefore, results in reduced agricultural diversity and increased reliance on chemical inputs, raising concerns about the spread of pests and diseases, contamination of surface and groundwater and the degradation of soil health and biodiversity. Additionally, this practice increases the overall economic risk for farmers (Alcon et al. 2020). To handle insect pests in an environment friendly manner, a pest-suppressive agroecosystem has to be considered (Thomine et al. 2022). A pest-suppressive agro-ecosystem can be built by choosing an appropriate intercrop with insect pest deterrence and a border crop improving natural enemy activity by acting as a refugium. Practising multi-cropping on a farm enhances plant and insect diversity compared to monoculture. The continuous availability of floral rewards and alternative food sources for natural predators supports their abundance, diversity and conservation (Christmann et al. 2021). Additionally, chemicals released by intercrops can mask the scent of the main crop, making it harder for pests to locate their host, which leads to a reduced occurrence of insect pests in the primary crop. So, crop diversity through intercropping in widely separated crops minimizes weed infestation and boosts overall productivity. Cultivations of crops with diverse botanical associations in the same piece of land lessen weed and pest/disease stresses.

12.5.3 Nutritional Benefits

Malnutrition and child health are major policy issues in developing countries. Crop diversification offers a potential solution to improve health and child nutrition. Increasing production diversity at the household level leads to greater dietary variety and better nutrient intake in many developing regions (Sibhatu and Qaim 2018). Additionally, crop diversity particularly with vegetables and root crops is linked to a reduced likelihood of child stunting (Shively and Sununtnasuk 2015). In the face of climate change, the nutritional benefits of crop diversity can also serve as a coping mechanism to enhance human health, dietary nutrition and food security (Lovo and Veronesi 2019). Food security is another significant aspect of crop diversification. To ensure an adequate supply of healthy and sufficient food in a future that is hotter, drier and more densely populated, agriculture must be resilient to various risks. Crop diversity is central to global food security, as it supports a resilient agricultural system and preserves the potential for securing food in an increasingly unpredictable world (Massawe et al. 2016). Agricultural diversification boosts food grain production and increases the per capita food supply. For small landholders, crop diversification is a vital strategy to secure food for themselves and their households. By regularly diversifying crops, they can ensure both short and long-term nutritional food security and maintain a high-quality diet (Kumar et al. 2021).

12.5.4 Economic Advantages

The primary goals of diversification are to increase per capita income by creating opportunities for productive employment in both farm and non-farm sectors and to ensure that economic growth is broad-based and sustainable over the long term (Babu et al. 2016). The level of agricultural diversity in a region is largely influenced by its geo-climatic conditions, socio-economic factors, and technological advancements. Ensuring an adequate food supply to meet rising demand, coupled with increasing weather variability and the growing societal preference for low-input agriculture, makes farming increasingly vulnerable to risk and uncertainty. This challenge is further compounded by increasing price volatility, which adds another layer of risk to farmers' incomes. These challenges are especially pronounced in industrial farming systems that rely heavily on monocropping and large amounts of synthetic inputs. Diversifying land use is widely seen as a key strategy to manage risk and uncertainty. Crop diversity promotes the use of agroecological practices and incorporating a variety of plant species in production and consumption can significantly enhance food security, health, nutrition and the income of farmers. Diversification towards new crops from the existing cropping systems will boost agricultural productivity, considering multiple economic benefits from the value-added crops with accompanying marketing scenarios is considered as crop diversification. Crop diversification aims to expand the range of crops grown on a farm, encouraging farmers to avoid reliance on a single crop (monoculture) for their income.

12.6 Crop Diversification for Environmental Sustainability

The rising impact of climate change globally is posing significant challenges to sustainable agriculture and livelihoods. Developing nations are highly vulnerable to climate change due to their small land holdings, heavy reliance on agriculture, and limited access to essential resources and infrastructure. Crops or their varieties that are less vulnerable to climatic difficulties are getting a larger share of production portfolios. Crop responses to climatic shocks vary, encouraging risk-averse growers to choose a mix of crops with low-correlated earnings as a way to diversify risk. Crop diversification is a preemptive technique to lessen the impact of environmental shocks in developing nations (Macours et al. 2012). Crop diversification demonstrates considerable potential in mitigating these challenges by withstanding weather anomalies, guaranteeing the security of food and nutrition, generating income, and enhancing overall health (Fig. 12.3). This could lead to a strengthened capacity to adapt to and mitigate the impact of these challenges, ultimately fostering greater

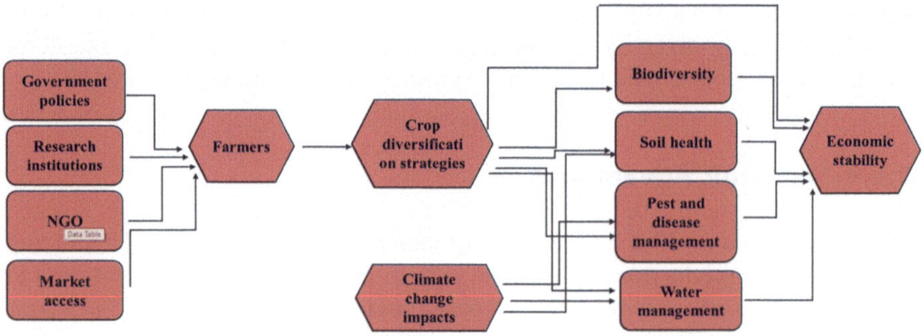

Fig. 12.3 Crop diversification for mitigating climate change

stability and sustainability in rural areas. The implementation of crop rotation diversification leads to a substantial increase in maize yields, with an average improvement of 28.1%. Moreover, this practice results in a noteworthy reduction in yield losses ranging from 14.0% to 89.9% in drought years, thereby making a significant contribution to environmental sustainability (Bowles et al. 2020).

Diversity within and among species protects against environmental changes. Agrobiodiversity promotes ecological resilience, while species diversity buffers against crop failure (Schroff and Ramos 2020). The strategy of cultivating crops and varieties that are adapted to altered rainfall patterns and seasons as a means of drought mitigation is referred to as climate-resilient cropping. Crop diversification in addition to reducing GHG emissions, improves soil carbon sequestration. Techniques like intercropping in woody cropping systems are among the most promising sustainable agriculture management solutions for maintaining current soil organic carbon reserves and revitalizing carbon-depleted soils. These approaches address food security, climate change and soil degradation simultaneously. Intercropping mitigates the loss of soil organic carbon that occurs when natural ecosystems in Mediterranean regions. Intercropping mitigates the risk of crop failure and enhances dietary diversity and food security for small-scale growers (Raseduzzaman and Jensen 2017). Diversified cropping systems featuring longer crop rotations have been shown to yield a 12% increase in crop production, outperforming monoculture (Francaviglia et al. 2020).

12.7 Major Setbacks of Crop Diversification

Crop diversification even though a promising practice yet impeded by a range of challenges which includes risk aversion, land suitability, social norms, income disparities and interactions with extension officers (Ijaz et al. 2019). Market and policies often prioritize economic benefits over comprehensive farming strategies, marginalizing the importance of diversification (Keller et al. 2024). Additionally, knowledge and technology gaps further hinder adoption, as many farmers lack the technical expertise and equipment required for diverse farming systems. Crop diversification in India is shaped by several influencing factors including rural literacy, irrigation intensity, road infrastructure, per capita GDP and the nation's burgeoning urban populace (Neogi and Ghosh 2022). The complexity of managing multiple crops fosters a general reluctance to embrace new perceived risky approaches. The susceptibility to weather fluctuations and pests along with inadequate postharvest infrastructure, pose significant obstacles to the economic viability of a wide range of crops. Other challenges include limited connectivity between research institutions, extension services and farmers; restricted access to financial resources, governmental assistance and ancillary services and climate-related data, among others. These factors collectively impose severe limitations on the capacity of cultivators to effectively embrace crop diversification and harness associated advantages.

12.8 Conclusion

Crop diversification is one of the most significant measures for driving land restoration for food security under diverse agroecosystems. Shifting away from monoculture and toward a more diverse crop portfolio can significantly improve soil health by reconstituting soil nutritional content and, in general, increasing agricultural resilience. Diversified farming, which includes grains, pulses, oilseeds, fruits and vegetables, could reduce reliance on chemical inputs, combat soil deterioration and promote sustainable land management methods. On the economic front, diversified cropping systems generate many income streams, reducing reliance on single crops and thereby mitigating market volatility risks. High-value crops, combined with traditional and nutritious legumes and millets increase economic returns while meeting the growing demand for diversified and nutritious diets caused by urbanization. Crop diversification allows growers to adapt to climate unpredictability by dispersing risks across crops, hence improving food system stability. Given the twin issues of land degradation and rising food demand, adopting diverse and sustainable farming practices will be critical to achieving long-term agricultural production, restoring damaged lands, and ensuring food security for future generations. Government incentives for market access and funding for agricultural research are crucial for the promotion of crop diversification.

References

Alcon F, Marín-Miñano C, Zabala JA, de Miguel MD, Martínez-Paz JM (2020) Valuing diversification benefits through intercropping in Mediterranean agroecosystems: a choice experiment approach. Ecol Econ 171:106593. https://doi.org/10.1016/j.ecolecon.2020.106593

Altieri M, Funes-Monzote F, Petersen P (2012) Agroecologically efficient agricultural systems for smallholder farmers: contributions to food sovereignty. Agron Sustain Dev 32(1):1–13. https://doi.org/10.1007/s13593-011-0065-6

Ansari MA, Babu S, Choudhary J, Ravisankar N, Panwar AS (2022) Soil quality restoration and yield stabilization in acidic soils of northeastern Himalayas: five years impact of green manuring and crop residue management. Front Environ Sci 10:940349. https://doi.org/10.3389/fenvs.2022.940349

Arafat Y, Tayyab M, Khan MU, Chen T, Amjad H, Awais S, Lin S (2019) Long-term monoculture negatively regulates fungal community composition and abundance of tea orchards. Agronomy 9(8):466. https://doi.org/10.3390/agronomy9080466

Avasthe RK, Devi SH, Bhupenchandra I, Kumar A, Chongtham SK, Babu S, Singh R, Das A, Gudade BA, Bora SS (2023) Agro-tactics for reducing carbon footprint in agricultural production systems: a review. Indian J Agron 68(2):115–125. https://doi.org/10.59797/ija.v68i2.332

Babu S, Singh R, Avasthe RK, Yadav GS, Rajkhowa DJ (2016) Intensification of maize (Zea mays)-based cropping sequence in rainfed ecosystem of Sikkim Himalayas for improving system productivity, profitability, employment generation and energy-use efficiency under organic management conditions. Indian J Agric Sci 86(6):778–784. https://doi.org/10.56093/ijas.v86i6.58980

Babu S, Mohapatra KP, Das A, Yadav GS, Tahasildar M, Singh R, Panwar AS, Yadav V, Chandra P (2020a) Designing energy-efficient, economically sustainable and environmentally safe cropping system for the rainfed maize–fallow land of the Eastern Himalayas. Sci Total Environ 722:137874. https://doi.org/10.1016/j.scitotenv.2020.137874

Babu S, Singh R, Avasthe RK, Yadav GS, Mohapatra KP, Selvan T, Das A, Singh VK, Valente D, Petrosillo I (2020b) Soil carbon dynamics in Indian Himalayan intensified organic rice-based cropping sequences. Ecol Indic 114:106292. https://doi.org/10.1016/j.ecolind.2020.106292

Babu S, Mohapatra KP, Yadav GS, Lal R, Singh R, Avasthe RK, Das A, Chandra P, Gudade BA, Kumar A (2020c) Soil carbon dynamics in diverse organic land use systems in North Eastern Himalayan ecosystem of India. Catena 194:104785. https://doi.org/10.1016/j.catena.2020.104785

Babu S, Singh R, Avasthe R, Rathore SS, Kumar S, Das A, Layek J, Sharma V, Wani OA, Singh VK (2023) Conservation tillage and diversified cropping enhance system productivity and eco-efficiency and reduce greenhouse gas intensity in organic farming. Front Sustain Food Syst 7:1114617. https://doi.org/10.3389/fsufs.2023.1114617

Banerjee GD, Banerjee S (2014) Crop diversification: an exploratory analysis. In: Diversification of agriculture in eastern India. Springer, New Delhi, pp 37–57. https://doi.org/10.1007/978-81-322-1997-2_4

Baptistella JLC, de Andrade SAL, Favarin JL, Mazzafera P (2020) Urochloa in tropical agroecosystems. Front Sustain Food Syst 4:119. https://doi.org/10.3389/fsufs.2020.00119

Beillouin D, Ben-Ari T, Malézieux E, Seufert V, Makowski D (2021) Positive but variable effects of crop diversification on biodiversity and ecosystem services. Glob Chang Biol 27:4697–4710. https://doi.org/10.1111/gcb.15747

Biswas A, Biswas A (2014) Comprehensive approaches in rehabilitating salt affected soils: a review on Indian perspective. Open Trans Geosci 1(1):13–24. https://doi.org/10.15764/GEOS.2014.01003

Bowles T, Bowles T, Mooshammer M, Schmer M, Strock J, Grandy A, Socolar Y, Calderón F, Cavigelli M, Culman S, Deen W, Drury C, Garcia A, Gaudin A, Harkcom W, Lehman R, Osborne S, Robertson G, Salerno J, Grandy S (2020) Long-term evidence shows that crop-rotation diversification increases agricultural resilience to adverse growing conditions in North America. One Earth 2(3):284–293. https://doi.org/10.1016/j.oneear.2020.02.007

Brar SK, Dhaliwal SS, Sharma V, Sharma S, Kaur M (2023) Replacement of rice-wheat cropping system with alternative diversified systems concerning crop productivity and their impact on soil carbon and nutrient status in soil profile of north-west India. Cogent Food Agric 9(1):2167483. https://doi.org/10.1080/23311932.2023.2167483

Cercioglu M, Anderson SH, Udawatta RP, Haruna SI (2018) Effects of cover crop and biofuel crop management on computed tomography-measured pore parameters. Geoderma 319:80–88. https://doi.org/10.1016/j.geoderma.2018.01.005. Get rights and content

Chapagain T, Riseman A (2015) Nitrogen and carbon transformations, water use efficiency and ecosystem productivity in monocultures and wheat-bean intercropping systems. Nutr Cycl Agroecosyst 101:107–121. https://doi.org/10.1007/s10705-014-9647-4

Choudhary M, Datta A, Jat H, Yadav A, Gathala M, Sapkota T, Das A, Sharma P, Jat MSR, Ladha JK (2018) Changes in soil biology under conservation agriculture based sustainable inten-

sification of cereal systems in Indo-Gangetic Plains. Geoderma 313:193–204. https://doi.org/10.1016/j.geoderma.2017.10.041
Christmann S, Bencharki Y, Anougmar S, Rasmont P, Smaili MC, Tsivelikas A (2021) Farming with alternative pollinators benefits pollinators, natural enemies, and yields, and offers transformative change to agriculture. Sci Rep 11(1):1–10. https://doi.org/10.1038/s41598-021-97695-5
Cowger C, Weisz R (2008) Winter wheat blends (mixtures) produce a yield advantage in North Carolina. Agron J 100(1):169–177. https://doi.org/10.2134/agronj2007.0128
Das A, Babu S, Kumar DM, Sanjeev SR, Avasthe RK, Rathore SS, Yadav SK, Singh VK (2022a) Restoring soil carbon in marginal land of Indian Himalayas: impact of crop intensification and conservation tillage. J Environ Manag 318:115603. https://doi.org/10.1016/j.jenvman.2022.115603
Das A, Babu S, Singh R, Kumar S, Rathore SS, Singh VK, Datta M, Yadav SK, Wani OA (2022b) Impact of live mulch- based conservation tillage on soil properties and productivity of summer maize in Indian Himalayas. Sustainability 14:12078. https://doi.org/10.3390/su141912078
Davis KF, Chhatre A, Rao ND, Singh D, Ghosh-Jerath S, Mridul A, Poblete-Cazenave M, Pradhan N, DeFries R (2019) Assessing the sustainability of post-green revolution cereals in India. Proc Natl Acad Sci 116(50):25034–25041. https://doi.org/10.1073/pnas.1910935116
Egea I, Estrada Y, Faura C, Egea-Fernández JM, Bolarin MC, Flores FB (2023) Salt-tolerant alternative crops as sources of quality food to mitigate the negative impact of salinity on agricultural production. Front Plant Sci 14:1092885. https://doi.org/10.3389/fpls.2023.1092885
El Chami D, Daccache A, El Moujabber M (2020) What are the impacts of sugarcane production on ecosystem services and human well-being? A review. Ann Agric Sci 65(2):188–199. https://doi.org/10.1016/j.aoas.2020.10.001
Fan S, Brzeska J (2016) Sustainable food security and nutrition: demystifying conventional beliefs. Glob Food Sec 11:11–16. https://doi.org/10.1016/j.gfs.2016.03.005
FAO (2022) The state of food security and nutrition in the world
FAO (2023) The State of Food and Agriculture 2023. http://www.fao.org
Fei L, Meijun Z, Jiaqi S, Zehui C, Xiaoli W, Jiuchun Y (2020) Maize, wheat and rice production potential changes in China under the background of climate change. Agric Syst 182:102853. https://doi.org/10.1016/j.agsy.2020.102853
Francaviglia R, Álvaro-Fuentes J, Bene C, Gai L, Regina K, Turtola E (2020) Diversification and management practices in selected European regions. A data analysis of arable crops production. Agronomy 10(2):297. https://doi.org/10.3390/agronomy10020297
Fu L, Feng A, Xiao J, Wu Q, Ye Q, Peng S (2021) Remediation of soil contaminated with high levels of hexavalent chromium by combined chemical-microbial reduction and stabilization. J Hazard Mater 403:123847. https://doi.org/10.1016/j.jhazmat.2020.123847
Gachene CK, Nyawade SO, Karanja NN (2020) Soil and water conservation: an overview. Zero Hunger:810–823. https://doi.org/10.1007/978-3-319-95675-6_91
Ghimire R, Bista P (2016) Crop diversification improves pH in acidic soils. J Crop Improv 30(6):657–667. https://doi.org/10.1080/15427528.2016.1219894
Ghosh BN, Dogra P, Sharma NK, Bhattacharyya R, Mishra PK (2015) Conservation agriculture impact for soil conservation in maize–wheat cropping system in the Indian sub-Himalayas. Int Soil Water Conserv Res 3(2):112–118. https://doi.org/10.1016/j.iswcr.2015.05.001
Himmelstein J, Ares A, Gallagher D, Myers J (2016) A meta-analysis of intercropping in Africa: impacts on crop yield, farmer income, and integrated pest management effects. Int J Agric Sustain 15(1–10):1080
Hombegowda HC, Adhikary PP, Jakhar P, Madhu M (2022) Alley cropping agroforestry system for improvement of soil health. In: Soil health and environmental sustainability: application of geospatial technology. Springer, Cham, pp 529–549. https://doi.org/10.1007/978-3-031-09270-1_23
Ijaz M, Nawaz A, Ul-Allah S, Rizwan M, Ullah A, Hussain M, Sher A, Ahmad S (2019) Crop diversification and food security. Agron Crops:607–621. https://doi.org/10.1007/978-981-32-9151-5_26

Jain N, Singh H, Dashora L, Mundra S (2015) Maize (Zea mays) wheat (Triticum aestivum) cropping system: intensification through introduction of pulses. Indian J Agron 60(3):347–351. https://doi.org/10.59797/ija.v60i3.4483

Jaiswal SK, Naamala J, Dakora FD (2018) Nature and mechanisms of aluminium toxicity, tolerance and amelioration in symbiotic legumes and rhizobia. Biol Fertil Soils 54:309–318. https://doi.org/10.1007/s00374-018-1262-0

Jin Q, Wang C, Sardans J, Vancov T, Fang Y, Wu L, Liangquan W, Huang X, Garriga A, Penuelas J, Wang W (2022) Effect of soil degradation on the carbon concentration and retention of nitrogen and phosphorus across Chinese rice paddy fields. Catena 209:105810. https://doi.org/10.1016/j.catena.2021.105810

Jing T, Li J, He Y, Shankar A, Saxena A, Tiwari A, Maturi SM, Singh V, Eissa M, Ding Z, Xie J, Awasthi M (2024) Role of calcium nutrition in plant physiology: advances in research and insights into acidic soil conditions-a comprehensive review. Plant Physiol Biochem:108602. https://doi.org/10.1016/j.plaphy.2024.108602

Jones SK, Sánchez AC, Juventia SD, Estrada-Carmona N (2021) A global database of diversified farming effects on biodiversity and yield. Sci Data 8:1–6. https://doi.org/10.1038/s41597-021-01000-y

Kamau H, Roman S, Biber-Freudenberger L (2023) Nearly half of the world is suitable for diversified farming for sustainable intensification. Commun Earth Environ 4:446. https://doi.org/10.1038/s43247-023-01062-3

Keller C, Joshi S, Joshi T, Goldmann E, Riar A (2024) Challenges for crop diversification in cotton-based farming systems in India: a comprehensive gap analysis between practices and policies. Front Agron 6. https://doi.org/10.3389/fagro.2024.1370878

Kumar P, Sharma PK (2020) Soil salinity and food security in India. Front sustain Food Syst 4:533781. https://doi.org/10.3389/fsufs.2020.533781

Kumar V, Jain M, Rani V, Kumar A, Kumar S, Naresh J (2018) A review of soil compaction-concerns, causes and alleviation. Int J Plant Soil Sci 22:1–9. https://doi.org/10.9734/IJPSS/2018/40351

Kumar S, Bamboriya SD, Rani K, Meena RS, Sheoran S, Loyal A, Kumawat A, Jhariya MK (2021) Grain legumes: a diversified diet for sustainable livelihood, food and nutritional security. In: Advances in legumes for sustainable intensification. Academic, pp 157–178. https://doi.org/10.1016/B978-0-323-85797-0.00007-0

Kumar A, Avasthe R, Singh R, Babu S, Das SK, Ravisankar N, Panwar AS, Qureshi AA, Laha, Mishra VK, Bhupenchandra I, Prasad SK (2023) Productivity and profitability of maize (Zea mays) cultivars under organic management condition in mid-hills of Sikkim. Indian J Agron 68(1):26–29. https://doi.org/10.59797/ija.v68i1.198

Kumawat A, Bamboriya SD, Meena RS, Yadav D, Kumar A, Kumar S, Raj A, Pradhan G (2022) Legume-based inter-cropping to achieve the crop, soil, and environmental health security. In: Advances in legumes for sustainable intensification. Academic, pp 307–328. https://doi.org/10.3389/fsufs.2020.533781

Li C, He X, Zhu S, Zhou H, Wang Y, Li Y, Yang J, Fan J, Yang J, Wang G, Long Y (2009) Crop diversity for yield increase. PLoS One 4(11):e8049. https://doi.org/10.1371/journal.pone.0008049

Liang J, Shi W (2021) Cotton/halophytes intercropping decreases salt accumulation and improves soil physicochemical properties and crop productivity in saline-alkali soils under mulched drip irrigation: a three-year field experiment. Field Crop Res 262:108027. https://doi.org/10.1016/j.fcr.2020.108027

Lin BB (2011) Resilience in agriculture through crop diversification: adaptive management for environmental change. Bioscience 61(3):183–193. https://doi.org/10.1525/bio.2011.61.3.4

Lovo S, Veronesi M (2019) Crop diversification and child health: empirical evidence from Tanzania. Ecol Econ 158:168–179. https://doi.org/10.1016/j.ecolecon.2019.01.005

Macours K, Premand P, Vakis R (2012) Transfers, diversification and household risk strategies: experimental evidence with lessons for climate change adaptation, Policy research working paper no. 6053. World Bank, Washington, DC

Maitra S, Hossain A, Brestic M, Skalicky M, Ondrisik P, Gitari H, Brahmachari K, Shankar T, Bhadra P, Palai JB, Jena J (2021) Intercropping-a low input agricultural strategy for food and environmental security. Agronomy 11(2):343. https://doi.org/10.3390/agronomy11020343

Massawe F, Mayes S, Cheng A (2016) Crop diversity: an unexploited treasure trove for food security. Trends Plant Sci 21(5):365–368. https://doi.org/10.1016/j.tplants.2016.02.006

Maximillian J, Brusseau ML, Glenn EP, Matthias AA (2019) Pollution and environmental perturbations in the global system. In: Environmental and pollution science. Academic, pp 457–476. https://doi.org/10.1016/B978-0-12-814719-1.00025-2

Mirdoraghi M, Farahani SM, Rezazadeh A (2024) Oilseeds in intercropping systems: strategies to increase oil quality and fatty acid profile, a review. J Agric Food Res 101229. https://doi.org/10.1016/j.jafr.2024.101229

Mishra AK, Yeggina PK, Sinha DD, Srivastava AK, Sharma S (2022) Scope for production of pulses in rice fallow lands in South Asia. In: Advances in legumes for sustainable intensification. Academic, pp 9–26. https://doi.org/10.1016/B978-0-323-85797-0.00015-X

Moharana PC, Naitam RK, Verma TP, Meena RL, Kumar S, Tailor BL, Samal SK (2017) Effect of long-term cropping systems on soil organic carbon pools and soil quality in western plain of hot arid India. Arch Agron Soil Sci 63(12):1661–1675. https://doi.org/10.1080/03650340.2017.1304637

Morugán-Coronado A, Linares C, Gómez-López MD, Faz Á, Zornoza R (2020) The impact of intercropping, tillage and fertilizer type on soil and crop yield in fruit orchards under Mediterranean conditions: a meta-analysis of field studies. Agric Syst 178:102736. https://doi.org/10.1016/j.agsy.2019.102736

Moustakidis I, Schilling K, Weber L (2024) Soil texture and vegetation root density assessment on regulating erosion across river floodplains. Int J Sediment Res. https://doi.org/10.1016/j.ijsrc.2024.06.004

Musa I, Samuel J, Adams M, Abdulsalam M, Nathaniel V, Maude A, Abdulsalam N, Maude A, Adedayo O, Tiamiyu A (2024) Soil erosion, mineral depletion and regeneration. In: Prospects for soil regeneration and its impact on environmental protection. Springer, Cham, pp 159–172. https://doi.org/10.1007/978-3-031-53270-2_7

Neogi S, Ghosh B (2022) Evaluation of crop diversification on Indian farming practices: a panel regression approach. Sustainability 14(24):16861. https://doi.org/10.3390/su142416861

Ni K, Shi YZ, Yi XY, Zhang QF, Fang L, Ma LF, Ruan J (2018) Effects of long-term nitrogen application on soil acidification and solution chemistry of a tea plantation in China. Agric Ecosyst Environ 252:74–82. https://doi.org/10.1016/j.agee.2017.10.004

Njeru EM (2013) Crop diversification: a potential strategy to mitigate food insecurity by smallholders in sub-Saharan Africa. J Agric Food Syst Commun Dev 3(4):63–69. https://doi.org/10.5304/jafscd.2013.034.006

Nyawade SO, Gachene CK, Karanja NN, Gitari HI, Schulte-Geldermann E, Parker ML (2019) Controlling soil erosion in smallholder potato farming systems using legume intercrops. Geoderma Reg 17:e00225. https://doi.org/10.1016/j.geodrs.2019.e00225

Onet A, Dincă LC, Grenni P, Laslo V, Teusdea A, Vasile D, Enescu R, Crisan V (2019) Biological indicators for evaluating soil quality improvement in a soil degraded by erosion processes. J Soils Sediments 19:2393–2404. https://doi.org/10.1007/s11368-018-02236-9

Paroda R (2022) Crop diversification for sustainable agriculture. Ecol Econ Soc INSEE J 5(1):15–21. https://doi.org/10.37773/ees.v5i1.611

Pingali P (2015) Agricultural policy and nutrition outcomes–getting beyond the preoccupation with staple grains. Food Secur 7:583–591. https://doi.org/10.1007/s12571-015-0461-x

Pretty J, Bharucha ZP (2014) Sustainable intensification in agricultural systems. Ann Bot 114(8):1571–1596. https://doi.org/10.1093/aob/mcu205

Qian X, Zhao L, Fang Q, Fan C, Zi R, Fang F (2024) Rill formation and evolution caused by upslope inflow and sediment deposition on freshly tilled loose surfaces. Soil Tillage Res 235:105868. https://doi.org/10.1016/j.still.2023.105868

Rahman M, Lee S, Ji H, Kabir A, Jones C, Lee K (2018) Importance of mineral nutrition for mitigating aluminium toxicity in plants on acidic soils: current status and opportunities. Int J Mol Sci 19(10):3073. https://doi.org/10.3390/ijms19103073

Rahman S, Han J, Ahmad M, Ashraf M, Khaliq M, Yousaf M, Wang Y, Yasin G, Nawaz M, Khan K, Du Z (2024) Aluminium phytotoxicity in acidic environments: a comprehensive review of plant tolerance and adaptation strategies. Ecotoxicol Environ Saf 269:115791. https://doi.org/10.1016/j.ecoenv.2023.115791

Raseduzzaman MD, Jensen ES (2017) Does intercropping enhance yield stability in arable crop production? A meta-analysis. Eur J Agron 91:25–33. https://doi.org/10.1016/j.eja.2017.09.009

Rathore SS, Subhash B, Shekhawat K, Singh R, Yadav SK, Singh VK, Singh C (2022) Designing energy cum carbon-efficient environmentally clean production system for achieving green economy in agriculture. Sustain Energy Technol Assess. https://doi.org/10.1016/j.seta.2022.102190

Sánchez AC, Kamau HN, Grazioli F, Jones SK (2022) Financial profitability of diversified farming systems: a global meta-analysis. Ecol Econ 201:107595. https://doi.org/10.1016/j.ecolecon.2022.107595

Schreinemachers P, Howard J, Turner M, Groot SN, Dubey B, Mwadzingeni L, Wopereis MC (2021) Africa's evolving vegetable seed sector: status, policy options and lessons from Asia. Food Secur 13(3):511–523. https://doi.org/10.1007/s12571-021-01146-y

Schroff R, Ramos C (2020) The biodiversity paradigm: building resilience for human and environmental health. Development 63:172–180. https://doi.org/10.1057/s41301-020-00260-2

Semmartin M, Cosentino D, Poggio SL, Benedit B, Biganzoli F, Peper A (2023) Soil carbon accumulation in continuous cropping systems of the rolling Pampa (Argentina): the role of crop sequence, cover cropping and agronomic technology. Agric Ecosyst Environ 347:108368. https://doi.org/10.1016/j.agee.2023.108368

Shaheb MR, Venkatesh R, Shearer S (2021) A review on the effect of soil compaction and its management for sustainable crop production. Journal of. Biosyst Eng 1–23. https://doi.org/10.1007/s42853-021-00117-7

Shankar V, Evelin H (2019) Strategies for reclamation of saline soils. In: Microorganisms in saline environments: strategies and functions, pp 439–449. https://doi.org/10.1007/978-3-030-18975-4_19

Sharma N, Singh R, Mandal D, Kumar A, Alam N, Keesstra S (2017) Increasing farmer's income and reducing soil erosion using intercropping in rainfed maize-wheat rotation of Himalaya, India. Agric Ecosyst Environ 247:43–53. https://doi.org/10.1016/j.agee.2017.06.026

Sharma T, Mrabet R, Singh A, Kumar N, Baghla D, Kumar S, Rana BB, Chauhan G (2024) Global hunger: the need for smart and sustainable agriculture. In: Sustainable green nanotechnology. CRC Press, pp 129–151

Shively G, Sununtnasuk C (2015) Agricultural diversity and child stunting in Nepal. J Dev Stud 51(8):1078–1096. https://doi.org/10.1080/00220388.2015.1018900

Shukla S, Sharma L, Jaiswal V, Dwivedi A, Yadav S, Pathak A (2022) Diversification options in sugarcane-based cropping systems for doubling farmers' income in subtropical India. Sugar Tech 24(4):1212–1229. https://doi.org/10.1007/s12355-022-01127-1

Sibhatu KT, Qaim M (2018) Farm production diversity and dietary quality: linkages and measurement issues. Food Secur 10:47–59. https://doi.org/10.1007/s12571-017-0762-3

Singh R, Babu S, Avasthe R, Yadav G, Das A, Mohapatra K, Chandra P (2021a) Crop productivity, soil health, and energy dynamics of Indian Himalayan intensified organic maize-based systems. Int Soil Water Conserv Res 9(2):260–270. https://doi.org/10.1016/j.iswcr.2020.11.003

Singh R, Subhash B, Avasthe RK, Meena RS, Yadav GS, Das A, Mohapatra KP, Rathore SS, Kumar A, Singh C (2021b) Conservation tillage and organic nutrients management improve soil properties, productivity, and economics of the maize-vegetable pea system in the eastern Himalayas. Land Degrad Dev 32(16):4637–4654. https://doi.org/10.1002/ldr.4066

Singh S, Yadav P, Singh D, Shukla S, Tripathi M, Bahadur L, Kumar S (2021c) Intercropping in sugarcane improves functional diversity, soil quality and crop productivity. Sugar Tech 23:794–810. https://doi.org/10.1007/s12355-021-00955-x

Su K, Mu L, Zhou T, Kamran M, Yang H (2024) Intercropped alfalfa and spring wheat reduces soil alkali-salinity in the arid area of northwestern China. Plant Soil 499(1):275–292. https://doi.org/10.1007/s11104-022-05846-y

Tallou A, Erraji H, Al Masmoudi Y, Abbatantuono F, Ali S, Mabrouki J, Moustaqim K, Salcedo F, Vivaldi G (2024) Phytoremediation of contaminated environments using halophytes: general overview. In: Integrated solutions for smart and sustainable environmental conservation. Springer, pp 143–163. https://doi.org/10.1007/978-3-031-55787-3_11

Tamburini G, Bommarco R, Wanger TC, Kremen C, Van Der Heijden MG, Liebman M, Hallin S (2020) Agricultural diversification promotes multiple ecosystem services without compromising yield. Sci Adv 6(45):eaba1715. https://doi.org/10.1126/sciadv.aba1715

Thilakarathna M, McElroy M, Chapagain T, Papadopoulos Y, Raizada M (2016) Belowground nitrogen transfer from legumes to non-legumes under managed herbaceous cropping systems. A review. Agron Sustain Dev 36:1–16. https://doi.org/10.1007/s13593-016-0396-4

Thomine E, Mumford J, Rusch A, Desneux N (2022) Using crop diversity to lower pesticide use: socio-ecological approaches. Sci Total Environ 804:150156. https://doi.org/10.1016/j.scitotenv.2021.150156

Timsina J, Jat ML, Majumdar K (2010) Rice-maize systems of South Asia: current status, future prospects and research priorities for nutrient management. Plant Soil 335(1):65–82. https://doi.org/10.1007/s11104-010-0418-y

Udawatta RP, Rankoth LM, Jose S (2019) Agroforestry and biodiversity. Sustainability 11(10):2879. https://doi.org/10.3390/su11102879

Uteau D, Pagenkemper S, Peth S, Horn R (2013) Root and time dependent soil structure formation and its influence on gas transport in the subsoil. Soil Tillage Res 132:69–76. https://doi.org/10.1016/j.still.2013.05.001

Van Dijk M, Morley T, Rau ML, Saghai Y (2021) A meta-analysis of projected global food demand and population at risk of hunger for the period 2010–2050. Nat Food 2:494–501. https://doi.org/10.1038/s43016-021-00322-9

Wezel A, Casagrande M, Celette F, Vian J, Ferrer A, Peigné J (2014) Agroecological practices for sustainable agriculture. A review. Agron Sustain Dev 34(1):1–20. https://doi.org/10.1038/s43016-021-00322-9

World Bank (2023) World development indicators. World Bank, Washington, DC

Wu X, Yang J, Ruan H, Wang S, Yang Y, Naeem I, Wang L, Liu L, Wang D (2021) The diversity and co-occurrence network of soil bacterial and fungal communities and their implications for a new indicator of grassland degradation. Ecol Indic 129:107989. https://doi.org/10.1016/j.ecolind.2021.107989

Yadav GS, Das A, Kandpal B, Babu S, Lal R, Datta M, Das B, Singh R, Singh V, Mohapatra K, Chakraborty M (2021) The food-energy-water-carbon nexus in a maize-maize-mustard cropping sequence of the Indian Himalayas: an impact of tillage-cum-live mulching. Renew Sust Energ Rev 151:111602. https://doi.org/10.1016/j.rser.2021.111602

Yadav D, Singh D, Madhu M, Singh DV, Islam S, Kumar G, Gupta AK, Mandal D, Babu S, Barh A, Kumawat A, Dubey R, Pati K, Bhargavi B, Yadav DK (2024) Interplanted crops and rainfall variability regulate soil erosion in rainfed maize of the Indian Himalayas. Land Degrad Dev. https://doi.org/10.1002/ldr.5221

Yang X et al (2021) Diversified crop rotations enhance groundwater and economic sustainability of food production. Food Energy Secur 10:e311. https://doi.org/10.1002/fes3.311

Yang R, Wang Z, Fahad S, Geng S, Zhang C, Harrison M, Adnan M, Saud S, Zhou M, Liu K, Wang X (2022) Rice paddies reduce subsequent yields of wheat due to physical and chemical soil constraints. Front Plant Sci 13:959784. https://doi.org/10.3389/fpls.2022.959784

Yang X, Xiong J, Du T, Ju X, Gan Y, Li S, Xia L, Shen Y, Pacenka S, Steenhuis TS, Siddique KHM, Kang S, Butterbach-Bahl K (2024) Diversifying crop rotation increases food production, reduces net greenhouse gas emissions and improves soil health. Nat Commun 15:198. https://doi.org/10.1038/s41467-023-44464-9

Yu R, Yang H, Xing Y, Zhang W, Lambers H, Li L (2022) Belowground processes and sustainability in agroecosystems with intercropping. Plant Soil 476(1):263–288. https://doi.org/10.1007/s11104-022-05487-1

Zhang Z, Peng X (2021) Bio-tillage: a new perspective for sustainable agriculture. Soil Tillage Res 206:104844. https://doi.org/10.1016/j.still.2020.104844

Zhang Z, Cao L, Zhu Z, He C, Xiang H, Xu L, Sun C, Lin C, Li K (2019) Evaluation on soil bioengineering measures in agricultural areas: poorer durability of wooden structures and better aboveground habitat improvements. Ecol Eng 129:1–10. https://doi.org/10.1016/j.ecoleng.2019.01.004

Zhang N, Zhang Q, Li Y, Zeng M, Li W, Chang C, Xu Y, Huang C (2020) Effect of groundcovers on reducing soil erosion and non-point source pollution in citrus orchards on red soil under frequent heavy rainfall. Sustain For 12(3):1146. https://doi.org/10.3390/su12031146

Zhang H, Liu G, Wu J (2024) The effect of land degradation and restoration on particulate and mineral-associated organic carbon. Appl Soil Ecol 196:105322. https://doi.org/10.1016/j.apsoil.2024.105322

Zhao J, Chen J, Beillouin D, Lambers H, Yang Y, Smith P, Zeng Z, Olesen JE, Zang H (2022) Global systematic review with meta-analysis reveals yield advantage of legume-based rotations and its drivers. Nat Commun 13:4926. https://doi.org/10.1038/s41467-022-32464-0

Zhu X, Shen R (2023) Towards sustainable use of acidic soils: deciphering aluminium-resistant mechanisms in plants. Fundamental Res. https://doi.org/10.1016/j.fmre.2023.03.004

Zhu Z, Zhu YG, Chan FKS, Xu M, Li G, Feng M (2024) A systematic review of factors affecting the performance of the rice–aquaculture system and regional perspectives. Crit Rev Environ Sci Technol 1–20. https://doi.org/10.1080/10643389.2024.2361505

Precision Farming for Food Security and Agricultural Land Restoration

13

Kadagonda Nithinkumar, Abhishek Panchadi, Shreyas Bagrecha, Twinkle Jena, and Megha Kumari

Abstract

Precision farming utilizes precise input application to increase crop yields over traditional methods, focusing on the 5Rs: right time, right amount, right place, right source, and right manner. It aims to manage spatial and temporal variability to enhance productivity while minimizing negative effects. Technologies such as Global Positioning Systems (GPS); Geographic Information Systems (GIS); Remote Sensing; Internet of Things (IoT); Drones and Unmanned Aerial Vehicles (UAVs); Data Analytics (DA) and Machine Learning (ML) are crucial for precision farming, enabling precise nutrient, water, weed, and pest management. Precision farming techniques, including laser land leveling, and precise nutrient, water, weed, and pest management, have the potential to significantly impact agricultural production and food security. Land degradation is a serious concern that affects the soil health and productivity. The agricultural lands experience 40% moderate degradation and 9% severe degradation adversely impacting global food production by 13%. Precision farming also addresses land degradation through agronomic measures like conservation tillage and cover cropping, mulching, organic matter, contour farming, etc. Despite its benefits, challenges like data management, security, and connectivity remains. Future innovations integrating AI, IoT, and UAVs hold promise for advancing agricultural practices and sustainability. The progress in technology and the utilization of big data are enhancing the connectivity, precision, efficiency, and broad applicability of pre-

K. Nithinkumar (✉) · T. Jena · M. Kumari
Division of Agronomy, ICAR-Indian Agricultural Research Institute, New Delhi, India

A. Panchadi
Department of Plant and Soil Sciences, Mississippi State University, Starkville, MS, USA

S. Bagrecha
Agronomy Section, ICAR-National Dairy Research Institute, Karnal, Haryana, India

© The Author(s), under exclusive license to Springer Nature Singapore Pte Ltd. 2025

S. Babu et al. (eds.), *Ecological Solutions to Agricultural Land Degradation*,
https://doi.org/10.1007/978-981-96-3392-0_13

cision farming tools. Improvements in technological infrastructure and regulatory frameworks can enhance access to precision farming, hence magnifying its socioeconomic advantages.

Keywords

Global positioning system · Internet of Things · Laser land leveling · Data management

Abbreviation

AL	Active Learning
ATPs	Agricultural Technology Providers
CNNs	Convolutional neural networks
DA	Data Analytics
DBN	Deep Belief Networks
DL	Deep Learning
DSSAT	Decision Support System for Agrotechnology Transfer
FAO	Food and Agriculture Organisation
FP	Farmer practice
FSL	Few-shot Learning
GA	Genetic Algorithms
GAN	Generative Adversarial Networks
GEE	Google Earth Engine
GIS	Geographic Information Systems
GPS	Global Positioning Systems
IoT	Internet of Things
LCC	Leaf Color Chart
LIDAR	Light Detection and Ranging
ML	Machine Learning
NDVI	Normalized Difference Vegetation Index
NE	Nutrient Expert®
NUE	Nutrient Use Efficiency
RDN	Recommended Dose of Nitrogen
RF	Random Forest
RNNs	Recurrent Neural Networks
SPAD	Soil Plant Analysis Development
SSL	Semi-Supervised Learning
TL	Transfer Learning
UAVs	Unmanned Aerial Vehicles
VRT	Variable Rate Technology

13.1 Introduction

The global population is expected to strike 11.0 billion marks by 2100 (Babu et al. 2022), necessitating a minimum 70% shoot-up in food production to fulfill the demand of the increasing population (Cravero and Sepúlveda 2021; Lakshmi et al. 2023). This growth strains existing resources, and expanding agricultural areas alone cannot address this issue due to resource mismanagement (Karthikeyan et al. 2020). Even expansion in areas under agriculture is unable to satisfy the increasing population load due to mismanagement of resources. Various intensive agricultural practices including overuse of fertilizers, irrigation, and synthetic pesticides, aim to increase productivity but severely impact the environment and soil biodiversity, which reduces resource use efficiency (Hakkim et al. 2016). Food security being a major concern to feed the increasing population with limited resources. Food security refers to the state or condition when individuals have access to food according to their needs. Food supply chains in the short run are affected by weather factors and ultimately impacted by climate change and farm management factors. Severe occurrences such as drought, flood, salinity, and pest and disease outbreaks also affect food security by affecting availability, access, sustainable utilization, and food prices, thereby influencing food security (Brown et al. 2015). Sustainably managing natural resources are going to be crucial in providing food security, while adapting to the climate change scenario (FAO 2016; Das et al. 2022). Precision farming technology is emerging as a solution in agriculture to address these challenges. Precision farming is a resource-use efficiency technology that accesses spatial and temporal variability (McFadden et al. 2023) and helps in decision-making in various aspects of field management. Utilizing state-of-the-art technology like is central to this field (Javaid et al. 2022; Karunathilake et al. 2023). It enhances crop yield and efficiency by the help of precise implementation of inputs like water, fertilizers, seeds, pesticides, etc. minimizing waste and reducing the harm on the climate.

Healthy soil primarily a forerunner of good crop production, promoting food security (Avasthe et al. 2023). However, a third of the farmland is already degraded (FAO 2022) necessitating urgent practices, and technologies that promote land restoration. It necessitates restoring degraded agricultural land by controlling soil erosion, watershed management, afforestation, etc. Precision farming helps in the identification of degraded land while collecting precise information regarding soil properties, crop conditions, and environmental factors. This information helps in the targeted restorative methods, tailored to that specific area. This amalgamation on the one hand boosts up the restoration process while keeping in view that the restored lands maintain their productivity and sustainability over time. In precision farming, regular monitoring of crop growth stages allows early intervention in the occurrence of unfavorable events (Sahoo et al. 2023a). Laser land leveling, precision nutrient, water, weed, and insect-pest management are some of the management practices that help in maintaining food security, aimed at enhancing productivity by increasing input/resource use efficiency while reducing the negative environmental impact, economic costs, and agricultural energy input (Sharma and

Srushtideep 2022). It aids farmers in the optimization of resources, improving crop yields, and thereby reducing environmental footprint.

13.2 Concepts and Principles of Precision Farming

Precision farming, as initially defined in 1997, refers to an integrated farming system that combines information and production methods. Its purpose is to enhance long-term, site-specific, and overall farm production efficiency, productivity, and profitability. At the same time, it aims to minimize any unintended negative effects on wildlife and the environment (US House of Representatives 1997). The focus of the definition is reducing the negative impact on the environment while simultaneously improving productivity. Dr. Pierre C. Robert, father of modern precision farming is one of the pioneers working in this field. It relies that the inputs should be based on 5Rs i.e. right time, right amount, right place, right source, and right manner, thereby reducing inputs and improving efficiency. The main aim being to manage spatial and temporal variability related to all the dimensions of agricultural management to enhance productivity while minimizing the negative effects. It focuses not only on crop cultivation but also on other aspects such as aquaculture, agroforestry, and livestock (Ma et al. 2021). Major steps include accessing variations, managing variations, and evaluating them.

The advent of GPS and GIS in the 1900s gave a boost to precision agriculture. Remote sensing monitors crop growth stages, pest and disease incidence, and yield assessment. While drones and unmanned aerial vehicles assist in fertilizer, pesticide spraying as well as crop monitoring. Variable rate technology with the assistance of maps interpolated by GIS delivers precise inputs in the field reducing loss. Agriculture 4.0 with the introduction of new technologies like the IoT, DA, and ML helped to get better idea from the data collected. IoT stands for the network of interconnected appliances, aiding the sharing of data between themselves. IoT-based monitoring systems for animal reproductive health have been successfully developed (Raj et al. 2021). ML leverages algorithms and data, and aids in agricultural automation. Both farmers and the environment will be having lots of benefits from these improved technologies, helping to assess the soil parameters via, composition, nutrient status, temperature, moisture as well as plant physicochemical properties. Reducing input use (pesticide, fertilizers, irrigation), such as by applying it with the help of variable rate technology, thereby reducing cost and making it more profitable. Effective crop pest (diseases, insects, weed) management, crop stress evaluation, field monitoring via remote sensing, and weather forecasting with ML, allows taking early action in case of adverse conditions while reducing environmental footprint. Assists in decision-making by collecting real-time data produced from different sources, analyzing and centralizing it for effective farm planning (Javaid et al. 2022; Belal et al. 2021). These technologies aim for sustainable agriculture, preventing soil degradation, enhancing crop production eventually maintaining food security.

13.3 Technologies in Precision Farming

Precision farming relies on a pool of advanced technologies to collect, analyze, and apply data for optimized agricultural practices (Fig. 13.1). Variable Rate Technology (VRT) is a precision agriculture technique used to apply fertilizers, pesticides, or water at varying rates across a field based on the specific needs of different areas (He 2022). The primary goal of VRT is to optimize the use of resources, improve crop yields, and minimize environmental impact (Jovarauskas et al. 2021). Core components of VRT are creating management zones based on soil properties, topography, and yield data using GIS and remote sensing (Fabiani et al. 2020), precisely applying inputs (fertilizers, seeds, pesticides) at varying rates according to the prescribed management zones and utilizing GPS-guided equipment and advanced control systems to execute VRT operations. Some of the key technologies for assisting VRT are GPS; GIS; Remote Sensing; IoT; Drones and UAVs; DA and ML (Saleem et al. 2023; Sharma et al. 2014). These tools helps in improving agricultural yields, resource use efficiency, protecting the environment and facilitating land restoration.

13.3.1 Global Positioning Systems (GPS)

Anywhere on our planet, anyone can obtain up-to-the-minute position and time data via this satellite navigation system. It helps determine precise coordinates (latitude, longitude and height) of a location (US Department of Commerce NO and AA 2024). GPS provide the foundational framework for precision agriculture by offering precise location coordinates (Sharma et al. 2014). This satellite-based technology is

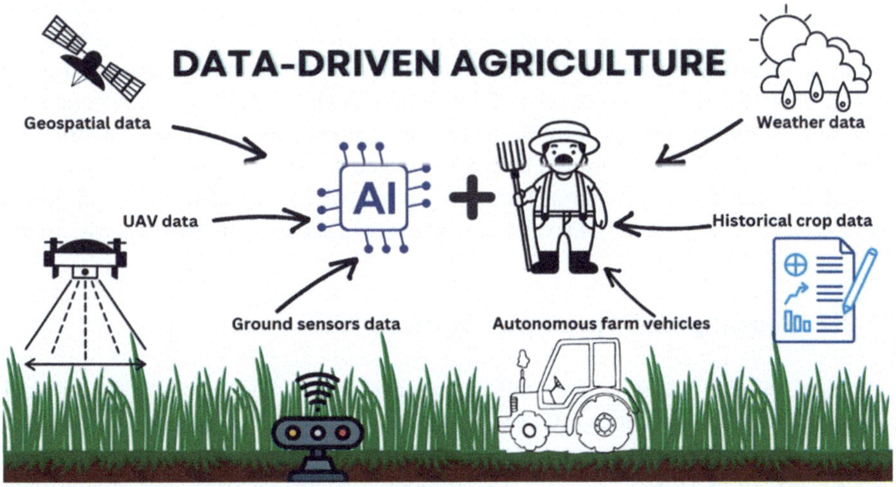

Fig. 13.1 Components of data-driven agriculture

instrumental in guiding farm equipment, creating accurate field boundaries, and collecting essential data for informed decision-making (Ghazal et al. 2024). This allows for detailed mapping of fields, identifying variations in soil composition, topography, and crop yields helping farmers identify high-performing areas and optimize planting strategies (Somashekar et al. 2024). Beyond simple mapping, GPS data forms the backbone of complex data management systems (Brisco et al. 2014). These systems integrate information from various sources (soil sensors, weather stations, yield monitors) to create comprehensive farm profiles. By integrating GPS with remote sensing data, farmers can track their crops' health in real-time, which helps them identify stress conditions quickly and take appropriate action (Brisco et al. 2014). GPS-based soil mapping and precision tillage practices enhances structure of the soil, reduce erosion, and enhance infiltration of water (Kroulík et al. 2011). Soil moisture sensors and irrigation systems that employ GPS technology enable more accurate water application, which in turn improves water usage efficiency and decreases water waste (Aslam et al. 2023). GPS can be used to monitor land use changes and assess the carbon sequestration potential of different agricultural practices (Mgalula 2016). GPS-guided tractors and implements enhance operational efficiency by enabling automated steering, reducing overlaps, and ensuring consistent application of inputs (e.g., fertilizers, seeds, pesticides) (Tyagi et al. 2024; Schimmelpfennig 2016). By integrating GPS with sensors and mapping systems, farmers can apply inputs at varying rates based on site-specific conditions. This method increases harvest yields while decreasing negative effects on the environment. GPS-equipped harvesters generate high-resolution yield maps, providing valuable insights into crop performance and identifying areas with high and low productivity (Mishra 2022; Pedersen and Lind 2017). These maps inform subsequent management decisions, such as variable rate fertilization and soil sampling. GPS technology is used to accurately delineate field boundaries, creating a digital representation of the farm (Belal et al. 2021; Persello et al. 2019). It is critical to have data for field-specific management practices comprising of soil sampling, crop planning, and equipment guidance. Several studies have shown that adopting GPS technology can lead to increased farm profitability due to improved resource efficiency, higher yields, and reduced costs (Belal et al. 2021; Aslam et al. 2023). Factors impacting the adoption of GPS technology are cost, technical expertise, and access to infrastructure (Fadeyi et al. 2022). Addressing challenges related to infrastructure development is crucial for widespread adoption. Protecting sensitive farm data from unauthorized access is a growing concern. Developing robust data security measures is essential.

13.3.2 Geographic Information Systems (GIS)

The purpose of this system is to collect, organize, store, and process geographical and spatial data (ESRI, 2024). GIS uses different types of data, comprised of GPS data, to create maps, analyze spatial relationships, and rely on geographic data when making decisions (ESRI, 2024). GIS provides a robust platform for spatial data management and analysis, crucial for precision farming (Sharma et al. 2014). Farmers may generate comprehensive field maps to guide management decisions by combining data layers including topography, yield maps, and soil characteristics (Khanal et al. 2018). GIS facilitates precise soil sampling and analysis, enabling the creation of detailed soil maps. These maps identify areas with varying soil characteristics, allowing for targeted nutrient management and lime applications. Khanal et al. (2018) observed the effectiveness of GIS-based soil mapping and improving crop yields. By overlaying historical yield data with soil and environmental factors, GIS models can predict crop performance and identify areas with high yield potential (Onyango et al. 2021). This information assists in crop planning, input optimization, and risk management. Research by Li et al. (2021) showcased the accuracy of GIS-based yield prediction models in guiding crop management decisions. GIS enables the establishing of field-based management zones according to soil characteristics, topography, and yield variability (Fabiani et al. 2020). This approach allows for tailored crop rotations, fertilization, and pest control practices, enhancing resource efficiency and crop productivity (Nahry et al. 2011; Onyango et al. 2021). Prescriptions for the variable rate application of fertilizers, seeds, and pesticides may be made using maps created by GIS. GIS-based models can simulate soil erosion processes, helping farmers identify areas prone to erosion and implement conservation measures (Senanayake et al. 2020). GIS is used to assess water availability, quality, and distribution for efficient irrigation planning and management (Nahry et al. 2011). GIS can evaluate the suitability of different crops for specific areas based on climatic, soil, and market factors (Sahoo et al. 2023b; Singha and Swain 2016). GIS is useful in mapping the spatial distribution of diseases and pests, aiding in early detection and control measures (Ghosh et al. 2022). Elements impacting the acceptance of GIS in agriculture, such as cost, technical expertise, and data accessibility, need to be addressed.

13.3.3 Internet of Things (IoT)

The IoT refers to physical objects that are enhanced with sensors, software, and other technologies, allowing them to communicate and share data over the internet (Paul and Jeyaraj 2019). In order to maximize the efficiency of watering, fertilizing, and pest control, sensors monitor, soil moisture, temperature, pH, and other indicators of crop health (Rehman et al. 2022; Yin et al. 2021). IoT devices track equipment location, usage, and performance, enabling predictive maintenance and fuel optimization (Elijah et al. 2018). Sensors track animal health, location, and behavior, facilitating early disease detection, reproductive management, and grazing optimization (Gehlot et al. 2022; Hassan et al. 2023; Neethirajan 2017). IoT-enabled systems control feed and water distribution, ensuring optimal animal nutrition and well-being (Gehlot et al. 2022). IoT generates vast amounts of data (Paul and Jeyaraj 2019) that can be analyzed to identify patterns, trends, and correlations. IoT data used to build predictive models for crop yields, disease outbreaks, and equipment failures (Elijah et al. 2018). IoT enables automated systems to respond to real-time data, optimizing resource allocation and reducing labor costs (Maraveas et al. 2022). Protecting sensitive agricultural data from cyberattacks is crucial. Dependable internet connection is crucial in remote regions for IoT applications. Ensuring seamless Inter-device communication within the IoT devices and platforms is a challenge. One obstacle that many farmers, particularly smallholders, face is the high initial cost of using the IoT. Ensuring equitable access to IoT technologies for farmers in different regions is crucial. Combining IoT with AI, ML, and robotics will create new opportunities agricultural innovation.

13.3.4 Drones and Unmanned Aerial Vehicles (UAVs)

UAV is a broader term that refers specifically to any aircraft operated without a human pilot onboard (Stöcker et al. 2017). This includes both drones and other types of unmanned aircraft. The word "Drone" is often synonymously use for UAVs. In recent years, they have become increasingly useful in farming, offering unique perspectives and capabilities for data collection, monitoring, and management. Gathering data and high-resolution photos for evaluating crop health, nutrient deficiencies, and pest and disease detection, creating detailed digital maps of fields for soil analysis, topography, and yield variability (Mutanga et al. 2017). Applying pesticides and fertilizers with precision, reducing chemical usage and environmental impact (Sahoo et al. 2023b). They help in identifying and locating problem areas for targeted interventions (Khan et al. 2021). UAVs capture various data types, including high-resolution imagery, thermal data, and multispectral data (Sahoo et al. 2023a). Data collected by UAVs may be enriched with the use of modern image processing methods and ML algorithms (Amarasingam et al. 2024). Combining UAV data with GIS and IoT systems provides a comprehensive view of agricultural operations (Zhao et al. 2023). Navigating complex drone regulations is essential for safe and legal operations. Limited flight time restricts the operational range of UAVs. Beyond visual inspection, UAVs equipped with multispectral sensors can detect nutrient deficiencies, diseases, and water stress at early stages, allowing for timely interventions (Abbas et al. 2023). By analyzing NDVI (Normalized Difference Vegetation Index) data from UAV imagery, farmers can estimate crop yields and optimize resource allocation (Gano et al. 2024). UAVs may fly above water sources to check for flooding or drought, measure how well irrigation is working, and keep an eye on overall water quality (Maes and Steppe 2019). Increasing payload capacity is essential for carrying heavier sensors and equipment, expanding the capabilities of UAVs. Developing autonomous flight capabilities will improve efficiency and reduce reliance on human operators. Advancements in data processing and analysis techniques are needed to fully unlock the potential of UAV data.

13.3.5 Remote Sensing

Remote sensing data is integrated with ground-based data (soil samples, yield maps, weather data) to create comprehensive farm information systems (Brisco et al. 2014; Maes and Steppe 2019). Meaningful information may be extracted from remote sensing data using advanced image processing techniques like change detection and classification (Shafique et al. 2022). Remote sensing data feeds into decision support systems that provide recommendations for farmers on the basis of condition of the crops, soil properties, and weather projections (Mustafa et al. 2011). The vast volume of data produced by remote sensing requires efficient processing and storage capabilities. Cloud-based platforms offer opportunities for processing and analyzing remote sensing data on a large scale (Yan et al. 2018). Advancements in sensor technology, such as hyperspectral and lidar sensors, will expand the capabilities of remote sensing in agriculture (Amarasingam et al. 2024; Gano et al. 2024). High-resolution data can be expensive and not always available at the frequency needed for timely agricultural decisions (Carletto 2021). Atmospheric interference, such as cloud cover, can obstruct satellite imagery, affecting data quality. The cost of acquiring remote sensing data, maintaining technology, and the computational resources needed for data analysis can be prohibitive, especially for small-scale farmers. Data quality, frequency and affordable technology can provide light into use of remote sensing at farmer level.

13.3.6 Data Analytics and Artificial Intelligence

DA and Artificial Intelligence (AI) are powerful tools that play a critical role in interpreting and leveraging data for various applications (Fig. 13.2). DA is looking at raw data to make inferences about that data. It is the process of utilizing statistical, computational, and graphical methods to examine data and derive important insights. AI, short for Artificial intellect, is the process of replicating human intellect in computers through programming, enabling them to think and learn in a similar manner to humans. AI comprises a wide array of technologies and strategies that

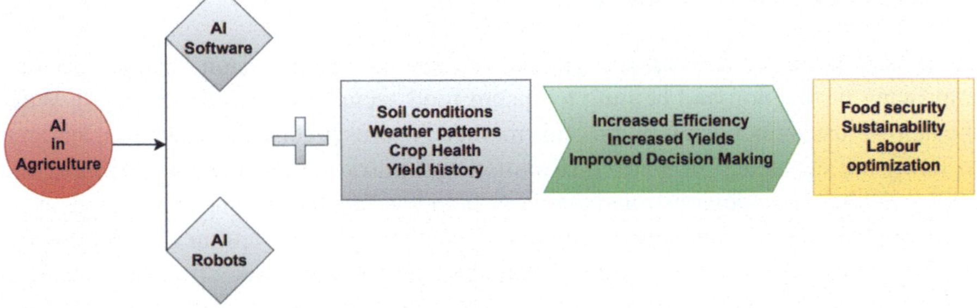

Fig. 13.2 Implications of Artificial Intelligence in Agriculture (Panchadi et al. 2024)

are designed to enable robots to carry out activities that usually need human intellect (Bhat and Huang 2021). AI consists of sub-divisions namely ML and Deep Learning (DL). ML and DL algorithms help at every stage of farming by analyzing climate data to suggest suitable crops and using AI sensors to monitor soil moisture for efficient irrigation (Panchadi et al. 2024). In agriculture, common ML algorithms like decision trees, random forests, support vector machines, etc. are used for crop yield prediction, while DL algorithms such as convolutional neural networks (CNNs) (Persello et al. 2019) and recurrent neural networks (RNNs) are employed for tasks like image-based pest detection and weather forecasting. Data privacy is a major concern for AI adoption in agriculture. A study found that most farmers are hesitant to adopt AI in farms due to the ambiguity about privacy and ownership of the data (Tzachor et al. 2022). A study found that 74% of farmers are unaware of these agreements (Wiseman et al. 2019), while a report stated that 55% of individuals enter into data contracts with Agricultural Technology Providers (ATPs) without obtaining clarity on crucial conditions related to data security and usage (Schuster 2017).

13.4 Precision Farming for Agricultural Production and Food Security

Precision farming is an approach that improves resource use efficiency leading to determining agricultural production and potentially enhancing food security by adopting techniques like laser land leveling, precision nutrient, water, and weed management. Precision farming utilizes advanced technology such as satellite images, drones, and IoT sensors to enhance resource use, monitor crop health, and improve pest and disease detection for farmers. This leads to improved yields, nutritional quality, and climate resilience. Precision farming also promotes sustainable soil health and nutrient management through the targeted application of fertilizers and soil amendments. Additionally, it can help reduce post-harvest losses by enabling real-time monitoring and optimizing storage conditions. Schematic precision farming approaches in agriculture is depicted in Fig. 13.3.

13.4.1 Laser Land Leveling

Laser land leveling has been recognized as a key tool in precision farming, greatly improving production and helping to ensure food security (Lohan et al. 2014). With the help of laser-guided equipment, this method ensures precise field leveling, leading to improved efficiency in water distribution and nutrient management (Wang et al. 2021). Studies have demonstrated that laser field leveling has the capacity to considerably improve water utilization efficiency and increase agricultural productivity. Studies have documented a notable enhancement in water usage efficiency, ranging from 25% to 30%. This improvement has led to an expansion of agricultural land by an extra 5–10%, resulting in a significant boost in crop yields by 10–20% (Jat et al.

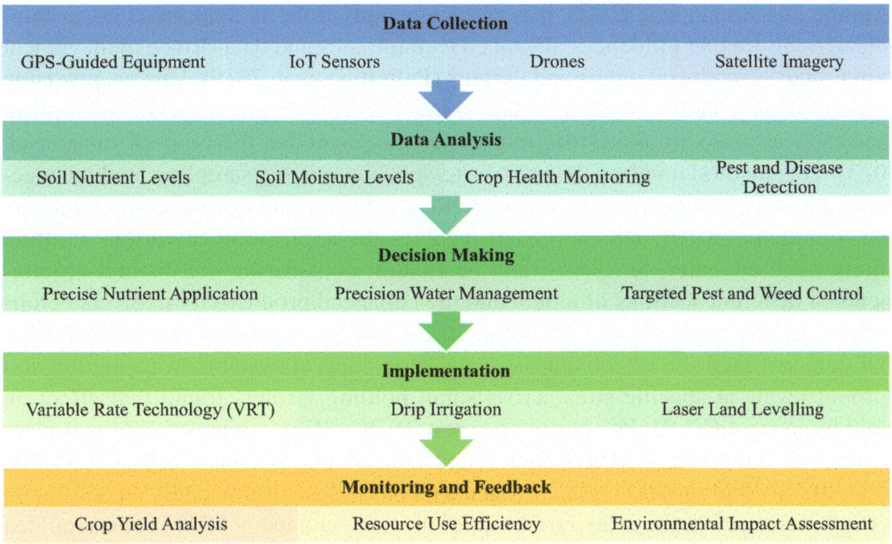

Fig. 13.3 Precision farming procedure for agricultural production

2006; Rickman 2002). Aryal et al. (2015) reported that laser land leveling increased the wheat yield by 6.6% over traditional land leveling. In addition, it promotes consistent crop growth by ensuring equal distribution of water, which helps to reduce issues such as waterlogging and dry patches. The consistency in field conditions helps to improve the distribution of nutrients, reducing the occurrence of nutrient runoff and soil erosion that often plague uneven fields (Yadav et al. 2017). The technology significantly reduces the labour and time required for irrigation, leading to decreased cultivation expenses (Jat et al. 2015). Farmers experience advantages because of decreased fuel consumption and operational expenses as a result of improved water utilization (Chen et al. 2024). In addition, laser land leveling has been found to reduce weed growth by eliminating uneven depressions that are often favourable for weed growth (Jat et al. 2006). By minimizing weed growth, crop productivity is boosted and the reliance on herbicides is reduced, leading to more sustainable farming practices. The use of laser land leveling has been found to have positive environmental impacts, such as decreased greenhouse gas emissions resulting from improved irrigation practices and reduced soil disruption (Tomar et al. 2020). In general, laser land leveling plays a crucial role in contemporary precision farming, providing significant advantages in terms of efficient resource utilization, increased crop productivity, and improved environmental sustainability (Aryal et al. 2015).

13.4.2 Precision Nutrient Management

Precision farming is a contemporary method of agricultural management that enables the customization of decisions based on specific locations and time periods

(Ammal and Rajakumar 2023). It is an increasingly important approach to sustainable agriculture that influences advanced technologies and data-driven management to optimize agricultural inputs and outputs (Pallottino et al. 2019). Various precision nutrient management tools such as leaf color chart (LCC), green seeker, and optical sensors were used to assess the nutrient requirement for the crop (Kumar et al. 2023). This method involves the use of site-specific sensing, sampling, and management, allowing farmers to address each field as a diverse and distinct entity (Gebbers and Adamchuk 2010). This approach involves the application of various farming inputs, such as water, fertilizers, and pesticides, considering the distinct requirements of different sections of a field and the enhanced productivity in diverse crops (Table 13.1) (Anbumozhi and Shanthini 2023). The meta-analysis provided findings that demonstrated the effectiveness of decision support systems in managing soil nutrient levels at specific sites across small holding farmers found that effect on yield by using LCC 81.3%, Nutrient expert 78.2%, Rice Advice 85.7%, soil plant analysis development (SPAD) 58.3% and overall improvement by 78.5% (Sida et al. 2023). Applying 60 kg/ha nitrogen as a basal dose, followed by an additional 60 kg/ha nitrogen at the crown root initiation stage, and using NDVI sensor-guided nitrogen delivery, resulted in enhanced wheat production and nutrient utilization efficiency (Mitra et al. 2023). Leaf colour chart guided nitrogen management improving the yield and saving the nitrogen requirement which has an LCC threshold value of 4.0 for rice and wheat (Singh et al. 2014) and 5.0 for maize (Naik et al. 2019; Naik and Hemalatha 2021). Likewise, Greenseeker-guided nitrogen management was saved the nitrogen by 46% and also improved the productivity of wheat (Kumar 2020). However, precision farming lowers the variable cost and reduces greenhouse gas emissions into the atmosphere (Finger et al. 2019; Ramankutty et al. 2018). The integration of technological improvements and big data has been helpful in enhancing the connectivity, efficiency, and broad applicability of precision farming technologies.

13.4.3 Precision Water Management

Water scarcity is a serious concern at the global level that influences sustainable development and climate change. Across the world, a total of 18.4% of renewable freshwater resources were used by various economic activities (United Nations 2024); however, the agriculture sector utilizes 85% of freshwater for irrigation purposes (Lakshmi et al. 2023). In this view, unsustainable management of water resources leads to a reduction of groundwater levels and lowers water use efficiency (Barman et al. 2021). Hence, 4.0 technology such as ML, DL, the IoT, and other computer applications emerging as a promising solution (Preite et al. 2023). These technologies optimize water resource usage by collection of big data, storage, and analysis (Et-taibi et al. 2024). Crop simulation algorithm-based models a cost-effective and optimize water management and save irrigation water in agriculture (Table 13.2). Decision Support System for Agrotechnology Transfer (DSSAT) (Hoogenboom et al. 2019), CROPWAT (Smith 1992), Root Zone Water Quality

Table 13.1 Precision nutrient management strategies in various crop

Precision nutrient management strategy	Crop	Nutrient/grain yield increase in percentage	Country	References
Drip fertigation	Maize	Yield: 8% Nutrient use efficiency (NUE):18%	Northeast China	Fan et al. (2020)
Sub-surface drip fertigation (75% RDN 4 splits)	Maize	20–25.3% higher yield than conventional irrigation	India	Singh et al. (2024a)
Sub-surface drip fertigation (75% RDN 4 splits)	Maize	25–50% N save	India	Singh et al. (2024b)
Postponing nitrogen application	Spring maize	Yield:27–69.4% Agronomic efficiency: 9.4–24.8%	China	Hou et al. (2019)
Soil test-based N application	Zero-till direct-seeded rice	Yield: 12%	India	Sadhukhan et al. (2023)
Nutrient expert® (+LCCN) based NPK	Zero-till direct-seeded rice	Agronomic efficiency of N: 7% Agronomic efficiency of P: 35%	India	Sadhukhan et al. (2023)
Green seeker guided N management	Wheat	Yield 22–33%	India	Mitra et al. (2023)
Decision support technique for site specific soil nutrient management	–	Meta-analysis Yield: 74.7%	Ethiopia	Sida et al. (2023)
Nutrient expert® -Rice (NE) recommendation	Hybrid Rice	Yield: 49.2%	Nepal	Shrestha et al. (2018)
Deep placement of urea briquette	Rice	Yield: 25.9% NUE: 90.8%	India	Chatterjee et al. (2018)
GreenSeeker-based SSNM	Finger millet	NUE: 21–36% Yield: 12%	India	Pramanick et al. (2024)
50% of N, full of P + K at basal and 50% of N SPAD based	Soybean	Biological yield: 26.8%	India	Sachin et al. (2023)
SPAD based N management	Soybean	Saves 30–45 kg N ha^{-1}	India	Dass et al. (2014)
Point placement of N fertilizer	Wheat	Yield: 10%–12% compared to surface application	India	Nayak et al. (2022)
Preliminary integrated precision rice management system	Rice	Grain yield by 10% and NUE compared to 51–97% over farmer practice (FP)	China	Zhao et al. (2013)

(continued)

Table 13.1 (continued)

Precision nutrient management strategy	Crop	Nutrient/grain yield increase in percentage	Country	References
High-efficiency system optimizing nutrient and water management	Rice	Yield: 16% NUE: 46–63% over FP	China	Zhao et al. (2013)
High-yield system optimizing transplanting density and water management	Rice	Yield: 11% NUE: 19–89% over FP	China	Zhao et al. (2013)

Table. 13.2 Precision water management strategies effects on water saving in various crops

Precision water management strategy	Crop	Water saving	Country	References
Monte Carlo simulation	Tomato	27%	Italy	Preite and Vignali (2024)
IoT, a LoRa-based ML	Tomato	46%	India	Lakshmi et al. (2023)
Water-driven AquaCrop model.	Wheat	1000–1100 m^3/ha	China	Li et al. (2024)
Decision support system developed for precision irrigation management (DSSPIM)	Orange trees	20%	Spain	Cayuela et al. (2022)
Long-range wide-area network (LoRaWAN)	Tomato	Irrigation water use efficiency: 22.7%	USA	Zhang (2022)
Severe deficit strategy in drip plot using	Soybean	Water productivity: 7.57 kg/m^3	Austria	Morales-Santos et al. (2023)
Precision irrigation strategy based on daily trunk growth (precision).	Olive	31%	Spain	Arbizu-Milagro et al. (2023)
Sub-surface drip irrigation	Maize-wheat-green gram	58.5%	India	Dhayal et al. (2023)
Sub-surface drip irrigation with HYDRUS-2D	Maize	41.6%	India	Patra et al. (2021)
Surface drip irrigation with residue retention	Maize-wheat	168 mm in maize 88 mm in wheat	India	Sandhu et al. (2019)
Sub-surface fertigation through drip based on conservation agriculture	Rice-wheat	48–53% in rice 42–53% in wheat	India	Sidhu et al. (2019)
Sprinkler irrigation with 60% ETc	Soybean	Irrigation water productivity: 1.08 kg ha^{-1} m^{-3}	India	Sachin et al. (2023)

Model (RZWQM2) (Ahuja et al. 2000; Ma et al. 2012) and AquaCrop (Raes et al. 2017) models were successfully used for irrigation scheduling (Ale et al. 2023). A Monte Carlo simulation model achieved water and energy savings of 57% and 27%

respectively (Preite and Vignali 2024). AquaBamboo utilized data-driven techniques to enhance water management and promote sustainable growth in the growing of bamboo (Mahule et al. 2024). Likewise, the idCROP app is a simple and powerful tool, helping to efficiently and limit the use of water (Ale et al. 2023). The application of DL-based thermal imagery helped to predict the moisture stress in cotton crops in drip irrigation with mulched plots (Jin et al. 2024). The integration of AI of Things (AIoT) advocates the precise application of water and optimization of yields (Karn et al. 2024). However, precise irrigation application methods such as drip (surface and sub-surface) and sprinkler irrigation methods save the irrigation water and improve the water use efficiency ultimately affecting sustainable productivity (Aydinsakir et al. 2021). Sub-surface drip irrigation with conservation agricultural practices saved the irrigation water 54% in maize (Patra et al. 2023) and 50.1% in rice (Rana et al. 2022). Adoption of these precision water management strategies significantly improves productivity and leads to food security and land restoration.

13.4.4 Precision Weed Management

Weeds pose a persistent threat to agricultural ecosystems, significantly impacting crop yields (Darbyshire et al. 2023). Traditional weed control methods including indiscriminate herbicide spraying, lead to increased costs, potential crop damage, the development of herbicide-resistant weeds, and environmental pollution (Partel et al. 2019). Similarly, globally, insect pests cause 30–40% of yield losses, making their identification, classification, and management crucial to avoiding significant agricultural losses (Kariyanna and Sowjanya 2024). To address these challenges, precision weed management strategies, including AI, DL, ML, IoT, robots, and various UAVs such as drones, as well as sensors and imagery technologies, are being adopted to accurately identify and detect weeds and insect pests. These technologies enable the precise application of herbicides and insecticides (Sharma and Shivandu 2024). The use of UAVs and DL techniques allows for the identification and classification of weed patches through spatial and temporal resolution imagery (Ajayi and Ashi 2023). UAV-based spraying is a fast and efficient method (Pranaswi et al. 2024), with drone sprayers using less herbicide and reducing environmental impacts compared to conventional knapsack and boom sprayers in soybean crops (Hiremath et al. 2024). Moreover, RTK GPS smart technology, when connected to smart sprayers, can develop algorithms to automatically generate weed maps for targeted herbicide applications (Partel et al. 2019). Digital cameras equipped with multispectral and hyperspectral capabilities, along with Artificial Neural Networks, achieve high accuracy in smart spraying (Andujar and Martinez-Guanter 2022). Similarly, AI, predictive modeling algorithms, sensors, and imagery technologies are utilized to identify and classify insect pests, leading to effective pest control while reducing the indiscriminate use of insecticides (Kariyanna and Sowjanya 2024). Overall, the inclusion of precision farming tools can save up to 97% of herbicides, 90% of insecticides, and reduce weed density by 89% (Anastasiou et al. 2023). Precision weed management, attained with intelligent sprayers that utilize machine vision and AI, is essential for improving agricultural productivity and food security, ensuring sustainable agriculture by mitigating losses due to weeds and insect pests (Vijayakumar et al. 2023).

13.5 Precision Farming for Agricultural Land Restoration/ Soil Health

13.5.1 Soil Erosion Control Practices

Soil erosion can be reduced by many agronomic practices, including conservation tillage, organic matter application, mulching, cover crops, intercropping, contour farming etc (Ahmad et al. 2020). Plots treated with downslope tillage showed a 10% reduction in runoff while plots treated with contour ridge tillage saw a 49% reduction in runoff under heavy rains (Dai et al. 2018). Soil erosion may be effectively controlled by utilizing mulching in conjunction with various tillage techniques (Wang et al. 2019). Physical soil characteristics, including as microbial and biotic activities, are enhanced by the no-tillage method. As a result, infiltration, bulk density, wilting point, and field capacity are all increased (Ahn and Kim 2016) and the no-tillage plot had a 64.9% lower runoff ratio as compare to conventional plots (Park et al. 2016). Further minimum tillage along with vegetative barrier reduced runoff and soil loss (Ghosh et al. 2016b). Cowpea and okra intercropping decreased soil loss by 43% and runoff by 26%, respectively (Sharma et al. 2017). The runoff loss was 166.3 mm lower and the soil erosion loss was 5.6 times lesser in the maize-sweet potato-wheat system compared to the maize-wheat system individually (Yadav et al. 2024). Legume crops are very effective cover crops due to their closer spacing and development of good crop canopy (Sharma and Dadhwal 2011). Soil erosion can be reduced by as much as 68% when cover crops are planted in the spaces between rows instead of using standard tillage methods (Novara et al. 2011). The crops used as a canopy utilizing cover crops resulted in a 95% decrease in soil erosion and a runoff reduction of over 60% (Eshel et al. 2015; Maharjan et al. 2016). Reduced runoff ranges from 15.5% to 81% and sediment yield from 40.7% to 97% when mulching is performed (Pan et al. 2018; Dai et al. 2018). When applied alone, organic matter reduces soil erodibility by 14%; when applied in conjunction with straw, it reduces it by 30% (Zhang et al. 2016). Soil erosion may be efficiently mitigated by employing biochar, either with or without polyacrylamide (Wani et al. 2021; Babu et al. 2023). The use of Palmarosa and panicum grass as vegetative barriers, together with organic amendment, resulted in a reduction in both runoff and soil erosion (Ghosh et al. 2016a). Agroforestry is a method of managing land in a sustainable way by combining the growth of trees or shrubs with agricultural crops and animal production on the same piece of land at the same time (Yadav et al. 2021). Various agroforestry systems, including Agri silviculture, Agri horticulture, Alley cropping, and Silvi pasture, have the potential to reduce soil erosion by 10% (Udawatta et al. 2002; Gupta et al. 2007). According to Xu et al. (2018), the terraced plot had a 47.2% drop in average runoff coefficient and an 83.9% decrease in erosion rate compared to the unterraced plots. The use of micro-irrigation tillage resulted in a significant reduction of 63% and 96% in runoff and sediment rates, respectively (Sui et al. 2016). According to Singh et al. (2019), the combination of conservation tillage with *Arundo donax* geotextile resulted in a 24% reduction in runoff and an 8.22 t ha^{-1} reduction in soil loss compared to using simply conservation tillage. Geotextiles have the ability to absorb up to 40% of moisture and contribute to the stability of soil (Vishnudas et al. 2018).

13.5.2 Reforestation and Afforestation in Degraded Land

Restoring forests with a focus on multiple ecosystem services in addition to climate change mitigation is a more promising approach that considers the species and functional diversity both above and below ground (Wani et al. 2023a, b). Ecological restoration has the capacity to restore some functions and elements of the original biodiversity found in damaged environments (Chazdon 2008; Aerts and Honnay 2011). Degraded land can provide various ecosystem services at a reed level than that of the original landscape (Hobbs et al. 2006). Restoring forests may significantly enhance biodiversity, increasing it by 15–84%, and improve vegetation structure by 36–77% when compared to degraded land (Crouzeilles et al. 2016). Earthworms and termites which are known as soil engineers due to their ability to accelerate organic matter decomposition and nutrient turnover can improve soil physical and biological properties.

13.5.3 Land Restoration in Watershed Area

Land degradation and water usage and management are closely interconnected. Land restoration may enhance water usage efficiency and promote effective water utilization. Additionally, it can avoid secondary salinization and soil erosion, so enhancing soil structure and water storage capacity, ultimately increasing productivity. The watershed approach is the cornerstone of the development of rainfed areas which account for about 65% area contributing to 45% of food production, inclusive of various problems such as saline and waterlogged areas, hill areas, ravines, and gullies which make the land degraded (Bhan 2013). Various Soil management practices can improve water use efficiency to the tune of 25–40% by increasing infiltration and soil water storage capacity whereas efficient fertilizer management strategies can enhance water use efficiency by 15–25% (Hatfield et al. 2001). An increase in water productivity at both farm and system scales can be achieved from 70% to 100% and 15–30% (Pretty et al. 2006) in rainfed and irrigated situations, respectively, by employing resource-conserving agriculture strategies. Various resource conservation technologies such as precision nutrient management, irrigation management, and agroforestry can be practiced widely by precision farming technologies.

13.5.4 The Use of Precision Agriculture in Land Degradation Prevention

13.5.4.1 Precision Farming Technologies for Land Restoration/Soil Health

Soil health is vital for preserving biodiversity, mitigating climate change, and maintaining ecosystem stability, all of which are crucial for global food security (Kok et al. 2024). Healthy soil supplies essential nutrients to plants, sequesters carbon, enhances soil macro and microflora and supports vital ecosystem services (Karlen et al. 2019; Lal et al. 2021). However, factors such as waterlogging, salinity, alkalinity, erosion, and poor agricultural practices lead to soil degradation (Dhyani et al. 2023). Thus, precision farming technologies are essential for land restoration and improving soil health. Precision farming technologies, including deep learning (DL), modified U-Net (MU-NET), artificial neural networks (ANN) (El-Rawy et al. 2024), convolutional neural networks (CNN) in 1D, 2D, and 3D configurations, recurrent neural networks (RNN), remote sensing, deep belief networks (DBN), generative adversarial networks (GAN), genetic algorithms, transfer learning (TL), semi-supervised learning (SSL), few-shot learning (FSL), and active learning (AL) (Guerri et al. 2024), are crucial for monitoring and assessing land restoration. Ecological stability is key to sustainable development on both regional and global scales. An integrated Google Earth Engine (GEE) platform utilizing the random forest (RF) machine learning method has been applied to assess ecological stability predictors such as precipitation, salinity, soil organic carbon index, and elevation (Amindin et al. 2024). Agroforestry, a sustainable agriculture management practice, involves growing trees alongside oat crops, which increases soil organic carbon, mitigates greenhouse gases (N_2O and CH_4), and improves soil physical, chemical, and biological properties in Southern Finland (Heimsch et al. 2023). Buthelezi et al. (2024) reported that the Light Gradient Boosting Machine (LightGBM) and Landsat 7 were effectively used for mapping and classification, guiding targeted forest rehabilitation and restoration, thereby preserving and enhancing forest ecosystem services. The deep learning modified U-Net (MU-NET) algorithm and ANN have been employed to generate salinity maps, identifying saline areas and aiding in addressing salinity issues at Siwa Oasis, Egypt (El-Rawy et al. 2024). Similarly, using machine learning algorithms and genetic algorithms (GA) to precisely forecast soil available phosphorus (SAP) and generate mapping SAP improves understanding of SAP dynamics and enhances soil nutrient status (Guo et al. 2024). Thus, adopting precision farming technologies, including machine learning and deep learning, is crucial for predicting and assessing degraded soil areas, facilitating land restoration, and improving soil health.

13.5.4.2 Precision Conservation Agriculture

Conservation Agriculture that adheres to three fundamental principles of low soil disturbance, surface residue coverage, and crop rotation need tailored methodologies to determine their suitability in various contexts (Giller et al. 2011) and precision conservation agriculture such as hand-dug planting holes, precise lime application around the root zone and precise spatial placing of plants (Jerich 2011) can help in increasing benefits of conservation agriculture. Remote sensing technologies, in conjunction with GIS, facilitate the subdivision of the field into smaller management zones. This division is based on specific soil characteristics, including soil types, soil pH, soil EC, soil biological activity, nutrient availability, fertility requirement, and soil compaction. These characteristics are geographically referenced within a GIS database. Precision deep tillage at varying depths is more beneficial than deep tillage at 400 mm depth (Wells et al. 2005). Controlled farming technologies (Godwin et al. 2015), reducing tire pressure systems (Godwin et al. 2017), and site-specific deep tillage are used for soil compaction reduction. Remote sensing and UAV drones can be used on a large scale to monitor the large-scale mapping of land. Variable Rate Technology, one of the components of precision farming can be used for the optimization of different inputs such as fertilizer, chemicals, lime, gypsum, and irrigation water by applying at differential rates across a field thus minimizing land degradation due to excessive application of chemicals and secondary salinization (Onyango et al. 2021). Kriged maps and inverse distance weighting interpolation (Lomeling and Lasu 2015) can be used for targeted land improvement and characterization of spatial pattern of soil compaction respectively. Light detection and ranging (LIDAR) analyses forests in a 3-D format by using laser sensors and provides information related to topography along with the vertical structure of forest canopies. It helps in estimating tree diameter and biomass in remote areas thereby enabling the measurement of the degree of deforestation in a certain region (Takoutsing et al. 2017).

13.6 Challenges and Limitations of Precision Farming

Precision farming is an important resource-efficient technology that offers significant benefits, such as increased nutrient and water use efficiency, minimized cost of resources, and improved yields, however, has several limitations and challenges. Developing efficient methods for collecting, storing, and evaluating high volumes of GPS data is a significant challenge and research area. Reliable GPS signals and high-speed internet connectivity are essential for precision agriculture. The increasing concern is around safeguarding confidential agricultural data from illegal intrusion. Developing robust data security measures is essential. Integrating data from various sources (GPS, remote sensing, field data) and performing complex spatial analyses remains a challenge. Protecting sensitive agricultural data from cyberattacks is crucial. Uninterrupted internet access in remote regions is crucial for the functioning of IoT applications. Facilitating uninterrupted connectivity across various IoT devices and platforms is a challenge. The substantial upfront investment required for implementing IoT might provide a significant obstacle for several farmers, particularly those with little resources, such as smallholders. Ensuring equitable access to IoT technologies for farmers in different regions is crucial. Access to the necessary technology and infrastructure for remote sensing can be limited in some regions, affecting the ability to implement and benefit from remote sensing. To ensure the widespread use of precision agriculture, clear data practices and digital literacy education are essential, along with cooperation among researchers, developers, policymakers, and farmers for sustainable growth.

13.7 Future Directions and Innovations in Precision Farming

Ongoing research focuses on combining GPS with additional technologies, such as drones, AI, and IoT to further enhance agricultural practices. Subsequent investigations should prioritize the resolution of existing gaps in information and understanding the implications of precision farming adoption in different agricultural systems, particularly in terms of promoting food security in developing countries. Combining IoT with AI, ML, and robotics will create new opportunities for agricultural innovation. Increasing payload capacity is essential for carrying heavier sensors and equipment, expanding the capabilities of UAVs. Developing autonomous flight capabilities will improve efficiency and reduce reliance on human operators. Further developments in data processing and analysis techniques are required to fully use the potential of UAV data. Need to focus on developing methodologies to optimize nutrient and water management, weed and pest control, and yield estimation while minimizing the environmental impact. Overall, precision farming holds promise for improving agricultural production and food security, and further research and advancements in technology can contribute to its widespread adoption and impact.

13.8 Conclusion

Precision farming connects with advanced technologies like VRT, GPS, GIS, IoT, drones, and data analytics to transform agricultural practices. This modern approach enables the precise application of inputs including water, fertilizers, and pesticides; thereby improving crop yields, improving resource efficiency, and minimizing environmental impacts. By treating each field as a unique entity, precision farming allows for targeted interventions that reduce waste and support sustainable agriculture. Advanced precision farming techniques, such as laser land leveling and precise control of nutrients, water, and pesticides, effectively tackle inefficiencies, enhance production, and reduce environmental expenses. Land degradation, driven by both natural and human factors, severely impacts global food security and agricultural productivity. Effective mitigation strategies include soil erosion control practices, reforestation, afforestation, and watershed restoration. Precision farming complements these efforts by optimizing input use and reducing land degradation through technologies like remote sensing, GIS, and variable rate technology. Subsequent investigations should prioritize expanding the implementation of precision farming, especially in small-scale and developing country contexts. Integrating innovations such as AI, drones, and IoT with precision farming holds promise for advancing agricultural practices and addressing food security challenges. Addressing current limitations and fostering technological advancements will be crucial for maximizing precision farming's benefits, leading to more sustainable agriculture and restored lands.

References

Abbas A, Zhang Z, Zheng H, Alami MM, Alrefaei AF, Abbas Q, Naqvi SAH, Rao MJ, Mosa WFA, Abbas Q, Hussain A, Hassan MZ, Zhou L (2023) Drones in plant disease assessment, efficient monitoring, and detection: a way forward to smart agriculture. Agronomy 13(6):Article 6. https://doi.org/10.3390/agronomy13061524

Aerts R, Honnay O (2011) Forest restoration, biodiversity and ecosystem functioning. BMC Ecol 11:29

Ahmad NSBN, Mustafa FB, Didams G (2020) A systematic review of soil erosion control practices on the agricultural land in Asia. Int Soil Water Conserv Res 8(2):103–115

Ahn SR, Kim SJ (2016) The effect of rice straw mulching and no-tillage practice in upland crop areas on nonpoint-source pollution loads based on HSPF. Water 8(3):106

Ahuja L, Rojas J, Hanson M, Shaffer M, Ma L (2000) Root zone water quality model. Water Resources Publications, Highlands Ranch

Ajayi OG, Ashi J (2023) Effect of varying training epochs of a faster region-based convolutional neural network on the accuracy of an automatic weed classification scheme. Smart Agric Technol 3:100128

Ale S, Su Q, Singh J, Himanshu S, Fan Y, Stoker B, Gonzalez E, Sapkota BR, Adams C, Biggers K, Kimura E, Wall J (2023) Development and evaluation of a decision support mobile application for cotton irrigation management. Smart Agric Technol 5:100270

Amarasingam N, Vanegas F, Hele M, Warfield A, Gonzalez F (2024) Integrating artificial intelligence and UAV-acquired multispectral imagery for the mapping of invasive plant species in complex natural environments. Remote Sens 16(9):Article 9. https://doi.org/10.3390/rs16091582

Amindin A, Siamian N, Kariminejad N, Clague JJ, Pourghasemi HR (2024) An integrated GEE and machine learning framework for detecting ecological stability under land use/land cover changes. Glob Ecol Conserv 53(3):e03010

Ammal UB, Rajakumar R (2023) Precision nutrient management in agriculture for sustaining crop productivity. J Nat Resour Conserv Manage 4(2):147–160

Anastasiou E, Fountas S, Voulgaraki M, Psiroukis V, Koutsiaras M, Kriezi O, Lazarou E, Vatsanidou A, Fu L, Bartolo FD, Barreiro-Hurle J, Gómez-Barbero M (2023) Precision farming technologies for crop protection: a meta-analysis. Smart Agric Technol 5:100323

Anbumozhi A, Shanthini A (2023) Leaf diseases identification and classification of self-collected dataset on groundnut crop using progressive convolutional neural network (PGCNN). Int J Adv Comput Sci Appl 14(2). https://doi.org/10.14569/ijacsa.2023.0140244

Andujar D, Martinez-Guanter J (2022) An overview of precision weed mapping and management based on remote sensing. Remote Sens 14(15):3621

Arbizu-Milagro J, Castillo-Ruiz FJ, Tascón A, Peña JM (2023) Effects of regulated, precision and continuous deficit irrigation on the growth and productivity of a young super high-density olive orchard. Agric Water Manag 286:108393

Aryal JP, Mehrotra MB, Jat ML, Sidhu HS (2015) Impacts of laser land leveling in rice–wheat systems of the north–western indo-gangetic plains of India. Food Secur 7(3):725–738

Aslam MA, Cheema MJM, Saleem S, Basit A, Hussain S, Waqas MS (2023) Application of sensor-based precision irrigation methods for improving water use efficiency of maize crop. Environ Sci Proc 23(1):Article 1. https://doi.org/10.3390/environsciproc2022023038

Avasthe RK, Devi SH, Bhupenchandra I, Amit K, Chongtham SK, Subhash B, Raghavendra S, Anup D, Gudade BA, Bora SS (2023) Agro-tactics for reducing carbon footprint in agricultural production systems: a review. Indian J Agron 68(2):115–125. https://doi.org/10.59797/ija.v68i2.332

Aydinsakir K, Dinc N, Buyuktas D, Kocaturk M, Ozkan CF, Karaca C (2021) Water productivity of soybeans under regulated surface and subsurface drip irrigation conditions. Irrig Sci 39(6):773–787

Babu S, Singh R, Yadav D, Rathore SS, Raj R, Avasthe R, Yadav SK, Das A, Yadav V, Yadav B, Shekhawat K, Upadhyay P K, Yadav D K, Singh V K (2022) Nanofertilizers for agricultural and environmental sustainability. Chemosphere, 292:133451. https://doi.org/10.1016/j.chemosphere.2021.133451

Babu S, Singh R, Kumar S, Rathore SS, Yadav D, Yadav SK, Yadav V, Ansari MA, Das A, Rajanna GA, Wani O A, Raj R, Yadav DK and Singh VK (2023) Biochar implications in cleaner agricultural production and environmental sustainability. Environmental Science: Advance, 2(8):1042–1059. https://doi.org/10.1039/D2VA00324D

Barman A, Singh VK, Rathore SS, Subhash B, Sharma VK (2021) Deficit-irrigation management and Agri-horti system compatibility using drip for enhancing water-use efficiency and profitability in Indian mustard (*Brassica juncea*). Indian J Agron 66(3):378–381

Belal AA, El-Ramady H, Jalhoum M, Gad A, Mohamed ES (2021) Precision farming technologies to increase soil and crop productivity. In: Abu-hashim M, Khebour Allouche F, Negm A (eds) Agro-environmental sustainability in MENA regions. Springer, Cham, pp 117–154. https://doi.org/10.1007/978-3-030-78574-1_6

Bhan S (2013) Land degradation and integrated watershed management in India. Int Soil Water Conserv Res 1(1):49–57

Bhat SA, Huang NF (2021) Big data and AI revolution in precision agriculture: survey and challenges. IEEE Access 9:110209–110222

Brisco B, Brown R, Hirose T, McNairn H, Staenz K (2014) Precision agriculture and the role of remote sensing: a review. Can J Remote Sens 24. https://doi.org/10.1080/07038992.1998.10855254

Brown ME, Antle JM, Backlund P, Carr ER, Easterling WE, Walsh MK, Ammann C, Attavanich W, Barrett CB, Bellemare MF, Dancheck V, Funk C, Grace K, Ingram JSI, Jiang H, Maletta H, Mata T, Murray A, Ngugi M, Ojima D, Neill BO, Tebaldi C (2015) Climate change, global food security, and the U.S. food System. https://www.usda.gov/sites/default/files/documents/CCFS_Executive_Summary.pdf

Buthelezi MNM, Lottering R, Peerbhay K, Mutanga O (2024) Optimising forest rehabilitation and restoration through remote sensing and machine learning: mapping natural forests in the eThekwini municipality. Remote Sens Appl Soc Environ 2024:101335

Carletto C (2021) Better data, higher impact: improving agricultural data systems for societal change. Eur Rev Agric Econ 48(4):719–740. https://doi.org/10.1093/erae/jbab030

Cayuela CMF, Perea RG, Poyato EC, Montesinos P (2022) An ICT-based decision support system for precision irrigation management in outdoor orange and greenhouse tomato crops. Agric Water Manag 269:107686

Chatterjee D, Mohanty S, Guru PK, Swain CK, Tripathi R, Shahid M, Kumar U, Kumar A, Bhattacharyya P, Gautam P, Lal B, Dash PK, Nayak AK (2018) Comparative assessment of urea briquette applicators on greenhouse gas emission, nitrogen loss and soil enzymatic activities in tropical lowland rice. Agric Ecosyst Environ 252:178–190

Chazdon RL (2008) Beyond deforestation: restoring forests and ecosystem services on degraded lands. Science 320(5882):1458–1460

Chen G, Hu L, Luo X, Wang P, He J, Huang P, Zhao R, Feng D, Tu T (2024) A review of global precision land-leveling technologies and implements: current status, challenges and future trends. Comput Electron Agric 220:108901

Cravero A, Sepúlveda S (2021) Use and adaptations of machine learning in big data applications in real cases in agriculture. Electronics 10:552

Crouzeilles R, Curran M, Ferreira MS, Lindenmayer DB, Grelle CE, Rey Benayas JM (2016) A global meta-analysis on the ecological drivers of forest restoration success. Nat Commun 7(1):11666

Dai C, Liu Y, Wang T, Li Z, Zhou Y (2018) Exploring optimal measures to reduce soil erosion and nutrient losses in southern China. Agric Water Manag 210:41–48

Darbyshire M, Salazar-Gomez A, Gao J, Sklar EI, Parsons S (2023) Towards practical object detection for weed spraying in precision agriculture. Front Plant Sci 14:1183277. https://doi.org/10.3389/fpls.2023.1183277

Das A, Subhash B, Kumar DM, Sanjeev SR, Avasthe RK, Rathore SS, Yadav SK, Singh VK (2022) Restoring soil carbon in marginal land of Indian Himalayas: impact of crop intensification and conservation tillage. J Environ Manag. https://doi.org/10.1016/j.jenvman.2022.115603

Dass A, Suri VK, Choudhary AK (2014) Site specific nutrient management approaches for enhanced nutrient-use efficiency in agricultural crops. RRJoCST 3(3):1–6

Dhayal D, Lal K, Khanna M, Sudhishri S, Brar AS, Sindhu VK, Singh M, Bhattacharyya R, Rajath E, Rosin KG, Chakraborty D (2023) Performance of surface and subsurface drip fertigated wheat-moongbean-maize cropping system under different irrigation schedules and nutrient doses. Agric Water Manag 284:108338

Dhyani S, Santhanam H, Dasgupta R, Bhaskar D, Murthy IK, Singh K (2023) Exploring synergies between India's climate change and land degradation targets: lessons from the Glasgow climate COP. Land Degrad Dev 34(1):196–206

Elijah O, Rahman TA, Orikumhi I, Leow CY, Hindia MN (2018) An overview of internet of things (IoT) and data analytics in agriculture: benefits and challenges. IEEE Internet Things J 5(5):3758–3773. https://doi.org/10.1109/JIOT.2018.2844296

El-Rawy M, Sayed SY, AbdelRahman MA, Makhloof A, Al-Arifi N, Abd-Ellah MK (2024) Assessing and segmenting salt-affected soils using in-situ EC measurements, remote sensing, and a modified deep learning MU-NET convolutional neural network. Eco Inform 81:102652

Eshel G, Egozi R, Goldwasser Y, Kashti Y, Fine P, Hayut E, Kazukro H, Rubin B, Dar Z, Keisar O, DiSegni DM (2015) Benefits of growing potatoes under cover crops in a Mediterranean climate. Agric Ecosyst Environ 211:1–9

ESRI (2024) What is GIS? Geographic Information System Mapping Technology. (n.d.). https://www.esri.com/enus/what-is-gis/overview

Et-taibi B, Abid MR, Boufounas EM, Morchid A, Bourhnane S, Hamed TA, Benhaddou D (2024) Enhancing water management in smart agriculture: a cloud and IoT-based smart irrigation system. Results Eng 22:102283

Fabiani S, Vanino S, Napoli R, Zajíček A, Duffková R, Evangelou E, Nino P (2020) Assessment of the economic and environmental sustainability of Variable Rate Technology (VRT) application in different wheat intensive European agricultural areas. A water energy food nexus approach. Environ Sci Policy 114:366–376. https://doi.org/10.1016/j.envsci.2020.08.019

Fadeyi OA, Ariyawardana A, Aziz AA (2022) Factors influencing technology adoption among smallholder farmers: a systematic review in Africa. https://doi.org/10.17170/kobra-202201195569

Fan J, Lu X, Gu S, Guo X (2020) Improving nutrient and water use efficiencies using water-drip irrigation and fertilization technology in Northeast China. Agric Water Manag 241:106352

FAO (2016) The state of food and agriculture climate change, agriculture and food security. Food and Agriculture Organization of the United Nations, Rome

FAO (2022) Healthy soils for a healthy people and planet: FAO calls for reversal of soil degradation. https://www.fao.org/newsroom/detail/agriculture-soils-degradation-FAO-GFFA-2022/. Accessed 29 July 2024

Finger R, Swinton S, Benni N, Walter A (2019) Precision farming at the nexus of agricultural production and the environment. Ann Rev Resour Econ. https://doi.org/10.1146/ANNUREV-RESOURCE-100518-093929

Gano B, Bhadra S, Vilbig JM, Ahmed N, Sagan V, Shakoor N (2024) Drone-based imaging sensors, techniques, and applications in plant phenotyping for crop breeding: a comprehensive review. Plant Phenome J 7(1):e20100. https://doi.org/10.1002/ppj2.20100

Gebbers R, Adamchuk VI (2010) Precision agriculture and food security. Science 327(5967):828–831. https://doi.org/10.1126/science.1183899

Gehlot A, Malik PK, Singh R, Akram SV, Alsuwian T (2022) Dairy 4.0: intelligent communication ecosystem for the cattle animal welfare with Blockchain and IoT enabled technologies. Appl Sci 12(14):Article 14. https://doi.org/10.3390/app12147316

Ghazal S, Munir A, Qureshi WS (2024) Computer vision in smart agriculture and precision farming: techniques and applications. Artif Intell Agric 13:64–83. https://doi.org/10.1016/j.aiia.2024.06.004

Ghosh BN, Dogra P, Sharma NK, Alam NM, Singh RJ, Mishra PK (2016a) Effects of resource conservation practices on productivity, profitability and energy budgeting in maize–wheat cropping system of Indian sub-Himalayas. Proc Natl Acad Sci India B Biol Sci 86:595–605

Ghosh BN, Meena VS, Alam NM, Dogra P, Bhattacharyya R, Sharma NK, Mishra PK (2016b) Impact of conservation practices on soil aggregation and the carbon management index after seven years of maize–wheat cropping system in the Indian Himalayas. Agric Ecosyst Environ 216:247–257

Ghosh P, Kumpatla SP, Zhang Y, Cheng Q (2022) GIS applications in agriculture. In: Geographic information systems and applications in coastal studies. IntechOpen. https://doi.org/10.5772/intechopen.104786

Giller KE, Corbeels M, Nyamangara J, Triomphe B, Affholder F, Scopel E, Tittonell P (2011) A research agenda to explore the role of conservation agriculture in African smallholder farming systems. Field Crop Res 124:468–472

Godwin R, Misiewicz P, White D, Smith E, Chamen T, Galambošová J, Stobart R (2015) Results from recent traffic systems research and the implications for future work. Acta Technol Agric 18(3):57–63

Godwin RJ, Misiewicz PA, Smith EK, Millington WAZ, White DR, Dicken ET et al (2017) Summary of the effects of three tillage and three traffic systems on cereal yields over a four-year rotation. In: 2017 ASABE annual international meeting, Spokane. ASABE. 16–19 July, 1701652, St. Joseph, pp 1–8

Guerri MF, Distante C, Spagnolo P, Bougourzi F, Taleb-Ahmed A (2024) Deep learning techniques for hyperspectral image analysis in agriculture: a review. ISPRS Open J Photogramm Remote Sens 12:100062

Guo J, Jin S, Wang K (2024) Machine learning and genetic algorithm for mapping soil available phosphorus in coastal provinces in Southeast China. Ecol Indic 166:112294

Gupta SK, Raina NS, Sehgal S (2007) Potential of silvi-pastoral systems in improving the forage production in the hills of Jammu and Kashmir. J Res SKUAST–J 6(2):160–168

Hakkim VA, Joseph EA, Gokul AA, Mufeedha K (2016) Precision farming: the future of Indian agriculture. J Appl Biol Biotechnol 4(6):68–72. https://doi.org/10.7324/JABB.2016.40609

Hassan M, Park JH, Han MH (2023) Enhancing livestock management with IoT-based wireless sensor networks: a comprehensive approach for health monitoring, location tracking, behavior analysis, and environmental optimization. J Sustain Urban Futures 13(6):34–46

Hatfield JL, Sauer TJ, Prueger JH (2001) Managing soils to achieve greater water use efficiency: a review. Agron J 93(2):271–280

He L (2022) Variable rate technologies for precision agriculture. In: Zhang Q (ed) Encyclopedia of smart agriculture technologies. Springer, pp 1–9. https://doi.org/10.1007/978-3-030-89123-7_34-3

Heimsch L, Huusko K, Karhu K, Mganga KZ, Kalu S, Kulmala L (2023) Effects of a tree row on greenhouse gas fluxes, growing conditions and soil microbial communities on an oat field in Southern Finland. Agric Ecosyst Environ 352:108525

Hiremath C, Khatri N, Jagtap MP (2024) Comparative studies of knapsack, boom, and drone sprayers for weed management in soybean (*Glycine max* L.). Environ Res 240:117480

Hobbs RJ, Arico S, Aronson J, Baron JS, Bridgewater P, Cramer VA, Epstein PR, Ewel JJ, Klink CA, Lugo AE, Norton D (2006) Novel ecosystems: theoretical and management aspects of the new ecological world order. Glob Ecol Biogeogr 15(1):1–7

Hoogenboom G, Porter CH, Boote KJ, Shelia V, Wilkens PW, Singh U, White JW, Asseng S, Lizaso JI, Moreno LP, Pavan W, Ogoshi R, Hunt LA, Tsuji GY, Jones JW (2019) The DSSAT crop modeling ecosystem. Burleigh Dodds Science Publishing, Cambridge

Hou YP, Yang J, Yin CX, Qin YB, Li Q, Yu L, Liu ZQ (2019) Effect of postponing nitrogen application on the yield, nitrogen absorption and utilization and soil nitrogen supply in spring maize. J Maize Sci 27:146–154

Jat ML, Sharma SK, Gill MA, Yadvinder-Singh (2006) Laser land leveling: a precursor technology for resource conservation. Rice-Wheat Consortium for the Indo-Gangetic Plains, New Delhi

Jat ML, Singh Y, Gill G, Sidhu H, Aryal JP, Stirling C, Gerard B (2015) Laser assisted precision land leveling: impacts in irrigated intensive production systems of South Asia. Adv Soil Sci 2014:323–352

Javaid M, Haleem A, Singh RP, Suman R (2022) Enhancing smart farming through the applications of agriculture 4.0 technologies. Int J Intell Netw 3:150–164

Jerich M (2011) Potential of precision conservation agriculture as a means of increasing productivity and incomes for smallholder farmers. J Soil Water Conserv 66:171A–174A

Jin K, Zhang J, Wang Z, Zhang J, Liu N, Li M, Ma Z (2024) Application of deep learning based on thermal images to identify the water stress in cotton under film-mulched drip irrigation. Agric Water Manag 299:108901

Jovarauskas D, Steponavičius D, Kemzūraitė A, Zinkevičius R, Venslauskas K (2021) Comparative analysis of the environmental impact of conventional and precision spring wheat fertilization under various meteorological conditions. J Environ Manag 296:113150. https://doi.org/10.1016/j.jenvman.2021.113150

Kariyanna B, Sowjanya M (2024) Unravelling the use of artificial intelligence in management of insect pests. Smart Agric Technol 8:100517

Karlen DL, Veum KS, Sudduth KA, Obrycki JF, Nunes MR (2019) Soil health assessment: past accomplishments, current activities, and future opportunities. Soil Tillage Res 195:104365

Karn S, Kotecha R, Pandey RK (2024) Towards sustainable farming: leveraging AIoT for precision water management and crop yield optimization. Proc Comp Sci 233:772–781

Karthikeyan L, Chawla I, Mishra AK (2020) A review of remote sensing applications in agriculture for food security: crop growth and yield, irrigation, and crop losses. J Hydrol 586:124905

Karunathilake EMBM, Le AT, Heo S, Chung YS, Mansoor S (2023) The path to smart farming: innovations and opportunities in precision agriculture. Agriculture 13:1593

Khan S, Tufail M, Khan MT, Khan ZA, Iqbal J, Wasim A (2021) Real-time recognition of spraying area for UAV sprayers using a deep learning approach. PLoS One 16(4):e0249436. https://doi.org/10.1371/journal.pone.0249436

Khanal S, Fulton J, Klopfenstein A, Douridas N, Shearer S (2018) Integration of high resolution remotely sensed data and machine learning techniques for spatial prediction of soil properties and corn yield. Comput Electron Agric 153:213–225. https://doi.org/10.1016/j.compag.2018.07.016

Kok M, Sarjant S, Verweij S, Vaessen SF, Ros GH (2024) On-site soil analysis: a novel approach combining NIR spectroscopy, remote sensing and deep learning. Geoderma 446:116903

Kroulík M, Kvíz Z, Kumhála F, Hůla J, Loch T (2011) Procedures of soil farming allowing reduction of compaction. Precis Agric 12(3):317–333. https://doi.org/10.1007/s11119-010-9206-1

Kumar BN (2020) Precision nitrogen management in irrigated wheat (*Triticum dicoccum* L.) using optical sensor. J Farm Sci 33(04):555–558

Kumar B, Shaloo BH, Meena MC, Dey A, Dass A, Paramesh V, Subhash B, Upadhyay PK, Prajapati VK, Chandanshive A, Suna T, Yadav SK, Saini AK, Dwivedi N, Brahmanand PS, Jha AK (2023) Nitrogen management sensor optimization, yield, economics, and nitrogen use efficiency of dierent wheat cultivars under varying nitrogen levels. Front Sustain Food Syst 7:1228221. https://doi.org/10.3389/fsufs.2023.1228221

Lakshmi GP, Asha PN, Sandhya G, Sharma SV, Shilpashree S, Subramanya SG (2023) An intelligent IOT sensor coupled precision irrigation model for agriculture. Meas Sens 25:100608

Lal R, Bouma J, Brevik E, Dawson L, Field DJ, Glaser B, Hatano R, Hartemink AE, Kosaki T, Lascelles B (2021) Soils and sustainable development goals of the United Nations: an international union of soil sciences perspective. Geoderma Reg 25:e00398

Li R, Wei C, Afroz MD, Lyu J, Chen G (2021) A GIS-based framework for local agricultural decision-making and regional crop yield simulation. Agric Syst 193:103213. https://doi.org/10.1016/j.agsy.2021.103213
Li M, Zhou S, Shen S, Wang J, Yang Y, Wu Y, Lei Y (2024) Climate-smart irrigation strategy can mitigate agricultural water consumption while ensuring food security under a changing climate. Agric Water Manag 292:108663
Lohan SK, Sidhu HS, Singh M (2014) Laser guided land leveling and grading for precision farming. In: Ram T, Lohan SK, Singh R, Singh P (eds) Precision farming: a new approach, 1st edn. Astral International Pvt Ltd, New Delhi, pp 148–158
Lomeling D, Lasu DM (2015) Spatial patterns of penetration resistance and soil moisture distribution in a Sandy loam soil (*Eutric leptosol*). Int J Soil Sci 10:130–141
Ma LW, Trout TJ, Ahuja LR, Bausch WC, Saseendran SA, Malone RW, Nielsen DC (2012) Calibrating RZWQM2 model for maize responses to deficit irrigation. Agric Water Manag 103:140–149
Ma M, Liu J, Liu M, Zeng J, Li Y (2021) Tree species classification based on sentinel-2 imagery and random forest classifier in the eastern regions of the qilian mountains. Forest 12:1736
Maes WH, Steppe K (2019) Perspectives for remote sensing with unmanned aerial vehicles in precision agriculture. Trends Plant Sci 24(2):152–164. https://doi.org/10.1016/j.tplants.2018.11.007
Maharjan GR, Ruidisch M, Shope CL, Choi K, Huwe B, Kim SJ, Tenhunen J, Arnhold S (2016) Assessing the effectiveness of split fertilization and cover crop cultivation in order to conserve soil and water resources and improve crop productivity. Agric Water Manag 163:305–318
Mahule A, Sawarkar AD, Pakle G, Pachlor R, Singh L (2024) AquaBamboo data-driven suggested system for water management and sustainable growth of bamboo: a review. Adv Bamboo Sci 7:100072
Maraveas C, Piromalis D, Arvanitis KG, Bartzanas T, Loukatos D (2022) Applications of IoT for optimized greenhouse environment and resources management. Comput Electron Agric 198:106993. https://doi.org/10.1016/j.compag.2022.106993
McFadden J, Njuki E, Griffin T (2023) Precision agriculture in the digital era: recent adoption on US farms, EIB-248. US Department of Agriculture, Economic Research Service. https://www.ers.usda.gov/webdocs/publications/105894/eib-248.pdf. Accessed 28 July 2024
Mishra S (2022) Emerging technologies—principles and applications in precision agriculture. In: Reddy GPO, Raval MS, Adinarayana J, Chaudhary S (eds) Data science in agriculture and natural resource management. Springer, pp 31–53. https://doi.org/10.1007/978-981-16-5847-1_2
Mitra B, Singha P, Roy Chowdhury A, Sinha AK, Skalicky M, Laing AM, Hossain A (2023) Optical sensor-based nitrogen management: an environmentally friendly and cost-effective approach for sustainable wheat (*Triticum aestivum* L.) production on eastern plains of India. Front Sustain Food Syst 7:1153575
Mohame E, Belal AA, Ali RR, Saleh A, Hendawy EA (2019) Land degradation. In: The soils of Egypt. Springer, Cham, pp 159–174
Morales-Santos A, García-Vila M, Nolz R (2023) Assessment of the impact of irrigation management on soybean yield and water productivity in a subhumid environment. Agric Water Manag 284:108356
Mustafa A, Singh M, Sahoo R, Ahmed N, Khanna M, Sarangi A, Mishra AK (2011) Land suitability analysis for different crops: a multi criteria decision making approach using remote sensing and GIS. Researcher 3:61–84
Mutanga O, Dube T, Galal O (2017) Remote sensing of crop health for food security in Africa: potentials and constraints. Remote Sens Appl Soc Environ 8:231–239. https://doi.org/10.1016/j.rsase.2017.10.004
Nahry AHE, Ali RR, Baroudy AAE (2011) An approach for precision farming under pivot irrigation system using remote sensing and GIS techniques. Agric Water Manag 98(4):517–531. https://doi.org/10.1016/j.agwat.2010.09.012
Naik MR, Hemalatha S (2021) Precision nitrogen management using leaf colour chart. Chronicle Bioresour Manag 5(Jun, 2):059–062

Naik MR, Hemalatha S, Reddy APK, Madhuri KV, Umamahesh V (2019) Need based nitrogen management on growth attributes of Rabi maize using leaf colour chart under varied plant density. Int J Chem Stud 7(3):2591–2596

Nayak HS, Parihar CM, Mandal BN, Patra K, Jat SL, Singh R, Singh VK, Jat ML, Garnaik S, Nayak J, Abdallah AM (2022) Point placement of late vegetative stage nitrogen splits increase the productivity, N-use efficiency and profitability of tropical maize under decade long conservation agriculture. Eur J Agron 133:126417

Neethirajan S (2017) Recent advances in wearable sensors for animal health management. Sens Bio Sens Res 12:15–29. https://doi.org/10.1016/j.sbsr.2016.11.004

Novara A, Gristina L, Saladino SS, Santoro A, Cerdà A (2011) Soil erosion assessment on tillage and alternative soil managements in a Sicilian vineyard. Soil Tillage Res 117:140–147

Onyango CM, Nyaga JM, Wetterlind J, Söderström M, Piikki K (2021) Precision agriculture for resource use efficiency in smallholder farming systems in sub-Saharan Africa: a systematic review. Sustain For 13(3):Article 3. https://doi.org/10.3390/su13031158

Pallottino F, Antonucci F, Costa C, Bisaglia C, Figorilli S, Menesatti P (2019) Optoelectronic proximal sensing vehicle-mounted technologies in precision agriculture: a review. Comput Electron Agric 162:859–873. https://doi.org/10.1016/j.compag.2019.05.034

Pan D, Zhao X, Gao X, Song Y, Dyck M, Wu P, Li Y, Ma L (2018) Application rate influences the soil and water conservation effectiveness of mulching with chipped branches. Soil Sci Soc Am J 82(2):447–454

Panchadi A, Jha PK, Mehan S (2024) The essential role of AI in agriculture. Resource 31(4):20–21

Park SW, Shin Y, Choi Y, Kim H and Kim S (2016) Effects of tillage practices on runoff and soil loss in a sloping upland field. Soil & Tillage Research, 155:233–240. https://doi.org/10.1016/j.still.2015.08.0121

Partel V, Kakarla SC, Ampatzidis Y (2019) Development and evaluation of a low-cost and smart technology for precision weed management utilizing artificial intelligence. Comput Electron Agric 157:339–350

Patra K, Parihar CM, Nayak HS, Rana B, Singh VK, Jat SL, Panwar S, Parihar MD, Singh LK, Sindhu HS, Gerard B, Jat ML (2021) Water budgeting in conservation agriculture-based subsurface drip irrigation in tropical maize using HYDRUS-2D in South Asia. Sci Rep 11(1):16770

Patra K, Parihar CM, Nayak HS, Rana B, Sena DR, Anand A, Reddy KS, Chowdhury M, Pandey R, Kumar A, Singh LK, Ghatala MK, Sidhu HS, Jat ML (2023) Appraisal of complementarity of subsurface drip fertigation and conservation agriculture for physiological performance and water economy of maize. Agric Water Manag 283:108308

Paul A, Jeyaraj R (2019) Internet of Things: a primer. Human Behav Emerg Technol 1(1):37–47. https://doi.org/10.1002/hbe2.133

Pedersen SM, Lind KM (2017) Precision agriculture—from mapping to site-specific application. In: Pedersen SM, Lind KM (eds) Precision agriculture: technology and economic perspectives. Springer, pp 1–20. https://doi.org/10.1007/978-3-319-68715-5_1

Persello C, Tolpekin VA, Bergado JR, de By RA (2019) Delineation of agricultural fields in smallholder farms from satellite images using fully convolutional networks and combinatorial grouping. Remote Sens Environ 231:111253. https://doi.org/10.1016/j.rse.2019.111253

Pramanick B, Choudhary S, Kumar M, Singh SK, Jha RK, Singh SK, Jha RK, Singh SK, Salem SH, Ansar MJ, Hossain A (2024) Can site-specific nutrient management improve the productivity and resource use efficiency of climate-resilient finger millet in calcareous soils in India? Heliyon 10(12):e32774. https://doi.org/10.1016/j.heliyon.2024.e32774

Pranaswi D, Jagtap MP, Shinde GU, Khatri N, Shetty S, Pare S (2024) Analyzing the synergistic impact of UAV-based technology and knapsack sprayer on weed management, yield-contributing traits, and yield in wheat (Triticum aestivum L.) for enhanced agricultural operations. Comput Electron Agric 219:108796

Preite L, Vignali G (2024) Artificial intelligence to optimize water consumption in agriculture: a predictive algorithm-based irrigation management system. Comput Electron Agric 223:109126

Preite L, Solari F, Vignali G (2023) Technologies to optimize the water consumption in agriculture: a systematic review. Sustainability 15:7. MDPI. https://doi.org/10.3390/su15075975

Pretty J, Noble A, Bossio D, Dixon J, Hine R, Penning de Vries FTW, Morison J (2006) Resource-conserving agriculture increases yields in developing countries. Environ Sci Technol 40(4):1114–1119

Raes D, Steduto P, Hsiao TC, Fereres E (2017) AquaCrop reference manual (Version 6.0)

Raj M, Gupta S, Chamola V, Elhence A, Garg T, Atiquzzaman M, Niyato DA (2021) Survey on the role of internet of things for adopting and promoting agriculture 4.0. J Netw Comput Appl 187:103107

Ramankutty N, Mehrabi Z, Waha K, Larissa J, Kremen C, Herrero M, Rieseberg L (2018) Trends in global agricultural land use: implications for environmental health and food security. Annu Rev Plant Biol 69:789–815. https://doi.org/10.1146/annurev-arplant-042817-040256

Rana B, Parihar CM, Nayak HS, Patra K, Singh VK, Singh DK, Pandey R, Abdallah A, Gupta N, Sidhu HS, Gerard B, Jat ML (2022) Water budgeting in conservation agriculture-based subsurface drip irrigation using HYDRUS-2D in rice under annual rotation with wheat in Western Indo-Gangetic Plains. Field Crop Res 282:108519

Rehman A, Saba T, Kashif M, Fati SM, Bahaj SA, Chaudhry H (2022) A revisit of Internet of Things technologies for monitoring and control strategies in smart agriculture. Agronomy 12(1):Article 1. https://doi.org/10.3390/agronomy12010127

Rickman JF (2002) Manual for laser land leveling. International Rice Research Institute

Sachin KS, Dass A, Dhar S, Rajanna GA, Singh T, Sudhishri S, Sannagoudar MS, Choudhary AK, Kushwaha HL, Praveen BR, Prasad S, Sharma VK, Pooniya V, Krishnan P, Khanna M, Singh R, Varatharajan T, Kumari K, Nithinkumar K, San A-A, Devi AD (2023) Sensor-based precision nutrient and irrigation management enhances the physiological performance, water productivity, and yield of soybean under system of crop intensification. Front Plant Sci 14:1282217. https://doi.org/10.3389/fpls.2023.1282217

Sadhukhan R, Kumar D, Sen S, Sepat S, Ghosh A, Shivay YS, Meena MC, Anand A, Kumar R, Sharma LD, Patra K, Pratap V, Alsuhaibani AM, Gaber A, Hossain A (2023) Precision nutrient management in zero-till direct-seeded rice influences the productivity, profitability, nutrient, and water use efficiency as well as the environmental footprint in the Indo Gangetic plain of India. Agriculture 13(4):784

Sahoo RN, Rejith RG, Gakhar S, Ranjan R, Meena MC, Dey A, Mukherjee J, Dhakar RK, Meena A, Daas A, Subhash B, Upadhyay PK, Sekhawat K, Kumar S, Kumar M, Chinnusamy V, Khanna M (2023a) Drone remote sensing of wheat N using hyperspectral sensor and machine learning. Precis Agric. https://doi.org/10.1007/s11119-023-10089-7

Sahoo RN, Gakhar S, Rejith RG, Ranjan R, Meena MC, Dey A, Mukherjee J, Dhakar RK, Arya S, Dass A, Subhash B, Upadhyay PK, Sekhawat K, Kumar S, Kumar M, Chinnusamy V, Khanna M (2023b) Unmanned aerial vehicle (UAV)–based imaging spectroscopy for predicting wheat leaf nitrogen. Photogramm Eng Remote Sens 89(2):107–116. 0099-1112/22/107–116

Saleem SR, Zaman QU, Schumann AW, Abbas Naqvi SMZ (2023) Variable rate technologies: development, adaptation, and opportunities in agriculture. In: Zaman Q (ed) Precision Agriculture. Academic, pp 103–122. https://doi.org/10.1016/B978-0-443-18953-1.00010-6

Sandhu OS, Gupta RK, Thind HS, Jat ML, Sidhu HS (2019) Drip irrigation and nitrogen management for improving crop yields, nitrogen use efficiency and water productivity of maize-wheat system on permanent beds in north-west India. Agric Water Manag 219:19–26

Schimmelpfennig D (2016) Farm profits and adoption of precision agriculture (ERR-217). U.S. Department of Agriculture, Economic Research Service

Schuster J (2017) Big data ethics and the digital age of agriculture. Resour Magaz 24(1):20–21

Senanayake S, Pradhan B, Huete A, Brennan J (2020) A review on assessing and mapping soil erosion Hazard using geo-informatics technology for farming system management. Remote Sens 12(24):Article 24. https://doi.org/10.3390/rs12244063

Shafique A, Cao G, Khan Z, Asad M, Aslam M (2022) Deep learning-based change detection in remote sensing images: a review. Remote Sens 14(4):Article 4. https://doi.org/10.3390/rs14040871

Sharma NK, Dadhwal KS (2011) In–situ sun hemp green manure mulching in rainfed maize based cropping system for higher productivity. Tech Bull. Published by Central Soil and Water Conservation Research and Training Institute, Dehradun

Sharma K, Shivandu SK (2024) Integrating artificial intelligence and internet of things (IoT) for enhanced crop monitoring and management in precision agriculture. Sens Int 5:100292

Sharma S, Srushtideep A (2022) Precision agriculture and its future. Int J Plant Soil Sci 34(24):200–204

Sharma S, Sharma R, Manhas S, Lohan SK (2014) Potential of variable rate application technology in India. AMA, Agricultural Mechanization in Asia, Africa & Latin America 45:74–89

Sharma NK, Singh RJ, Mandal D, Kumar A, Alam NM, Keesstra S (2017) Increasing farmer's income and reducing soil erosion using intercropping in rainfed maize-wheat rotation of Himalaya, India. Agric Ecosyst Environ 247:43–53

Shrestha S, Amgain LP, Subedi R, Shrestha P, Shahi S (2018) Productivity and profitability analysis of old aged hybrid rice seedling using nutrient-Rice expert and other precision nutrient management practices at Lamjung, Nepal. Int j Appl Sci Biotechnol 6(3):232–237

Sida TS, Gameda S, Chamberlin J, Andersson JA, Getnet M, Woltering L, Craufurd P (2023) Failure to scale in digital agronomy: an analysis of site-specific nutrient management decision-support tools in developing countries. Comput Electron Agric 212:108060

Sidhu HS, Jat ML, Singh Y, Sidhu RK, Gupta N, Singh P et al (2019) Sub-surface drip fertigation with conservation agriculture in a rice-wheat system: a breakthrough for addressing water and nitrogen use efficiency. Agric Water Manag 216:273–283

Singh V, Singh B, Thind HS, Singh Y, Gupta RK, Singh S, Singh M, Kaur S, Singh M, Brar JS, Singh A, Singh J, Kumar A, Singh S, Kaur A, Balasubramanian V (2014) Evaluation of leaf colour chart for need-based nitrogen management in rice, maize and wheat in north-western India. J Res Punjab Agric Univ 51(3 & 4):239–245

Singh RJ, Deshwal JS, Sharma NK, Ghosh BN, Bhattacharyya R (2019) Effects of conservation tillage based agro-geo-textiles on resource conservation in sloping croplands of Indian Himalayan region. Soil Tillage Res 191:37–47

Singh AA, Dass A, Dhar S, Sudhishri S, Shekhawat K, Meena M, Nithinkumar K, Devi AD (2024a) Sub-surface drip fertigation of nitrogen coupled with crop residue incorporation enhanced the growth and yield of maize in an alluvial soils. Indian J Agron 69(2):144–150

Singh AB, Dass A, Sudhishri S, Singh VK, Shekhawat K, Meena MC, Sahoo RN, Soora NK, Upadhyay PK, Dhar S, Nithinkumar K (2024b) Sub-surface drip-fertigation and legume residue improved maize yield and nitrogen use. Nutr Cycl Agroecosyst 129:1–17

Singha C, Swain KC (2016) Land suitability evaluation criteria for agricultural crop selection: a review. Agric Rev. https://arccjournals.com/journal/agricultural-reviews/R-1588

Smith M (1992) CROPWAT: a computer program for irrigation planning and management. Food and Agriculture Organization, Rome, p 1992

Somashekar KS, Moinuddin Belagalla N, Srinatha TN, Abhishek GJ, Kumar V, Tiwari A (2024) Revolutionizing agriculture: innovative techniques, applications, and future prospects in precision farming. J Sci Res Rep 30(8):Article 8. https://doi.org/10.9734/jsrr/2024/v30i82263

Stöcker C, Bennett R, Nex F, Gerke M, Zevenbergen J (2017) Review of the current state of UAV regulations. Remote Sens 9(5):Article 5. https://doi.org/10.3390/rs9050459

Sui Y, Ou Y, Yan B, Xu X, Rousseau AN, Zhang Y (2016) Assessment of micro-basin tillage as a soil and water conservation practice in the black soil region of Northeast China. PLoS One 11(3):e0152

Takoutsing B, Martín JAR, Weber JC, Shepherd K, Sila A, Tondoh J (2017) Landscape approach to assess key soil functional properties in the highlands of Cameroon: repercussions of spatial relationships for land management interventions. J Geochem Explor 178:35–44

Tomar SS, Singh YP, Naresh RK, Dhaliwal SS, Gurjar RS, Yadav R, Sharma D, Tomar S (2020) Impacts of laser land levelling technology on yield, water productivity, soil health and profitability under arable cropping in alluvial soil of north Madhya Pradesh. J Pharm Phytochem 9(4):1889–1898

Tyagi V, Nagargade M, Govindasamy P, Babu S, Singh MK, Kumar A, Singh P (2024) Precision nitrogen management strategies and high yielding genotypes for enhanced growth, yield, economics, and nitrogen use efficiency in wheat. J Plant Nutr 1–20. https://doi.org/10.1080/01904167.2024.2380788

Tzachor A, Devare M, King B, Avin S, Ó hÉigeartaigh S (2022) Responsible artificial intelligence in agriculture requires systemic understanding of risks and externalities. Nat Mach Intell 4(2):104–109. https://doi.org/10.1038/s42256-022-00440-4

Udawatta RP, Krstansky JJ, Henderson GS, Garrett HE (2002) Agroforestry practices, runoff, and nutrient loss: a paired watershed comparison. J Environ Qual 31(4):1214–1225

United Nations (2024) Integrated Monitoring Initiative for SDG 6. Retrieved from: United Nations UnWater

US Department of Commerce NO and AA (2024) The Global Positioning System: Global Positioning Tutorial. Retrieved August 5, 2024, from https://oceanservice.noaa.gov/education/tutorial_geodesy/geo09_gps.html

US House of Representatives (1997) An Act to reform, extend and repeal certain agricultural research, extension and education programs and for other purposes. Title IV—New research, extension and education initiatives. Subtitle B—Precision Agriculture Bill No. H.R.2534. https://www.congress.gov/105/plaws/publ185/PLAW-105publ185.pdf. Accessed 30 July 2024

Vijayakumar V, Ampatzidis Y, Schueller JK, Burks T (2023) Smart spraying technologies for precision weed management: a review. Smart Agric Technol 6:100337

Vishnudas S, Anil KR, Savenije HH, Van Der Zaag P (2018) Use of coir geotextiles for soil and water conservation: case studies from India. In: Technological interventions in management of irrigated agriculture. Apple Academic, pp 161–209

Wang L, Yuan X, Liu C, Li Z, Chen F, Li S, Wu L, Liu Y (2019) Soil C and N dynamics and hydrological processes in a maize-wheat rotation field subjected to different tillage and straw management practices. Agric Ecosyst Environ 285:106616

Wang S-Q, Bai X, Xiaoling Z, Reis S, Deli C, Jianming X, Gu B (2021) Urbanization can benefit agricultural production with large-scale farming in China. Nat Food 2:183–191. https://doi.org/10.1038/s43016-021-00228-6

Wani I, Narde SR, Huang X, Remya N, Kushvaha V and Garg A (2021) Reviewing role of biochar in controlling soil erosion and considering future aspect of production using microwave pyrolysis process for the same. Biomass Conversion and Biorefinery, 13:11543–11569. https://doi.org/10.1007/s13399-021-02060-1

Wani OA, Akhter F, Kumar SS, Subhash B, Kanth RH, Mir SA, Mahdi SS, Malik AR, Bangroo S, Gaafar A-RZ, Popescu SM, Rathore SS (2023a) Mitigating soil erosion through biomass-derived biochar: exploring the influence of feedstock types and pyrolysis temperature. Land 12(12):2111. https://doi.org/10.3390/land12122111

Wani OA, Sharma V, Kumar SS, Subhash B, Sharma KR, Rathore SS, Marwaha S, Ganai NA, Dar SR, Yeasin M, Raghavendra S, Tomar J (2023b) Climate plays a dominant role over land management in governing soil carbon dynamics in North Western Himalayas. J Environ Manag. https://doi.org/10.1016/j.jenvman.2023.117740

Wells LG, Stombaugh TS, Shearer SA (2005) Crop yield response to precision deep tillage. Trans ASAE 48(3):895–901

Wiseman L, Sanderson J, Zhang A, Jakku E (2019) Farmers and their data: an examination of farmers' reluctance to share their data through the lens of the laws impacting smart farming. NJAS: Wageningen J Life Sci 90–91(1):1–10. https://doi.org/10.1016/j.njas.2019.04.007

Xu QX, Wu P, Dai JF, Wang TW, Li ZX, Cai CF, Shi ZH (2018) The effects of rainfall regimes and terracing on runoff and erosion in the three gorges area, China. Environ Sci Pollut Res 25:9474–9484

Yadav RL, Dwivedi BS, Prasad K (2017) Use of precision land leveling for increasing water and nutrient use efficiency in the Indo-Gangetic Plains. Indian J Agric Sci 87(3):305–310

Yadav GS, Kandpal BK, Das A, Subhash B, Mohapatra KP, Devi AG, Devi HL, Chandra P, Singh R, Barman KK (2021) Impact of 28-year-old agroforestry systems on soil carbon dynamics in Eastern Himalayas. J Environ Manag. https://doi.org/10.1016/j.jenvman.2021.111978

Yadav D, Singh D, Babu S, Madegowda M, Singh D, Mandal D, Rathore AC, Sharma VK, Singhal V, Kumawat A, Yadav DK (2024) Intensified cropping reduces soil erosion and improves rainfall partitioning and soil properties in the marginal land of the Indian Himalayas. Int Soil Water Conserv Res 12(3):521–533

Yan J, Ma Y, Wang L, Choo KKR, Jie W (2018) A cloud-based remote sensing data production system. Futur Gener Comput Syst 86:1154–1166. https://doi.org/10.1016/j.future.2017.02.044

Yin H, Cao Y, Marelli B, Zeng X, Mason AJ, Cao C (2021) Soil sensors and plant wearables for smart and precision agriculture. Adv Mater 33(20):2007764. https://doi.org/10.1002/adma.202007764

Zhang H, He L, Di Gioia F, Choi D, Elia A, Heinemann P (2022) LoRaWAN based internet of things (IoT) system for precision irrigation in plasticulture fresh-market tomato. Smart Agric Technol 2:100053

Zhang Q, Liu D, Cheng S, Huang X (2016) Combined effects of runoff and soil erodibility on available nitrogen losses from sloping farmland affected by agricultural practices. Agric Water Manag 176:1–8

Zhao G, Miao Y, Wang H, Su M, Fan M, Zhang F, Jiang R, Zhang Z, Liu C, Liu P, Ma D (2013) A preliminary precision rice management system for increasing both grain yield and nitrogen use efficiency. Field Crop Res 154:23–30

Zhao W, Wang M, Pham VT (2023) Unmanned aerial vehicle and geospatial analysis in smart irrigation and crop monitoring on IoT platform. Mob Inf Syst 1:4213645. https://doi.org/10.1155/2023/4213645

14 Watershed Management for Soil Restoration and Environmental Sustainability

Shamal Shasang Kumar, Owais Ali Wani, Sanjeev Kumar, and Khushboo Devi

Abstract

Watershed management is a holistic and multifaceted approach towards management of water, land, and natural resources within specific geographical boundaries, with the objective of promoting environmental preservation and sustainable development. This chapter explores various facets of watershed management, including soil erosion, soil restoration principles, sustainable land management practices, and hydrological management for soil health. The significance of policy and governance with watershed management efforts is emphasized, particularly in view of the challenges caused by changing climate. Studies from diverse global contexts suggest practical applications and the outcomes of distinctive watershed management strategies. The chapter also identifies future research necessities, highlighting the importance of integrating climate change (CC) adaptation, advancing hydrological modeling, promoting sustainable land and water use, and enhancing community engagement. By addressing these aspects, there is critical need of managing watersheds to maintain ecosystem functioning, ensuring water security, and fostering socio-economic development.

S. S. Kumar
Crop Research Division, Ministry of Agriculture & Waterways (MOA & W), Suva, Fiji

O. A. Wani (✉)
Division of Soil Science and Agricultural Chemistry, Sher-e-Kashmir University of Agricultural Sciences and Technology of Kashmir, Srinagar, India

S. Kumar
Division of Soil Science and Agricultural Chemistry, Sardar Vallabhbhai Patel University of Agriculture and Technology, Meerut, India

K. Devi
Division of Agronomy, ICAR-Indian Agricultural Research Institute, New Delhi, India

© The Author(s), under exclusive license to Springer Nature Singapore Pte Ltd. 2025

S. Babu et al. (eds.), *Ecological Solutions to Agricultural Land Degradation*,
https://doi.org/10.1007/978-981-96-3392-0_14

Keywords

Climate change · Ecosystem health · Soil erosion · Sustainable land management

Abbreviations

ARPA	Amazon Region Protected Areas
BC	Before Christ
C	Carbon
CC	Climate change
CWA	Clean Water Act
EbA	Ecosystem-based adaptation
GIS	Geographic information system
IPM	Integrated Pest Management
IWRM	Integrated Water Resources Management
LU	Land-use
MRC	Mekong River Commission
NB-DSS	Nile Basin Decision Support System
NBI	Nile Basin Initiative
OM	Organic matter
pH	Potential of Hydrogen
PES	Payments for ecosystem services
SLM	Sustainable Land Management
SOM	Soil organic matter
TBS	Tarun Bharat Sangh
US	United States
USA	United States of America
WFD	Water Framework Directive

14.1 Introduction

Watershed management plays an important role in managing the water, land, and natural resources within a geographic area defined by the drainage basin that channels precipitation to rivers, lakes, and other water bodies (Hassan et al. 2024). This approach is critical for sustaining both the environment and the human population that rely on these resources. Watersheds are dynamic systems where natural processes and human activities are intricately linked, making the management of these areas essential for ensuring water quality, flood control, biodiversity conservation, and the overall health of ecosystems (Pandey and Ghosh 2023). A watershed is an area of land where all the water flows down through a network of streams to a single point, often called the outlet (Sheng 1990). It is more than just a geographic area which defines where water flows, where it also functions as a socioeconomic and political unit helping to manage and sustain natural resources (Blomquist and Schlager 2005). Analysing a watershed reveals how actions can be taken with upstream and downstream conditions, showing the interconnected cause-and-effect relationship within the area. The goals of watershed management focus on raising awareness and involving relevant agencies and communities to enhance management practices. These efforts aim to create sustainable, productive land and ensure that water resources are maintained in the best possible quantity and quality for long-term sustainability (Narendra et al. 2021). The watershed management concept dates to 2000 Before Christ (BC) and continues to evolve. It is an ongoing process that involves balancing the management of natural resources with the complexities of human activities. However, the challenges related to watershed management are persistent and complex (Wang et al. 2016).

The importance of watershed management has grown in response to escalating environmental challenges such as deforestation, soil erosion, urbanization, pollution, and CC (Mondal and Palit 2022). These challenges have resulted in the degradation of water resources causing significant impact on both ecosystems and human communities. For instance, deforestation and poor agricultural practices can lead to increased sedimentation in rivers, reducing water quality and causing habitat loss for aquatic species (Ogidi and Akpan 2022). Urbanization, with its associated increase in impervious surfaces, exacerbates runoff and flooding, while also introducing pollutants into water systems (Nazari-Sharabian et al. 2019). CC adds another layer of complexity to these challenges by shifting precipitation patterns, increasing the frequency of extreme weather events, and affecting the availability and distribution of water resources (Sivakumar 2011). Watershed management seeks to tackle these challenges through integrated, sustainable practices that account for the entire water cycle. This approach involves protecting and restoring natural landscapes like forests and wetlands, which are necessary for regulating water flow, filtering pollutants, and providing habitats for wildlife (Swanson et al. 2017). Soil conservation techniques like contour farming and terracing are crucial for reducing erosion and preserving soil fertility (Sarvade et al. 2019). Additionally, watershed management includes sustainable land-use (LU) planning, which carefully balances the needs of agriculture, urban development, and conservation to reduce environmental impacts (Sadeghi et al. 2009).

A key principle of watershed management is Integrated Water Resources Management (IWRM), which advocates for the coordinated development and management of water, land, and related resources (Martins et al. 2017). IWRM strives to enhance economic and social welfare fairly while ensuring the sustainability of essential ecosystems. This approach highlights the importance of involving stakeholders, acknowledging that the varied needs and interests of different sectors and communities within a watershed must be factored into planning and decision-making (Benson et al. 2020). Effective watershed management relies on collaboration among a range of stakeholders, including government agencies, local communities, non-governmental organizations, and private sectors. Community involvement is particularly crucial in watershed management, as local populations often have a deep understanding of their environment and a vested interest in its sustainable management. Involving communities in the planning, implementation, and monitoring of watershed management programs helps build a sense of ownership and responsibility, resulting in more effective and lasting outcomes (Pahl-Wostl 2005). Moreover, community-based approaches can strengthen the resilience of local population to environmental changes by empowering them to adapt to and mitigate the impacts of these changes (Ensor et al. 2018).

In conclusion, watershed management is a comprehensive and integrated approach that tackles the complex interactions between water, land, and human activities within a watershed. By emphasizing sustainable practices, community involvement, and the integration of diverse management strategies, watershed management seeks to protect and enhance water resources while ensuring the long-term resilience of ecosystems and human communities. As environmental challenges intensify, the urgency for effective watershed management has never been greater.

14.2 Understanding Watershed Dynamics

Watersheds are intricate systems where land, water, and living organisms interact in complex ways, making them crucial units for ecological balance and human activity. Watershed dynamics refer to the complex interactions between the physical, biological, and human components within a watershed. These dynamics govern how water moves through the landscape, how it interacts with the environment, and how it affects and is affected by human activities. The dynamics within a watershed are shaped by various factors such as topography, climate, soil type, vegetation, and LU (Dessalegn et al. 2014). These factors affect how water moves across the landscape, shaping its distribution, storage, and use within the watershed. Table 14.1 highlights key challenges in watershed management.

Understanding these dynamics is crucial for managing water resources, preventing environmental degradation, and ensuring long-term use of natural resources. Water movement within a watershed is driven by the hydrological cycle, which includes precipitation, infiltration, runoff, evaporation, and transpiration (Edwards et al. 2015). When rain falls within a watershed, it can either infiltrate into the soil, contributing to groundwater, or it can run off the surface, eventually flowing into rivers, lakes, or other water bodies (Bhardwaj 2019). The rate and volume of runoff depend on factors like soil type, vegetation cover, and LU. For example, areas with dense vegetation and porous soils allow more water to infiltrate, reducing runoff, while urbanized areas with impervious surfaces like roads and buildings increase runoff, leading to higher risks of flooding (Ferreira et al. 2021). The shape and slope of the land within a watershed significantly influence water movement. Steep slopes cause water to move quickly, increasing the potential for erosion and transporting sediments downstream (Yu et al. 2016). Conversely, flatter areas have slow movement of water, which enhances infiltration and lowers the likelihood of erosion. The topography also determines the drainage patterns within a watershed, influencing where water collects and how it flows towards the main water bodies (Daniel et al. 2011). Soil composition and vegetation cover play critical roles in watershed

Table 14.1 Major challenges in watershed management and their impacts

Challenge	Description	Impact on watershed	Country	References
Soil erosion	Loss of topsoil due to wind and water action	Reduced soil fertility, increased sedimentation	Loess Plateau, China	Zhao et al. (2013)
Deforestation	Removal of forest cover	Altered water cycles, loss of biodiversity	Amazon Rainforest, Brazil	Ometto et al. (2011)
Urbanization	Expansion of cities and impervious surfaces	Increased runoff, pollution, and flooding	Greater Los Angeles Area, USA	Tong et al. (2016)
Agricultural practices	Unsustainable farming practices like monoculture and overgrazing	Soil degradation, water pollution	Punjab, India	Singh (2013)
CC	Altered precipitation patterns and temperature shifts	Increased flooding, droughts, and water scarcity	Nile River basin, Africa	Coffel et al. (2019)
Policy and governance	Ineffective or lack of integrated management policies	Fragmented management, resource conflicts	Murray-Darling Basin, Australia	Wallis and Ison (2011)
Overextraction	Excessive withdrawal of water for irrigation and industries	Reduced streamflow, depletion of aquifers	Indus Basin, Pakistan	Lytton et al. (2021)
Pollution	Contamination from agricultural runoff and industrial discharge	Decreased water quality, harm to aquatic life	Ganges River, India	Dwivedi et al. (2018)
Invasive species	Introduction of non-native species	Disruption of local ecosystems, competition for resources	Great Lakes, USA	Mason et al. (2024)
Land degradation	Long-term loss of productive land due to misuse	Decreased agricultural productivity, erosion	Sahel Region, Africa	Dregne (2002)

dynamics. Soils rich in organic matter (OM) and with good structure absorb water more effectively; decreasing runoff and supporting groundwater recharge (Seiler and Gat 2007). Vegetation, such as forests and grasslands, stabilizes the soil, prevents erosion, and helps filter pollutants from runoff before they reach water bodies (Zafirah et al. 2017). Clearing vegetation through deforestation or agricultural activities can result in higher erosion rates, poor water quality, and changes in hydrological processes (Comte et al. 2012). Anthropogenic activities such as agricultural management, urban development, and industrial operations significantly affect its dynamics. Agriculture can introduce fertilizers and pesticides into the water system, leading to pollution (Moss 2008). Urban development increases impervious surfaces, reducing infiltration and increasing runoff, which can lead to flooding and water pollution. Industrial activities may discharge pollutants into water bodies, affecting water quality and aquatic life (Ferreira et al. 2018). Effective watershed management must account for these activities and their impacts, promoting practices that minimize negative effects on the watershed's health.

Watersheds encompass a variety of interconnected ecosystems, including forests, wetlands, rivers, and lakes. These ecosystems offer essential services like water purification, flood regulation, and wildlife habitat. The health of these ecosystems is closely linked to the dynamics within the watershed. For example, wetlands act as natural filters, trapping sediments and pollutants, while forests regulate water flow and maintain soil stability (DeBusk and DeBusk 2001). Disruptions to these ecosystems, such as wetland drainage or deforestation, can have cascading effects throughout the watershed, leading to degraded water quality, biodiversity loss, and lead to an increase in vulnerability to natural disasters (Nichols et al. 2019). Climate and weather patterns have a major impact on watershed dynamics. Seasonal changes in temperature and precipitation influence the availability and distribution of water throughout the watershed (Raymondi et al. 2013). For instance, in wet seasons, rivers might overflow, and groundwater recharge could rise, whereas dry seasons may result in decreased water flow and heightened competition for resources. CC is shifting these patterns, causing more frequent and severe extreme weather events, such as floods and droughts. These changes can disturb the natural balance within watersheds and create challenges for managing water resources (Erol and Randhir 2012). Activities in the upper parts of a watershed can have significant downstream impacts. For example, deforestation or construction upstream can increase sediment load in rivers, leading to siltation and reduced water quality downstream (Kidane and Alemu 2015). Similarly, water extraction or pollution upstream can diminish both the quantity and quality of water available to users downstream (Yoon et al. 2015). Understanding these relationships is essential for effective and coordinated watershed management, ensuring that actions in one part of the watershed do not harm other areas.

14.3 Soil Erosion and Degradation

Soil erosion and degradation are significant concerns within watersheds, where they have far-reaching impacts on both land and water resources. Within a watershed, soil erosion occurs when the protective topsoil is stripped away by water runoff or wind, often as a result of deforestation, overgrazing, or unsustainable agricultural practices (Kulkarni 2021). This erosion is not just a loss of fertile soil, it also contributes to the sedimentation of rivers, streams, and reservoirs within the watershed. As sediment builds up, it can lower water quality, disturb aquatic habitats, and reduce the storage capacity of reservoirs. This accumulation can increase the risk of flooding and decrease water availability during dry periods (Schleiss et al. 2016). Degradation of soil within a watershed further exacerbates these issues by weakening the land's ability to retain water, support vegetation, and resist further erosion

(Lal 2001). Degraded soils, which suffer from nutrient depletion, compaction, or contamination, are less effective at absorbing rainfall, leading to increased surface runoff (Osman and Osman 2013). This runoff can carry extra sediments and pollutants into water bodies, worsening water quality across the watershed. Additionally, the loss of vegetation from soil degradation diminishes the natural filtering and buffering functions of the watershed, making it more susceptible to extreme weather events and CC. Since soil and water within a watershed are interconnected, erosion and degradation in one area can lead to widespread effects throughout the entire system (Browning and Sawyer 2021). For example, soil loss on hillsides can lead to landslides and increased sediment load downstream, while degraded agricultural land can contribute to nutrient runoff, leading to eutrophication in downstream lakes and rivers (Nelson and Booth 2002). Addressing soil erosion and degradation requires integrated land management practices that stabilize soil, restore vegetation, and improve the overall health of the watershed (Gregersen et al. 2007). Techniques such as reforestation, contour farming, and the establishment of buffer zones along waterways are vital in maintaining the integrity of both the land and water resources (Abebe and Minale 2017). By effectively managing soil health, we can safeguard water quality, support biodiversity, and ensure the long-term sustainability of the entire watershed ecosystem.

14.4 Principles of Soil Restoration

Soil restoration is a critical process aimed at reversing the degradation of soil and restoring its health, productivity, and ecological functions (Aradottir and Hagen 2013). The principles of soil restoration focusses on improving soil structure, enhancing nutrient availability, increasing OM content, and promoting biodiversity within the soil ecosystem (Lal 2015). These principles are crucial for restoring damaged lands, promoting sustainable farming practices, and guaranteeing the enduring health of ecosystems. Enhancing soil structure is a core component of successful restoration initiatives (Bronick and Lal 2005). Well-structured soil has a balanced composition of sand, silt, and clay, and contains sufficient pore spaces to ensure proper air and water circulation (Radcliffe et al. 2002). Restoration techniques like reduced tillage, adding OM, and using cover crops contribute to rebuilding soil structure. These practices enhance the soil's capacity to retain water, resist erosion, and promote healthy root development (Baumhardt et al. 2015). Soil organic matter (SOM) is the backbone of healthy soil (Hussain et al. 2023). It improves soil fertility by supplying vital nutrients, improving soil structure, and increasing the soil's ability to retain water and nutrients (Gurmu 2019). Restoration practices such as composting, applying manure, and incorporating crop residues and green manure crops are effective methods for enhancing the SOM content (Tejada et al. 2008). This in turn fosters a more vibrant and resilient soil ecosystem. Restoring soil fertility requires careful management of soil nutrients. This involves the balanced application of fertilizers, both organic and inorganic, to replenish nutrients that have been depleted due to erosion, leaching, or intensive farming (Campbell 2017). Soil restoration also emphasizes the importance of maintaining potential of hydrogen (pH) at an optimal level, as this influences the availability of nutrients to plants. Crop rotation and planting legumes can improve soil nutrient levels by fixing nitrogen and adding variety to the cropping system. (Jena et al. 2022).

Healthy soils are teeming with life, from microorganisms like bacteria and fungi to earthworms and insects. These organisms are important for nutrient cycling, breaking down OM, and developing soil structure (Yatoo et al. 2020). Soil restoration efforts aim to promote this biodiversity by reduced chemical inputs, which can harm beneficial soil organisms, and by creating conditions that support diverse microbial and macrofaunal communities (Menta 2012). Practices such as mulching, maintaining ground cover, and incorporating organic amendments can boost soil biodiversity and overall soil health (Zahed et al. 2022). Preventing further soil loss is a key aspect of soil restoration. Erosion control strategies like planting cover crops, building terraces, and creating windbreaks are essential for stabilizing the soil and safeguarding it from erosion caused by wind and water (Nair et al. 2021). These practices not only prevent further degradation but also contribute to the gradual rebuilding of soil depth and quality over time. Effective water management is essential for soil restoration, as water is crucial for soil processes and plant growth (Hobley et al. 2018). Restoration practices focus on enhancing the soil's capacity to absorb and hold water, which helps to decrease runoff and enhance groundwater recharge. Techniques such as contour plowing, the creation of swales, and the restoration of wetlands can help to manage water effectively within the soil, promoting a healthier and more resilient soil system (Blanco and Lal 2023a). By adhering to these principles, soil restoration efforts can rehabilitate degraded lands, making them productive and sustainable once again. These practices not only improve soil health but also support broader environmental objectives including better water quality, greater biodiversity, and reduced impacts of CC.

14.5 Sustainable Land Management Practices

Sustainable land management (SLM) practices are essential strategies designed at preserving and enhancing land productivity, environmental health, and socioeconomic stability over the long-term (Baskent 2021). These practices strive to harmonize the demands of agriculture, forestry, and other land uses with the conservation of natural resources. They ensure that land remains both productive and resilient in the face of environmental challenges like CC, soil degradation, and biodiversity loss. Table 14.2 presents various SLM practices employed in watersheds. Agroforestry, for instance, involves incorporating trees and shrubs into agricultural landscapes. This approach improves biodiversity, enriches soil fertility through OM from leaf litter, and mitigates soil erosion by stabilizing the soil with tree roots (Fahad et al. 2022). For example, a study found that smallholder farmers practicing agroforestry experienced improved soil fertility, reduced erosion, and increased farm productivity compared to those using conventional farming methods (Awazi and Tchamba 2019). Trees in agroforestry systems offer shade and wind protection,

Table 14.2 SLM practices in watersheds

Practice	Description	Benefits	Country	References
Agroforestry	Integrating trees with crops and livestock	Improves soil structure, reduces erosion, enhances biodiversity	Niger, West Africa	Gupta et al. (2020)
Conservation tillage	Minimal soil disturbance during planting	Reduces soil erosion, enhances water retention	Midwestern United States of America (USA)	Blevins et al. (2018)
Contour farming	Plowing along the contour lines of a slope	Reduces runoff, prevents soil erosion	Terraced Rice Fields, Philippines	Mag-Aso et al. (2024)
Riparian buffer strips	Vegetated areas along water bodies	Filters pollutants, stabilizes streambanks	Chesapeake Bay Watershed, USA	MacKenzie et al. (2024)
Reforestation	Planting trees to restore forest cover	Enhances water infiltration, reduces erosion	Loess Plateau, China	Guan et al. (2024)
Rainwater harvesting	Collecting and storing rainwater for agricultural use	Reduces reliance on groundwater, improves water availability	Rajasthan, India	Mehta et al. (2024)
Cover cropping	Growing cover crops to protect and enrich the soil	Improves soil health, prevents erosion, increases biodiversity	Southern Brazil	Souza et al. (2024)
Terracing	Building terraces to create flat areas on slopes	Reduces soil erosion, conserves water	Ethiopian Highlands	Gashure and Wana (2023)
IPM	Combining biological, cultural, mechanical and chemical practices to manage pests	Reduces pesticide use, promotes environmental health	Central Java, Indonesia	Sujianto et al. (2023)

which can improve crop yields and decrease the reliance on chemical inputs. Additionally, agroforestry systems can diversify farm income by supplying timber, fruits, nuts, and other valuable products (Kuyah et al. 2020). Conservation tillage, including no-till and reduced-till methods, minimizes soil disturbance, preserving soil structure and reducing erosion (Blanco-Canqui and Wortmann 2020). A long-term study in USA demonstrated that no-till farming significantly reduced soil erosion, improved SOM, and increased water infiltration, while maintaining or even increasing crop yields (Islam and Reeder 2014). Conservation tillage helps to maintain soil moisture, reduce fuel and labor costs, and promote carbon (C) sequestration, contributing to CC mitigation (Baye and Aamaa 2019). Rotating crops and diversifying plant species in agricultural systems are important for maintaining soil health and reducing pest and disease pressures (Yang et al. 2020). A study showed that crop rotation, especially with legumes, improved soil structure and fertility, reduced pest outbreaks, and decreased reliance on synthetic fertilizers and pesticides (Kremen and Miles 2012). Crop rotation breaks pest-disease cycles, hence reducing the use of chemical pesticides, and helps manage soil nutrients by alternating deep-rooted and shallow-rooted plants (Boincean et al. 2019).

Integrated Pest Management (IPM) is a holistic approach towards pest control that blends biological, cultural, physio-chemical methods to manage pests in an environmentally and economically sustainable manner (Trivedi and Ahuja 2011). A study on rice cultivation found that IPM strategies, including the use of natural predators and resistant crop varieties, significantly reduced pest incidence, leading to higher yields and lower pesticide costs (Fahad et al. 2015). Terracing involves reshaping steep land into a series of steps to reduce soil erosion and runoff, while contour farming aligns planting rows with the natural contours of the land. A study in the Tigray region of Ethiopia found that terracing effectively reduced soil erosion, increased water retention, and improved crop yields, contributing to food security in local communities (Nyssen et al. 2008). These methods are especially useful in hilly or mountainous areas where erosion is a major issue. Sustainable water management practices, such as rainwater harvesting and efficient irrigation systems, are essential for conserving water resources and ensuring crops receive the necessary moisture without depleting natural water sources (Li et al. 2000). A study demonstrated that rainwater harvesting techniques, such as Zai pits, significantly improved water infiltration and soil moisture retention, leading to higher crop yields (Kebenei and Mucheru-Muna 2023). Techniques like drip irrigation and the use of mulches to retain soil moisture can reduce water use, enhance crop productivity, and mitigate the impacts of drought (Zhang et al. 2022). Reforestation (replanting trees in previously deforested areas) and afforestation (planting trees in areas where trees have not previously grown) are key strategies for restoring degraded lands, improving soil health, and sequestering C (Di Sacco et al. 2021). China's "Grain for Green" program, one of the world's largest reforestation initiatives has been shown to substantially reduce soil erosion, enhance water quality, and increase C sequestration. Additionally, it offers economic incentives to farmers, supporting environmental and economic benefits (Capodaglio and Callegari 2018). These practices help in restoring ecosystems, enhancing biodiversity, and delivering essential ecosystem services, including water regulation, soil stabilization, and climate moderation.

Sustainable grazing management focuses on regulating the timing, intensity, and duration of grazing to keep pasturelands healthy and avoid overgrazing. Overgrazing can cause soil compaction, erosion, and loss of vegetation (Blanco and Lal 2023b). A study found that rotational grazing allowed pastures to recover between grazing periods, leading to increased vegetation cover, improved soil structure, and enhanced biodiversity, while also improving livestock productivity (Teague and Kreuter 2020). By implementing these SLM practices, land users can safeguard and improve the productivity of their lands, ensure the long-term health of natural resources, and support broader environmental goals like biodiversity conservation, CC mitigation, and the preservation of ecosystem services. These practices are essential for balancing human needs with environmental sustainability, ensuring that land resources remain available for future generations.

14.6 Hydrological Management for Soil Health

Hydrological management is integral to maintaining soil health. This approach involves managing water resources to enhance soil quality, prevent degradation, and support sustainable agricultural practices (Lal 2012). By adopting practices like efficient irrigation, rainwater harvesting, flood control, drainage systems, wetland restoration, riparian buffers, and IWRM, we can protect and enhance soil quality, ensuring sustainable agricultural production and broader ecosystem services (Russo et al. 2014). Table 14.3 outlines hydrological management techniques aimed at enhancing soil health. Efficient irrigation techniques like drip irrigation method helps preserve water and prevent soil degradation (Fazliev et al. 2019). For instance, a study demonstrated that drip fertigation notably increases crop yield, water productivity, and nitrogen use efficiency compared to traditional irrigation and fertilization methods (Li et al. 2021). Drip irrigation minimizes water loss from evaporation and runoff, helps maintain consistent soil moisture levels, and reduces both erosion and nutrient leaching (Gabriel and Quemada 2017). It directly delivers water to the root zone, helps preserve soil structure and fertility, promoting healthier soils. In regions with irregular rainfall, rainwater harvesting is a crucial practice for maintaining soil moisture and preventing erosion (Yosef and Asmamaw 2015). An example from Burkina Faso shows how farmers used Zai pits i.e., small, circular pits dug into the soil to capture and store rainwater, improving water infiltration and reducing runoff. This technique increased in crop yields by 50%, as the stored water provided a reliable source of moisture during dry periods, enhancing soil health and productivity (Reij et al. 2009). Rainwater harvesting also helps to recharge groundwater, which is essential for long-term soil and water resource sustainability (Kumar

Table 14.3 Hydrological management techniques for soil health

Technique	Description	Impact on soil health	Country	References
Terracing	Creating stepped levels on slopes	Reduces soil erosion, increases water retention	Ethiopian Highlands, Ethiopia	Mekonnen (2021)
Check dams	Small barriers built across streams to slow water flow	Reduces runoff, promotes groundwater recharge	India	Saha et al. (2022)
Wetland restoration	Restoring natural wetlands	Filters pollutants, stabilizes water table	Everglades, USA	Riedinger-Whitmore (2015)
Subsurface drainage	Installing drains below the soil surface	Prevents waterlogging, enhances root growth	Netherlands	Ritzema and Stuyt (2015)
Managed aquifer recharge	Directing surface water into groundwater systems	Replenishes aquifers, prevents soil salinization	Western Australia	Ross (2018)

et al. 2006). In flood-prone areas, managing floodwaters is essential to protect soil health. The Mississippi River Basin in the USA is a prime example, where flood management strategies, such as the construction of levees and the restoration of floodplains, have been implemented to reduce soil erosion and sediment deposition. These measures have contributed to preserving nutrient-rich topsoil, safeguarding agricultural lands, and sustaining the overall health of the soil ecosystem (Jacobson et al. 2015). Flood management not only safeguards soil quality but also supports sustainable agricultural practices by mitigating the adverse effects of flooding (Kundzewicz 1999). Proper drainage is important in areas with high rainfall or poorly drained soils to prevent waterlogging, which can lead to soil degradation (Nayakekorale 2020). In the Netherlands, where much of the land is below sea level, extensive drainage systems have been developed to manage excess water and prevent soil saturation (Oude Essink et al. 2010). These systems include a combination of canals, ditches, and pumping stations that keep the soil aerated and suitable for agriculture. As a result, the Netherlands has been able to maintain productive agricultural soils despite its challenging hydrological conditions (Oweis 2018). Wetlands act as natural buffers against flooding and are crucial for maintaining water quality within a watershed (Klemas 2014). The restoration of the Everglades in Florida, USA, is a notable example of wetland management aimed at improving hydrological function and soil health (Sklar et al. 1999). The project involved re-establishing natural water flow patterns which helped reduce soil erosion, trap sediments, and filter out pollutants. This restoration has led to the recovery of soil health in surrounding agricultural areas, demonstrating that wetlands are essential for sustaining land and water (Curtis 2002). IWRM integrates the management of water, land, and related resources to ensure sustainability. An example of successful IWRM is the Murray-Darling Basin in Australia, where coordinated efforts have been made to balance water use for agriculture, urban needs, and environmental conservation. Through IWRM, the basin has seen improvements in water use efficiency, soil health, and ecosystem resilience. This approach has helped maintain soil fertility and agricultural productivity, while protecting the basin's water resources for future generations (Docker and Robinson 2014). These highlight the diverse strategies and their tangible benefits, emphasizing the importance of integrated approaches to managing water and soil.

14.7 Policy and Governance in Watershed Management

Effective watershed management depends on robust policy frameworks and governance structures that support the sustainable management of water and land resources. These policies and governance systems are essential for coordinating stakeholder actions, enforcing regulations, and promoting practices that protect and improve watershed health. Central to this approach is the adoption of IWRM policies, which aims to balance social, economic, and environmental priorities. IWRM policies emphasize the holistic management of water resources, integrating LU, water supply, and environmental protection. For example, South Africa's National

Water Act of 1998 is based on the IWRM principles, aiming to manage water resources sustainably and equitably by considering the entire watershed in decision-making. The Act promotes protection of water ecosystems, equitable access to water, and efficient resource utilization, providing a comprehensive policy framework for watershed management (van Koppen and Schreiner 2014). Decentralized governance, which involves delegating decision-making authority to local governments and communities, is essential for effective watershed management. Local participation ensures that community needs and knowledge are integrated into management plans. An example of successful decentralized watershed governance is found in the Indian state of Maharashtra, where the watershed development programme has empowered local communities to manage their water resources. By involving local stakeholders in planning and implementation, the program has led to improved water availability, soil conservation, and agricultural productivity, demonstrating the effectiveness of community-based governance (Samuel et al. 2007). Regulatory frameworks and environmental laws are crucial for enforcing watershed management practices. In USA, the Clean Water Act (CWA) provides a regulatory framework for protecting water quality in rivers, lakes, and wetlands. The Act requires states to develop water-quality standards and implement pollution control programs within their watersheds. The CWA has been instrumental in reducing point-source pollution, restoring impaired water bodies, and protecting aquatic ecosystems, highlighting the importance of strong legal frameworks in watershed management (Walsh and Ward 2022). Watersheds often cross political boundaries, requiring coordinated governance among different jurisdictions. The Mekong River Basin, which spans multiple countries in Southeast Asia, serves as an example of transboundary watershed management. The Mekong River Commission (MRC) was established to facilitate cooperation among riparian countries, focusing on sustainable development, water resource management, and environmental protection. The MRC's work has led to collaborative efforts in managing the basin's resources, addressing challenges such as hydropower development, water allocation, and CC impacts (Sok et al. 2019). Economic incentives, like payments for ecosystem services (PES), are increasingly employed to encourage sustainable watershed management. PES programs provide compensation to landowners or communities for managing their land in ways that deliver ecological benefits, such as better water quality or enhanced biodiversity. In Costa Rica, the government's PES program has successfully encouraged forest conservation and reforestation in watersheds, leading to improved water regulation and reduced soil erosion (Wallbott et al. 2019). Such market-based instruments align economic and environmental goals, incentivizing sustainable practices within watersheds. Effective watershed management necessitates the integration of policies across multiple sectors, including agriculture, forestry, urban planning, and environmental protection. Cross-sectoral coordination ensures that actions in one sector do not undermine the goals of watershed management. For instance, the European Union's Water Framework Directive (WFD) mandates that member states develop river basin management plans that integrate water management with LU, agriculture, and industrial activities. The WFD has enabled the adoption of integrated approaches to managing Europe's water resources, enhancing water quality, and encouraging sustainable LU practices within watersheds (Moss 2004). Ongoing monitoring and evaluation are essential components of governance in watershed management. These processes enable the evaluation of management results and the modification of strategies in response to evolving conditions. Adaptive management, which involves learning from experience and revising policies as necessary, is especially crucial in addressing CC and other dynamic environmental challenges. For example, the Chesapeake Bay Program in USA employs an adaptive management approach to restore and protect the Chesapeake Bay watershed. Continuous monitoring and the flexibility to modify management strategies have been key to the program's success in improving water quality and ecosystem health (Boesch et al. 2001). Effective policy and governance are essential for successful watershed management. By implementing IWRM policies, fostering decentralized governance, establishing regulatory frameworks, promoting transboundary cooperation, providing economic incentives, integrating policies, and utilizing adaptive management, governments and stakeholders can ensure the sustainable management of water and land resources.

14.8 Climate Change and Watershed Management

CC presents major challenges to watershed management by altering precipitation patterns, increasing the frequency of extreme weather events, and affecting water availability, which can worsen existing vulnerabilities in watersheds. To tackle these challenges, adaptive strategies are needed to bolster watershed resilience against climate impacts, while ensuring sustainable management of water and land resources. CC is causing shifts in precipitation patterns, with some areas experiencing heavier rainfall and others enduring prolonged droughts. These changes can disrupt the hydrological balance within watersheds, leading to higher runoff, increased soil erosion, and diminished groundwater recharge. For example, in the western United States (US), reduced snowpack and earlier snowmelt are impacting the timing and availability of water resources, making it difficult to manage water supplies for agriculture, municipal use, and ecosystem needs (Siirila-Woodburn et al. 2021). Watershed management strategies must account for these changing precipitation patterns by enhancing water storage, improving soil moisture retention, and restoring wetlands that can buffer against extreme rainfall events. CC is expected to increase the frequency and intensity of extreme weather events, such as floods, hurricanes, and droughts. These events can severely damage watersheds, resulting in vegetation loss, soil degradation, and disruptions to water supplies. In Bangladesh, for example, the increased occurrence of cyclones and storm surges has led to severe erosion in coastal watersheds, threatening agricultural lands and freshwater resources (Rezaie et al. 2019). To mitigate these impacts, watershed management practices must focus on disaster-risk reduction, including construction of flood defenses, reforestation, and the creation of buffer zones to protect critical water resources and infrastructure. CC can also affect water quality within watersheds by increasing the frequency of algal blooms, altering nutrient cycling, and raising water temperatures. Warmer temperatures can lead to stratification in lakes and reservoirs, reducing oxygen levels and creating conditions that favor harmful algal blooms. In the Great Lakes region, rising temperatures and increased nutrient runoff have contributed to more frequent and severe algal blooms, which threaten water quality and aquatic ecosystems (McKindles et al. 2020). Watershed management strategies must include measures to reduce nutrient inputs, such as implementing best management practices in agriculture, restoring riparian buffers, and enhancing wastewater treatment facilities. CC can alter time and distribution of water availability, affecting both surface and groundwater resources. Regions such as Sahel in Africa, where rainfall is becoming more erratic, water scarcity is becoming a critical issue, leading to conflicts over water use and increased pressure on limited resources (Epule et al. 2014). Watershed management in these areas must focus on improving water use efficiency, promoting rainwater harvesting, and developing alternative water sources. To address the effects of CC on watersheds, strategic management strategies are essential. This involves incorporating climate projections into watershed management plans, enhancing monitoring and data collection, and developing flexible management techniques. In the Colorado River Basin, for example, stakeholders are using climate models to inform water allocation decisions, ensuring that water management strategies are resilient to future climate scenarios (Gerlak et al. 2021). Enhancing watershed resilience also depends on community engagement, as local knowledge and practices are vital in adapting to CC. Ecosystem-based adaptation (EbA), which leverages biodiversity and ecosystem services, is a key strategy to help communities cope with the negative impacts of CC. In watershed management, this can involve restoring wetlands, reforesting degraded areas, and protecting natural floodplains to improve the ability of ecosystems to regulate water-flow, reduce erosion, and maintain water quality. A successful example of EbA is the restoration of mangroves in Vietnam's Mekong Delta, which has helped protect coastal communities from storm surges while enhancing fishery productivity and C sequestration (Wolf et al. 2021). Integrating CC considerations into watershed management policies is critical for long-term sustainability. Climate-smart watershed management involves planning and implementing actions that mitigate greenhouse gas emissions and improve adaptive capacity of watersheds. For instance, in the European Union, the WFD has been complemented by climate adaptation strategies that address the impacts of CC on water resources, ensuring that watershed management remains effective under changing climatic conditions (Quevauviller 2011). CC presents significant challenges for watershed management, requiring adaptive strategies that enhance the resilience of watersheds to changing conditions. Integrating climate projections, advancing ecosystem-based adaptation, and embedding climate considerations into policy frameworks are key strategies in watershed management. These approaches help protect water resources, sustain ecosystem services, and bolster the livelihoods of communities that rely on well-functioning watersheds. With the growing challenges posed by CC, adopting forward-thinking, climate-resilient strategies in watershed management will be essential to maintaining the health and sustainability of water and land resources.

14.9 Case Studies and Practical Applications

Watershed management provide valuable insights into how various strategies are implemented in different contexts and the outcomes they produce. Several examples highlight the importance of tailored strategies which consider the unique environmental, social, and economic conditions of each watershed. Nile River Basin is one of the most complex and politically sensitive watersheds in the world, spanning 11 countries in northeastern Africa. The Nile Basin Initiative (NBI), established in 1999, is a cooperative framework aimed at sustainable management of the river's resources. NBI focuses on equitable water sharing, joint development projects, and the prevention of conflicts over water resources. A notable success of the NBI is the development of the Nile Basin Decision Support System (NB-DSS), which integrates data from across the basin to help countries make informed decisions about water allocation, hydropower development, and agricultural irrigation (Jonker and Seid 2014). The initiative has fostered cooperation among the riparian states, leading to improved water management practices and reduced tensions over water use. The Loess Plateau in China was once one of the most degraded regions in the world, with severe soil erosion, deforestation, and poverty. The Chinese government launched an integrated watershed management project in the 1990s to address these issues. The project combined reforestation, terracing, water conservation, and sustainable agricultural practices to restore the landscape. As a result, soil erosion was significantly reduced, agricultural productivity increased, and local incomes improved. The Loess Plateau project is often cited as a model of how integrated watershed management can achieve both environmental restoration and socio-economic development (Yi and Wang 2013). The Chesapeake Bay in the US is the largest estuary in North America which has faced various environmental challenges as a result of nutrient pollution, habitat loss, and overfishing. The Chesapeake Bay Program, established in 1983, is a regional partnership that brings together federal, state, and local governments, along with NGOs and other stakeholders, to restore the bay's health. Key strategies include reducing nutrient runoff through agricultural best management practices, restoring wetlands and riparian buffers, and improving wastewater treatment. The initiative has resulted in significant enhancements in water quality, a rise in the populations of important species such as blue crabs, and the rejuvenation of habitats vital to the bay's ecosystem (Boesch et al. 2001). The Chesapeake Bay Program demonstrates the importance of long-term, collaborative efforts in watershed management. In Ethiopia's highlands, the deterioration of watersheds has caused significant soil erosion, decreased agricultural yields, and heightened susceptibility to drought. The Ethiopian government, with support from international organizations, has implemented community-based watershed management projects to address these issues. These initiatives highlight the crucial role of local communities in planning and executing soil and water conservation efforts, including building terraces, reforestation, and managing water resources. A successful example is the Tigray region, where community-led watershed management has resulted in the restoration of degraded lands, improved water availability, and enhanced food security (Duale 2024). This case study highlights the effectiveness of engaging local communities in watershed management efforts. The Murray-Darling Basin is Australia's largest river system, and it has faced significant challenges related to water over-extraction, salinity, and ecosystem degradation. In response, the Australian government implemented the Murray-Darling Basin Plan in 2012, which aimed to restore the ecological health of the basin while ensuring sustainable water use for agriculture and communities. The plan includes measures such as water buybacks to reduce over allocation, investments in irrigation efficiency, and the restoration of wetlands and floodplains. The implementation of the plan has been challenging, but it has led to improvements in water quality, increased flows to critical ecosystems, and better management of water resources (Grafton et al. 2014). Murray-Darling Basin Plan illustrates the complexities of balancing environmental sustainability with economic needs in watershed management. In Rajasthan's arid region, India, persistent issues of water scarcity and land degradation have posed significant challenges. The Tarun Bharat Sangh (TBS), a grassroots organization, has worked with local communities to revive traditional water harvesting techniques and implement soil conservation practices. The construction of small check dams (known as johads) has recharged groundwater, improved water availability, and increased agricultural productivity. This community-driven approach has transformed the landscape, turning previously barren areas into fertile farmland and providing a sustainable water supply for drinking and irrigation (Staddon and Everard 2017). The success of TBS in Rajasthan underscores the importance of traditional knowledge and community involvement in watershed management. The Amazon Basin, home to the world's largest tropical rainforest, is a crucial watershed essential for global climate regulation and biodiversity preservation. However, deforestation and the conversion of land for agricultural purposes have significantly threatened the integrity of this vital ecosystem. Various initiatives have been implemented to sustainably manage the Amazon's resources, including establishing protected areas, promoting sustainable forestry practices, and engaging indigenous communities in conservation efforts. The Amazon Region Protected Areas (ARPA) program, launched in 2002, is one of the largest conservation projects in the world, aiming to protect 60 million hectares of rainforest. This program has helped reduce deforestation rates and maintain the ecological integrity of the Amazon watershed (Robinson et al. 2024). The ARPA program demonstrates the importance of large-scale, coordinated efforts in managing complex and ecologically significant watersheds. These case studies and practical applications illustrate the diverse approaches and strategies employed in watershed management across different regions and contexts. From community-based initiatives in Ethiopia and India to large-scale restoration efforts in Australia and the Amazon, these examples highlight the importance of tailored, context-specific approaches that address the unique challenges of each watershed. Successful watershed management requires collaboration among stakeholders, the blending of traditional and contemporary practices, and a dedication to long-term sustainability.

14.10 Future Directions and Research Needs

As challenges such as CC, population growth, and LU pressures make watershed issues more complex, the field of watershed management must adapt and advance to address these evolving demands. Future directions and research needs in watershed management will focus on advancing scientific understanding, improving management practices, and fostering innovative solutions that ensure the sustainability of water resources and ecosystems. Research should focus on integrating CC adaptation and mitigation strategies into watershed management. This involves creating models to predict how CC will affect hydrological cycles, water availability, and ecosystem health. Research is needed in exploring the effectiveness of EbA approaches, such as reforestation, wetland restoration, and the preservation of natural floodplains, in enhancing watershed resilience to CC. Additionally, studies should investigate the role of watersheds in C sequestration and how watershed management practices can contribute to climate mitigation efforts. There is a growing need for more accurate and comprehensive hydrological models that can account for the complex interactions within watersheds, including surface and groundwater dynamics, LU changes, and climate variability. Future research should also be focussed towards refining these models for greater accuracy and accessibility for practitioners. Advances in remote sensing technology, geographic information systems (GIS), and big data analytics can greatly improve the monitoring and management of watersheds. Research should also explore the integration of traditional knowledge with modern scientific methods to develop more holistic approaches to watershed management. Sustainable land and water use practices are essential for maintaining the health of watersheds. Research should explore innovative agricultural practices such as agroforestry, precision farming, and conservation tillage to reduce soil erosion, improve water retention, and boost crop yields while minimizing environmental impact. Additionally, the development of water-saving technologies and policies that encourage efficient water use across agriculture, industry, and urban areas should be a focus. Investigating the role of PES schemes in promoting sustainable watershed practices is also essential. Community involvement remains crucial for the success of watershed management efforts. Future research should explore ways to strengthen community participation in watershed management, particularly in marginalized and vulnerable communities. This includes developing tools and frameworks that facilitate stakeholder engagement, enhance local capacity for resource management, and ensure that the benefits of watershed management are equitably distributed. Research should also address the social and cultural aspects of watershed management, examining how local knowledge and practices can be effectively incorporated into management strategies. Effective policy and governance frameworks are essential for successful watershed management. Research should examine the role of multi-level governance in managing transboundary watersheds, where cooperation among different jurisdictions is necessary. Studies should also explore the effectiveness of existing watershed management policies and identify gaps that need to be addressed. Research on adaptive governance, which provides flexibility and responsiveness to evolving conditions, is essential, especially in the face of CC and other emerging challenges. Sustainable financing is key to the long-term success of watershed management projects. Future research should explore innovative financing mechanisms, such as green bonds, environmental trust funds, and public-private partnerships, to support the implementation of watershed management initiatives. Additionally, research should explore the potential of market-based instruments, such as water trading and PES schemes, in generating revenue for watershed conservation and restoration efforts. Watershed management is inherently interdisciplinary, requiring the integration of knowledge from hydrology, ecology, agriculture, economics, sociology, and other fields. Future research should promote interdisciplinary collaboration to address the complex and interconnected challenges facing watersheds. This involves building partnerships between academic institutions, government agencies, non-governmental organizations, and the private sector. Collaborative research can lead to the creation of more holistic and effective watershed management strategies that incorporate diverse viewpoints and priorities. As CC heightens the frequency and intensity of extreme weather events, research is needed on designing and implementing climate-resilient infrastructure within watersheds. This includes developing green infrastructure solutions, such as permeable pavements, rain gardens, and constructed wetlands, which can help manage flooding, reduce runoff, and enhance water quality. Research should also focus on the retrofitting of existing infrastructure to enhance its resilience to climate-related stresses. Equity and social justice are increasingly recognized as critical components of sustainable watershed management. Future research should explore how watershed management practices can be designed and implemented to promote social equity, focusing on equitable access to water resources and fair distribution of benefits. This involves examining the impacts of these practices on vulnerable populations, including indigenous communities, women, and low-income groups, and developing strategies to safeguard their needs and rights. Additionally, ongoing monitoring and evaluation are crucial for assessing the effectiveness of watershed management strategies and making timely adjustments. Future research should focus on developing robust monitoring frameworks that can track changes in water quality, biodiversity, soil health, and other key indicators over time. Long-term impact assessments are essential for comprehending the cumulative effects of watershed management practices. These evaluations help to inform adaptive management approaches by providing insights into the sustained outcomes and effectiveness of various strategies over time. The future of watershed management will require a combination of scientific innovation, community engagement, policy reform, and interdisciplinary collaboration. Addressing the research needs outlined above will enable watershed management to adapt to the challenges of the twenty-first century, ensuring the sustainable management of water and land resources for future generations.

14.11 Conclusion

Effective watershed management is crucial for sustaining natural resources and ensuring ecosystem resilience as we face escalating environmental challenges. This paper has examined the intricate aspects of watershed management, including soil erosion, restoration efforts, sustainable LU practices, and hydrological techniques, highlighting its complex nature. Strong governance and well-crafted policies are vital for steering these initiatives, especially as CC exerts greater stress on our water and land resources. The case studies discussed reveal that successful watershed management requires strategies tailored to specific contexts, active involvement from local communities, and a steadfast commitment to sustainable practices. These examples illustrate the need for approaches that adapt to local conditions and engage all stakeholders to achieve effective results. Looking ahead, future research must address several crucial areas. It's important to integrate CC adaptation into management strategies and to enhance the precision and usability of hydrological models. Research should also investigate advanced agricultural and land management techniques, such as agroforestry, precision farming, and conservation tillage, which can improve soil health, reduce erosion, and optimize water usage. Furthermore, ensuring social equity and justice in watershed management is essential, so that marginalized groups have fair access to resources and benefits. Collaboration among researchers, practitioners, policymakers, and local communities will be key to advancing watershed management strategies. Exploring new financing options, like green bonds, environmental trust funds, and public-private partnerships, will also be critical for implementing effective management practices. By prioritizing these areas, we can protect watershed health, promote sustainable development, and ensure that vital water resources are preserved for future generations. A proactive, inclusive, and adaptable approach will be essential to overcoming the evolving challenges in watershed management and achieving lasting sustainability.

References

Abebe WB, Minale AS (2017) Land use and watershed management practices in Lake Tana Basin. In: Social and ecological system dynamics: characteristics, trends, and integration in the Lake Tana Basin, Ethiopia, pp 479–521. https://doi.org/10.1007/978-3-319-45755-0_28

Aradottir AL, Hagen D (2013) Ecological restoration: approaches and impacts on vegetation, soils and society. Adv Agron 120:173–222. https://doi.org/10.1016/B978-0-12-407686-0.00003-8

Awazi NP, Tchamba NM (2019) Enhancing agricultural sustainability and productivity under changing climate conditions through improved agroforestry practices in smallholder farming systems in sub-Saharan Africa. Afr J Agric Res 14(7):379–388

Baskent EZ (2021) Assessment and improvement strategies of sustainable land management (SLM) planning initiative in Turkey. Sci Total Environ 797:149183. https://doi.org/10.1016/j.scitotenv.2021.149183

Baumhardt RL, Stewart BA, Sainju UM (2015) North American soil degradation: processes, practices, and mitigating strategies. Sustain For 7(3):2936–2960. https://doi.org/10.3390/su7032936

Baye KN, Aamaa B (2019) Role of conservation tillage as climate change mitigation. Fuel 11(1):12–23. https://doi.org/10.7176/CER
Benson D, Gain AK, Giupponi C (2020) Moving beyond water centricity? Conceptualizing integrated water resources management for implementing sustainable development goals. Sustain Sci 15(2):671–681. https://doi.org/10.1007/s11625-019-00733-5
Bhardwaj A (2019) Watershed hydrology and management. In: Watershed hydrology, management and modeling. CRC Press, pp 1–17
Blanco H, Lal R (2023a) Management of Grazing Lands. In: Soil conservation and management. Springer, Cham, pp 443–469. https://doi.org/10.1007/978-3-031-30341-8_18
Blanco H, Lal R (2023b) Soil conservation and management. Springer. https://doi.org/10.1007/978-3-031-30341-8
Blanco-Canqui H, Wortmann CS (2020) Does occasional tillage undo the ecosystem services gained with no-till? A review. Soil Till Res 198:104534. https://doi.org/10.1016/j.still.2019.104534
Blevins RL, Lal R, Doran JW, Langdale GW, Frye WW (2018) Conservation tillage for erosion control and soil quality. In: Advances in soil and water conservation. Routledge, pp 51–68
Blomquist W, Schlager E (2005) Political pitfalls of integrated watershed management. Soc Nat Resour 18(2):101–117. https://doi.org/10.1080/08941920590894435
Boesch DF, Brinsfield RB, Magnien RE (2001) Chesapeake Bay eutrophication: scientific understanding, ecosystem restoration, and challenges for agriculture. J Environ Qual 30(2):303–320. https://doi.org/10.2134/jeq2001.302303x
Boincean B, Dent D, Boincean B, Dent D (2019) Crop rotation. In: Farming the black Earth: sustainable and climate-smart management of Chernozem soils, pp 89–124. https://doi.org/10.1007/978-3-030-22533-9
Bronick CJ, Lal R (2005) Soil structure and management: a review. Geoderma 124(1–2):3–22. https://doi.org/10.1016/j.geoderma.2004.03.005
Browning TN, Sawyer DE (2021) Vulnerability to watershed erosion and coastal deposition in the tropics. Sci Rep 11(1):885. https://doi.org/10.1038/s41598-020-79402-y
Campbell LC (2017) Managing soil fertility decline. In: Nutrient use in crop production. CRC Press, pp 29–52
Capodaglio AG, Callegari A (2018) Can payment for ecosystem services schemes be an alternative solution to achieve sustainable environmental development? A critical comparison of implementation between Europe and China. Resources 7(3):40. https://doi.org/10.3390/resources7030040
Coffel ED, Keith B, Lesk C, Horton RM, Bower E, Lee J, Mankin JS (2019) Future hot and dry years worsen Nile Basin water scarcity despite projected precipitation increases. Earth's Future 7(8):967–977. https://doi.org/10.1029/2019EF001247
Comte I, Colin F, Whalen JK, Grünberger O, Caliman JP (2012) Agricultural practices in oil palm plantations and their impact on hydrological changes, nutrient fluxes and water quality in Indonesia: a review. Adv Agron 116:71–124. https://doi.org/10.1016/B978-0-12-394277-7.00003-8
Curtis J (2002) Establishing agricultural partnerships in the context of everglades ecosystem restoration: an assessment of potential partners in South Florida an Internal Report for the World Wildlife Fund
Daniel EB, Camp JV, LeBoeuf EJ, Penrod JR, Dobbins JP, Abkowitz MD (2011) Watershed modeling and its applications: a state-of-the-art review. Open Hydrol J 5(1). https://doi.org/10.2174/1874378101105010026
DeBusk TA, DeBusk WF (2001) Wetlands for water treatment. In: Applied wetlands science and technology. Lewis Publishers, Boca Raton, pp 241–280
Dessalegn D, Beyene S, Ram N, Walley F, Gala TS (2014) Effects of topography and land use on soil characteristics along the toposequence of Ele watershed in Southern Ethiopia. Catena 115:47–54. https://doi.org/10.1016/j.catena.2013.11.007
Di Sacco A, Hardwick KA, Blakesley D, Brancalion PH, Breman E, Cecilio Rebola L et al (2021) Ten golden rules for reforestation to optimize carbon sequestration, biodiversity recovery and livelihood benefits. Glob Chang Biol 27(7):1328–1348. https://doi.org/10.1111/gcb.15498

Docker B, Robinson I (2014) Environmental water management in Australia: experience from the Murray-Darling basin. Int J Water Resour Dev 30(1):164–177. https://doi.org/10.1080/07900627.2013.792039

Dregne HE (2002) Land degradation in the drylands. Arid Land Res Manag 16(2):99–132. https://doi.org/10.1080/153249802317304422

Duale MM (2024) Resilience and transformation: examining pastoral development initiatives in Ethiopia—article review. Open Access Library J 11(2):1–17. https://doi.org/10.4236/oalib.1111215

Dwivedi S, Mishra S, Tripathi RD (2018) Ganga water pollution: a potential health threat to inhabitants of Ganga basin. Environ Int 117:327–338. https://doi.org/10.1016/j.envint.2018.05.015

Edwards PJ, Williard KW, Schoonover JE (2015) Fundamentals of watershed hydrology. J Contemp Water Res Educ 154(1):3–20. https://doi.org/10.1111/j.1936-704X.2015.03185.x

Ensor JE, Park SE, Attwood SJ, Kaminski AM, Johnson JE (2018) Can community-based adaptation increase resilience? Clim Dev 10(2):134–151. https://doi.org/10.1080/17565529.2016.1223595

Epule ET, Peng C, Lepage L, Chen Z (2014) The causes, effects and challenges of Sahelian droughts: a critical review. Reg Environ Chang 14:145–156. https://doi.org/10.1007/s10113-013-0473-z

Erol A, Randhir TO (2012) Climatic change impacts on the ecohydrology of Mediterranean watersheds. Clim Chang 114:319–341. https://doi.org/10.1007/s10584-012-0406-8

Fahad S, Nie L, Hussain S, Khan F, Khan FA, Saud S et al (2015) Rice pest management and biological control. Sustain Agric Rev Cereals:85–106. https://doi.org/10.1007/978-3-319-16988-0_4

Fahad S, Chavan SB, Chichaghare AR, Uthappa AR, Kumar M, Kakade V et al (2022) Agroforestry systems for soil health improvement and maintenance. Sustain For 14(22):14877. https://doi.org/10.3390/su142214877

Fazliev J, Khaitova I, Atamurodov B, Rustamova K, Ravshanov U, Sharipova M (2019) Efficiency of applying the water-saving irrigation technologies in irrigated farming. Интернаука 21(103 часть 3):35

Ferreira CS, Walsh RP, Ferreira AJ (2018) Degradation in urban areas. Curr Opin Environ Sci Health 5:19–25. https://doi.org/10.1016/j.coesh.2018.04.001

Ferreira CS, Kalantari Z, Seifollahi-Aghmiuni S, Ghajarnia N, Rahmati O, Solomun MK (2021) Rainfall-runoff-erosion processes in urban areas. In: Precipitation. Elsevier, pp 481–498. https://doi.org/10.1016/B978-0-12-822699-5.00018-5

Gabriel JL, Quemada M (2017) Water management for enhancing crop nutrient use efficiency and reducing losses. In: Advances in research on fertilization management of vegetable crops, pp 247–265. https://doi.org/10.1007/978-3-319-53626-2_9

Gashure S, Wana D (2023) Sustainability of the long-term indigenous soil and water conservation practices in the UNESCO designated cultural landscapes of Konso, Ethiopia. Curr Res Environ Sustain 5:100221. https://doi.org/10.1016/j.crsust.2023.100221

Gerlak AK, Jacobs KL, McCoy AL, Martin S, Rivera-Torres M, Murveit AM et al (2021) Scenario planning: embracing the potential for extreme events in the Colorado River basin. Clim Chang 165:1–21. https://doi.org/10.1007/s10584-021-03013-3

Grafton RQ, Pittock J, Williams J, Jiang Q, Possingham H, Quiggin J (2014) Water planning and hydro-climatic change in the Murray-Darling basin, Australia. Ambio 43:1082–1092. https://doi.org/10.1007/s13280-014-0495-x

Gregersen HM, Ffolliott PF, Brooks KN (2007) Integrated watershed management: connecting people to their land and water. CABI

Guan N, Bi H, Song Y, Lu S, Lin D, Han J (2024) Vegetation restoration is affecting the characteristics and patterns of infiltration in the loess plateau. Catena 243:108190. https://doi.org/10.1016/j.catena.2024.108190

Gupta SR, Dagar JC, Teketay D (2020) Agroforestry for rehabilitation of degraded landscapes: achieving livelihood and environmental security. In: Agroforestry for degraded landscapes: recent advances and emerging challenges, vol 1, pp 23–68. https://doi.org/10.1007/978-981-15-4136-0_2

Gurmu G (2019) Soil organic matter and its role in soil health and crop productivity improvement. For Ecol Manag 7(7):475–483. https://doi.org/10.14662/ARJASR2019.147

Hassan MR, Alam MD, Haque MN, Khan MAQ, Talukder MAK (2024) A review of watershed Management in Bangladesh: options, challenges and legal framework. J Mater Environ Sci 15(2):225–241

Hobley E, Garcia-Franco N, Hübner R, Wiesmeier M (2018) Reviewing our options: managing water-limited soils for conservation and restoration. Land Degrad Dev 29(4):1041–1053. https://doi.org/10.1002/ldr.2849

Hussain A, Bashir H, Zafar SA, Rehman RS, Khalid MN, Awais M et al (2023) The importance of soil organic matter (som) on soil productivity and plant growth. Biol Agric Sci Res J 2023(1):11–11. https://doi.org/10.54112/basrj.v2023i1.11

Islam R, Reeder R (2014) No-till and conservation agriculture in the United States: an example from the David Brandt farm, Carroll, Ohio. Int Soil Water Conserv Res 2(1):97–107. https://doi.org/10.1016/S2095-6339(15)30017-4

Jacobson RB, Lindner G, Bitner C (2015) The role of floodplain restoration in mitigating flood risk, lower Missouri River, USA. In: Geomorphic approaches to integrated floodplain management of lowland fluvial systems in North America and Europe, pp 203–243. https://doi.org/10.1007/978-1-4939-2380-9_9

Jena J, Maitra S, Hossain A, Pramanick B, Gitari HI, Praharaj S et al (2022) Role of legumes in cropping system for soil ecosystem improvement. In: Ecosystem services: types, management and benefits. Nova Science Publishers, Inc, p 415

Jonker V, Seid A (2014) A decision support system for integrated water resource management in the Nile River basin: environmental engineering. Civil Eng/Siviele Ingenieurswese 2014(7):17–22

Kebenei MC, Mucheru-Muna M (2023) The potential of Zai pit technology and integrated soil fertility management to enhance crop productivity in semi-arid regions of Sub-Sahara Africa: a review. Int J Environ Agric Biotechnol 8(1). https://doi.org/10.22161/ijeab

Kidane D, Alemu B (2015) The effect of upstream land use practices on soil erosion and sedimentation in the upper Blue Nile Basin, Ethiopia. Res J Agric Environ Manag 4(2):55–68

Klemas V (2014) Remote sensing of riparian and wetland buffers: an overview. J Coast Res 30(5):869–880. https://doi.org/10.2112/JCOASTRES-D-14-00013.1

Kremen C, Miles A (2012) Ecosystem services in biologically diversified versus conventional farming systems: benefits, externalities, and trade-offs. Ecol Soc 17(4). https://doi.org/10.5751/ES-05035-170440

Kulkarni S (2021) Sustainable soil conservation and management: principles, issues, and strategies. In: Field practices for wastewater use in agriculture. Apple Academic, pp 133–178

Kumar MD, Ghosh S, Patel A, Singh OP, Ravindranath R (2006) Rainwater harvesting in India: some critical issues for basin planning and research. In: Land use and water resources research, vol 6. https://doi.org/10.22004/ag.econ.47964

Kundzewicz ZW (1999) Flood protection—sustainability issues. Hydrol Sci J 44(4):559–571. https://doi.org/10.1080/02626669909492252

Kuyah S, Sileshi GW, Luedeling E, Akinnifesi FK, Whitney CW, Bayala J et al (2020) Potential of agroforestry to enhance livelihood security in Africa. In: Agroforestry for degraded landscapes: recent advances and emerging challenges, vol 1, pp 135–167. https://doi.org/10.1007/978-981-15-4136-0_4

Lal R (2001) Soil degradation by erosion. Land Degrad Dev 12(6):519–539. https://doi.org/10.1002/ldr.472

Lal R (2012) Climate change and soil degradation mitigation by sustainable management of soils and other natural resources. Agric Res 1:199–212. https://doi.org/10.1007/s40003-012-0031-9

Lal R (2015) Restoring soil quality to mitigate soil degradation. Sustain For 7(5):5875–5895. https://doi.org/10.3390/su7055875

Li F, Cook S, Geballe GT, Burch WR Jr (2000) Rainwater harvesting agriculture: an integrated system for water management on rainfed land in China's semiarid areas. AMBIO J Human Environ 29(8):477–483. https://doi.org/10.1579/0044-7447-29.8.477

Li H, Mei X, Wang J, Huang F, Hao W, Li B (2021) Drip fertigation significantly increased crop yield, water productivity and nitrogen use efficiency with respect to traditional irrigation and fertilization practices: a meta-analysis in China. Agric Water Manag 244:106534. https://doi.org/10.1016/j.agwat.2020.106534

Lytton L, Ali A, Garthwaite B, Punthakey JF, Saeed B (2021) Groundwater in Pakistan's Indus Basin. World Bank. Accessed at file: Groundwater-in-Pakistan-s-Indus-Basin-Present-and-Future-Prospects, 20(1)

MacKenzie K, Auger S, Beitollahpour S, Gharabaghi B (2024) The role of stream restoration in mitigating sediment and phosphorous loads in urbanizing watersheds. Water 16(2):363. https://doi.org/10.3390/w16020363

Mag-Aso JR, Vista AB, Garcia FGN (2024, February) Estimation of agricultural productivity loss due to soil loss in Buayan-Malungon River basin, Philippines: a foundation for watershed protection and agricultural development. In: IOP conference series: earth and environmental science, vol 1302, No. 1. IOP Publishing, p 012130. https://doi.org/10.1088/1755-1315/1302/1/012130

Martins PD, Poleto C, Jerônimo GJ, Batista MD (2017) Study of water resources management plans as a tool for integrated watershed management. Int J River Basin Manag 15(4):445–452. https://doi.org/10.1080/15715124.2017.1372444

Mason DM, Redinger J, Elgin AK (2024) The Great Lakes' most unwanted: characterizing the impacts of the top ten Great Lakes aquatic invasive species. J Great Lakes Res 50:102365. https://doi.org/10.1016/j.jglr.2024.102365

McKindles K, Frenken T, McKay RML, Bullerjahn GS (2020) Binational efforts addressing cyanobacterial harmful algal blooms in the Great Lakes. In: Contaminants of the Great Lakes, pp 109–133. https://doi.org/10.1007/698_2020_513

Mehta P, Jangra MS, Baweja PK, Srivastav AL (2024) Impact of climate change on rural water resources and its management strategies. In: Water resources management for rural development. Elsevier, pp 45–54. https://doi.org/10.1016/B978-0-443-18778-0.00018-0

Mekonnen M (2021) Impacts of soil and water conservation practices after half of a generation age, northwest highlands of Ethiopia. Soil Tillage Res 205:104755. https://doi.org/10.1016/j.still.2020.104755

Menta C (2012) Soil fauna diversity-function, soil degradation, biological indices, soil restoration. In: Biodiversity conservation and utilization in a diverse world, vol 284

Mondal S, Palit D (2022) Challenges in natural resource management for ecological sustainability. In: Natural resources conservation and advances for sustainability. Elsevier, pp 29–59. https://doi.org/10.1016/B978-0-12-822976-7.00004-1

Moss T (2004) The governance of land use in river basins: prospects for overcoming problems of institutional interplay with the EU water framework directive. Land Use Policy 21(1):85–94. https://doi.org/10.1016/j.landusepol.2003.10.001

Moss B (2008) Water pollution by agriculture. Philos Trans R Soc B Biol Sci 363(1491):659–666. https://doi.org/10.1098/rstb.2007.2176

Nair PR, Kumar BM, Nair VD, Nair PR, Kumar BM, Nair VD (2021) Soil conservation and control of land-degradation. In: An introduction to agroforestry: Four decades of scientific developments, pp 445–474. https://doi.org/10.1007/978-3-030-75358-0_18

Narendra BH, Siregar CA, Dharmawan IWS, Sukmana A, Pratiwi, Pramono IB et al (2021) A review on sustainability of watershed management in Indonesia. Sustain For 13(19):11125. https://doi.org/10.3390/su131911125

Nayakekorale HB (2020) Soil degradation. In: The soils of Sri Lanka, pp 103–118. https://doi.org/10.1007/978-3-030-44144-

Nazari-Sharabian M, Taheriyoun M, Karakouzian M (2019) Surface runoff and pollutant load response to urbanization, climate variability, and low impact developments–a case study. Water Supply 19(8):2410–2421. https://doi.org/10.2166/ws.2019.123

Nelson EJ, Booth DB (2002) Sediment sources in an urbanizing, mixed land-use watershed. J Hydrol 264(1–4):51–68. https://doi.org/10.1016/S0022-1694(02)00059-8

Nichols CR, Zinnert J, Young DR (2019) Degradation of coastal ecosystems: causes, impacts and mitigation efforts. In: Tomorrow's coasts: complex and impermanent, pp 119–136. https://doi.org/10.1007/978-3-319-75453-6_8

Nyssen J, Naudts J, De Geyndt K, Haile M, Poesen J, Moeyersons J, Deckers J (2008) Soils and land use in the Tigray highlands (Northern Ethiopia). Land Degrad Dev 19(3):257–274. https://doi.org/10.1002/ldr.840

Ogidi OI, Akpan UM (2022) Aquatic biodiversity loss: impacts of pollution and anthropogenic activities and strategies for conservation. In: Biodiversity in Africa: potentials, threats and conservation. Springer, Singapore, pp 421–448. https://doi.org/10.1007/978-981-19-3326-4_16

Ometto JP, Aguiar APD, Martinelli LA (2011) Amazon deforestation in Brazil: effects, drivers and challenges. Carbon Manag 2(5):575–585. https://doi.org/10.4155/cmt.11.48

Osman KT, Osman KT (2013) Soil resources and soil degradation. In: Soils: principles, properties and management, pp 175–213. https://doi.org/10.1007/978-94-007-5663-2_12

Oude Essink GH, Van Baaren ES, De Louw PG (2010) Effects of climate change on coastal groundwater systems: a modeling study in The Netherlands. Water Resour Res 46(10). https://doi.org/10.1029/2009WR008719

Oweis T (2018) Drainage systems to support sustainable water use Henk Ritzema, Wageningen University, The Netherlands. In: Water management for sustainable agriculture. Burleigh Dodds Science Publishing, pp 489–516

Pahl-Wostl C (2005) Information, public empowerment, and the management of urban watersheds. Environ Model Softw 20(4):457–467. https://doi.org/10.1016/j.envsoft.2004.02.005

Pandey B, Ghosh A (2023) Urban ecosystem services and climate change: a dynamic interplay. Front Sustain Cities 5:1281430. https://doi.org/10.3389/frsc.2023.1281430

Quevauviller P (2011) Adapting to climate change: reducing water-related risks in Europe–EU policy and research considerations. Environ Sci Pol 14(7):722–729. https://doi.org/10.1016/j.envsci.2011.02.008

Radcliffe DE, Rasmussen TC, Warrick AW (2002) Soil water movement. In: Soil physics companion, pp 85–126. https://doi.org/10.1201/9781420041651

Raymondi RR, Cuhaciyan JE, Glick P, Capalbo SM, Houston LL, Shafer SL, Grah O (2013) Water resources: implications of changes in temperature and precipitation. In: Climate Change in the Northwest: implications for our landscapes, waters, and communities, pp 41–66. https://doi.org/10.5822/978-1-61091-512-0_3

Reij C, Tappan G, Smale M (2009) Agroenvironmental transformation in the Sahel: another kind of "Green Revolution". Intl Food Policy Res Inst

Rezaie AM, Ferreira CM, Rahman MR (2019) Storm surge and sea level rise: threat to the coastal areas of Bangladesh. In: Extreme hydroclimatic events and multivariate hazards in a changing environment. Elsevier, pp 317–342. https://doi.org/10.1016/B978-0-12-814899-0.00013-4

Riedinger-Whitmore MA (2015) Using palaeoecological and palaeoenvironmental records to guide restoration, conservation and adaptive management of Ramsar freshwater wetlands: lessons from the Everglades, USA. Mar Freshw Res 67(6):707–720. https://doi.org/10.1071/MF14319

Ritzema HP, Stuyt LCPM (2015) Land drainage strategies to cope with climate change in The Netherlands. Acta Agric Scand B Soil Plant Science 65(sup1):80–92. https://doi.org/10.1080/09064710.2014.994557

Robinson JG, LaBruna D, O'Brien T, Clyne PJ, Dudley N, Andelman SJ et al (2024) Scaling up area-based conservation to implement the global biodiversity Framework's 30×30 target: the role of Nature's strongholds. PLoS Biol 22(5):e3002613. https://doi.org/10.1371/journal.pbio.3002613

Ross A (2018) Speeding the transition towards integrated groundwater and surface water management in Australia. J Hydrol 567:e1–e10. https://doi.org/10.1016/j.jhydrol.2017.01.037

Russo T, Alfredo K, Fisher J (2014) Sustainable water management in urban, agricultural, and natural systems. Water 6(12):3934–3956. https://doi.org/10.3390/w6123934

Sadeghi SHR, Jalili K, Nikkami D (2009) Land use optimization in watershed scale. Land Use Policy 26(2):186–193. https://doi.org/10.3390/w6123934

Saha D, Sikka AK, Goklani R (2022) Artificial recharge endeavours in India: a review. Water Secur 16:100121. https://doi.org/10.1016/j.wasec.2022.100121
Samuel A, Joy KJ, Paranjape S, Peddi S, Adagale R, Deshpande P, Kulkarni S (2007) Watershed development in Maharashtra: present scenario and issues for restructuring the programme. SOPPECOM, Pune
Sarvade S, Upadhyay VB, Kumar M, Imran Khan M (2019) Soil and water conservation techniques for sustainable agriculture. In: Sustainable agriculture, forest and environmental management, pp 133–188. https://doi.org/10.1007/978-981-13-6830-1_5
Schleiss AJ, Franca MJ, Juez C, De Cesare G (2016) Reservoir sedimentation. J Hydraul Res 54(6):595–614. https://doi.org/10.1080/00221686.2016.1225320
Seiler KP, Gat JR (2007) Groundwater recharge from run-off, infiltration and percolation, vol 55. Springer
Sheng TC (1990) Watershed management field manual. Watershed survey and planning. In: FAO conservation guide. Food and Agriculture Organization of The United Nations, Rome
Siirila-Woodburn ER, Rhoades AM, Hatchett BJ, Huning LS, Szinai J, Tague C et al (2021) A low-to-no snow future and its impacts on water resources in the western United States. Nat Rev Earth Environ 2(11):800–819. https://doi.org/10.1038/s43017-021-00219-y
Singh J (2013) Depleting water resources of Indian Punjab agriculture and policy options: a lesson for high potential areas. Glob J Sci Front Res Agric Vet 13(4):16–23
Sivakumar B (2011) Global climate change and its impacts on water resources planning and management: assessment and challenges. Stoch Env Res Risk A 25:583–600. https://doi.org/10.1007/s00477-010-0423-y
Sklar F, McVoy C, Van Zee R, Gawlik D, Swift D, Park W et al (1999) Hydrologic needs: the effects of altered hydrology on the Everglades, Everglades interim report. SFWMD, West Palm Beach
Sok S, Meas S, Chea S, Chhinh N (2019) Regional cooperation and benefit sharing for sustainable water resources management in the lower Mekong basin. Lakes Reserv Res Manag 24(3):215–227. https://doi.org/10.1111/lre.12277
Souza VS, da Silva Vanolli B, Schiebelbein BE, de Souza Bortolo L, Carvalho ML, Mendes IC, Cherubin MR (2024) Cover crops and soil health in Brazilian agricultural systems. In: Soil health and sustainable agriculture in Brazil, Soil health series: volume 3, pp 103–144. https://doi.org/10.1002/9780891187448.ch4
Staddon C, Everard M (2017) Competing epistemologies of community-based groundwater recharge in semi-arid North Rajasthan: progress and lessons for groundwater-dependent areas. In: Water, knowledge and the environment in Asia. Routledge, pp 87–107
Sujianto S, Ermiati E, Sudjarmoko B (2023) Agricultural practice assessment to accelerate organic Rice farming adoption: a case study in West Java, Indonesia. In: BIO web of conferences, vol 69. EDP Sciences, p 04002. https://doi.org/10.1051/bioconf/20236904002
Swanson S, Kozlowski D, Hall R, Heggem D, Lin J (2017) Riparian proper functioning condition assessment to improve watershed management for water quality. J Soil Water Conserv 72(2):168–182. https://doi.org/10.1051/bioconf/20236904002
Teague R, Kreuter U (2020) Managing grazing to restore soil health, ecosystem function, and ecosystem services. Front Sustain Food Syst 4:534187. https://doi.org/10.3389/fsufs.2020.534187
Tejada M, Gonzalez JL, García-Martínez AM, Parrado J (2008) Application of a green manure and green manure composted with beet vinasse on soil restoration: effects on soil properties. Bioresour Technol 99(11):4949–4957. https://doi.org/10.1016/j.biortech.2007.09.026
Tong ST, Yang H, Chen H, Yang JY (2016) Hydrologic impacts of climate change and urbanization in the Las Vegas Wash Watershed, Nevada. J Water Clim Chang 7(3):598–620. https://doi.org/10.2166/wcc.2016.038
Trivedi TP, Ahuja DB (2011) Integrated pest management: approaches and implementation. Indian J Agric Sci 81(11):981–993
van Koppen B, Schreiner B (2014) Moving beyond integrated water resource management: developmental water management in South Africa. Int J Water Resour Dev 30(3):543–558. https://doi.org/10.1080/07900627.2014.912111

Wallbott L, Siciliano G, Lederer M (2019) Beyond PES and REDD+: Costa Rica on the way to climate-smart landscape management? Ecol Soc 24(1):24. https://doi.org/10.5751/ES-10476-240124

Wallis PJ, Ison RL (2011) Appreciating institutional complexity in water governance dynamics: a case from the Murray-Darling basin, Australia. Water Resour Manag 25:4081–4097. https://doi.org/10.1007/s11269-011-9885-z

Walsh R, Ward AS (2022) An overview of the evolving jurisdictional scope of the US clean water act for hydrologists. Wiley Interdiscip Rev Water 9(5):e1603. https://doi.org/10.1002/wat2.1603

Wang G, Mang S, Cai H, Liu S, Zhang Z, Wang L, Innes JL (2016) Integrated watershed management: evolution, development and emerging trends. J For Res 27:967–994. https://doi.org/10.1007/s11676-016-0293-3

Wolf S, Pham M, Matthews N, Bubeck P (2021) Understanding the implementation gap: policy-makers' perceptions of ecosystem-based adaptation in Central Vietnam. Clim Dev 13(1):81–94. https://doi.org/10.1080/17565529.2020.1724068

Yang T, Siddique KH, Liu K (2020) Cropping systems in agriculture and their impact on soil health-a review. Glob Ecol Conserv 23:e01118. https://doi.org/10.1016/j.gecco.2020.e01118

Yatoo AM, Ali S, Hamid S, Hassan B, Zaheen Z, Ali MN et al (2020) Role of soil biota and associated threats. In: Bioremediation and biotechnology, Vol 4: techniques for noxious substances remediation, pp 143–165. https://doi.org/10.1007/978-3-030-48690-7_7

Yi X, Wang L (2013) Land suitability assessment on a watershed of loess plateau using the analytic hierarchy process. PLoS One 8(7):e69498. https://doi.org/10.1371/journal.pone.0069498

Yoon T, Rhodes C, Shah FA (2015) Upstream water resource management to address downstream pollution concerns: a policy framework with application to the N akdong R iver basin in S outh K Orea. Water Resour Res 51(2):787–805. https://doi.org/10.1002/2013WR014201

Yosef BA, Asmamaw DK (2015) Rainwater harvesting: an option for dry land agriculture in arid and semi-arid Ethiopia. Int J Water Resour Environ Eng 7(2):17–28

Yu S, Xu Z, Wu W, Zuo D (2016) Effect of land use types on stream water quality under seasonal variation and topographic characteristics in the Wei River basin, China. Ecol Indic 60:202–212. https://doi.org/10.1016/j.ecolind.2015.06.029

Zafirah N, Nurin NA, Samsurijan MS, Zuknik MH, Rafatullah M, Syakir MI (2017) Sustainable ecosystem services framework for tropical catchment management: a review. Sustain For 9(4):546. https://doi.org/10.3390/su9040546

Zahed Z, Mufti S, Kumar SS, Wani OA, Mushtaq F, Rasool R et al (2022) Organic and inorganic mulches combination improves the productivity, quality and profitability of rainfed potato in the temperate himalayan region. Gesunde Pflanzen 74(4):1109–1122. https://doi.org/10.1007/s10343-022-00650-1

Zhang W, Dong A, Liu F, Niu W, Siddique KH (2022) Effect of film mulching on crop yield and water use efficiency in drip irrigation systems: a meta-analysis. Soil Tillage Res 221:105392. https://doi.org/10.1016/j.still.2022.105392

Zhao G, Mu X, Wen Z, Wang F, Gao P (2013) Soil erosion, conservation, and eco-environment changes in the loess plateau of China. Land Degrad Dev 24(5):499–510. https://doi.org/10.1002/ldr.2246

Ecosystem Services in Agro-Ecosystems: A Step Towards Sustainability

15

Ananya Gairola, Vipin Kumar, Diksha Sharma, Sandeep Kumar, and Sanjay Singh Rathore

Abstract

The agro-ecosystem has evolved into a highly specialized framework for crop production, significantly influenced by human activity. Unlike natural ecosystems, which maintain a balance of various functions, agro-ecosystems prioritize crop yield, often compromising other indispensable services encompassing atmospheric regulation through gas modulation, the preservation of soil integrity, and the sustenance of biological diversity. Within these systems, entomophagous predators and parasitoids play a crucial role, providing about 33% of natural pest control. The total eradication of pollinators could result in a 3–8% decline in global agricultural output. Recognizing and valuing the wide array of essential ecosystem services provided by these natural agents is crucial for sustaining agro-ecosystem resilience amid global environmental pressures. This is particularly pertinent for India, which is dedicated to mitigating greenhouse gas emissions by 33–35% and restoring 26 million hectares of degraded land in accordance with Sustainable Development Goals (SDGs) by 2030. Stakeholders need to work together to protect these ecosystem services, ensuring that agro-ecosystems remain functional and productive despite growing environmental challenges.

Keywords

Agro-ecosystem · Biological diversity · Resilience · Sustainability

---

A. Gairola (✉) · V. Kumar · D. Sharma · S. S. Rathore
Division of Agronomy, ICAR-Indian Agricultural Research Institute, New Delhi, India

S. Kumar
ICAR-National Bureau of Soil Survey and Land Use Planning, Jorhat, Assam, India

© The Author(s), under exclusive license to Springer Nature Singapore Pte Ltd. 2025

S. Babu et al. (eds.), *Ecological Solutions to Agricultural Land Degradation*,
https://doi.org/10.1007/978-981-96-3392-0_15

Abbreviations

TEEB	The Economics of Ecosystems and Biodiversity
EU	European Union
IPBES	Intergovernmental Platform on Biodiversity and Ecosystem Services
PES	Payment for Ecosystem Services
ES	Ecosystem Services
SDGs	Sustainable Development Goals
MEA	Ministry of External Affairs
GHGs	Green House Gases
GIS	Geographic Information System
IPM	Integrated Pest Management
%	Percentage
RUE	Radiation Use Efficiency
$t\ ha^{-1}$	Tons per hectare
°C	Degree Centigrade
USA	United States of America

15.1 Introduction

Currently, the world is experiencing a concerning decline in biodiversity, with nearly 25% of entire floral and faunal species at risk of imminent extinction. This troubling trend is diminishing nature's productivity, resilience, and adaptability, posing serious risks to our economies, livelihoods, and overall well-being. However, there lies significant untapped potential within global biodiversity. It is essential to restore an equilibrium between the requisites and provisions of natural resources, so that we can progress towards sustainable production and consumption (Dasgupta 2021). Ecosystems are vital for human livelihoods, providing various services that contribute to and improve human life (Daily 1997; MA 2005). The notion that environment confers advantages upon human society has ancient origins. Over the course of time, this understanding has undergone progressive refinement into the concept of environmental services (Wilson and Matthews 1970). The term "Nature's service" was, for the very first time, used by Westman (1977), but it was Ehrlich and Ehrlich (1981) and Ehrlich and Mooney (1983) who renamed it as "ecosystem services". This concept gained significant attention starting in the year 1997 (Gómez–Baggethun et al. 2010). Originally conceived simply as metaphor (Norgaard 2010), it has since evolved into a foundational concept for extensive literature aimed at evaluating, quantifying, and appreciating human society's reliance on nature. This once-metaphorical idea has inspired a substantial body of work and influenced policy changes. Policymakers often seek assessments and economic evaluations so that they can get an idea about the direct link between loss of biodiversity and the diminutions in overall well-being, as seen in TEEB (The Economics of Ecosystems and Biodiversity) study mandated by the EU (European Union) (Sukhdev 2008). As a result, several governments have established Intergovernmental Platform on

Biodiversity and Ecosystem Services (IPBES 2022) and additionally, various Payment for Ecosystem Services (PES) programs have also been commenced, addressing services like watershed management, preservation of biodiversity, the sequestration of atmospheric carbon, alongside a myriad of other ecological functions and services.

Several authors have observed that focusing on generation of just one ecosystem service can negatively impact others due to the intrinsic trade-offs (Bennett et al. 2009; Seppelt et al. 2013). So, it becomes crucial to decode the complex interactions between agricultural output and its related ecosystem services (Zhang et al. 2007) to develop strategies that minimize the trade-offs and promoting simultaneously (Dumont et al. 2013). The production of livestock, for instance, is highlighted for its significant impact on various ecosystem services (Herrero and Thornton 2013). The escalation of land use intensification aimed at enhancing food production worsens the already declining biodiversity related ecosystem services (Teillard et al. 2016). Grasslands play quite a significant role in rendering synergies among ecosystem services as they support agricultural practices and contribute to services like sequestration of C (Soussana et al. 2004) and enhancement of biodiversity (Dross et al. 2018). Additionally, grasslands offer the potential to increase livestock production without the need for pesticides used in fodder cultivation.

15.2 Ecosystem Services

Ecosystem services represent the advantages derived by humans from natural ecosystems (MA 2005). A framework focused on ecosystem services helps highlight how much society depends on these systems for its own welfare. It shows connection between the physical and biological structures, their functional roles, and the advantages they confer. This approach represents significance of measuring and evaluating these benefits in order to embed the valuation of natural capital and ecosystem services within decision-making frameworks for public as well as private sectors (TEEB 2010; Burkhard et al. 2012). Ecological economists frequently view human societies as integral components of ecosystems, arguing that natural capital is not easily substituted with constructed or manufactured capital (van den Belt et al. 2011; Daly and Farley 2010, Farley 2012). Consequently, valuation of many ecosystem services is frequently seen from perspective of the human subsystem (Costanza and Liu 2014). While few definitions of "ecosystem services" focus on functional processes of ecosystems, via which, humans derive benefits (Costanza et al. 1997; Daily 1997), many other definitions emphasize their utilitarian aspects, aligning them with principles of economic accounting (Boyd and Banzhaf 2007). There are also definitions that highlight well-being of humans (Fisher et al. 2009) or underscore importance of values (TEEB 2010). Over time, these definitions have transformed into comprehensive frameworks that guide our thoughts, establish the organizing principles, and even enhance our ability to progress towards the sustainable development. Few frameworks adopt linear approach, often for accounting purposes, whereas others stress the interdependence of human systems and ecosystems, considering aspects such as biodiversity and culture. This perspective requires managing human activities in a way that aligns with the ecosystems' carrying capacity. Adopting an ecosystem services perspective to decode how well-being of humans rely on the natural systems generally requires integration of existing knowledge from diverse networks to guide policy and management choices, while also pinpointing and addressing areas where understanding is lacking.

In past 10 years, notion of ecosystem services (ES) achieved considerable scientific prominence (Potschin et al. 2016) recognizing biodiversity and the ecosystems' functioning as the foundation for supplying ES. In 2016, the United States released a memorandum directing federal agencies to integrate ecosystem services into their strategic planning and decision-making processes. Furthermore, ES are acknowledged for their role in global sustainability, especially in Goal 15 (Geijzendorffer et al. 2017). Ecosystem services are generally described as different types of advantages that mother nature offers to the humankind, that are crucial for survival and well-being of human race (Costanza et al. 1997). These encompass provisioning services such as food materials, and energy; regulating and maintenance services, including climate and water regulation; and last but not least cultural services like recreation (Haines–Young and Potschin 2013). Provisioning ES, particularly those from managed ecosystems, are of significant interest due to their direct impact on human activities like nutrition and material supply (Plieninger et al. 2016). Different classification frameworks for provisioning ecosystem services concentrate on the production of biomass, genetic resources, medicinal and ornamental resources, and freshwater (TEEB 2010). Provisioning services are distinct because they usually take the form of those ecosystem 'goods' that can be directly consumed/traded, unlike other ES with less functional markets (Boyd and Banzhaf 2007). A new definition of ecosystem services encompasses the combined effects of the ecosystem structure and function, along with other contributing factors, in order to improve human well-being (Burkhard et al. 2012). The MEA (2005) recognized that human race is an important part of ecosystems; European Landscape Convention views landscapes as areas shaped by natural and human factors (European Council 2000). This is especially true for intensively used landscapes like agro-ecosystems, which depend on modifications of the natural ecosystems and human-induced inputs (Pérez–Soba et al. 2012). Figure 15.1 illustrates various categories of the ecosystem services.

Fig. 15.1 Classification of the ES

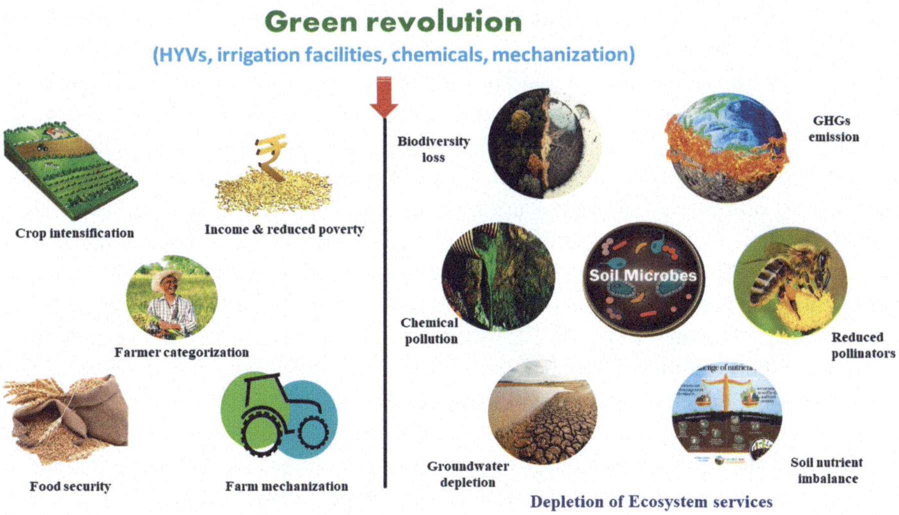

Fig. 15.2 Sustainability challenges after green revolution era

15.3 Agricultural Transformation and Sustainability Challenges

Agricultural ecosystems in India are indispensable for sustaining human well-being and livelihoods. The Green Revolution of the 1960s, characterized by introduction of advanced agricultural technologies such as high-yielding variety seeds, sophisticated irrigation systems, mechanization, synthetic fertilizers, and pest control measures, catalyzed a significant transformation in Indian agriculture. This revolution not only enhanced farm productivity and ensured financial stability for farmers but also played a pivotal role in poverty alleviation and economic advancement. This agricultural transformation reshaped rural livelihoods and reinforced the nation's economic resilience and self-sufficiency in food production. While it helped meet food demands without expanding farmland into fragile ecosystems like forests and wetlands, it also led to significant environmental challenges, including soil degradation and water resource depletion (Fig. 15.2).

15.4 Agro-Ecosystems

Agro-ecosystems are defined as managed ecosystems that rely on human inputs and are able to affect multiple other ES, leading to both positive and negative impacts on environment (Petz and van Oudenhoven 2012). The concurrent production of agro-ecosystem services requires the integration of natural as well as anthropogenic capital (Jones et al. 2016). Primary objective of agro-ecosystems: to produce food, forage, fuel, pharmaceuticals, and bioenergy crops (Power 2010). They also provide regulating services like climate regulation and water regulation; cultural services like aesthetics of landscape (Huang et al. 2015). Nevertheless, agro-ecosystems may also result provision of ecosystem disservices such as soil erosion and leaching of nitrogen as nitrate (Fridman and Kissinger 2018). Agro-ecosystems are globally recognized mainly due to their distinctive capacity to offer all four ES (Zhang et al. 2007; Power 2010). The provision of ES by the agro-ecosystems relies on how these systems are managed (Foley et al. 2005). The production of crops and other agriculture-related outputs is contingent upon other ecosystem services (ES), especially regulating services like pollination, pest and disease control, and nutrient regulation (Power 2016). Boosting regulating ecosystem services and human welfare while enhancing agricultural competitiveness necessitates advancing landscape multi-functionality (Huang et al. 2015). Adopting efficient agricultural management approaches can amplify the beneficial impacts of agro-ecosystem services along with mitigating adverse environmental effects (Kanter et al. 2018). The comparative significance of various ESs by different ecosystems is provided in Fig. 15.3.

Grasping the intricate dynamics of agricultural systems and the ultimate need to achieve various goals within sustainable agro-ecosystems requires a multidisciplinary approach that combines insights and methods from a wide range of fields. In order to address this complexity, systems approaches have been developed, and their use has expanded considerably in past few decades. The agronomic systems possess unique geographical and temporal characteristics, with spatial dimensions evident at levels ranging from individual crops and fields to farms, regions, and beyond, each accompanied by its specific temporal elements. Systems analysis in agronomic systems involves utilizing expert insights, stakeholder experience, and information obtained from scientific observations and model-based simulations.

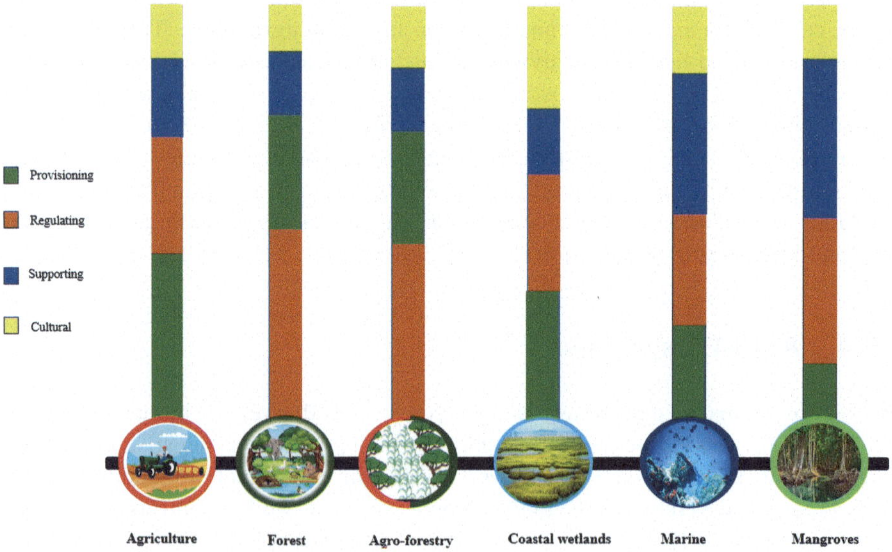

Fig. 15.3 Ecosystem services under different agro-ecosystems

These scientific measurements and simulations can range from simple, quick, and inexpensive procedures (relatively unreliable) to complex, time-consuming, and costly data-intensive procedures. Selecting appropriate issue specific procedures, in terms of both measurements and simulation models, requires careful consideration, as every issue requires a distinct research methodology. The choice of the appropriate systems approach should be based on the necessary outputs and the accessibility of data. To tackle issues in the agronomic systems at a specific scale, various approaches can be used. This usually starts with an in-depth system analysis, followed by different types of studies such as exploratory, protective, predictive, or those aimed at supporting decision-making, with relevant examples provided. Nowadays, the systems approach often involves engaging stakeholders, such as farmers and local community members, to refine the problem, carry out research, and apply the results effectively. A thorough interdisciplinary examination of production systems is crucial for devising innovative and sustainable solutions for the future.

Research in agronomy served an important role in substantial rise of agricultural output in the latter part of the twentieth century. This expansion was mainly fuelled by the development of enhanced crop varieties and advanced management techniques that incorporated external inputs, in response to the increasing global food demand. The global population increased from approximately 1.6 billion in the year 1900 to over 8.2 billion at present, which spurred this demand. As a result, the food supply system has developed into a complex network involving production, land utilization, industrial activities, trade, market dynamics, and national as well as international policies. The rise in output per unit area, due to the application of fertilizers and biocides alongside extensive agricultural transformations, has resulted in irreversible alterations in landscapes, environmental diversity, and natural resource quality. These developments have sparked global concerns regarding the agricultural sustainability. In highly industrialized nations such as the European Union, there is an increasing demand for organic agricultural practices that strictly eliminate the utilization of synthetic chemical inputs. To tackle the challenge of augmenting food production, it becomes essential to raise the productivity, economic viability, and ultimately, the long-term sustainability of agricultural systems. This is particularly urgent in regions like Asia, where food demand is highest, and the increase in production must be attained with less land, labour, water, and pesticides, while conserving scarce natural resources. Historically, agronomic research concentrated on single-discipline studies in fields like plant breeding, plant nutrition, plant growth physiology, and pest control through pesticide use. Field studies established dose-effect relationships to determine optimal fertilization, irrigation scheduling, residual effects of agrochemicals used for pest and disease control. However, it is quite evident that the efficiency of single factors is significantly influenced by other production factors, necessitating an integrated approach. There is now a greater emphasis on analysing entire agronomic production systems to develop agro-ecosystems that achieve multiple goals. These interdisciplinary evaluations necessitate contributions from diverse fields to more precisely delineate and comprehend the entire agronomic production system. Systems approaches, which systematically and quantitatively analyse agricultural systems and synthesize comprehensive, functional concepts, have been developed to facilitate these cross-disciplinary research. These methods utilize techniques including simulation modelling, expert systems, databases, linear programming, and geographic information systems (GIS). Mathematical models and computer simulations grounded in scientific principles offer unbiased methods for assessing the biophysical impacts of different resource management strategies at the levels of fields, farms, and regions. To conduct a comprehensive systems analysis, biophysical evaluations must be paired with socio-economic assessments to generate benefits at various scales, thereby requiring an integration of biophysical and social sciences. Systems approaches have advanced from research to application, evidenced by numerous high-quality interdisciplinary research teams worldwide. At field level, resource use optimization is important for achieving goals related to food supply, income, and environmental protection. This necessitates deciphering how genotype-environment-management interactions work to more effectively customize choosing genotypes and developing management strategies tailored to specific local conditions and targeted objectives. System-specific approaches are now being adopted to enhance breeding programs, evaluate yield potential in various environments, optimize water and nitrogen usage, and improve crop protection by emphasizing prevention and using natural predators to lessen reliance on pesticides (Kropff et al. 1997). Over the past decade, water and nutrient management-based research has advanced in response to environmental challenges, including nitrate contamination and competition for water resources among industries, households, and agriculture. Majority of this research is centred on aligning the seasonal availability of resources with crop requirements at each growth stage to optimize resource use efficiency and reduce nutrient losses. Thus, resource-use optimization often serves multiple, seemingly contradictory objectives. Figure 15.4 illustrates the core ecological principles that contribute to the endurance and resilience of agro-ecosystems.

In crop protection, the focus has transitioned from simply eliminating or eradicating pests to implementing IPM (Integrated Pest Management) strategies. Problems associated with pesticides, including gradual pest resistance development, environmental consequences, and high expenses, have prompted creation of innovative systems and methods for managing pests. These new methods are founded on an improved understanding of the complex pest dynamics, their natural enemies and the pest-crop interactions. Improved system understanding helps identify new control techniques and supports various decision-making processes in the management of pests. At the field level, decisions often involve navigating trade-offs among various biophysical and socio-economic objectives. Integrated approaches have been developed to assist in farm-level decision-making by combining the biophysical and socio-economic strategies (Rossing et al. 1997). On a regional scale, systems approaches have facilitated eco-regional analyses, with innovative tools such as GIS aiding in the organization and application of extensive databases, which gain added value through system models for analysis. Novel approaches are being developed to address scaling issues in these studies (Teng et al. 1995).

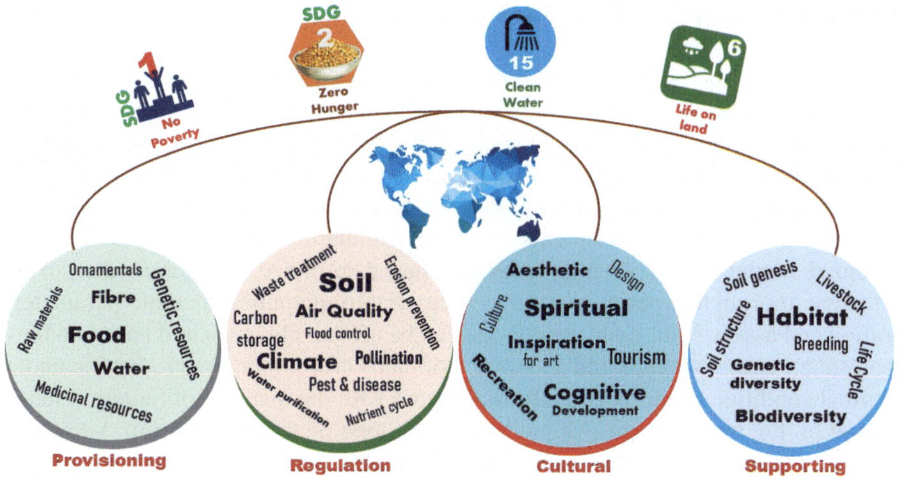

Fig. 15.4 Ecological foundation for agro-ecosystem endurance

15.4.1 Strategies for Enhancing Crop Production

Historically, field experiments at 'representative sites' were used to evaluate how different genotypes performed and identify ways to enhance production efficiency. The results from these studies were then applied to other similar sites (Nix 1983). In recent years, agronomic research has transitioned to methodology based, driven by knowledge and centred on systems analysis. Simulation tools have become essential for evaluating and implementing management strategies. Consequently, systems approaches have gained prominence in agricultural research in areas like breeding (Cooper and Hammer 1996) and agri-business management (Teng et al. 1995). The current focus at ground level is on optimizing genotype-environment interactions to meet the goals of farmers and societal needs (economic, environmental, and socio-economic). The various systems approaches facilitate the evaluation of potential yields for specific genotypes across various input levels and environmental conditions. The simplest scenario involves evaluating crop production potential based on factors such as genotype, radiation, and temperature. Including additional factors like water, nutrients, pests, diseases, and weeds adds complexity to system. Historical weather data can be used for simulating potential production and analysing yield gaps (benchmarking). Incorporating elements like water, nutrients, and management into simulations supports decision-making and the evaluation of strategies for achieving targeted yields and various other environmental goals. Those models that connect crop growth with the dynamics of pests, diseases, and weeds can aid in developing management scenarios, risk assessments, and plant resistance designs. These models are key for integrating scientific understanding of physiological processes to offer a thorough examination of the system and set research priorities.

15.4.2 Refining Breeding Strategies

In field crops, the actual yields rarely reach their calculated potential, ranging between less than 5% and up to 60% of the capability (Oerke et al. 1994). Potential yield, which is the productivity influenced solely by the growth factors as well as yield determining factors, varies across different agro-ecological zones due to temperature and solar radiation variability. Muchow and Kropff (1997) emphasized significance of the field experiments and crop simulation models in assessing the potential output across diverse systems. They determined that for models to be effective, thorough experimentation is necessary to guarantee accurate parameterization and assessment. Crop production can be effectively measured based on how well solar radiation is captured and used. For many crops, biomass production under optimal conditions is proportionally associated with the quantity of radiation intercepted by the crop canopy (Monteith 1977). Slope of this relationship, known as RUE (radiation use efficiency), measures the effectiveness of converting energy into biomass. Variations in benchmark potential yields across different environments are influenced by both temperature and solar radiation (Kropff et al. 1997). For instance, Muchow and Kropff (1997) demonstrated that potential yields for maize, rice, and wheat differ significantly, being lower in the tropical regions and higher in temperate zones at greater latitudes and altitudes. Kropff et al. (1995) discovered that the potential yield of rice varied from 6 t ha^{-1} in tropical wet season environments (Los Banos, Philippines) to 15 t ha^{-1} at higher latitudes (Yanco, Australia). Temperature mainly affects growth duration, with cooler temperatures extending the radiation interception period. Potential biomass accumulation is directly related to intercepted solar radiation, and grain yield has a direct correlation with biomass for a specific harvest index. Thus, higher yields were linked to cooler temperatures and higher solar radiation in the studied areas. For maize, a simple mechanistic model successfully simulated yields ranging from 9.5 to 17.1 t ha^{-1}, with the main variation attributed to growth duration (84–153 days) resulting from temperature differences (28.7 °C to 18 °C; Muchow and Kropff 1997). These findings suggest that sunlight along with temperature define the upper limits of potential yield in any given environment.

15.4.3 Agroecology in Terms of Sustainability

The concept of sustainable agriculture has emerged as a reaction to the natural resources degradation resulting from contemporary farming methods (McIsaac and Edwards 1994). Currently, agricultural production is seen not only as a technical challenge but as one that involves social, cultural, political, and economic factors. Although definition of sustainability is often debated and varies, it remains a useful term as it addresses the interconnectedness of agriculture with socioeconomic and natural systems (Reijntjes et al. 1992). Comprehending the wider context of agriculture entails investigating its interplay with global environment and social systems, while acknowledging that agricultural growth and development is resulted from multiple interacting factors. This comprehensive understanding can lead to new management strategies aligned with SDGs. The sustainability concept has spurred discussions and highlighted the need for significant changes in traditional agriculture to turn it sustainable environmentally, socially, and economically. A range of solutions is being investigated to address the environmental challenges associated with capital and technology-intensive agricultural practices, focusing on reducing or eliminating agrochemical use through organic nutrients and integrated pest management (Gliessman et al. 1998). Despite numerous environmentally-oriented research projects and technological advancements, the predominant approach remains highly technological, focusing on addressing limiting factors or symptoms rather than systemic issues within agroecosystems. This perspective views pests and nutrient deficiencies as primary causes of low productivity rather than symptoms of an imbalanced agroecosystem (Carrol et al. 1990). As a result, the main goal has been to overcome these limiting factors through new technologies, rather than understanding and addressing the root causes of agricultural problems (Altieri 1993). Agroecology, defined as applying ecological principles to designing and managing sustainable agroecosystems, offers a framework for understanding agroecosystem complexity (Altieri 1995). It seeks to create agricultural systems that depend minimally on high levels of agrochemicals and energy inputs, emphasizing ecological interactions and synergies among biological components to maintain soil health, crop yield, and pest management (Altieri and Rosset 1996).

15.4.4 Principles of Agroecology

In the pursuit of incorporating ecological principles into agricultural practices, scientists and developers have frequently missed an essential element: a deep comprehension of agroecosystems and their operational dynamics. This oversight has led to the emergence of agroecology, a field of study that offers fundamental ecological principles for the analysis, design, and management of agroecosystems that are productive, conserve natural resources, and support cultural, social, and economic sustainability (Altieri 1995). Agroecology transcends a simplistic perspective of agroecosystems by incorporating aspects like genetics, agronomy, and soil science, to understand their ecological and social evolution, structure, and function. Rather than concentrating on individual components, agroecology highlights the interconnectedness of all elements within an agroecosystem and the not so simple interplay of ecological processes (Vandermeer 1995). Agroecosystems, which consist of floral and faunal interactions with their physical and chemical environments, are shaped by humans to produce various products such as food, fiber, fuel and timber. Agroecology represents a comprehensive study of these systems, integrating all environmental and human factors. It explores structure, complexity, functions, and processes of their interconnections. For example, an agricultural field is understood as a complex system where natural ecological processes like nutrient cycling, predator/prey relationships, competition, symbiosis, and succession take place. Agroecological research operates on the principle that a thorough comprehension of

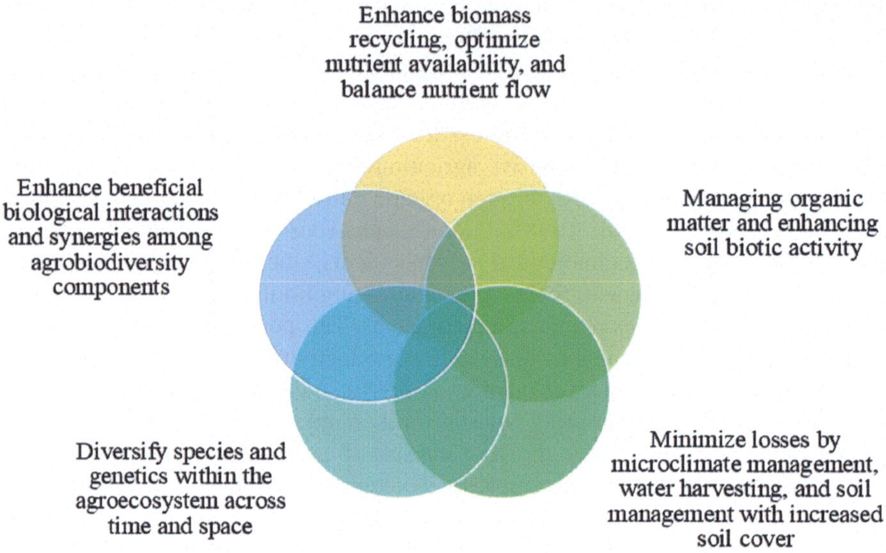

Fig. 15.5 Ecological principles

ecological interactions and processes allows for the manipulation of agroecosystems to enhance production and sustainability, while minimizing adverse environmental and social effects and decreasing dependence on external inputs (Altieri 1995).

The ecological principles can be applied using a variety of techniques and approaches (Fig. 15.5); influencing the productivity, stability, and resilience in distinct ways based on local opportunities, resource limitations, and market conditions. The primary aim of agroecological design is to combine various elements to enhance overall biological efficiency, protect biodiversity, sustain the productivity and self-regulating ability of agroecosystem. This objective involves creating a mosaic of agroecosystems within a landscape that mirrors structure and functions of natural ecosystems.

15.4.5 Agroecosystems Diversification

From a management viewpoint, agroecology seeks to create well-balanced environments that sustain agricultural output, support soil fertility through natural biological activities, and naturally control pests using a mix of diverse agroecosystems and minimal-input practices (Gleissman 1998). Present day agroecologists acknowledge sustainable practices like intercropping and agroforestry that emulate natural ecological functions. By creating agricultural systems that replicate natural processes, optimal utilization of solar radiation, soil nutrients, and rainfall can be realized (Pretty 1994). Agroecological management focuses on enhancement of nutrient

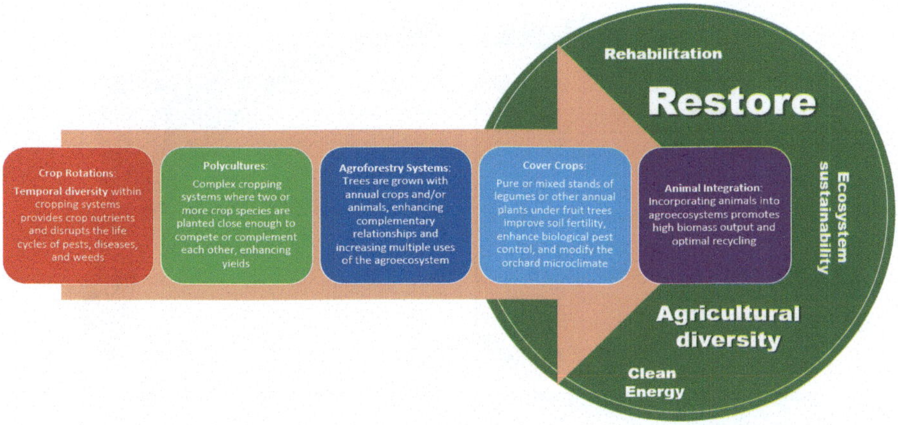

Fig. 15.6 Approaches to reintroduce agricultural diversity

recycling, decomposition of organic matter, closed energy cycles, and preservation of water and soil, while also ensuring a balance between pests and their natural predators. This approach takes advantage of the synergistic effects and complementarities of various crop, tree, and animal combinations arranged in both spatially and temporally (Altieri 1994). The effectiveness of agroecosystems is dependent on the dynamics between living and non-living components (Fig. 15.6). By incorporating a variety of functional biodiversity, it is possible to create synergies that support agroecosystem processes and deliver essential ecological services, such as boosting soil biological activity, nutrient recycling, and encouraging beneficial arthropods and natural enemies (Altieri and Nicholls 1999). Currently, a wide array of practices and technologies is accessible, each varying in effectiveness and strategic value (Fig. 15.7).

Increasing the diversity offers various benefits (Altieri 1994; Gleissman 1998):

- Species co-existence and positive interactions, fostering the sustainability of agroecosystems.
- Improved resource-use efficiency.
- In diverse systems, the abundance and variety of natural enemies increase resistance to herbivores.
- Creation of diverse microclimates within cropping systems, supporting beneficial predators, parasites, pollinators, soil fauna, and antagonists.
- Support for preserving biodiversity in adjacent natural habitats.
- Delivery of ecological benefits, including nutrient recycling and the breakdown of harmful chemicals in the soil.
- Reduced risk for farmers, especially in unpredictable environments, as diversity ensures that if one crop fails, others may compensate.

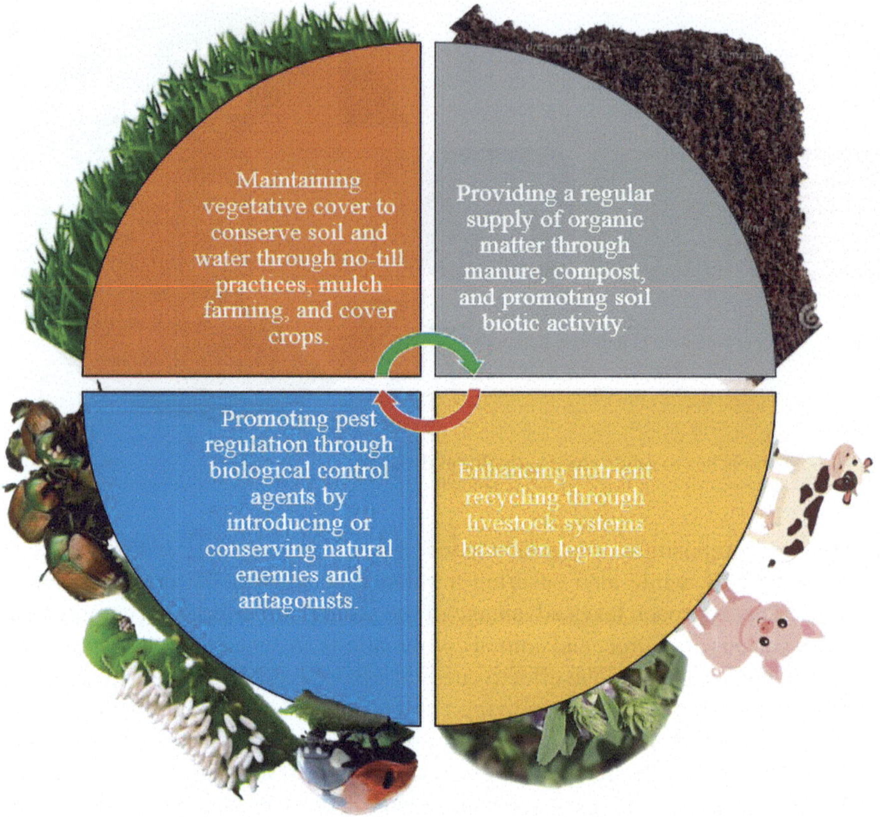

Fig. 15.7 Common features shared by diversified agroecosystems

15.5 Development of Sustainable Agroecosystems

Most advocates of sustainable agriculture aim to build farming systems that maintain long-term productivity by (Pretty 1994; Vandermeer 1995)

- Integrating various elements of the farming system—flora, fauna, soil, water, climate, and human factors- to complement one another and optimize synergistic benefits.
- Reducing external inputs: Minimizing utilization of off-farm, external, and non-renewable inputs that may harm environment, and using remaining inputs more strategically to lower variable costs.
- Reliance on local resources: Substituting the external inputs with nutrient cycling, improved conservation methods, and a greater reliance on local resources.

Table 15.1 Strategies for agroecosystem optimization and resilience mechanism (Altieri 2000)

Ecological processes for agroecosystems optimization	Mechanisms to achieve agroecosystem resilience
Boost and maintain productivity over the long term.	Expansion of floral species and genetic diversity across different time periods and spatial contexts.
Reduce toxicity by removing agrochemicals.	Augmentation of functional biodiversity (natural enemies, antagonists, etc.)
Optimize metabolic functions (organic matter decomposition and nutrient cycling)	Improvement in soil organic matter + biological activity
Achieve equilibrium in regulatory systems such as nutrient cycles, soil-water balance, energy flow, and pest population control.	Enhancement of soil coverage and the competitive capacity of crops
Improve the preservation and restoration of soil and water resources, as well as biodiversity.	Removal of harmful inputs and residues

- Aligning cropping pattern: Adjusting cropping patterns to match productive potential and environmental challenges of the climate and landscape thus ensuring long-term sustainability.
- Valuing and conserving biodiversity: Protecting biological diversity in both wild and domesticated landscapes; optimizing biological as well as innate potential of plant and animal species.
- Utilizing local knowledge: Leveraging local expertise and techniques, encompassing innovative methods widely utilized by farmers but not yet thoroughly understood by researchers.

Agroecology offers the insights and methodologies required to create an agricultural system which is ecologically sustainable, exceptionally productive, socially equitable, and economically viable (Table 15.1). By implementing the agroecological principles, sustainable agriculture can better utilize internal resources, minimize external inputs, and renew internal resources. The main goal of designing agroecosystems is to integrate these elements in order to improve overall biological efficiency, protect biodiversity, and maintain the productivity and self-regulating capabilities of the agroecosystem. This involves creating agroecosystems that mimic local natural environments, which have high species diversity and healthy soils. These systems support natural pest control, nutrient recycling, and good soil coverage to prevent resource depletion.

15.6 Biodiversity-Driven Services in Agroecosystems

15.6.1 Crop, Forage, and Timber Yields

It has been widely recognized for a long time that combining grasses and legumes in mixtures and rotations results in greater biomass production compared to monocultures. This beneficial effect of species interactions has been recognized for centuries and still applied in modern agricultural systems (Harper 1977; Hector and Hooper 2002; Sanderson et al. 2004; Trenbath 1974; Vandermeer 1989). Several research have proven that the interactions between different species can also have a beneficial impact on crop yields (Kiaer et al. 2009), forage (Bullock et al. 2007), and timber (Zhang et al. 2012). Previous studies primarily compared a limited high-yielding mixtures and monocultures (Sanderson et al. 2004), but current thoroughly explore the effects of richness of species versus composition of species (O'Connor et al. 2017). Furthermore, substantial evidence from both experimental and observational studies indicates that enhancing the variety of tree species can result in a substantial increase in wood yield (Chamagne et al. 2017; Paquette and Messier 2011; Zhang et al. 2012) and also enhance carbon storage in trees (Hulvey et al. 2013). The silviculture-based experiments often reveal better growth in mixed-species stands compared to monocultures, especially when including nitrogen-fixing species (Piotto 2008). Research in both natural and managed forests increasingly supports the idea that greater tree diversity leads to improved growth and wood production among all major forest biomes (Liang et al. 2016; Scherer–Lorenzen 2014). Despite these generally positive effects, it is evident that the relationship between tree diversity and production is not uniform and can be influenced by climate, with the strongest benefits observed in areas where the climate significantly restricts the ability of trees to grow (Jucker et al. 2016).

15.6.2 Yield Stabilization

Plant communities that are diverse in terms of taxonomy (Gross et al. 2014; Isbell et al. 2015), function (Polley et al. 2013), otherwise in genetics (Prieto et al. 2015) tend to demonstrate increased stability in their overall yields over time and demonstrate better resilience to climate disruptions compared to less diverse plant communities. Plant diversity contributes to yield stability by providing protective effects that help to reduce or mitigate the impact of stress or disturbances that reduce variations in yield over time (Yachi and Loreau 1999). Buffering happens when different species have distinct responses to environmental disturbances, leading to a drastic reduction in yield variance (Yachi and Loreau 1999), or when species compensate for each other's losses by flourishing when others decline (Lehman and Tilman 2000), or achieved by averaging out the fluctuations of individual species over time (Doak et al. 1998). These processes help stabilize yields in grasslands (De Mazancourt et al. 2013; Gross et al. 2014) and forests (Jucker et al. 2016; Tuck et al. 2016). Additionally, increased genetic diversity improves the resilience of frequently harvested grasslands by fostering variations in seasonal growth and development trends among species, a concept referred to as phenological niche partitioning (Prieto et al. 2015).

Plant diversity's influence on stability may fluctuate depending on factors such as (i) land use practices, (ii) structure of plant communities, and (iii) spatial differences in environmental conditions.

15.6.3 Pollinator Diversity

Despite the widespread management of honeybees for crop pollination globally, other insect pollinators—such as various bee species, moths, flies, beetles, butterflies, ants, and wasps—often play a larger role in crop pollination than previously recognized, sometimes even surpassing the impact of honeybees on crop productivity (Rader et al. 2016). Floral diversity plays a crucial role in sustaining wild pollinator communities over an extended period and across different areas by (i) sustaining a diverse range of pollinator species with varying floral choices and level of specialization, and also because (ii) Pollinator species with long flight periods that last longer than the blooming period of any one plant species (Kremen et al. 2007). Thus, floral richness and diversity are essential for fostering and maintaining crop pollination services provided by unmanaged pollinators (Garibaldi et al. 2014; Kremen et al. 2007; Scherber et al. 2010). One research showed that an increase in richness of plant species was directly linked to greater functional diversity among pollinators and various other pollination service metrics in a positive manner (Orford et al. 2016).

Due to their high mobility, pollinator communities within agricultural environments exhibit positive responses to increased floral diversity at numerous spatial scales (Kremen and Miles 2012; Kennedy et al. 2013). At local level, practices like planting strips of flowering plants, increasing their diversity, creating perennial hedgerows, using intercrops and polycultures, and adopting organic farming methods have all been observed to positively affect pollinators and their services (Garibaldi et al. 2014), despite the limited number of studies have specifically examined the impact of polyculture or intercropping. A synthesis of 39 studies at the landscape level demonstrated that greater distances from natural habitats notably decreased both the average and stability of wild bee diversity, visitation rates, and fruit set within fields over spatial and temporal scale (Garibaldi et al. 2011). The beneficial effects of adjacent natural habitats are likely attributable to the floral and/or nesting resources they supply, which support local biodiversity (Williams et al. 2012; Kremen et al. 2007).

15.6.4 Weed Regulation

Implementing diverse cropping systems across time as well as space play quite an essential role in successfully managing weed populations (Bennett et al. 2012; Liebman et al. 2001; Levine et al. 2004; Finn et al. 2013; Balvanera et al. 2006; Quijas et al. 2010). Temporal diversification can be achieved by implementing crop rotations, a time-honoured practice that dates back to ancient Greece, Rome, and China (Bennett et al. 2012). Diverse rotations usually involve crops with diverse planting and harvest schedules, varying growth patterns and competitive traits, as well as differing tillage and weed management practices. This approach exposes weeds to various kinds of stresses and mortality risks, limiting their ability for uncontrolled growth and reproduction. A review conducted on different crop species revealed that crop rotations led to lower weed densities in 21 cases when compared to continuous monoculture systems (Liebman and Dyck 1993). Diversification in time can be implemented through use of cover crops, that can be cultivated before or after the major crop seasons. Cover crops help smother weeds by altering the micro-climatic conditions that influence weed germination and subsequent establishment, competing for growth resources, and releasing phytotoxins. This reduces weed seeds and vegetative propagules present in following crops (Teasdale et al. 2007). Additionally, combination of forage species with rapid and slow growth rates enhances weed suppression (Finn et al. 2013). Spatial diversification can be achieved through intercropping. The synergistic use of resources and beneficial interactions among intercrop components improve the uptake of light, water, and nutrients (Liebman et al. 2001). As a result, intercropping is often more effective than monoculture in outcompeting weeds for resources and inhibiting their growth (Liebman and Dyck 1993).

15.6.5 Pest Regulation

Monocultures, where only one type of plant is grown, create dense resource pockets that favour certain herbivores while lacking the "associational resistance" that occurs when a variety of plants compete for these resources and simultaneously increase those available to natural predators (Tahvanainen and Root 1972; Root 1973). By introducing diversity into agricultural systems—whether through mixed cropping, crop rotations, varietal mixtures, or planting non-crop species—pest colonization and growth can be curbed by (i) disrupting the resource availability for herbivores and (ii) boosting the abundance and variety of their natural enemies (Bianchi et al. 2006; Quijas et al. 2010; Poveda et al. 2008; Chaplin-Kramer et al. 2011; Letourneau et al. 2011; Kremen and Miles 2012; Rusch et al. 2016). Research synthesizing multiple cropping experiments indicates that diversified planting systems generally outperform simpler, monocultural setups in terms of lower insect herbivore populations, reduced crop damage, higher herbivore mortality, and improved crop yields (Letourneau et al. 2011). A prominent illustration of this is the push-pull strategy, which involves planting insect-repellent species within fields to repel pests and lure crops outside the fields to attract and trap them, effectively disrupting pest life cycles (Midega et al. 2015). This method was shown to reduce pest damage by 138% and boost crop yields by 84% across 14 studies, while other diversification methods sometimes led to lower yields (Letourneau et al. 2011).

Genetic variation within a crop species can help reduce the growth and survival of herbivorous pests (Wetzel et al. 2016). In absence of crucial resources for predators and parasitoids, mixed plantings and non-crop borders can supplement them by providing nectar, pollen, alternative prey, and refuges, thereby increasing the number and variety of natural enemies. Higher diversity among natural enemies generally leads to higher pest mortality rates, although instances of interference and intraspecific predation can sometimes interfere with the effectiveness of biological pest control (Letourneau et al. 2009). Gaining deeper insights into the behavioural and evolutionary mechanisms involved will help in designing plant communities that suppress pests by supporting their natural enemies. The complexity of the surrounding landscape is often linked to more natural enemies and greater diversity in fields, as well as higher herbivore diversity, though it doesn't always lead to more pests (Chaplin-Kramer et al. 2011). Simplified landscapes have been shown to reduce the effectiveness of aphid biological control by 46% in Europe as well as in the States (Rusch et al. 2016). Establishing clear connections between landscape diversity, populations of natural enemies, and pest dynamics is a vital yet complex challenge for future research.

15.6.6 Nutrient Enrichment and Carbon Storage in Soil

Expanding the diversity of agroecosystems can markedly improve the internal processes of nutrient generation and recycling, which diminishes the requirement for fertilizers and lessens the associated social costs. Enhancing plant diversity can result in greater accumulation and retention of soil nitrogen (N) (Fornara and Tilman 2008) thus improving nitrogen use efficiency (van Ruijven and Berendse 2005). Rotational cropping systems that incorporate legumes such as alfalfa and clover, in conjunction with grains like maize, wheat, and barley, have been recognized for their capacity to consistently uphold and ensure the maintenance of elevated agricultural productivity over time by supplying N_2 (Robertson and Vitousek 2009). In soils lacking sufficient phosphorus, the phosphorus released by one crop has the potential to augment the productivity of a subsequent crop cultivated in alternating rows (Li et al. 2007). Therefore, crop diversity plays a quite a crucial role in improving the uptake and recycling of more than one essential nutrient simultaneously. While increasing the diversity of soil organisms from very low to moderate levels may also enhance nutrient cycling, the effectiveness of nutrient cycling often hinges more on the presence and abundance of specific soil food web components and their interactions than on overall soil organism diversity (Bardgett and Van Der Putten 2014).

To ensure agricultural productivity and climate stability, it is crucial to preserve or enhance soil carbon stocks. The cultivation practices led to an estimated loss of 78 ± 12 gigatonnes of carbon from global soils between 1850 and 1998 (Lal 2004). Greater diversity among plant species can lead to increased accumulation of SOM in the temperate grasslands (Fornara and Tilman 2008; Steinbeiss et al. 2008) and intercropping systems (Cong et al. 2014) over relatively short periods (4–12 years). Plant communities with greater diversity are likely to sequester greater amounts of soil carbon due to the higher production of root biomass. This enhanced biomass contributes to soil carbon reserves through the decomposition of plant debris and the addition of microbial biomass and its byproducts (Fornara and Tilman 2008; Cong et al. 2014; Lange et al. 2015). This process may even improve soil aggregation, thus preventing organic matter decomposition by microbes (Gould et al. 2016). However, increasing the diversity of soil organisms could potentially accelerate decomposition and respiration rates, particularly in intensively managed

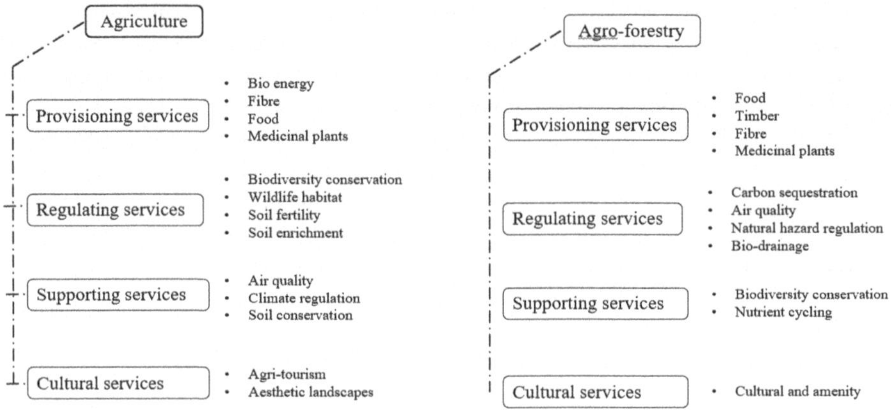

Fig. 15.8 Inventory of ecosystem services from agro-ecosystems

agricultural soils with low biodiversity (Nielsen et al. 2011). The effect mainly relies on the species composition and the specific traits of soil microbes and fauna (Bardgett and Van Der Putten 2014). Since plant diversity has the ability to influence both the accumulation and loss of soil carbon, further research is required to gain a deeper understanding of how it affects soil carbon storage and overall ecosystem carbon balance (Wilsey and Polley 2004). Figure 15.8 depicts the various ES offered by agro-ecosystems.

15.7 Tackling Challenges and Proposing Solutions

Although there have been notable advancements in comprehending the advantages of diversifying agroecosystems, numerous challenges remain in applying these strategies effectively. Economic and cultural obstacles are prominent barriers to diversification in agriculture. For example, diversifying often requires substantial investments in procuring new harvesting equipment and high-yielding seed varieties. Additionally, finding and maintaining the best species mixtures to optimize yields can be difficult, especially as the abundance of species relatively changes over time (Connolly et al. 2013; Finn et al. 2013). Identifying species combinations that not only optimize yields but also improve additional ecosystem services is even more challenging, as different species contribute various functions in grasslands (Hector and Bagchi 2007; Allan et al. 2015; Lefcheck et al. 2015; Storkey et al. 2015) and forests (Gamfeldt et al. 2013; Castagneyrol et al. 2014; Van Der Plas et al. 2016). Despite these significant struggles, diversification is both attainable and advantageous, and we propose two complementary strategies to facilitate it.

1. Agroecosystems that are currently managed with low diversity—such as high-input cropping systems, pastures, hay meadows, and timber plantations—should be diversified. Enhancing the genetic diversity of crops and addition of species-

rich hedgerows around fields are practical strategies among many. However, there lies a trade-off between achieving high yields and preserving high diversity, as inputs aimed at boosting yields may reduce diversity. (e.g., Isbell et al. 2013). Nonetheless, it is possible to achieve high yields with moderate inputs and diversity, with diversity occasionally acting as a partial replacement for certain inputs. A major constraint in diversifying intensive production systems is that many crops have been specifically bred for monoculture, which may limit the potential for yield improvements in mixtures. However, there is significant potential to improve mixture yields by choosing appropriate crop species that complement each other (Zuppinger-Dingley et al. 2014). Diversification holds significant potential for advancing sustainable intensification, offering a promising strategy to enhance productivity while promoting environmental sustainability (Brooker et al. 2015; Pywell et al. 2015), also as part of a broader land-sparing strategy.

2. In systems with minimal management (low-input, high-diversity croplands, rangelands, hay meadows, and production forests), implementing diversification strategies—such as overseeding or promoting natural forest regeneration—can be highly effective. For example, introducing a diverse mixture of grassland plant species increased both biodiversity and yield by 43% in southern English hay meadows, in contrast to using a more limited blend of seven grass species (Bullock et al. 2007). This approach is particularly valuable in these systems, where the dual objectives are to enhance productivity while also securing ecosystem services such as carbon sequestration and habitat creation (Nelson et al. 2009; Gamfeldt et al. 2013), fitting within the broader context of a land-sharing paradigm.

The greatest gains of agroecosystem diversification are likely where the social costs of agricultural inputs are high and where single-crop systems fail to optimize environmental conditions for maximum yield (Table 15.2). Under variable environmental conditions, niche differentiation among plant species can be crucial for ensuring biomass production (Turnbull et al. 2016; Isbell et al. 2011; Tuck et al. 2016). Although diversification may not invariably improve all the ESs (Allan et al. 2013), increasing floral diversity has the potential to partially or entirely substitute for various agricultural inputs—fertilizers, pesticides, increased pollinator species, and irrigation (Tilman et al. 2006; Weigelt et al. 2009). These farm inputs are expensive for users and for those who are indirectly affected by their negative impacts, such as water and air pollution from nitrogen (Keeler et al. 2016). Rather than focusing solely on whether diverse mixtures yield more than current systems, we encourage adopting diverse mixtures even if they produce similar or slightly lower yields than monocultures, due to the multiple additional benefits they offer. We also advocate for research that evaluates the complete spectrum of costs and benefits related to diversification strategies. As monoculture yields plateau or decline and the demand for ES grows because of population rise and higher consumption rates, diversification could become crucial for maintaining agricultural productivity and ES in croplands, rangelands, and production forests over the coming decades.

Table 15.2 Addressing issues and best practices in agroecosystem Diversification

Barriers	Enhanced diversification framework	Comprehensive diversification strategy
Determine the best combination of species and their respective proportions	Perform experimental screenings to determine the best combinations of crops or trees	Introduce a range of plant species and allow the most competitive ones to outgrow the less competitive varieties
Ensure the best possible balance and proportions of different species	Regularly adjust the ideal crop combinations and apply moderate inputs to manage weeds or regulate the abundance of species (through thinning)	Periodically introduction of rare species or promote natural regeneration to increase their prevalence, especially if evolving environmental conditions make them more suitable
Allocate resources towards purchasing harvest machinery and a variety of seed mixtures	Create new machinery designed for harvesting a variety of crop mixtures	Increase production to lower the costs of various seed mixtures or propagules
Apply the approach to different environmental conditions or services	Determine the best crop or tree combinations for a broad array of relevant conditions	Evaluate how well the most suitable species meet ecosystem service goals and modify seeding rates as needed

15.8 Conclusion

Agroecology offers principles for creating diverse agricultural systems that integrate both plant and animal biodiversity. This integration boosts intricate interactions and synergistic effects, which improve ecosystem functions and processes like controlling harmful organisms, recycling nutrients, and producing and accumulating biomass, thereby enabling agroecosystems to maintain their own health. Implementing agroecological principles leads to enhanced economic and environmental sustainability, with management practices adapted to the local resources and aligned with current environmental and social conditions. An agroecological approach focuses on preserving and enhancing local agricultural resources—such as seeds, soil, beneficial animals, and plant diversity—by promoting a development strategy that promotes participation of farmers, utilizes traditional knowledge, and adapts farming practices to align with local needs and conditions. Balancing biodiversity conservation with sustainability objectives represents the most effective approach to achieving genuine ecological intensification, offering enhanced benefits for both wildlife and agricultural producers.

References

Allan E, Weisser WW, Fischer M, Schulze ED, Weigelt A, Roscher C, Schmid B (2013) A comparison of the strength of biodiversity effects across multiple functions. Oecologia 173:223–237

Allan E, Manning P, Alt F, Binkenstein J, Blaser S, Blüthgen N, Fischer M (2015) Land use intensification alters ecosystem multifunctionality via loss of biodiversity and changes to functional composition. Ecol Lett 18(8):834–843

Altieri MA (1993) Ethnoscience and biodiversity: key elements in the design of sustainable pest management systems for small farmers in developing countries. Agric Ecosyst Environ 46(1–4):257–272

Altieri MA (1994) Biodiversity and pest management in agroecosystems. Hayworth Press, New York

Altieri MA (1995) Agroecology: the science of sustainable agriculture. Westview Press, Boulder

Altieri MA (2000) Agroecology: principles and strategies for designing sustainable farming systems. Agroecol Action

Altieri MA, Nicholls CI (1999) Biodiversity, ecosystem function, and insect pest management in agricultural systems. In: Collins WW, Qualset CO (eds) Biodiversity in agroecosystems. CRC Press, Boca Raton

Altieri MA, Rosset R (1996) Agroecology and the conversion of large–scale conventional systems to sustainable management. Int J Environ Stud 50:165–185

Balvanera P, Pfisterer AB, Buchmann N, He JS, Nakashizuka T, Raffaelli D, Schmid B (2006) Quantifying the evidence for biodiversity effects on ecosystem functioning and services. Ecol Lett 9(10):1146–1156

Bardgett RD, Van Der Putten WH (2014) Belowground biodiversity and ecosystem functioning. Nature 515(7528):505–511

Bennett EM, Peterson GD, Gordon LJ (2009) Understanding relationships among multiple ecosystem services. Ecol Lett 12(12):1394–1404

Bennett AJ, Bending GD, Chandler D, Hilton S, Mills P (2012) Meeting the demand for crop production: the challenge of yield decline in crops grown in short rotations. Biol Rev 87(1):52–71

Bianchi FJ, Booij CJH, Tscharntke T (2006) Sustainable pest regulation in agricultural landscapes: a review on landscape composition, biodiversity, and natural pest control. Proc R Soc B Biol Sci 273(1595):1715–1727

Boyd J, Banzhaf S (2007) What are ecosystem services? The need for standardized environmental accounting units. Ecol Econ 63(2–3):616–626

Brooker RW, Bennett AE, Cong WF, Daniell TJ, George TS, Hallett PD, White PJ (2015) Improving intercropping: a synthesis of research in agronomy, plant physiology, and ecology. New Phytol 206(1):107–117

Bullock JM, Pywell RF, Walker KJ (2007) Long-term enhancement of agricultural production by restoration of biodiversity. J Appl Ecol 44(1):6–12

Burkhard B, De Groot R, Costanza R, Seppelt R, Jørgensen SE, Potschin M (2012) Solutions for sustaining natural capital and ecosystem services. Ecol Indic 21:1–6

Carrol CR, Vandermeer JH, Rosset PM (1990) Agroecology. McGraw Hill Publishing Company, New York

Castagneyrol B, Jactel H, Vacher C, Brockerhoff EG, Koricheva J (2014) Effects of plant phylogenetic diversity on herbivory depend on herbivore specialization. J Appl Ecol 51(1):134–141

Chamagne J, Tanadini M, Frank D, Matula R, Paine CT, Philipson CD, Hector A (2017) Forest diversity promotes individual tree growth in central European forest stands. J Appl Ecol 54(1):71–79

Chaplin-Kramer R, Tuxen-Bettman K, Kremen C (2011) Value of wildland habitat for supplying pollination services to Californian agriculture. Rangelands 33(3):33–41

Cong RG, Smith HG, Olsson O, Brady M (2014) Managing ecosystem services for agriculture: Will landscape-scale management pay?. Ecological Economics 99:53–62

Connolly J, Bell T, Bolger T, Brophy C, Carnus T, Finn JA, Weigelt A (2013) An improved model to predict the effects of changing biodiversity levels on ecosystem function. J Ecol 101(2):344–355

Cooper M, Hammer GL (1996) Plant adaptation and crop improvement. IRRI

Costanza R, Liu S (2014) Ecosystem services and environmental governance: comparing China and the US. Asia Pac Policy Stud 1(1):160–170

Costanza R, d'Arge R, De Groot R, Farber S, Grasso M, Hannon B, Van Den Belt M (1997) The value of the world's ecosystem services and natural capital. Nature 387(6630):253–260

Daily GC (1997) Introduction: what are ecosystem services? In: Nature's services: Societal dependence on natural ecosystems. Island Press, Washington, DC, pp 1–10
Daly HE, Farley J (2010) Ecological economics: principles and applications. Island Press
Dasgupta P (2021) The economics of biodiversity: the Dasgupta review. HM Treasury
De Mazancourt C, Isbell F, Larocque A, Berendse F, De Luca E, Grace JB, Haegeman B, Polley HW, Roscher C, Schmid B, Tilman D (2013) Predicting ecosystem stability from community composition and biodiversity. Ecol Lett 16(5):617–625
Doak DF, Bigger D, Harding EK, Marvier MA, O'malley RE, Thomson D (1998) The statistical inevitability of stability–diversity relationships in community ecology. Am Nat 151(3):264–276
Dross C, Princé K, Jiguet F, Tichit M (2018) Contrasting bird communities along production gradients of crops and livestock in French farmlands. Agric Ecosyst Environ 253:55–61
Dumont B, Fortun–Lamothe L, Jouven M, Thomas M, Tichit M (2013) Prospects from agroecology and industrial ecology for animal production in the 21st century. Animal 7(6):1028–1043
Ehrlich PR, Ehrlich AH (1981) Extinction. Ballantine, New York
Ehrlich PR, Mooney HA (1983) Extinction, substitution, and ecosystem services. Bioscience 33(4):248–254
European Council (2000) The European Landscape Convention of the Council of Europe. http://www.coe.int/en/web/landscape/home. Accessed 20 Mar 2019
Farley J (2012) Ecosystem services: the economics debate. Ecosyst Serv 1(1):40–49
Fisher B, Turner RK, Morling P (2009) Defining and classifying ecosystem services for decision making. Ecol Econ 68(3):643–653
Finn JA, Kirwan L, Connolly J, Sebastià MT, Helgadottir A, Baadshaug OH, Lüscher A (2013) Ecosystem function enhanced by combining four functional types of plant species in intensively managed grassland mixtures: a 3-year continental-scale field experiment. Journal of Applied Ecology 50(2):365–375.
Foley JA, DeFries R, Asner GP, Barford C, Bonan G, Carpenter SR, Snyder PK (2005) Global consequences of land use. Science 309(5734):570–574
Fornara DA, Tilman D (2008) Plant functional composition influences rates of soil carbon and nitrogen accumulation. J Ecol 96(2):314–322
Fridman D, Kissinger M (2018) An integrated biophysical and ecosystem approach as a base for ecosystem services analysis across regions. Ecosyst Serv 31:242–254
Gamfeldt L, Snäll T, Bagchi R, Jonsson M, Gustafsson L, Kjellander P, Bengtsson J (2013) Higher levels of multiple ecosystem services are found in forests with more tree species. Nat Commun 4(1):1340
Garibaldi LA, Steffan-Dewenter I, Kremen C, Morales JM, Bommarco R, Cunningham SA, Klein AM (2011) Stability of pollination services decreases with isolation from natural areas despite honey bee visits. Ecol Lett 14(10):1062–1072
Garibaldi LA, Carvalheiro LG, Leonhardt SD, Aizen MA, Blaauw BR, Isaacs R, Winfree R (2014) From research to action: enhancing crop yield through wild pollinators. Front Ecol Environ 12(8):439–447
Geijzendorffer IR, Cohen-Shacham E, Cord AF, Cramer W, Guerra C, Martín-López B (2017) Ecosystem services in global sustainability policies. Environmental Science & Policy 74: 40–48.
Gliessman SR, Engles E, Krieger R (1998) Agroecology: ecological processes in sustainable agriculture. CRC Press
Gómez–Baggethun E, De Groot R, Lomas PL, Montes C (2010) The history of ecosystem services in economic theory and practice: from early notions to markets and payment schemes. Ecol Econ 69(6):1209–1218
Gould IJ, Quinton JN, Weigelt A, De Deyn GB, Bardgett RD (2016) Plant diversity and root traits benefit physical properties key to soil function in grasslands. Ecol Lett 19(9):1140–1149
Gross K, Cardinale BJ, Fox JW, Gonzalez A, Loreau M, Wayne Polley H, van Ruijven J (2014) Species richness and the temporal stability of biomass production: a new analysis of recent biodiversity experiments. Am Nat 183(1):1–12

Haines–Young R, Potschin M (2013) Common international classification of ecosystem services (CICES): consultation on Version 4, August–December 2012. EEA framework contract no EEA/IEA/09/003

Harper JL (1977) Population biology of plants. New York, Academic

Hector A, Bagchi R (2007) Biodiversity and ecosystem multifunctionality. Nature 448:188–190

Hector A, Hooper R (2002) Darwin and the first ecological experiment. Science 295(5555):639–640

Herrero M, Thornton PK (2013) Livestock and global change: emerging issues for sustainable food systems. Proc Natl Acad Sci 110(52):20878–20881

Huang J, Tichit M, Poulot M, Darly S, Li S, Petit C, Aubry C (2015) Comparative review of multifunctionality and ecosystem services in sustainable agriculture. J Environ Manag 149:138–147

Hulvey KB, Hobbs RJ, Standish RJ, Lindenmayer DB, Lach L, Perring MP (2013) Benefits of tree mixes in carbon plantings. Nat Clim Chang 3(10):869–874

IPBES Secretariat. https://ipbes.net. Accessed on 23 Aug 2022

Isbell F, Calcagno V, Hector A et al (2011) High plant diversity is needed to maintain ecosystem services. Nature 477:199–202

Isbell F, Reich PB, Tilman D, Hobbie SE, Polasky S, Binder S (2013) Nutrient enrichment, biodiversity loss, and consequent declines in ecosystem productivity. Proc Natl Acad Sci 110(29):11911–11916

Isbell F, Craven D, Connolly J, Loreau M, Schmid B, Beierkuhnlein C, Eisenhauer N (2015) Biodiversity increases the resistance of ecosystem productivity to climate extremes. Nature 526(7574):574–577

Jones L, Norton L, Austin Z, Browne AL, Donovan D, Emmett BA, Willis GF (2016) Stocks and flows of natural and human–derived capital in ecosystem services. Land Use Policy 52:151–162

Jucker T, Avacariței D, Barnoaiea I, Duduman G, Bouriaud O, Coomes DA (2016) Climate modulates the effects of tree diversity on forest productivity. J Ecol 104:388–398

Kanter DR, Musumba M, Wood SL, Palm C, Antle J, Balvanera P, Andelman S (2018) Evaluating agricultural trade-offs in the age of sustainable development. Agricultural Systems 163:73–88.

Keeler BL, Gourevitch JD, Polasky S, Isbell F, Tessum CW, Hill JD, Marshall JD (2016) The social costs of nitrogen. Adv Sci 2:e1600219

Kennedy CM, Lonsdorf E, Neel MC, Williams NM, Ricketts TH, Winfree R, Kremen C (2013) A global quantitative synthesis of local and landscape effects on wild bee pollinators in agroecosystems. Ecol Lett 16(5):584–599

Kiaer LP, Skovgaard IM, Østergård H (2009) Grain yield increase in cereal variety mixtures: a meta-analysis of field trials. Field Crops Research 114(3):361–373.

Kremen C, Miles A (2012) Ecosystem services in biologically diversified versus conventional farming systems: benefits, externalities, and trade–offs. Ecol Soc 17(4):40

Kremen C, Williams NM, Aizen MA, Gemmill-Herren B, LeBuhn G, Minckley R, Ricketts TH (2007) Pollination and other ecosystem services produced by mobile organisms: a conceptual framework for the effects of land-use change. Ecol Lett 10(4):299–314

Kropff MJ, Williams RJ, Horie R, Angus JF, Singh U, Centeno HG, Cassman KG (1995) Predicting the yield potential of rice in different environments. In: Humphreys E, Murray EA, Clampett WS, Lewinn LG (eds) Temperate rice-achievements and potential. Proceedings of the temperate rice conference, Yanco, Australia, February 1994, pp 657–664

Kropff MJ, Muchow RC, Rabbinge R, Aggarwal PK (1997) Limits to intensive agricultural systems and opportunities for sustainable agricultural development. In: Keating B, Wilson J (eds) Intensive sugarcane production : meeting the challenge beyond 2000: proceedings of the Sugar 2000 symposium, Brisbane, Australia, 20–23 August 1996. CABI, pp 17–35

Lal R (2004) Soil carbon sequestration to mitigate climate change. Geoderma 123(1–2):1–22

Lange M, Eisenhauer N, Sierra CA, Bessler H, Engels C, Griffiths RI, Gleixner G (2015) Plant diversity increases soil microbial activity and soil carbon storage. Nat Commun 6:6707

Lefcheck JS, Byrnes JE, Isbell F, Gamfeldt L, Griffin JN, Eisenhauer N, Duffy JE (2015) Biodiversity enhances ecosystem multifunctionality across trophic levels and habitats. Nature communications 6(1):1–7.

Lehman CL, Tilman D (2000) Biodiversity, stability, and productivity in competitive communities. Am Nat 156(5):534–552

Letourneau DK, Jedlicka JA, Bothwell SG, Moreno CR (2009) Effects of natural enemy biodiversity on the suppression of arthropod herbivores in terrestrial ecosystems. Annu Rev Ecol Evol Syst 40(1):573–592

Letourneau DK, Armbrecht I, Rivera BS, Lerma JM, Carmona EJ, Daza MC, Trujillo AR (2011) Does plant diversity benefit agroecosystems? A synthetic review. Ecol Appl 21(1):9–21

Levine JM, Adler PB, Yelenik SG (2004) A meta-analysis of biotic resistance to exotic plant invasions. Ecol Lett 7(10):975–989

Li L, Li SM, Sun JH, Zhou LL, Bao XG, Zhang HG, Zhang FS (2007) Diversity enhances agricultural productivity via rhizosphere phosphorus facilitation on phosphorus–deficient soils. Proc Natl Acad Sci 104(27):11192–11196

Liang J, Crowther TW, Picard N, Wiser S, Zhou M, Alberti G, Reich PB (2016) Positive biodiversity–productivity relationship predominant in global forests. Science 354(6309):aaf8957

Liebman M, Dyck E (1993) Crop rotation and intercropping strategies for weed management. Ecol Appl 3(1):92–122

Liebman MA, Staver CP, Mohler CL (2001) Crop diversification for weed management. In: Ecological management of agricultural weeds. Cambridge University Press, New York, pp 322–374

MA (Millennium Ecosystem Assessment) (2005) Ecosystem and human well–being. Island Press, Washington, DC

McIsaac G, Edwards WR (1994) Sustainable agriculture in the American Midwest: lessons from the past, prospects for the future. University of Illinois Press, Urbana

MEA (Millennium Ecosystem Assessment) (2005) Ecosystems and human well–being: synthesis. Island Press/World Resources Institute, Washington, DC

Midega CA, Bruce TJ, Pickett JA, Khan ZR (2015) Ecological management of cereal stemborers in African smallholder agriculture through behavioural manipulation. Ecol Entomol 40:70–81

Monteith JL (1977) Climate and the efficiency of crop production in Britain. Philos Trans R Soc London B Biol Sci 281(980):277–294

Muchow RC, Kropff MJ (1997) Assessing the potential yield of tropical crops: role of field experimentation and simulation. In: Applications of systems approaches at the field level: volume 2 proceedings of the second international symposium on systems approaches for agricultural development, held at IRRI, Los Baños, Philippines, 6–8 December 1995. Springer, pp 101–112

Nelson E, Mendoza G, Regetz J, Polasky S, Tallis H, Cameron D, Shaw M (2009) Modeling multiple ecosystem services, biodiversity conservation, commodity production, and trade–offs at landscape scales. Front Ecol Environ 7(1):4–11

Nielsen UN, Ayres E, Wall DH, Bardgett RD (2011) Soil biodiversity and carbon cycling: a review and synthesis of studies examining diversity–function relationships. Eur J Soil Sci 62(1):105–116

Nix HA (1983) Minimum data sets for agrotechnology transfer. In: Proceedings of the international symposium on minimum data sets for agrotechnology transfer, March 1983, pp 181–188

Norgaard RB (2010) Ecosystem services: from eye–opening metaphor to complexity blinder. Ecol Econ 69(6):1219–1227

O'Connor MI, Gonzalez A, Byrnes JE, Cardinale BJ, Duffy JE, Gamfeldt L, Dolan KL (2017) A general biodiversity–function relationship is mediated by trophic level. Oikos 126(1):18–31

Oerke EC, Dehne HW, Schönbeck F, Weber A (1994) Crop production and crop protection: estimated losses in major food and cash crops. Elsevier, Amsterdam

Orford KA, Murray PJ, Vaughan IP, Memmott J (2016) Modest enhancements to conventional grassland diversity improve the provision of pollination services. J Appl Ecol 53(3):906–915

Paquette A, Messier C (2011) The effect of biodiversity on tree productivity: from temperate to boreal forests. Glob Ecol Biogeogr 20(1):170–180

Pérez–Soba M, Elbersen B, Kempen M, Braat L, Staristky I, Van Wijngaart R, Smith L (2012) Study on the role of agriculture as provisioning ecosystem service. Interim report to the Institute for Environment and Sustainability (JRC/IES). Ecologic Institute, University of Copenhagen, and EuroCARE, Alterra Wageningen UR

Petz K, van Oudenhoven AP (2012) Modelling land management effect on ecosystem functions and services: a study in The Netherlands. Int J Biodivers Sci Ecosyst Serv Manag 8(1–2):135–155

Piotto D (2008) A meta–analysis comparing tree growth in monocultures and mixed plantations. For Ecol Manag 255(3–4):781–786

Plieninger T, Raymond CM, Oteros–Rozas E (2016) Cultivated lands. In: Routledge handbook of ecosystem services. Routledge, pp 442–451

Polley HW, Isbell FI, Wilsey BJ (2013) Plant functional traits improve diversity–based predictions of temporal stability of grassland productivity. Oikos 122(9):1275–1282

Potschin M, Haines–Young R, Fish R, Turner RK (eds) (2016) Routledge handbook of ecosystem services. Routledge

Poveda K, Gómez MI, Martinez E (2008) Diversification practices: their effect on pest regulation and production. Rev Colomb Entomol 34(2):131–144

Power AG (2010) Ecosystem services and agriculture: trade–offs and synergies. Philo Trans R Soc B Biol Sci 365(1554):2959–2971

Power AG (2016) Can ecosystem services contribute to food security? In: Routledge handbook of ecosystem services. Routledge, pp 491–500

Pretty JN (1994) Alternative systems of inquiry for a sustainable agriculture. IDS Bull 25(2):37–49

Prieto I, Violle C, Barre P, Durand JL, Ghesquiere M, Litrico I (2015) Complementary effects of species and genetic diversity on productivity and stability of sown grasslands. Nature Plants 1(4):1–5

Pywell RF, Heard MS, Woodcock BA, Hinsley S, Ridding L, Nowakowski M, Bullock JM (2015) Wildlife–friendly farming increases crop yield: evidence for ecological intensification. Proc R Soc B Biol Sci 282(1816):20151740

Quijas S, Schmid B, Balvanera P (2010) Plant diversity enhances provision of ecosystem services: a new synthesis. Basic Appl Ecol 11(7):582–593

Rader R, Bartomeus I, Garibaldi LA, Garratt MP, Howlett BG, Winfree R, Woyciechowski M (2016) Non–bee insects are important contributors to global crop pollination. Proc Natl Acad Sci 113(1):146–151

Reijntjes C, Haverkort B, Waters–Bayer A (1992) Farming for the future. Macmillan Education

Robertson GP, Vitousek PM (2009) Nitrogen in agriculture: balancing the cost of an essential resource. Annu Rev Environ Resour 34(1):97–125

Root RB (1973) Organization of a plant–arthropod association in simple and diverse habitats: the fauna of collards (Brassica oleracea). Ecol Monogr 43(1):95–124

Rossing WA, Jansma JE, De Ruijter FJ, Schans J (1997) Operationalizing sustainability: exploring options for environmentally friendly flower bulb production systems. Eur J Plant Pathol 103:217–234

Rusch A, Chaplin–Kramer R, Gardiner MM, Hawro V, Holland J, Landis D, Bommarco R (2016) Agricultural landscape simplification reduces natural pest control: a quantitative synthesis. Agric Ecosyst Environ 221:198–204

Sanderson MA, Skinner RH, Barker DJ, Edwards GR, Tracy BF, Wedin DA (2004) Plant species diversity and management of temperate forage and grazing land ecosystems. Crop Sci 44(4):1132–1144

Scherber C, Eisenhauer N, Weisser WW, Schmid B, Voigt W, Fischer M, Tscharntke T (2010) Bottom–up effects of plant diversity on multitrophic interactions in a biodiversity experiment. Nature 468(7323):553–556

Scherer–Lorenzen M (2014) The functional role of biodiversity in the context of global change. In: Biodiversity and Climate Change: reports and guidance developed under the Bern convention, pp 83–97

Seppelt R, Lautenbach S, Volk M (2013) Identifying trade–offs between ecosystem services, land use, and biodiversity: a plea for combining scenario analysis and optimization on different spatial scales. Curr Opin Environ Sustain 5(5):458–463

Soussana JF, Loiseau P, Vuichard N, Ceschia E, Balesdent J, Chevallier T, Arrouays D (2004) Carbon cycling and sequestration opportunities in temperate grasslands. Soil Use Manag 20(2):219–230

Steinbeiss S, Bessler H, Engels C, Temperton VM, Buchmann N, Roscher C, Gleixner G (2008) Plant diversity positively affects short-term soil carbon storage in experimental grasslands. Glob Change Biol 14(12):2937–2949

Storkey J, Döring T, Baddeley J, Collins R, Roderick S, Jones H, Watson C (2015) Engineering a plant community to deliver multiple ecosystem services. Ecol Appl 25(4):1034–1043

Sukhdev P (2008) The economics of ecosystems and biodiversity: an interim report. European Union Commission for the Environment

Tahvanainen JO, Root RB (1972) The influence of vegetational diversity on the population ecology of a specialized herbivore, Phyllotreta cruciferae (Coleoptera: Chrysomelidae). Oecologia 10:321–346

Teasdale JR, Brandsaeter LO, Calegari A, Neto FS, Upadhyaya MK, Blackshaw RE (2007) Cover crops and weed management. In: Non–chemical weed management: principles, concepts and technology. C.A.B. International, Wallingford, pp 49–64

TEEB RO (2010) Mainstreaming the economics of nature. TEEB, Geneva

Teillard F, Doyen L, Dross C, Jiguet F, Tichit M (2016) Optimal allocations of agricultural intensity reveal win–no loss solutions for food production and biodiversity. Reg Environ Chang 17:1397–1408

Teng PS, Kropff MJ, ten Berge HFM, Dent JB, Lansigan FP, van Laar HH (1995) Applications of systems approaches at the farm and regional levels. In: Proceedings of the second international symposium on systems approaches for agricultural development, vol 1. IRRI, Los Baños

Tilman D, Hill J, Lehman C (2006) Carbon–negative biofuels from low–input high–diversity grassland biomass. Science 314(5805):1598–1600

Trenbath BR (1974) Biomass productivity of mixtures. Adv Agron 26:177–210

Tuck SL, O'Brien MJ, Philipson CD, Saner P, Tanadini M, Dzulkifli D, Hector A (2016) The value of biodiversity for the functioning of tropical forests: insurance effects during the first decade of the Sabah biodiversity experiment. Proc R Soc B Biol Sci 283(1844):20161451

Turnbull LA, Isbell F, Purves DW, Loreau M, Hector A (2016) Understanding the value of plant diversity for ecosystem functioning through niche theory. Proc R Soc B Biol Sci 283(1844):20160536

van den Belt M, Forgie V, Farley J (2011) Valuation of coastal ecosystem services. In: Wolanski E, McLusky DS (eds) Treatise on Estuarine and coastal science, vol 12. Academic, Waltham, pp 35–54

Van Der Plas F, Manning P, Allan E, Scherer–Lorenzen M, Verheyen K, Wirth C, Fischer M (2016) Jack–of–all–trades effects drive biodiversity–ecosystem multifunctionality relationships in European forests. Nat Commun 7(1):1–11

van Ruijven J, Berendse F (2005) Diversity–productivity relationships: initial effects, long–term patterns, and underlying mechanisms. Proc Natl Acad Sci USA 102(3):695–700

Vandermeer J (1989) The ecology of intercropping. Cambridge University Press, New York

Vandermeer J (1995) The ecological basis of alternative agriculture. Annu Rev Ecol Syst 26:201–224

Weigelt A, Weisser WW, Buchmann N, Scherer-Lorenzen M (2009) Biodiversity for multifunctional grasslands: equal productivity in high–diversity low–input and low–diversity high–input systems. Biogeosciences 6(8):1695–1706

Westman WE (1977) How much are nature's services worth? Measuring the social benefits of ecosystem functioning is both controversial and illuminating. Science 197(4307):960–964

Wetzel WC, Kharouba HM, Robinson M, Holyoak M, Karban R (2016) Variability in plant nutrients reduces insect herbivore performance. Nature 539(7629):425–427

Williams NM, Regetz J, Kremen C (2012) Landscape-scale resources promote colony growth but not reproductive performance of bumble bees. Ecology 93(5):1049–1058

Wilsey BJ, Polley HW (2004) Realistically low species evenness does not alter grassland species-richness–productivity relationships. Ecology 85(10):2693–2700

Wilson CM, Matthews WH (eds) (1970) Man's impact on the global environment: report of the study of critical environmental problems (SCEP). MIT Press, Cambridge

Yachi S, Loreau M (1999) Biodiversity and ecosystem productivity in a fluctuating environment: the insurance hypothesis. Proc Natl Acad Sci USA 96(4):1463–1468

Zhang W, Ricketts TH, Kremen C, Carney K, Swinton SM (2007) Ecosystem services and dis–services to agriculture. Ecol Econ 64(2):253–260

Zhang Y, Chen HY, Reich PB (2012) Forest productivity increases with evenness, species richness and trait variation: a global meta-analysis. J Ecol 100(3):742–749

Zuppinger-Dingley D, Schmid B, Petermann JS, Yadav V, De Deyn GB, Flynn DF (2014) Selection for niche differentiation in plant communities increases biodiversity effects. Nature 515(7525):108–111

The manufacturer's authorised representative in the EU is Springer Nature Customer Service Centre GmbH, Europaplatz 3, 69115 Heidelberg, Germany. If you have any concerns regarding our products, please contact ProductSafety@springernature.com

Printed and bound by CPI Group (UK) Ltd, Croydon, CR0 4YY
01/10/2026
02225930-0014